CAD/CAM Robotics and
Factories of the Future '90

Suren N. Dwivedi
Alok K. Verma
John E. Sneckenberger

Editors

CAD/CAM Robotics and Factories of the Future '90

Volume 2: Flexible Automation

5th International Conference on CAD/CAM,
Robotics and Factories of the Future
(CARS and FOF'90) Proceedings

International Society for Productivity Enhancement

With 245 Figures

Springer-Verlag Berlin Heidelberg GmbH

Suren N. Dwivedi
Department of Mechanical and
    Aerospace Engineering
West Virginia University
Morgantown, WV 26506-6101
USA

Alok K. Verma
Chairman
Department of Mechanical Engineering Technology
Old Dominion University
Norfolk, VA 23508
USA

John E. Sneckenberger
Concurrent Engineering Research Center
West Virginia University
Morgantown, WV 26506
USA

Library of Congress Cataloging-in-Publication Data
International Conference on CAD/CAM, Robotics, and Factories of the
    Future (5th: 1990: Norfolk, VA.)
        CAD/CAM, robotics, and factories of the future: 5th International
    Conference on CAD/CAM, Robotics, and Factories of the Future (CARS
    and FOF'90) proceedings /S.N. Dwivedi, Alok Verma, John
    Sneckenberger.
            p. cm.
        "International Society for Productivity Enhancement."
        Includes bibliographical references.
        Contents: v. 1. Concurrent engineering—v. 2. Flexible
    automation.
        ISBN 978-3-642-63504-5   ISBN 978-3-642-58214-1 (eBook)
        DOI 10.1007/978-3-642-58214-1
        1. CAD/CAM systems—Congresses. 2. Flexible manufacturing
    systems—Congresses. 3. Robotics—Congresses. 4. Manufacturing
    processes—Automation—Congresses. I. Dwivedi, Suren N.
    II. Verma, Alok. III. Sneckenberger, John. IV. International
    Society for Productivity Enhancement. V. Title.
    TS155.6.I5818 1990
    670.42'7—dc20   90-19478

Printed on acid-free paper.

Camera-ready prepared by the editors.

9 8 7 6 5 4 3 2 1

ISBN 978-3-642-63504-5

Conference Objective

The last decade has seen the emergence of a unified approach for product design which attempts to combine traditionally distinct tasks like design, management, marketing, analysis, manufacture and materials. Often called "Concurrent Engineering" or "Simultaneous Engineering", this new philosophy aims at improving cost competitiveness by reducing waste of time, money, and other resources inherent in the iterative traditional methods. In view of the importance of this new philosophy, Concurrent Engineering is selected as the theme for this conference.

The main objective of the conference is to bring together researchers and practitioners from government, industries and academia interested in the multi-disciplinary and inter-organizational productivity aspects of advanced manufacturing systems utilizing CAD/CAM, CAE, CIM, Parametric Technology, AI, Robotics, AGV Technology, etc.

Conference Organization

Sponsors

International Society for Productivity Enhancement (ISPE), USA

Center for Innovative Technology, Virginia, USA

Old Dominion University, Norfolk, Virginia, USA

Concurrent Engineering Research Center (CERC)
West Virginia University, West Virginia, USA

Department of Mechanical and Aerospace Engineering,
West Virginia University, West Virginia, USA

Committee Chairpersons

Conference General Chairperson:
Alok K. Verma, Old Dominion University, USA.

Conference co-chairs:
Suren N. Dwivedi, West Virginia University, USA.
Gary Crossman, Old Dominion University, USA.

Program Chairpersons:
John Sneckenberger, West Virginia University, USA.

Technical Chairperson:
Virendra Kumar, General Electric, USA.

International Chairperson:
Jean Marie Proth, INRIA, France.

Reception Chairperson:
John. M. Jeffords, Old Dominion University, USA.

Workshop Chairpersons:
Stewart Shen, Old Dominion University, USA.
Bharat Thacker, Universal Computer Services, USA.
Sumitra Reddy, CERC, West Virginia University, USA.
Hal Schall, Ford Motor Co., Dearborn, USA.

Printing and Publication Chairperson:
Sacharia Albin, Old Dominion University, USA.

Abstract and Paper Review Chairperson:
Jean Hou, Old Dominion University, USA.
Resit Unal, Old Dominion University, USA.

University Chairperson:
Robert Ash, Old Dominion University, USA.

Industrial Chairpersons:
Gary Crossman, Old Dominion University, USA.
Larry Richards, Old Dominion University, USA.

Plenary Chairperson:
Suren N. Dwivedi, West Virginia University, USA.

Exhibit Chairperson:
Thomas Houlihan, Jonathan Corporation, USA.

Student Chairpersons:
Drew Landman, Old Dominion University, USA.
Francis M. Williams, Old Dominion University, USA.

International Coordinators

H. Bera (U.K.)
M. Dominguez (Spain)
K. Ghosh (Canada)
V.M. Ponomaryov (USSR)
J.M. Proth (France)
R. Sagar (India)
T.-P. Wang (Taiwan)
T. Yamashita (Japan)

Committee Rosters

Abstract and Paper Review Committee

Dr. Ralph Wood
Dr. John Spears
Dr. Cheng Y. Lin
Dr. Duc Nguyen
Dr. Nageswara Rao

Program Committee

Donald W. Lyons
John Sneckenberger
Suren N. Dwivedi
Sumitra Reddy
Larry Banta
John Hackworth
Bob Creese
B. Gopalkrishnan
Waeik Iskander
Bruce Kang
Ken Means
Jacky Prucz
Nithi Sivaneri
Emil Steinhardt

Editorial Board and Publication Committee

Donald W. Lyons
Ralph Wood
John Spears
William Bentley
Zenon Kulpa
Michael Sobolewski
Sati Maharaj
Sisir Padhy
Bin Du
Prashanth Murthy
Deepak Kohli
Dandamudi Venugopal
Dhananjay Salunke

Industrial Committee

Thomas Houlihan
Moustafa R. Moustafa
Ed Wilson
Jim Fox
Larry Wilson

Reception Committee

Taj Mohieldin
Linda Vahala
Nancy Short

Professional Relations Committee

John Jurewicz
Donald W. Lyons
Ramana Reddy
Ralph Wood
Biren Prasad
Kumar Singh
John Spears

Conference Coordinators

William Bentley
Nancy Short
Georgette Ingram

Conference Staff

Sati Maharaj
Robin Johnson
Joette Claiborne
Marylin Host
Indira Dwivedi
Iva Dwivedi
Fern Wood
Vicki Grim
Pat Logar
Jean Shellito

Letter from the President, ISPE

The International Society for Productivity Enhancement (ISPE) is entering its seventh year. The Conference you are attending is our fifth of the international series on CAD/CAM, Robotics and Factories of the Future (CARS & FOF). The fourth conference was held at the Indian Institute of Technology, New Delhi, India in 1989. During the past seven years, we have expanded our activities significantly. The membership interest and international participation are also growing. During the past year alone, the Society has made tremendous progress in the following major frontiers:

JOURNAL: The Society now has its own journal entitled The International *Journal of Systems Automation Research and Applications* (SARA), an international, multidisciplinary research and applications-oriented journal to promote a better understanding of systems considerations in interdisciplinary automation using computers. The Journal contains important reading for design, engineering, and manufacturing persons as well as those with interest in research and development and applications of productivity tools, concepts and strategies to multidisciplinary systems environments. The Journal will only publish original, quality papers. To receive more information about this Journal, write to: Editor-in-Chief, ISPE, SARA Journal Department, P.O. Box 731, Bloomfield Hills, Michigan 48303-0731.

PROCEEDINGS: Starting this year (with the Fifth Conference), the Society is now making the Conference Proceedings available at the Conference. Selected papers from this Proceedings will also be considered for publication in *SARA*.

CONFERENCES: ISPE's annual conferences are now book until 1994. The Sixth International Conference will take place at South Bank Polytechnic, London from August 19-21, 1991. The Seventh and Eight International Conferences will be held in Leningrad, USSR and France, respectively.

COOPERATIVE PROGRAMS: In 1989, ISPE started a new cooperative program called the Indo-U.S. Forum for Cooperative Research and Technology Transfer (IFCRTT) in cooperation with West Virginia University and the National Science Foundation (NSF). The first joint meeting of the IFCRTT was held from December 17-18, 1989, in New Delhi, India. The meeting attracted a large body of scholars from industry, universities, and research institutions from both the United States and India. Similar cooperative programs are being arranged in the U.K., U.S.S.R. and France.

As you can see, we have made great strides, but significant changes are taking place in the manufacturing sectors due to global competitiveness and economic factors. Productivity enhancement needs are even larger than before, and such needs require us to be more dynamic and resourceful. ISPE is looking for a few good people to take leadership positions in its organization and committees for sponsored events. If you would like to help us build our technical program or if you would like to work on ideas of your own, please write to us. There are openings in the following areas:

* *SARA* Journal - Readers' Committee
* Productivity Directors
* Workshop and Tutorial Organizers
* CARS & FOF Conferences: University, Industry, International Representatives, Session Organizers, and Technical and Program Chairpersons.

We are still a very young organization and your leadership can play a significant role. Please do not hesitate to write us with your ideas and opinions.

Biren Prasad, Ph.D.
ISPE, P.O. Box 731, Bloomfield Hills, MI 48303-0731, USA.

Acknowledgments

The Fifth International Conference on CAD/CAM, Robotics, and Factories of the Future (Cars & FOF '90) was hosted by the College of Engineering and Technology at Old Dominion University and was endorsed by more than ten societies, associations, and international organizations. The conference was held in Norfolk, Virginia at the Omni International Hotel from December 2-5, 1990. Over 200 presentations organized into 40 specialty sessions, three plenary sessions, and eight workshops were conducted during the four days. Authors, plenary session speakers, and participants from 17 different countries around the world converged in Norfolk for this Conference. In view of the ever-increasing importance for integrating different facets of manufacturing with design process, the organizing committee selected "Concurrent Engineering" as the theme of the Conference.

I wish to acknowledge, with many thanks, the contributions of all the authors who presented their work at the Conference and submitted the manuscripts for publication. It is also my pleasure to acknowledge the role of banquet, luncheon, and plenary session speakers who shared their vision of the manufacturing industry and issues related to productivity. My sincere thanks to the session organizers, session chairs, and members of the Organizing Committee both at Old Dominion University and West Virginia University without whose cooperation this Conference would not be possible. Thanks are due to Ms. Georgette Ingram and other staff members in the MET Department for their patience and hard work. Financial support from the Center for Innovative Technology and industrial sponsors also made this Conference possible.

I acknowledge, with gratitude, the help and support received from Dr. James V. Kock, President, and Dr. Ernest J. Cross, Dean of the College of Engineering and Technology at Old Dominion University. From West Virginia University, I thank Dr. Donald W. Lyons, Chairman, MAE Department, for his support; Drs. Ralph Wood and John Spears for their help in reviewing conference papers and for allowing us to use the facilities of the Concurrent Engineering Research Center; and Ms. Sati Maharaj for her assistance in coordinating the conference. In addition, I extend my deepest gratitude to Dr. Suren N. Dwivedi for providing me with support and encouragement in organizing this conference. Furthermore, I express my sincere thanks to all my colleagues, friends, student volunteers, and family members who extended their help in organizing this conference.

I also acknowledge with great appreciation the excellent work done by Springer-Verlag in publishing both volumes of the proceedings.

Alok K. Verma
Conference General Chairperson

Preface

Flexibility is as acceptable an objective for today's industrial community as is automation. Thus, the title of this conference proceedings volume - Flexible Automation - reflects an added emphasis to the usual industrial automation. As with general automation that has impacted every component of the manufacturing office and plant, the identity of flexible automation can possess various forms and functions.

The papers in this volume have been grouped into two main categories. One category deals with implementation of so-called "intelligent manufacturing". This means use of algorithmic methods and artificial intelligence approaches to various problems encountered in practical factory automation tasks. The placement of papers into five chapters of this part cannot be very precise , due to multidisciplinary nature and constant rapid change of the field. The categories are arranged starting from problems of enhancement of current factory settings, and followed by the papers addressing more specific issues of production planning, process technology and product engineering. The fifth chapter contains papers on the very important aspects of factory automation - problems of design, simulation, operation and monitoring of manufacturing cells.

In the second category, papers dealing with practical developments in applied automation, especially robotics, have been grouped. The section starts from a parade of industrial applications of robotics, proceeding then to task performance analysis. The specific problems of robotics, namely motion specification and manipulator mechanics are thoroughly covered by the papers in next two chapters. The final chapter groups four papers addressing the topic of engineering education and training in the field of automated manufacturing - a problem of considerable importance to preapare people so that the idea of "factories of the future" may become a reality.

Suren N. Dwivedi

Contents, Volume 2

Section A: Implementation of Intelligent Manufacturing

Chapter V: Workcell Operations 303

Section B: Developments in Applied Robotics and Automation

Contents, Volume 1

SECTION A: IMPLEMENTATION OF INTELLIGENT MANUFACTURING

Chapter I

Factory Enhancements

Introduction

In this chapter, various approaches to enhance existing factories with flexible CAD/CAM/CIM tools and concepts are discussed from various points of view. The first paper is concerned with the transformation of existing enterprises into an integrated, though decentralized, manufacturing system, capable of reacting quickly to market demands - the difficult but necessary task to undertake in the authors' country. The second paper discusses concept, selection criteria and implementation problems of a Flexible Manufacturing System (FMS) for precision engineering components. The third paper presents results of a survey of successful U.S. manufacturers on their methods of strategic planning for Computer-Integrated Manufacturing (CIM).

In the fourth paper, the results of a simulation of a flexible manufacturing system are analyzed. Specifically, the optimization of the number of Automated Guided Vehicles (AGVs), machines needed, and job arrival rate is discussed. The fifth paper presents results of a research project on a computer-based safety system for a FMS, especially management logic necessary to avoid various danger situations resulting from the interaction of machines, people, and factory environment objects. The sixth paper deals with technical barriers existing in the course of implementation of software infrastructure for CIM and Concurrent Engineering (CE), especially in the support of integrated models in global repositories. The seventh paper deals with the requirements for a computerized CAD/CAM system for Diesel engine design and manufacturing in today's China, whereas the eighth one discusses the future of industrialized housing in the U.S., describing the 21st Century manufacturing facility employing advanced information and manufacturing technologies. Similarly, the next paper indulges in the future of the U.S. Postal Service, presenting some enabling technologies capable of implementing this vision.

The next three papers deal with formal mathematical analysis of common problems arising in advanced factory automation: scheduling optimization, modeling, and simulation. The last one discusses data analysis methods required to prepare data and inference rules for expert systems to be used in CIM.

From the Existing Manufacturing System to CIM

D.S. LVOV and E.I. ZAK
Central Economic and Mathematical Institute
of the Academy of Sciences of the USSR
Moscow, USSR

YU. M. ZYBAREV
Institute of Mathematics of the Academy of Sciences of the USSR
Novosibirsk, USSR

Abstract. The approach for transforming the existing manufacturing system in the machine-building industry into a new kind of decentralized manufacturing system is presented. The main idea is to replace existing enterprises (factories, plants , etc.) with an integrated manufacturing system (IMS) which combines a set of relatively independent (in an economic sense) specialized manufacturing modules (SMM) which interconnect with each other as supplier and consumer. This IMS is directly market-oriented, reacting quickly to market changes. We show the possibility of creating a programmable network of IMS such that market changes (i.e. changes in final product demand) as a rule lead only to new relations between SMMs.

1. Introduction

The Computer Integrated Manufacturing (CIM) concept provides better decision making capabilities which result in faster design and development cycles for products and more flexible manufacturing systems . A number of CIM studies have concentrated mainly on the cell, job-shop and plant levels [1, 3]. However, the computer integration and advanced information technology are clearly applicable to a higher level, the multi-plant enterprise. So the main goal of our research is building up a new kind of CIM in the metalworking machinery as a multi-plant enterprise which could flexibly reconfigured in response to changing a market.

2. A modular approach for manufacturing design

The integration should be considered in two dimensions: the first is an integration of all functions in a product life cycle including design, planning, production, distribution, and field service. The second dimension represents cooperation between manufacturers of the final products and intermediate products (components).

The manufacturing system such as a large enterprise is divided into specialized manufacturing modules (SMM) which are completely independent from the economic point of view. As a rule, every SMM manufactures a distinct kind of intermediate product: a part, unit or subassembly. The criteria for constructing a module are marketability of its products and similarity of production techniques for the various products. For example, one or more FMSs can compose an SMM.

The SMMs cooperate with one another. Every module may belong to a corporation such as an integrated manufacturing system (IMS) or keep its own independence. An IMS combines a set of SMMs which interconnects with each other as supplier and consumer. An IMS integrates information, material processing and transshipment of components from one SMM to other. An IMS include at least three elements:
- a number of SMMs ,
- a transport communication,
- an information network, and
- supervisory computer control.

An SMM involved in IMS becomes as semi-independent. One of the advantages of incorporating an SMM into a whole IMS is that it provides more continuity of flow of materials. It is more difficult for SMMs alone to achieve continuity of material flow. The overall effectiveness of the system is higher than the sum of the individual outputs from the subsystems.

Configuration of an IMS can vary. To switch the production from one kind of products to other it is necessary to replace software for SMMs and to set new relationships between SMMs only. The time needed for this reconstruction is greatly less than that in the case of a traditional enterprise.

There are two kinds of control here: a centralized control of IMS as management of IMS reconfiguration (long wave of control) and a decentralized control regarding each SMM (short wave of control). There is an analogy between management of an IMS and a programmable controller (PC). As well, in the PC case, the main advantage of IMS is the ease of changing the system logic (system configuration here). The creating a programmable network of IMS is becoming possible (Fig.1).

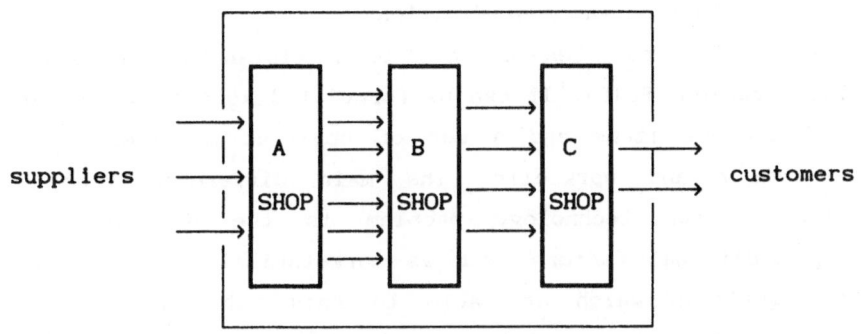

(a) a conventional enterprise

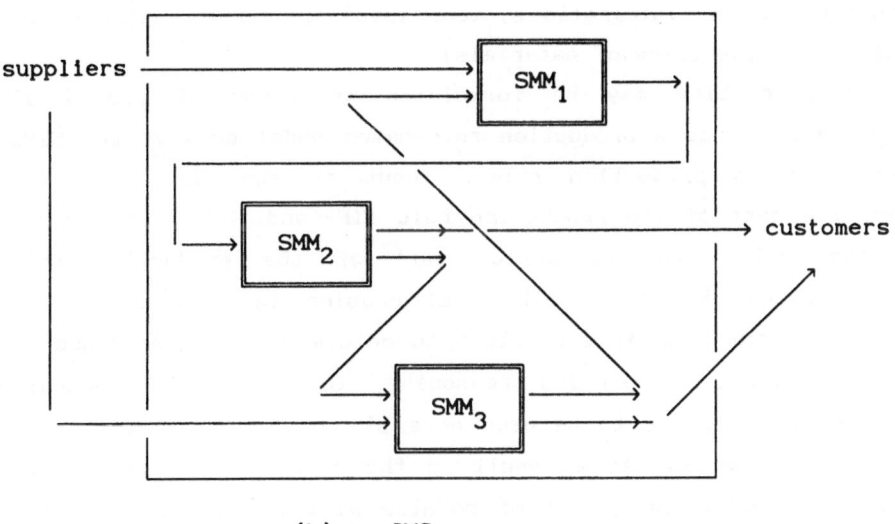

(b) an IMS

⟶ flow of materials (parts, components)

Figure 1.   Comparison of a conventional enterprise and an IMS

3. The problems of an IMS design

Besides typical CIM problems, there are some specific IMS design problems:
- Generation of SMMs,
- Finding an IMS structure,
- Balancing and synchronization, and
- Evaluation and justification.

The First Problem " Generation of SMMs " is an Extended Group Technology Problem (EGTP). It can be formulated as follows: Given a set of product items and a set of optional machines, find product family and workcells. The main difference from a conventional group technology problem is the necessity of involving additional factors such as marketability. There are various algorithms which are able to solve the usual group technology problem. The resolution of the EGTP implies increasing computational efforts.

The Second problem "Finding an IMS structure" may be expressed as follows. Given a potential set of SMMs, find the structure of the integrated system, involving modules itself and their relations (flow of materials).

This problem may be formulated in terms of procedural knowledge. It has a production rule-based model so that the SMMs correspond to production rules, input of the SMM - to the condition part of the production rule (IF-condition), and output of the SMM - to the action part of the production rule (THEN-action) (Fig.2). So the real problem in constructing a network of the SMMs is equivalent to determining the sequence of rule execution to provide reasoning. Obviously the backward chaining strategy is to be used here since the set of the final products are given. As a result in the context of a production rule-based model the subset of modules will be selected and the technological chain as related production processing stages will be fired.

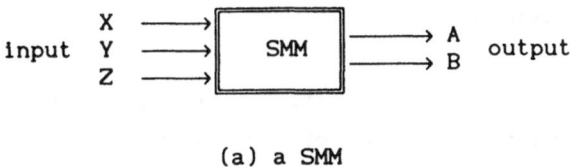

(a) a SMM

IF input is the set of components {X, Y, Z}
THEN output is the set of products {A, B}

(b) a production rule

Figure 2. Correspondence between a SMM and
an **IF** - **THEN** production rule

The Third problem "Balancing and synchronization" can be formulated as follows. It is necessary to provide equivalent capacity of different sections of the technological chain and balancing flow of materials.

A two phase algorithm is used here. On the first phase, the balancing problem as assignment operations to different SMMs is solved and after that the necessary number of every kind of SMMs to provide feasible flow of materials is determined so the capacity of each section of the technological chain would be almost the same. The procedure is based on an assembly line balancing algorithm for single-model and mixed-model cases [4].

The fourth problem "Evaluation and justification" may be formulated in the following way: Given a set of variants of manufacturing configuration obtained as a result of solving the previous problems (the case with a unique variant is also under consideration), find and evaluate the optimal variant from the economic point of view.

Effectiveness should be considered on two levels: SMM level and IMS level. For solving this problem the special modelling program can be used [2, 5].

4. Concluding remarks

A special case of CIM as IMS has been proposed. This approach is kindred with the strategy of "standardized design" where each final product is different but consists of basic standard parts.

The specific problems have been formulated for IMS design.

This concept is important with concurrent engineering point of view because it is possible to combine product design and an IMS reconfiguration.

We plan to use this tool to analyze a way of reconstructing real machine tool enterprises to CIM.

References

[1] ACACCIA, G.M., MICHELINI, R.C., MOLFINO, R.M., and Rossi, G.B. "Shopfloor logistics for flexible manufacturing based on distributed intelligence", *The International Journal of Advanced Manufacturing Technology* 4 (1989) 231-242.

[2] FEDORENKO, N.P., LVOV, D.S., et al. , *Formal method for complex evaluation and justification of innovations. Guide book* , Inform-electro, Moscow, 1989 (in Russian).

[3] VILLA, A., UKOVICH, W., and MURARI, G. "Modelling the incremental innovation process in an automated manufacturing plant", *The TC-7 IFIP International Conference on Modelling the Innovation: Communications, Automation and Information Systems*, Rome, Italy, March 1990.

[4] ZAK, E.I., "An assembly line synchronizing model for JIT system", in: A.A.Fridman, E.V.Levner (eds.), *Mathematical Modelling in Economics and Analysis of Discrete Systems*, Central Economic and Mathematical Institute, Moscow, 1988,138-145 (in Russian).

[5] ZYBAREV, Yu.M. , KAUROV V.M., Production Systems : *Interactive Modelling Technology*, Institute of mathematics, Novosibirsk, 1989 (in Russian).

Flexible Manufacturing System in Manufacture of Precision Engineering Components - Key Issues in Implementation

V.K. GUPTA
VXL India Ltd.,
Universal Engineering
Faridabad, India

R. SAGAR
Department of Mechanical Engineering
Indian Institute of Technology
New Delhi, India

ABSTRACT

Flexible manufacturing System's concept, design and implementation vary considerably and depend mainly on application and the environment under which these are required to operate. This paper brings out an integrated approach adopted for finalising concept, selection and implementation of flexible manufacturing system for manufacture of precision engineering components, restructuring of support infrastructure, services and tackling the environmental factors for its successful implementation.

FLEXIBLE MANUFACTURING SYSTEMS

FMS can be defined as the approach where the automation system is not dedicated throughout its life to the manufacture of a limited range of products, but has the versatility and adaptability to tackle both existing and future product needs.

Another definition gives ultimate goal for FMS systems as
CONTINUOUS - ZERO STOCK - BATCH MANUFACTURE.

PROBLEM AREAS IN EXISTING MANUFACTURING SYSTEMS

A detailed study of existing manufacturing system of a plant manufacturing precision engineering components, mainly for Defence, brought out following problem areas leading to low plant utilisation and high rejections.

a. Frequent and shifting bottlenecks in production due to a few types of complex components requiring several operations on several machines, thereby, disrupting production schedules in case of delay at any stage and machine breakdowns.

b. Rejection of components of each stage/machine setting resulting in high rejection of components requiring multiple machines and processes.

c. In view of stringent quality requirements, there was a need to inspect parts after each operation, thereby need for a large inspection department to cope up with inspection of products.

d. High inprocess inventory in a bid to smoothen the production process.

e.   Long lead time for manufacture of new complex components in view of elaborate tooling, gauge, fixtures and multiple machine settings.

f.   Large inventories of tools, gauges and fixtures requiring continuous maintenance and replacement putting excess strain on inhouse tool-room.

g.   Low moral amongst employees engaged mainly in handling crisis situations instead of actual planning and control.

FEASIBILITY STUDY

A feasibility study carried out to identify possible alternative solutions for this plant indicated a preference for flexible manufacturing systems as against conventional mass production technology being adopted by this plant in the past. This view was further supported by the following developments.

a.   As a result of technology innovations world wide, it has become possible to build flexibility in manufacturing systems and thus deliver a large variety of products in smaller quantities almost as efficiently as few types of products in large quantities.

b.   Increase in number of competitors of this organisation, forcing it to respond quickly to the requirements of the customers for a large variety of products in small quantities, which it could afford to ignore till recently.

c.   Increase in operating costs.

d.   Need to reduce development time for new products to keep competitors from catching up on high-tech products.

e.   Need to create flexible manufacturing facilities which would be adaptable to new product lines in shortest possible time and with negligible additional investment.

CRITERIA FOR SELECTION OF AREAS FOR INTRODUCING FLEXIBLE MANUFACTURING SYSTEMS:

After several rounds of discussions with the plant executives, the following criteria were adopted for selection of areas for introducing flexible manufacturing technology.

a.   New complex components requiring large number of tools, gauges, fixtures and setting on a number of production machines using conventional mass production technology.

b.   Complex components requiring several operations on several machines.

c. components requiring secondary operation to achieve desired accuracy or surface finish.

d. Components with short lead time as a competitive strategy.

CRITERIA FOR SELECTION OF FLEXIBLE MANUFACTURING SYSTEMS

Following criteria was finalised for selection of flexible manufacturing systems:

a. Flexibility
b. Total cost
c. Change over time for new setting
d. Modularity
e. Upgradibility
f. Connectivity
g. Integrtation
h. Adaptability

IMPLEMENTATION OF FMS

Selection of FMS solutions was made based on criteria stated above. Orders were placed for equipments finalised. Following problems were faced during implementation of FMS.

A. ORGANISATION CHANGE:

    i. During initial phase of implementation of FMS, it was strongly felt that the existing organisation, attitude of employees to mass manufacture using conventional technology, was not suited for new technology and initial adjustments were of little help in solving various problems.

    ii. A total reorganisation of the plant had to be done laying more emphasise on support functions like Tool Planning, Process Planning and Industrial Engineering.

    iii. A favourable environment was built by holding several co-ordination meetings, discussions and giving hands on experience to people at all levels, to build confidence in new technology and also to dispell fears amongst all sections of employees.

    iv. Widely publicised policy of 'Train from within' was practiced in true sense by training engineers, technicians and workers at machine manufacturer's plants in Japan and elsewhere. This led to filling most of the slots from within and hiring only a few key employees from outside.

B. Other issues involved in implementation of FMS:

   i. Raw Material Specifications:

   Physical specifications for most of the inputs (raw
   materials) for FMS systems were very tight compared to
   those being followed. A concentrated effort was required
   to improve quality of raw materials to meet the new
   specifications from existing sources. New sources had to
   be developed. This also led to marginal increase in cost
   of inputs in few cases (refer to Fig.1).

   ii. Power:

   Quality of power required for FMS systems is quite
   stringent in terms of frequency and voltage. Automatic
   voltage stabilizers, procured locally posed serious
   problems of frequent tripping etc. It look quite some
   time to resolve this problem by modifying these equipments
   installing isolation transformers and also ensuring stable
   power supply.

   iii. Tooling:

   a. Use of special tooling for automated manufacture such
      as quick change of tooling systems, use of standard
      tool holders and shanks, standard tool bits, use of new
      materials – coated carbides, ceramics etc. were a few
      steps taken to fully exploit high cutting speeds
      possible on FMS. Support services to manufacture and
      maintain special tools had to be set up.

   b. Computer Software package for Tool Planning and
      Management was developed to assist process engineers
      and production engineers to keep track of tools. This
      package maintains a data base on tooling and is
      designed to provide on line information on status of a
      specific tool, indent, expected date of availability
      etc. The data base is resident on a central computer
      and is accessible on terminals throughout the plant
      (refer to Fig 2.).

   iv. Cutting Oils:

   Cutting oils used on FMS posed several problems such as
   developing foul odour, growth of micro organism etc. These
   problems were resolved in consultation with machine
   manufacturers and cutting oil suppliers in India through
   intensive R&D efforts.

S No	Description	Specification	
		Existing	Proposed
1.	Tolerance on diameter (as per relevant Standard BS/IS)	Normal	Close (Class A) (approx 1/3rd of Normal).
2.	Tolerance on Length	–	1.85 + 0.05M – 0
3.	Straightness	–	0.5mm/mtr length
4.	Ovality & Taper	–	Max 1/3rd of total tolerance on dia.
5.	Bar to Bar variation in dia. within a lot.	–	Not exceed 1/3rd of total tolerance on dia.

PHYSICAL SPECIFICATIONS FOR ALUMINIUM/BRASS BARS

FIG. 1

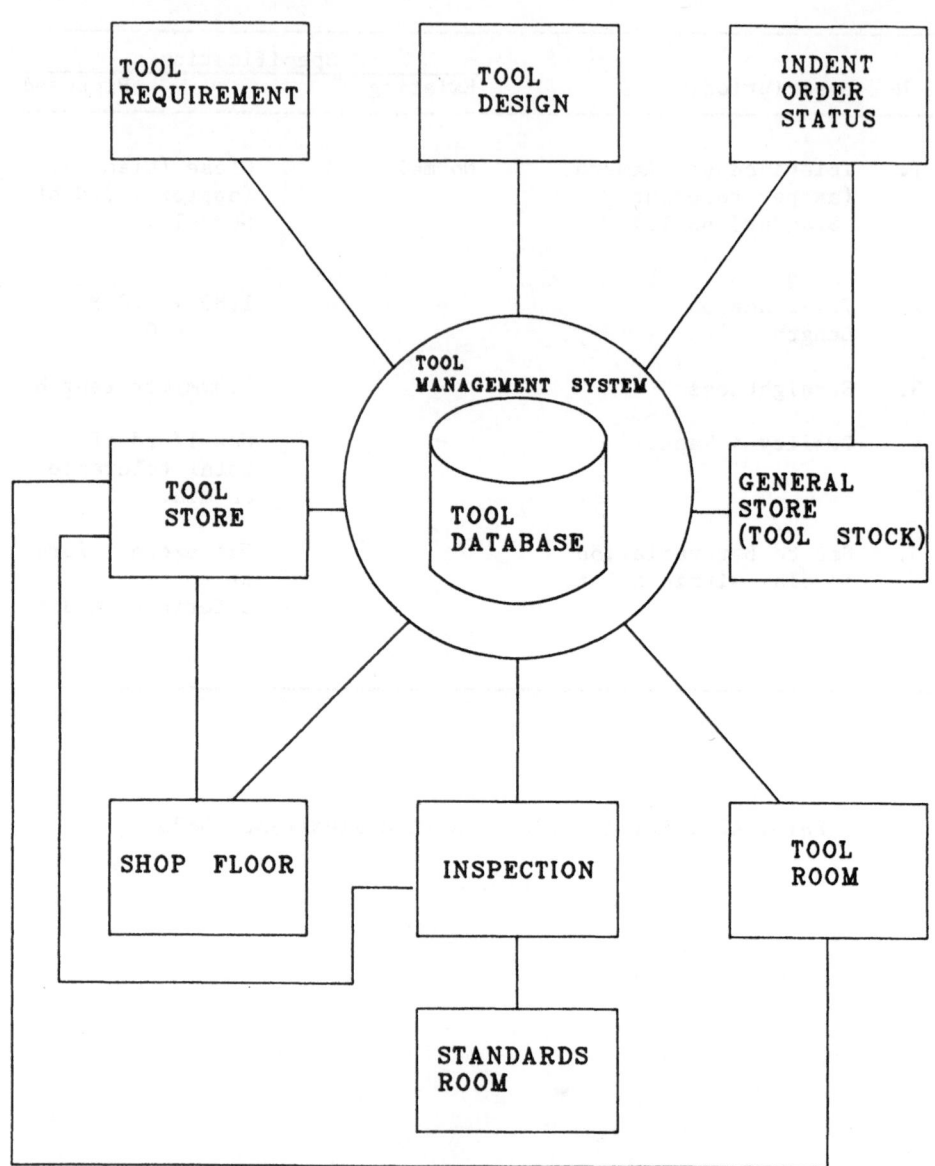

TOOL MANAGEMENT SYSTEM
FIG-2

v. Process Planning:

Process planning practices too had to be changed in view of entirely different approach required for manufacturing a finished complex component in one setting as against several settings on several machines previously. In addition to the task now being highly imaginative and innovative, process engineers had to get accustomed to the use of special complex tools, holding fixtures and also sequence of operations from an entirely different angle. A comparison of old and new manufacturing process of a component is given in Fig.3.

vi. Maintenance:

FMS systems needed a much higher level of maintenance skills hence, engineers and technicians had to be trained at machine manufacturer's plants for proper maintenance of these systems. A few problems like premature failure of main spindle of AC servo drives in quick succession and time lost in getting spare supply etc. forced system to remain down for a considerable period. These problems have now been overcome and alternate action taken to prevent reoccurances.

vii. Computer Aided QA System:

With the change in manufacturing processes from conventional systems to FMS; most of gauges available could not be used. To manufacture a new set of gauges for new process required considerable time and cost.

Computer aided QA systems were only answer to meet this.

To improve and provide quick feedback on quality of components and products, computer aided QA systems were installed. These systems improved storage and retrieval of quality information and also reduced dependence and hence judgement at every stage.

viii. Shop Floor Information System:

Shop Floor information recording and reporting systems were inadequate. Computer software package to provide information to production engineers and also to keep track of actual production as against planned production was developed. This package gives information on ;

a. Machinewise and operatorwise production of a component per shift.

b. No. of shifts required to finish balance quantity planned against actual produced till date.

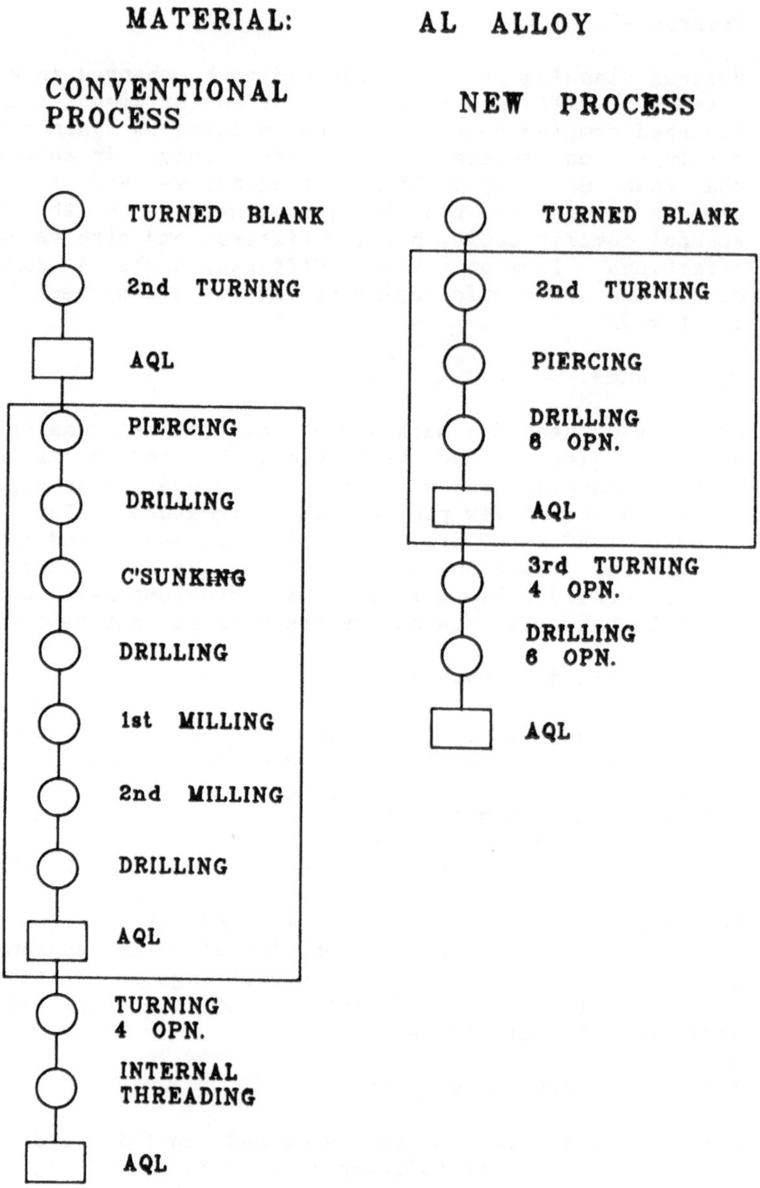

MATERIAL: AL ALLOY

CONVENTIONAL PROCESS

NEW PROCESS

CONVENTIONAL PROCESS	NEW PROCESS
TURNED BLANK	TURNED BLANK
2nd TURNING	2nd TURNING
AQL	PIERCING
PIERCING	DRILLING 6 OPN.
DRILLING	AQL
C'SUNKING	3rd TURNING 4 OPN.
DRILLING	DRILLING 6 OPN.
1st MILLING	AQL
2nd MILLING	
DRILLING	
AQL	
TURNING 4 OPN.	
INTERNAL THREADING	
AQL	

FIG-3

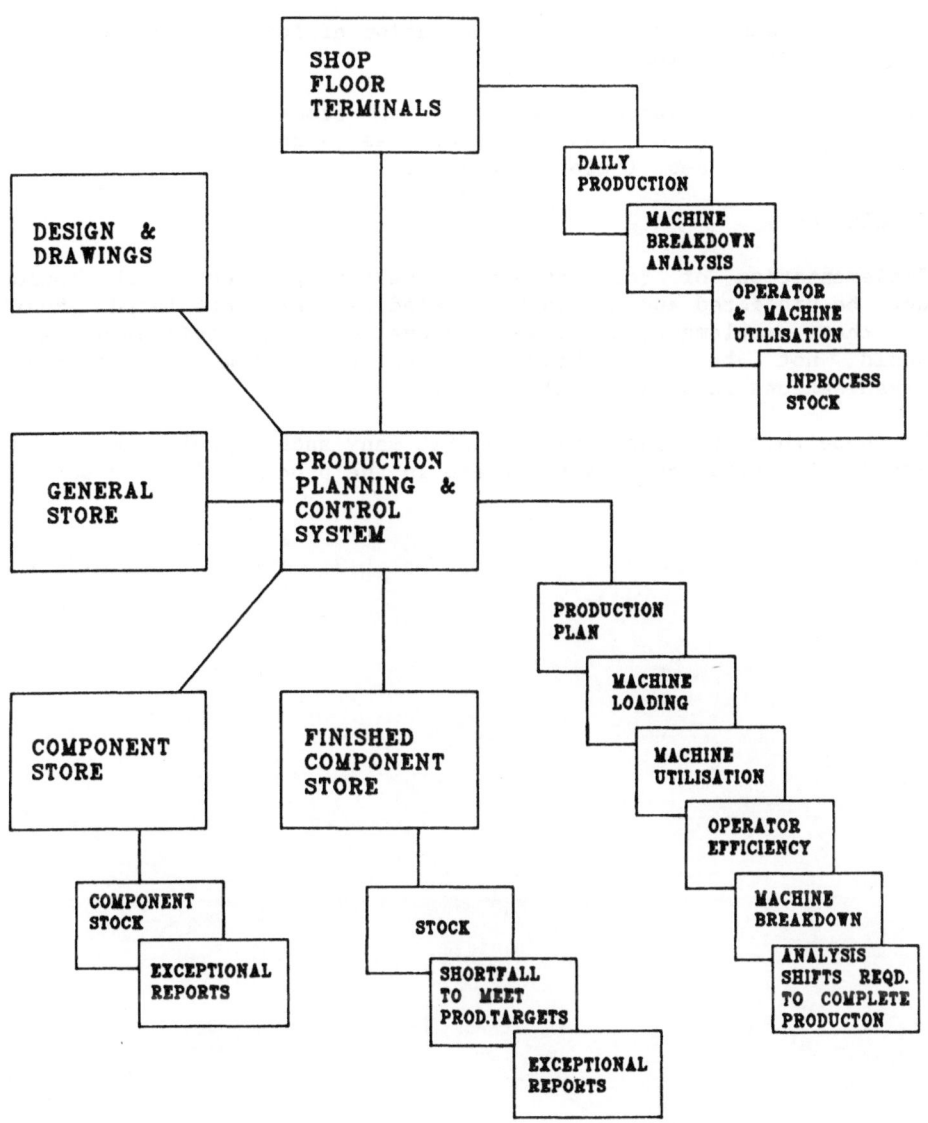

SHOP FLOOR INFORMATION SYSTEM
FIG-4

18

    c. Machine utilisation, operation efficiency and analysis of time lost.

In addition to above information on stock of specific raw material, status of indents etc. was also provided to the engineers on request (refer to Fig.4).

Conclusion

Implementation of FMS system posed several problems which could not be predicted and had to be tackled as these were faced. Most of these problems however, were unique to Indian Environment and could not be even thought of as a problem by similar organisations in Japan or elsewhere.

We hope that this paper will benefit many such organisations who are yet to instal FMS system or are in the process of doing so.

A Survey of CIM Strategic Planning in U.S. Industry

MARK. D. PARDUE
Old Dominion University
Norfolk, VA

FREDERICK J. MICHEL
Factory Operations and Automation
Alexandria, VA

ABSTRACT

This paper presents the results of a survey of successful U.S. manufacturers on their methods of strategic planning for Computer-integrated Manufacturing (CIM). The foundation for the survey was a literature search of relevant journals, trade publications, books, and industry and government documentation in the area of CIM. In addition, community-of-interest interviews were conducted to help determine successful companies with regard to CIM strategic planning. Those companies identified were interviewed to gather detailed information concerning their CIM strategic planning processes and the impacts of CIM on their companies. Information on CIM project justification, training, and organizational impacts was also obtained. This information is analyzed, and conclusions are offered by the authors as to what factors are important in CIM strategic planning.

INTRODUCTION

This report presents the results of a survey on the state of CIM strategic planning in the U.S. today. The survey consisted of three steps:

O   A literature search of relevant journals, trade publications, books, and government and industry documentation on CIM.

O   Community-of-interest   interviews   to   identify
     successful companies in the area of CIM strategic
     planning who have achieved superior results.

O   Interviews with the selected companies to gather
     detailed information concerning their CIM strategic
     planning processes, and the impact of CIM on their
     companies.

FINDINGS

Motivation to Implement CIM

The overriding motivation for the implementation of CIM
was survival, i.e. survival of product lines in the face of
increasing cost and quality competition, or even survival of
the company itself. CIM was viewed as the way to cut costs,
improve quality, shorten delivery times, and provide
flexibility.

Planning

The planning process used by most companies was a
combination top-down/bottom-up process. All respondents to
this survey indicated that their CIM planning team was made up
of   representatives   from   all   departments   including
representatives from the corporate CIM organization. For the
most part no outside expertise was used in their CIM planning.
The planning team reported to the Office of the CEO, Vice
President for Manufacturing, a CIM steering committee, or, in
the case of two defense contractors, to the Director for the
CIM organization.

All responding companies performed some type of financial
analysis for their CIM projects; however, the majority of

those companies did not use the results from these analyses as critical factors to justify a CIM project. Two companies indicated that payback was important except where a project was a foundation or an enabling project. Foundation or enabling projects are those projects that provide a basis upon which to implement other CIM projects, or are prerequisites for other CIM projects. Typically, the defense-related companies perform cost/benefit analyses, while other companies calculate return on investment (ROI), payback, or total economic impact on the company. One company simply targets the price for a new product and then determines how to meet or beat that target. Except for the companies receiving Department of Defense Industrial Modernization Incentives Program (IMIP) funds to subsidize their CIM efforts, all respondents stated that CIM projects are long-term investments, and that savings must include soft benefits.

All respondents indicated that top management commitment was a key factor in their success. Having all departments represented and having top management commitment seems to foster a great deal of cooperation, participation, and enthusiasm from everyone. Another primary measure cited was justification based upon company strategic objectives and mission needs, with cost justification a secondary for CIM projects in general, or at least for foundation or enabling CIM projects.

Implementation

Companies normally begin CIM implementation with a pilot program in a selected area. Although the respondents were not in agreement on which areas to implement first, the consensus was that the area piloted first should be strategically important and should have the potential of early success. The results of the literature search indicate that CAD/CAE/CAM has been by far the most prevalent CIM technology implemented; however, two companies that did implement this area as the first CIM project responded that the benefits from this initial implementation were marginal, and these benefits resulted from improved drafting productivity. Only after integration across the enterprise were the real benefits realized. These companies stated that in retrospect they would not select CAD/CAE/CAM as the pilot program. Establishing a top-level architecture is considered to be an important first step in a company's CIM implementation.

All respondents established a separate corporate CIM organization. This organization provides the CIM expertise for the company and participates in most CIM projects. Three companies most successful in implementing CIM, merged the Product Engineering and Manufacturing Engineering groups. This reorganization supports concurrent (simultaneous) engineering and avoids the "throwing it over the wall" syndrome.

Training programs for CIM are usually implemented throughout the organization, from shop floor to CEO. These programs are administered by in-house personnel except for specialized areas such as statistical process control (SPC) where contractors are brought in. Engineering and management staffs keep abreast of CIM technology through conferences, seminars, discussions with other companies and vendors in the industry, and through self-education. Most companies do not bring in outside expertise to plan and implement CIM. Usually a lead engineer is given the responsibility to become the expert in a CIM technology that the company may want to implement.

The relationships that companies have with their suppliers has changed with the implementation of CIM. They are now closer and involve fewer suppliers. Higher quality requirements and shorter delivery requirements to support JIT are passed on to the suppliers. With the advent of CIM, suppliers are viewed as venture partners.

All respondents to the question on accounting indicated that the standard labor burden accounting methods are unsatisfactory for CIM implementations, and that costing methods based on the contribution of each activity must be used. Some companies have already made changes to their accounting system long these lines, while the others stated that these types of changes will be forthcoming.

All respondents indicated that they had some type of quality program and that CIM was a major factor in improving quality. CIM reduces rework, scrap, and the number of inspectors. In their opinion, CIM assures that quality is built into the product and eliminates the need for a quality department. Some companies are pushing the responsibility for inspection of purchased parts back to the parts suppliers, while others are using the CIM data base to automate the inspection of incoming parts. When pushing the responsibility for inspection back to the supplier, the company typically continues sample inspection of an incoming purchased part until the quality performance of the supplier can be certified, sometimes on a part number basis. Several companies indicated that in-process inspection of parts has been automated using specification data in the CIM data base, and is now an integral part of the manufacturing process.

The most common implementation barrier mentioned by the respondents was lack of capital to adequately resource all of their needed CIM projects. Other problems were lack of technical talent, coordination between departments, and resistance to change.

Results

Several respondents gave an indication of the resulting cost-benefit or payback of CIM. Payback periods were from 2-7 years. The companies also indicated difficulties in quantifying benefits. CIM allowed all companies to either remain competitive or improve their competitive position. They were able to reduce costs, improve quality and reduce delivery times to customers.

SUMMARY

In reviewing and analyzing the data obtained from these world class companies, several factors emerged as being important to their success. They are:

O   Top management committment and understanding.

O   Strategic, long-term view.

O   Multi-disciplinary team composed of management, engineering, manufacturing, and shop floor personnel.

O   Top-level CIM architecture, with a set of hardware, software, and communications standards applied across the company but allowing flexibility for individual plant requirements.

O   Partnership with suppliers.

O   Establishing a corporate CIM organization.

O   Adapting a contribution-based accounting system.

O   Implementing a CIM training program for all levels of the company.

Modelling and Optimization of a Flexible Manufacturing System

R. N. CHAVALI, S. KESWANI and S. C. BOSE

Department of Mechanical and Manufacturing Engineering
Utah State University
Logan, UT

Abstract

A flexible Manufacturing System (FMS) is simulated using SIMSCRIPT with five workstations. Jobs are assumed to enter and leave an input/output station. Three different types of jobs arrive at the input/output station with interarrival times that are independent exponential random variables at a certain probability. Each job has a routing system before it leaves the manufacturing system. Service time for each task at a particular workstation is an independent exponential random variable. An Automated Guided Vehicle (AGV) is used to transport the jobs to the workstations. The AGV processes requests by jobs in a First In First Out (FIFO) basis. If the job transported by the AGV to a work station is found to be busy, the job will be placed in a single FIFO queue at that workstation. The FMS facility was simulated for 1000 hours and statistics were gathered for total delay for the jobs in queue, time average number of jobs in queue, average proportion of time that machines were idle, working, or blocked. Analysis were performed to optimize the number of AGVs required for transportation, job arrival rate in anticipation of the demand, and number of machines needed for uniform machine utilization. The results of the analysis are presented in the paper.

Introduction

A manufacturing cell consisting of five workstations and an input/output station is considered for simulation. Three types of jobs are manufactured. The jobs arrive at the input/output station with interarrival times that are independent exponential random variables. The probability associated with a given job which are of Type I, Type II, and Type III are 0.3, 0.5, and 0.2 respectively. Each job has different number of tasks to be performed at different workstations. The sequence of tasks, and hence the routing of the jobs is different. Every job enters the input/output station, goes through the routing, and finally exits at the input/output station. The job is moved from one workstation to the other on it's routing by an Automated Guided Vehicle (AGV). The AGVs process the job requests on First In First Out (FIFO) basis. If a workstation to which a job is transported is busy, then the job is placed in a single FIFO queue. Once the task has been completed the machine requests the AGV to unload the job and transport the job to the next workstation on that particular job's routing. Until an AGV is available to unload the machine, the machine is said to be blocked and cannot be used. The service times for the jobs of a given type at a particular workstation are independent random exponential variables with a constant mean value.

In a completely automated factory environment, where the components are being moved piece by piece rather than in batches, the effect of the transporter on delay time (the non value adding time) and machine block time is important. A balance between the number of transporters on one hand and the delay and block time on the other, has to be struck for economic viability or increase the profitability of the factory. Increasing the AGVs increase the capital investment while increase in delay time and block time adds to the cost of the product without adding any value to the product. A completely automated factory is capital intensive. Therefore all effort should be directed to keep the utilization at a higher level and at the same time ensure quick flow of jobs with minimal delay time. There should also be a built-in production expansion capability for an anticipated increase in demand. This necessitates the need to identify the potential bottleneck areas. Simulation plays an important role in this identification.

There are two approaches to simulation. First is the event scheduling simulation [1], and the other is process interaction approach [2]. In event scheduling approach it is required to concentrate on events (instantaneous occurrence which change the state of the system) and how they effect the system state. In the process interaction approach, it is required to concentrate on a single entity and the sequence of events and activities the entity undergoes as it moves through the system. In this paper the process interaction approach is followed. The process interaction approach is preferred as it describes the entire experience of a process entity and requires fewer lines of code than a comparable program using event scheduling approach.

Statement of the Problem

The manufacturing cell consists of an input/output station and five workstations. Presently the workstations 1, 2, ..., 5 consist of 3, 3, 4, 4, and 1 identical machines, respectively. There are three types of jobs with probabilities 0.3, 0.5, and 0.2. The routings and the mean service times for different job types are as follows.

Job Type	Workstations in routing	Mean service time for successive tasks (hours)
1	3, 1, 2, 5	0.5, 0.6, 0.85, 0.5
2	4, 1, 3	1.1, 0.8, 0.75
3	2, 5, 1, 4, 3	1.2, 0.25, 0.7, 0.9, 1.0

Automated Guided Vehicles (AGVs) are used to transport the jobs between the workstations along the job routings. If a workstation to which the job is brought is busy or blocked, then the job enters a single FIFO queue at that station. When a machine finishes processing a job, it requests the AGV to unload the job and transport it to the next station on the job routing. Till the time AGV comes and unloads the machine, the machine is said to be blocked and cannot be used.

The manufacturing cell is simulated for 1000 hours and the following statistics are obtained:

(1) The mean total delay in queue and the mean total transporter delay for each job.

(2) The time average number of jobs in queue and the mean delay in queue for each workstation.

(3) The average proportion of time that the machines are working, are blocked, and are idle for each workstation.

From the statistics gathered, analysis is performed to find the optimal number of transporters required, job arrival rate in anticipation of demand, and the number of machines required at each workstation.

<u>Analysis of Results</u>

SIMSCRPIT II.5 was used to program the simulation and runs were carried out on a VAX 8650. The program consists of basically two processes; one to generate the jobs, and the other to simulate the various tasks that a given job undergoes as it passes through the manufacturing system. Job types were considered permanent entities while the tasks in a given routing were treated as temporary entities. Simulation time was set to 1000 hours and the statistics were gathered as stated in the problem statement. Initial setup was designed for an expected demand rate of four jobs per hour. It was anticipated that in the future there would be an increase of 25% in demand over the current demand. Simulation was carried out for job arrival rates of two, three, four, five, and six jobs per hour. For each job arrival rate, the number of AGVs were increased from one to five and their effect on delay in queue, transporter delay, blocked time, and idle time were studied. Based on these values the bottleneck areas for increased production were identified. Having identified the bottleneck areas, the number of machines in the bottleneck workstation were increased to give a reduced delay time while maintaining the machine utilization close to 90%.

The delay for job type I as a function of job arrival rate and number of AGVs is given in Table 1. As expected, for demands below four jobs per hour, there is no change in the delay time for the job with the change in number of AGVs because the manufacturing cell is being under utilized. For a demand of four jobs per hour or more, the delay time decreases initially and then increases with the increase in number of AGVs. One would expect that the delay time to decrease or remain constant with the increase in the number of AGVs. This apparent anomaly is due to the fact that the jobs of type II which have higher process time than job types I and III at common workstations, enter the queues of these workstations earlier than job types I and III because of better transporter response.

Table 1. Delay time in queue (hours) for job type I					
Job rate	Number of AGVs				
	1.0	2.0	3.0	4.0	5.0
2	0.4	0.4	0.4	0.4	0.4
3	1.3	1.2	1.2	1.2	1.2
4	6.8	6.5	6.2	5.9	6.0
5	98.8	96.5	95.6	96.2	96.7
6	158.5	156.5	156.5	156.0	156.4

The average number in queue, average delay in queue, proportion of working time, proportion of block time, and proportion of idle time for job rates of 4 and 5 per hour are given in Tables 2 and 3 respectively. From these tables it can be observed that workstation 1 is the bottleneck area. By increasing the production capacity of workstation 1 (i.e the number of machines from 3 to 4), we shift the bottleneck area to workstation 5, which has only one machine. To improve the flow rate, it was necessary to increase the number of machines in workstation 5 from 1 to 2. The results after modification are given in Table 4. This reduces the delay time quite drastically and still maintains the machine utilization around 90% for all workstations except workstation 5. From the analysis it can be concluded that the number of AGVs required for transportation is 3 with an optimal job arrival rate of 5 per hour. The machine utilization in all workstations could be maintained at around 90% with 4 machines in workstation 1 and 2 machines in workstation 5. This avoids any pile up of jobs at any given workstation and maintains the flow of jobs.

Table 2. Statistics for job rate of four per hour					
Station	Avg. Num in queue	Avg. Delay in queue, hour	Prop. time working	Prop time blocked	Prop time idle
1	15.680	3.940	0.955	0.010	0.035
2	0.940	0.460	0.682	0.006	0.312
3	0.690	0.170	0.711	0.005	0.283
4	1.150	0.420	0.724	0.004	0.272
5	3.300	1.630	0.807	0.013	0.180

Table 3. Statistics for job rate of five per hour					
Station	Avg. Num in queue	Avg. Delay in queue, hour	Prop. time working	Prop time blocked	Prop time idle
1	458.050	91.890	0.986	0.010	0.004
2	1.380	0.610	0.769	0.006	0.225
3	0.980	0.220	0.764	0.006	0.231
4	3.960	1.210	0.867	0.005	0.128
5	7.460	3.310	0.883	0.014	0.103

Table 4. Statistics after modification (job rate of five per hour)					
Station	Avg. Num in queue	Avg. Delay in queue, hour	Prop. time working	Prop time blocked	Prop time idle
1	5.050	1.010	0.907	0.010	0.083
2	4.340	1.700	0.849	0.007	0.144
3	4.280	0.860	0.895	0.007	0.099
4	6.490	1.880	0.904	0.005	0.091
5	0.220	0.090	0.514	0.008	0.478

Conclusions

Simulation of a manufacturing cell was carried out by varying the number of AGVs, job arrival rates and number of machines in each workstation with an aim of finding a configuration that would increase the flow of jobs and maintain the machine utilization around 90%. The analysis shows that the delay time decreases initially and then increases with the increase in number of AGVs. The optimum was found to be 3 AGVs. Workstation 1 constitutes a bottleneck area with an increase in demand from 4 to 5 jobs per hour. Addition of one more machine to workstation 1 eliminates this problem at workstation 1 but shifts the bottleneck to workstation 5. Increasing the number of machines in workstation 5 from 1 to 2 reduces the delay time quite drastically while maintaining the job flow and uniform utilization of machines. Hence 4 machines are required in workstation 1 and 2 in workstation 5 for an increase in demand.

References

1. Gordon, G.: System Simulation. Prentice Hall, Inc. 1978.

2. Aburdene, M. F.: Computer Simulation of Dynamic Systems. Wm. C. Brown Publishers, Dubuque, Iowa 1988.

3. Banks, J; Carson, J. S, II;: Discrete-Event System Simulation. Prentice Hall, Inc. 1984

4. Law, A. M.; Larmey, C. S.: An Introduction to Simulation Using SIMCRIPT II.5. CACI, Inc. 1984.

Computer Based Safety System for the FMS - Management Logic

C.F. MARCIOLLI

Mechanical Department
Faculty of Engineering
University of Brescia
Brescia, Italy

Abstract

This paper refers to developments in a research project concerning a computer - based safety system for the FMS. The purpose of the project is to carry out an analysis of actual and potential danger situations, resulting from possible interactions between machine and machine, man and machine or machine and external objects present in the working area. The final objective is to modify the production plant so as to remove the above situations.
The aspect of the system discussed in the paper is that of its management logic.
As part of the work, certain algorithms necessary for the correct and rapid functioning of the system are put forward and discussed. The software architecture is also touched upon.

Note:
The work described was carried out under sponsorship of the C.N.R. (Mechanical Technologies Project).

==

1. INTRODUCTION

In previous works (ref. 1,2 and 3) it has been proposed to develop a safety system (SICUR) designed for mechanical plant for the production of piece-parts, in batches or singly, using computer assistance and normally referred to as FMS.
The idea xas bornn from the realisation that, in this type of plant, safety problems are radically different from those typically ecountered in conventional plant. Because of this, they should be faced by employing appropriate logic schemes. The solutions to the various problems, whilst subject to different laws and norms, are still substantially the same for each type of installation.
The principal function of SICUR is to prevent unwanted collisions between production system elements on one hand and people, objects and other elements of the same plant on the other.
In this way, a precise pattern of accidents, typical of FMS installations, has been arrived at.
For the time being, other accident patterns which are common in different types of plant have deliberately not been taken into consideration.
The SICUR system is based on the knowledge of the topological and kinetic characteristics of the machine tool forming the plant.
SICUR operates by acquiring data from the ambient space (environment).

When processed, these data make it possible to identify possible clashes between the system elements. After their processing, SICUR chooses and implements the best strategy for prevention of possible accidental collisions. Once the dangerous situation has passed, SICUR restores the system to normal operating conditions. The initial studies on the flessibility of the safety system have served to identify numerous aspects of the system requiring in-depth detailed investigation.

This paper presents the most recent results referring to the management logic of SICUR.

The logic operates essentially in three phases. The first of these consists of acquiring data on the motions of the controlled elements and of determining the maximum volumes which might be occupied by such elements as a result of any type of event. The second phase consists of indicating the potential dangers of a collision. The third and final phase consists of deciding on subsequent action designed to eliminate such a possibility.

Assuming thus that a system consisting of many elements has to be controlled and that all te necessary data referring to this system have been collected and processed, the SICUR management logic must necessarily proceed to determine, from time to time, the action strategy required.

It should be noted that, in order to simplify the presentation of the argumente, at least in this phase of the analysis, but without imparing its validity, it has been assumed that the spaces to be controlled are two-dimensional rather than three-dimensional.

In order to be able to carry out simulation tests on the management logic, it has been necessary to define certain algorithms, serving both for the determination of areas to be controlled and for taking decisions. This aspect of the problem has not, of course, been treated exhaustively.

2. DETAILED DEFINITION OF THE PROBLEM

As already mentioned, the problem is faced by assuming that SICUR is applied to a configuratoin of a known production system and that the operating environment is also known.

It is hence assumed that the geometrical, kinematic and dynamic characteristics of all the system elements (machine tools, robos, pallets, etc.) situated in the entire zone being controlled are known.

The possibility of the presence of human beings inside the operational area is also provided for. This is because, whilst normal operation is unmanned, the presence of human beings not only cannot be excluded, but generally happens rather frequently. In addition, it has been assumed that the interface with the ambient provides the instantaneous positions and instantaneous conditions of motion of all the elemnts present and controlled within the area being discussed.

Under the above conditions, the problem can be formulated in the following way:

- Given the geometrical, kinematic and dynamic parameters of all the elements present in the system, define the algorithms which permit the following:

a) to predict all the possible movements (including anomalous ones) of all the controlled elements at a defined time, and hence to determine the areas or volumes containing such movements,

b) to check wether the above areas/volumes overlap. If so, the movements could lead to collisions between the parts (elements) involved,

c) in case a possibility of collisions exists, to initiate action having the minimum impact on the rate of production of the system, while at the same time maintaning the safety level required,

d) to restore the normal functioning of the system as soon as the conditions, which have led to action on the part of SICUR modifying the planned succession of controls and the programmed characteristics of the movements, are no longer present.

3. SAFETY SYSTEM ARCHITECTURE

From the logical point of view, the optimal approach combines continuous control of the situation with the capacity for immediate action.

In practice, however, the two above requirements (continuous control and capacity for immediate action) appear to be impossible to attain.

This is because there always exists a delay between the acquisition of data on the situation by means of the interface with the environment (sensors) and their understanding as a result of suitable signal processing. In addition, there is also a delay between the moment of a decision, togheter with a command initiating action, and the carrying out of that action. The delay is caused both by physical limitations (inertia, actuation times etc.) and by causes related to the necessity to inform and synchronise all the parts of the system which work togheter to carry out the action decided upon.

It thus appears reasonable to abandon the assumptions of continuous control and immediate action and replace them by assumptions of discontinuous control (at predetermined time intervals) and of action delayed by the time essential for the carrying out of calculations, analysis, activation etc.

It is clear that, the more the time interva between successive controls is reduced and the shorter the action delays, the closer becomes control situation.

On the other hand, even if very fast processors are used, it is easy to see that the time variable is extremely critical. The architecture which has been developed (see fig. 1) has been essentially based on the necessity to reduce the times needed to carry out the controls to an absolute minimum. Because of this, the architecture is based on the subdivion of the system into logically independent local subsystems.

In each subsystem, control is exercised over a definite number of areas, determined on the basis of the input data. Some of these refer to elemnts which are always present in the subsystem, while others refer to entties which may enter into the subsystem. When not present they clearly have zero dimensions. All the subsystem are supplied with the same management logic.

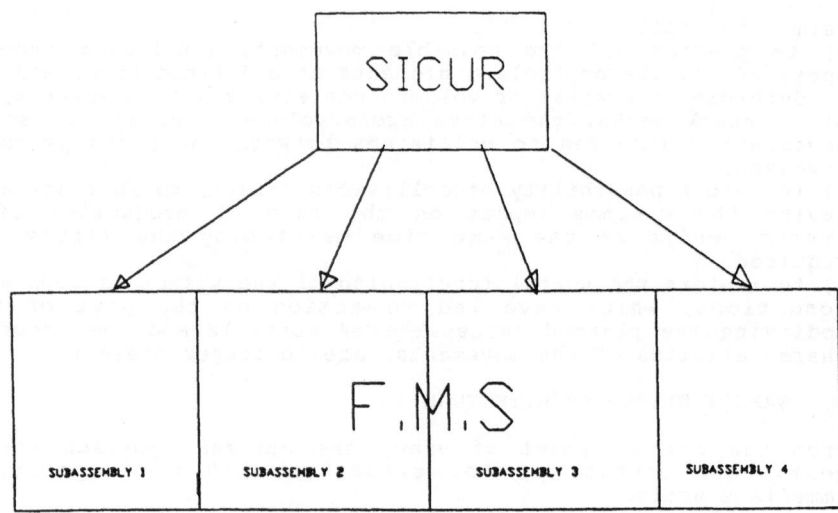

SUBDIVISION OF A PRODUCTION SYSTEM(FMS)INTO SEVERAL LOGICALLY INDIPENDENT SUBSYSTEM MANAGED BY SICUR

FIG.1 STRUCTURE OF THE SAFETY SYSTEM

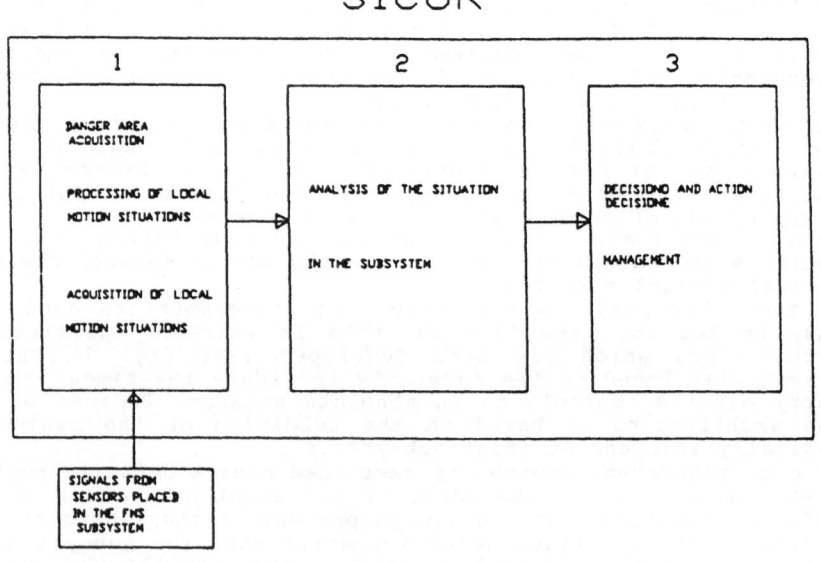

FIG.2 LOGICAL STRUCTURE OF SICUR

Given that, within an FMS, it is inevitable that certain elements (e.g. the transport system) form part of more than one subsystm, it is necessary to establish interconnections between all the SICUR units within the system. This is in order to hand over the parameters relating to the situation in each subsystem which many lead too changes inside any of the other subsystems. Needless to say, the interconnection of SICUR units creates problems of management, but at this stage of the analysis these are considered to be of minor relevance. In order to simplify the problem, at least in this stage of the analysis, it is assumed that the times needed to carry out the various activities are constant and equal to the maximum time occurring in practice.

On the above assumption, the times needed to sense the danger areas and to analyse the actions to be carried out appear to be substantially determined and irreducible, once the algorithms called into play have been optimised.

The problem can be divided into three phases, in strict time succession, which are reflected in a three-level hierarchical structure of the safety system.

Referring to fig. 2, the following three phases/levels have been indentified:

1 -acquisition of the situation within the subsystem,
2 -analysis of the situation within the subsystem,
3 -decision and the management of action taken.

The danger area acquisition level can, in turn, be subdivided into two cascaded sub-levels. The first of thse relates to the acquisition of the local motion situations, while the second relates to their processing up to a point where the danger areas are identified.

The first sub-level presents problems mainly concerned with sensing (interface with the environment), while the second calls for fast processors and optimised algorithms.

Because of the diversity of ways in which local data are acquired, the use of multiprocessor systems appears to be more effective at this level.

Th elevel of analysis of the situation within the subsystem (check on the overlapping between the danger areas and the definition of possible courses of action to remove it) also requires rapid processing.

As the interface analysis programme, however complex it may be, is single and common for all the areas being controlled, it must be managed by a single microprocessor. The same microprocessor might, alternatively, also manage the second sub-level of the first level.

The level of decision and action management constitutes, by its very nature, a non - divisible synthesis level.

Referring to the above considerations, one arrives at the following definitions of the hardware and software architectures:
HARDWARE (see fig. 3)

-a series of microprocessor - equipped units with the tasks of managing the sub-level of acquisition of the conditions of motion and hence of managing the interface with the environment (first sub-level of level 1),

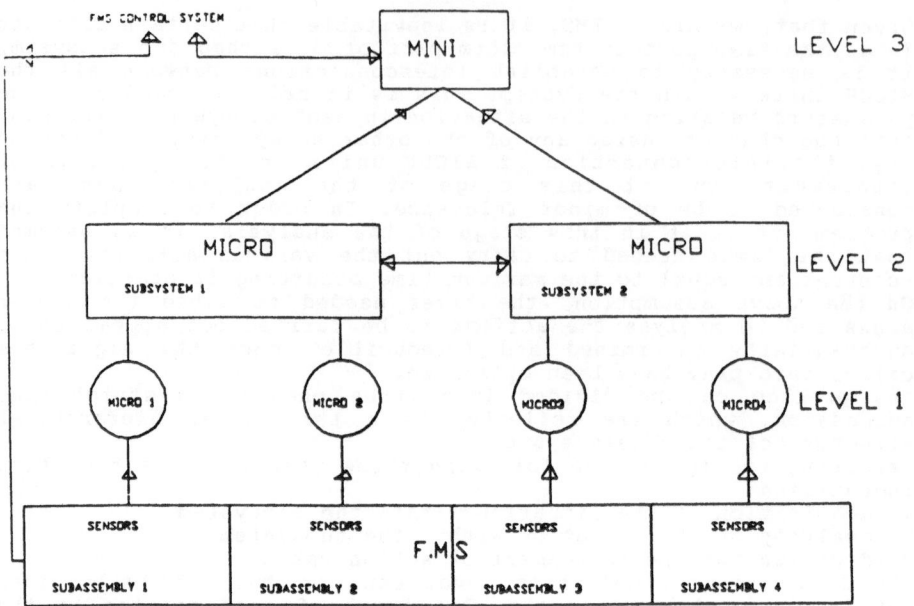

FMS CONTROL SYSTEM

MINI — LEVEL 3

MICRO — SUBSYSTEM 1   MICRO — SUBSYSTEM 2 — LEVEL 2

MICRO 1   MICRO 2   MICRO 3   MICRO 4 — LEVEL 1

SENSORS SUBASSEMBLY 1   SENSORS SUBASSEMBLY 2   F.M.S   SENSORS SUBASSEMBLY 3   SENSORS SUBASSEMBLY 4

FIG.3 SICUR HARDWARE ARCHITECTURE

POSSIBLE HARDWARE CONFIGURATION	SOFTWARE ARCHITECTURE	POSSIBLE HARDWARE CONFIGURATION	
M I N I	PROGRAMMES FOR DECISION-TAKING AND ACTION MANAGEMENT: 1) ACTION DECISION 2) INTERFACING WITH THE FMS COMPUTER	M I N I	LEVEL 1
	PROGRAMMES FOR MANAGEMENT OF THE SITUATION WITHIN THE SUBSYSTEM 1) VERIFICATION OF INTERFERENCES 2) COMPUTATION OF DANGER AREAS	MICRO	LEVEL 2
MICRO	PROGRAMMES FOR THE PROCESSING OF INCOMING SIGNALS FROM THE SENSORS	MICRO	LEVEL 3

FIG.4 SICUR SOFTWARE ARCHITECTURE

-a microprocessor for the analysis of the situation within the subsystem (level 2) and possibility with the task of processing of the data concerning danger areas (second sub-level of level 1),

-a minicomputer for the implementation of level 3 (decision and management of action taken). This mini could possibly be used also for the implementation of level 2. One could imagine that, for management of levels 1 and 2 might be employed.
This possibility is conditional upon the processing speed of the expert system or on its level of intervention.
An analysis along those lines is in the course of being carried out.
SOFTWARE (see fig. 4)
-a decision and action management module (level 3), implemented on the minicomputer, with the following tasks:
a) decision on the course of action to be taken,
b) management of the activation of the situation analysis module within the subsystem, in accordance with the various local degrees of danger,
c) interfacing with the contol units of other subsystemms and with the FMS control system, for passing information on the safety strategies adopted.
-Subsystem situation management module (level 2) implemented on a microprocessor or on the minicomputer, with the following functions:
a) check upon interferences between danger areas under control,
b) calculation of the effective degree of danger.
-Modules, implementedon microprocessor - equipped units, to determine danger areas by data processing. The units start by processing the condition of motion, then going on to work out the extreme limits which could be reached by all the subsystem elements/entities on the most pessimistic assumptions.
-Interface modules, implemented on microprocessor - equipped units, with the task of transforming th signals incoming from the environment interface into geometrical and kinematic parameters.
There exists also the possibility of synthesis for the identification of the nature of the intrusion.

4. ANALYSIS OF MOTION SITUATION AND DETERMINATION OF DANGER AREAS

Once the instantaneous conditions of motion have been acquired by means of interface with the environment, the determination of danger areas presents problems exclusively tied to the rapidity of execution of the algorithms which are formulated and optimised on a case by case basis.
It is interesting to note that, in order to determine danger areas, it may be generally convenient to refer to geometrically simple areas which circumscribe the effective danger area.
The dimensions of the potential danger area increase (clearly up to the limits imposed by the construction) as the scanning intervals between succesive check becomes greater. This intervals alsol depends directly on the time taken to carry out the necessary calculations.

It is thus clear that a reduction in the time taken for calculations needed to determine simplified areas and for the checks and action concernign possible clashes leads to a reduction in the scanning of the areas to be taken into consideration.

The above concepts can be clarified by referring to fig. 5. The minimum area of potential danger, which is morphologically closest to reality, is a function of the degrees of freedom and the kinematic characteristics of the element, of the time required for action by SICUR. This minimum area could be, at a given instant, the complex external one traced with short lines and made up of straight elements and of circular arcs.

Whenever we consider the rectangle (external, traced with a continuosu line) as the area of potential danger which circumscribes the area mentioned above, we certainly introduce an imporvement in the safety level, but only by departing from the goal of minimising the dimensions of potential danger areas.

Hovever, the ad hoc algorithm for the determination of the circumscribed rectangle uses much less processing time. It also shortens the calculation of interferences between rectangles rather than between areas of a more complex shape. Because of this, it is possible to reduce substantially the scanning interval, the number of elements/entities controlled being equal. In order to confirm the above statements, certain tests have been carried out using microvax.

It has been found that the ratio between the times needed for the determination of the complex area and of the rectangle containing it is 2.8 : 1,. The ratio between the times needed to check the interference between two areas and to evaluate the interfering entity is 30 to 1.

The overall time needed to detect interference between two areas of a complex shape is 3.2 times greater than the time needed for two rectangular areas.

The calculated rectangular danger area (internal solid line in fig. 5) will hence reasonably have a smaller surface than the minimum area determined using the other, more rigorous method. Despite this, it certainly contains the effective area of potential danger (internal, short lines). The latter can only be determined hypothetically, as it is still a function of the degrees of freedom and of the element, of the algorithms used for its determination and of the time required for action by SICUR. It is not, however a function of the new scanning interval.

5. MANAGEMENT OF A DANGER SITUATION

There is no logical bar to the interchange of information between the SICUR system, the central computer of the production unit and the various units controlling the machines involved. This is so in order to ensure that the kinematic parameters of the machines can be varied and that this can be taken into account in the overall management of the system.

For the time being, however, only the autonomous functioning of the SICUR system has been examined. This means that, when faced with the task of resolving an undesirable situation, SICUR relies solely on the use of resources intrinsic in its functioning logic

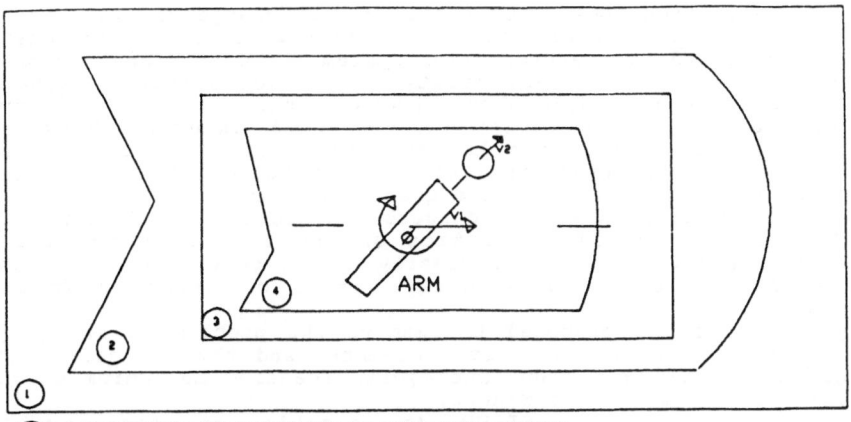

FIG.5 INFLUENCE OF THE SCANNING INTERVAL AND OF THE
COMPUTATION METHOD ON THE DANGER AREA

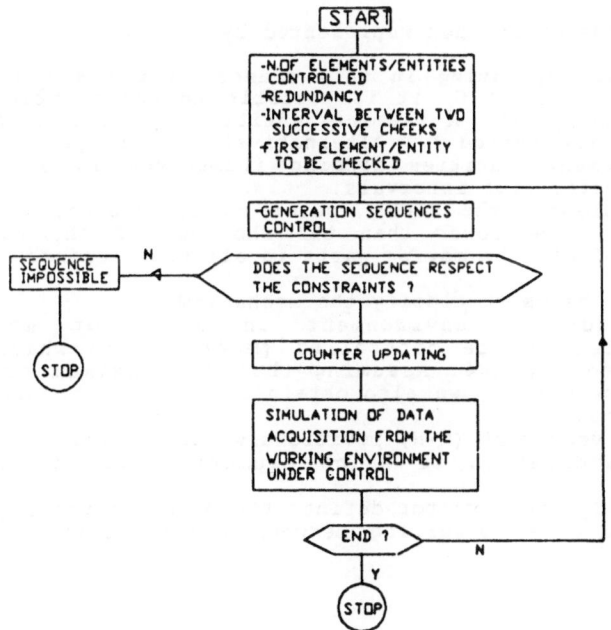

FIG.6 FLOW-CHART OF THE SEQUENCE ANALYSIS PROGRAM

On this assumption, the SICUR system cannot change the kinematic parameters of the machines. The only action which it can and must take with regard to the production system is to stop one or more machines, once the possibility of unacceptable clashes (interferences) is established. However, the degree of safety depends also on the lenght of the time interval between successive check (it should be noted from ref. 1 that, on the most probable assumption, the relationship between the movements and the time intervals follows a quadratic law).

Thus, when confronted with a dangerous situation, SICUR can act by "concentrating" its attention on the zone in which this situation has arisen. This is clearly at the cost of other zones where, in any case, an adequate degree of safety must be maintained.

In practice, it is proposed to resolve the problem by managing the analysis of different system elemnts, and thus of different danger areas, using a redundant cyclic scanning mechanism which operates in a competitive manner.

Let us assume that SICUR is capable of managing M elements/entities and that the longest time required for managing any element/entity is G. (The notion of management denotes here the acquisition of information from the environment interface, the processing of the above, the comparison with other elements/entities controlled and the activation of strategies required to maintain an adequate degree of safety).

If we conservatively assume that G is the time needed to manage each element, it is clear that SICUR is capable of carrying out check on the same element/entity at intervals of:

$I = MG$

Let us now suppose that the actual interval used between checks is:

$I' = (M+H)G$

The system redundancy is then represented by:

$R = HG$

It is clear that, operating in this manner in the H redundant time intervals of lenght G, it is possible to manage more than once all or some of the M elements/entities up to a total of (M+H) checks in the period I'. Alternatively, it is possible to recheck the elements/entities finding themselves in a danger situation at shorter time intervals.

In order to implement the SICUR system as postulated, I' must clearly have a value lower than the shortest of the minimum stopping times of all the moving parts in all the elements.

A simple example helps to clarify the mechanism

Let us consider an environment in which at most 4 elements/entities can be controlled (M=4) and in which the operating redundancy R=4. Assuming that the maximum control (management) time G is 1, we also obtain H=4, while the scanning interval I' becomes 8.

We can define a vector of (M+H) positions which contains, in each position, an indication of which element/entity is to be analysed.

Scanned cyclically, this vector defines the order of the analysis and the maximum delay occurring between two successive checks

(analyses) of the same element/entity.
Initially, with no danger situation insight, the vector is made up as follows:

Position no. :	1	2	3	4	5	6	7	8 :
Element :	A	-	B	-	C	-	D	- :

where A, B, C and D denote the elements/entities.
In the above table, the dashes represent the H time intervals during which SICUR can remain in a waiting mode. From this it can be noted that the interval I' between two successive scans is effectively 8 time units. However, if redundancy were removed, the scans could follow each other every 4 time units, according to the following table:

Position no. :	1	2	3	4	5	6	7	8 :
Element :	A	B	C	D	A	B	C	D :

When the analysis of area A indicates a danger situation, it is possible to modify the vector in such a way that the interval between two successive scans of element/entity A is reduced, for example, to a maximum value of 4 time units.
The vector configuration is then as follows:

Position no. :	1	2	3	4	5	6	7	8 :
Element :	A	-	B	C	A	-	D	- :

If all the G management ·intervals are utilised, it is possibile to scan element A every 2 or 3 units of time, while B, C and D are scanned every 5 units, according to the following arrangement:

Position no. :	1	2	3	4	5	6	7	8 :
Element :	A	B	C	A	D	A	B	C :

It is clear that, by using the redundancy of the mechanism to its maximum, the scanning of a single element can take place every 2 time units, while the other three elements are scanned every 6 time units. Alternatively, all the elements/entities can be scanned every 4 time units.
Taking this to the limit, if the other threee elements/entities can tolerate a scanning interval of 7 or 8 time units, there exists the possibility of carrying out, on the single element A, one or more sequences of two or three consecutive checks.
In the first of these two cases, the vector assumes the following configuration:

Position no. :	1	2	3	4	5	6	7	8 :
Element :	A	B	A	C	A	D	A	B :

It may be possible to manage the system so as not to arrive
necessarily to the stopping of machines in case of danger,
followed by a resumption of the normal sequence of check as soon
as possible, meanwhile maintaining the essential safety
requirements. This, however, requires a suitable management of
the vector referred to above, togheter with a protocol defining
the choice of scanning frequencies as a function of the degree of
danger, linked to the presence of an interference condition
between areas or of a reduction of the dangerous situation
(increase of the distance between the areas in question).
The redundancy - based competitive management mechanism makes
possible the return to the normal control sequence. It should be
understood, however, that the return parameter, being linked to
the minimum distance between the areas, has to be supplied with a
hysteresis provision. This is to avoid problems of resonance
between the danger situation and the normal one, which might
otherwise arise.
This means, in practice, that after the interference between the
two areas has ceased, the time interval between successive checks
on the element in question is maintained constant until the
distance between the two areas exceeds a predetermined limit. By
doing this, the possibility that the return to normal operation,
following only the separation between the areas, might lead to a
reappearance of the danger signs upon the successive scan. This
would aggravate the management situation of the control sequence.
It should finally be noted taht the cyclic method of scanning has
been chosen in preference to random access. This is in order to
make quite certain that, during the time interval I', all the M
elements/entities under control are always scanned.
The efficiency of this mechanism has been checked using a
computation programme which determines the checking sequence as a
function of the situations detected in the field and verifies its
agreement with the constraints imposed.
The programme requires, as input data, the number of
elements/entities to be kept under control, redundancy to be
provided, the interval between successive scans of the same
element/entity and the first element/entity to be checked. It
then establishes the first succession of checks necessary to
verify the agreement with the redundancy assigned.
At this point, an automatic procedure is started up, which
updates the sequence from time to time, making sure that it
always reamians within the redundancy limits and that constraints
relating to priority in checking are respected as the situation
evolves.
The piece of information which must be acquired each time for the
recalculation of the successive sequence is the interval between
two successive checks on the element/entity being examined. This
is determined on the basis of a simulation of the detection
effected on the environment. The intervals between two checks on
all other elements/entities are maintained constant unitl the
scan and analysis which follow.
For each sequence generated, the programme works out the residual
redundancy which could be considered an indication of the
criticality of the situation. The impossibility of determining a
control sequence respecting the constraints is clearly indicated

by the simulation programme, with a statement of the reason for
the failure.
A block schematic of such a programme is shown in fig. 6.

6. CONCLUSIONS

As pointed out already, the results obtained in thid phase of the
research are not definitve. They require a detailed verification,
in parallel with a complete set of tests on the safety system in
an environment which resembles the one to which the system is to
be applied.
In our opinion, however, the SICUR management logic need not be
changed substantially, unless this is required by the hardware
which must be employed to render the safety system efficient.
The continuation of the research is closely connected with the
analysis of many other aspects of the problem.
Some of those, on which work is proposed within the near future
with the aim of analysing certain possible plant solutions are:
a) definition of an interface (sensors and transducers) towards
the environment. The interface makes it possible to recongice who
or what is present in the space controlled. Should this body be
moving, it should be possible to determine the principal
parameters of the motion, preferably using a direct method,
b) clear discriminations between correct and anomalous
movements, as wll as between events which are tested and
foreseen, hence basically safety on the one hand and events which
are not foreseen or not recognised as normal within the system,
and hence potentially dangerous, on the other. For example, let
us consider the difference between a moving part getting close to
an unforeseen obstacle, with SICUR intervention required, and the
element getting closer according to a prearranged schedule, like
to a piece part on which the element has to carry out an
operation. In the second case, no SICUR intervention is
necessary.
c) definition of algorithms for the calculation of maximum
areas/volumes and for arriving at subsequent decisions. This
refers to the assumption that SICUR should act as soon as
interference occurs between the maximum areas/volumes which might
be occupied by the controlled element in a time interval which is
a multiple of a minimum time linked to the characteristics of the
production plant and of SICUR itself.
d) optimisation of such algorithms to ensure that they require
the minimum possible time for their execution. This is to
guarantee that the delays in SICUR action are not large enough to
impair its reliability.
A further specific subject in the continuation of the research is
the choice and evaluation of sensors and transducers forming the
interface with the environment.

7. REFERENCES

[1]Bonfioli M.-Marcolli C.F-Noe'C.:
    ''Safety system proposal for automated production''
    ''Robot safety''IFS Pubblication ltd,1985

[2]Bonfioli M-Marcolli C.F-Noe'C.:
    ''Maintenance and safety in FMS:problems and prospects''
    7th EFNMS Congress,Venice,may 1984

[3]Marcolli C.F-Noe'C-Turco F.:
    ''Safety,reliability,and maintenance in FMS:
    research near builders and industrial users
    of FMS''
    Pixel-ED.Rostro-jan 1986

[4]Bareggi F.-Federici A.-Marcolli C.F-Noe'C.:
    ''Methodology for planning safety system of automatic
    production plants''
    La rivista di meccanica- ED.Tecniche Nuove- oct 1986

[5]Hasegava-Sugimoto:
    ''Industrial safety and robots''
    11th ISIR 1981

[6]Safegguarding industrial robots-MTTA-1982

CIM Repositories

H. T. GORANSON

Science Applications International Corp.
Munden Point
Virginia Beach, VA

Summary
Software infrastructure for CIM and Concurrent Engineering is
now focusing heavily on the support of integrated models in
global repositories. However, major technical barriers still
exist. This paper reviews the problem, gives some context, and
indicates a promising approach.

Background

The basic engine of the world's economy is manufacturing. A
manufacturing enterprise typically includes a diversity of
activities: extraction of customer requirements; design and
engineering of the new product; devising and optimizing the
manufacturing process; making the item(s); and managing sales,
inventory and shipping. It is common for an enterprise to
include a number of other corporations in teaming or
subcontractor relationships. Sometimes the enterprise extends
through the life cycle of the product for continuing
maintenance, training and improvement. This is especially the
case with military and energy systems. In addition to
including this wide diversity of interests and functions,
manufacturing enterprises are frequently very large and/or very
complex.

Computerization has come to these enterprises in a manner that
is proving to be troublesome. Various functions within the
enterprise have automated as it has become appropriate for them
to do so. But for a variety of technical and business reasons,
many of these systems are distinct, representing *islands of
automation*. Industry is now learning the lesson that a

collection of suboptimized processes does not add up to an optimized enterprise; and this is more true as the enterprise becomes more intricate.

The Problem

In simple terms, the problem is how to integrate the enterprise. Clearly, this is the central problem faced by Concurrent Engineering (CE), Computer Integrated Manufacturing (CIM) and Computer aided Acquisition and Logistic Support (CALS). Any of these initiatives can lay claim to subsuming the others, depending on whether the initiating focus is design, manufacturing or support. For the sake of this paper, we use CIM as representative of the class of problems addressed by all.

The design, manufacture and support of military aircraft are among the most complex of enterprises, so it is no surprise that the Air Force Manufacturing Technology Directorate has taken the lead in addressing this problem. Their series of Integrated Computer Aided Manufacturing (ICAM) programs has now evolved into an Enterprise Integration Framework (EIF) effort with an associated Enterprise Integration Program (EIP). Planning for those programs has been supplemented by work done in Europe under the CIM Open Systems Architecture (CIMOSA) effort and in the US under sponsorship of the SEMATECH consortium. An unofficial consensus of the components of the problem is emerging as shown in Table 1.

POLICY AND CULTURE
TECHNOLOGY

  APPLICATION TECHNOLOGY
  INFRASTRUCTURE TECHNOLOGY

    CONNECTIVITY AND INTERFACING
    INTEGRATION

      COMMON EXECUTION ENVIRONMENT
      COMMON APPLICATION ENVIRONMENT

      CASE TOOLS AND METHODS
      INTEGRATED MODELS

TABLE 1

The table shows that at the highest level, the problem is as much one of business policy and culture as it is of technology. Resolution of high level issues simultaneously in both policy and technology areas is essential to any integration strategy. The technical issues are resolved into those which address *applications* and those which address *infrastructure*. Application technologies are concerned with the specific methods employed to support various processes in the enterprise. They are not a problem in this context; there appears to be a continuous supply of innovative application packages.

Infrastructure technologies support the applications. This includes traditional infrastructure services such as operating systems, communication and data management. However, new services are being demanded by users to bridge the islands of automation. The table resolves these into two tiers, *connectivity/ interfacing* and *integration*. The former is often the exclusive focus of an integration strategy. Such is currently the case with CALS Phase 1.

Connectivity is the ability of machines from different functions in an enterprise to connect to each other, even though they might look quite different internally. This problem has been the focus of numerous efforts, beginning with Manufacturing Automation Protocol/ Technical Office Protocol

(MAP/TOP), evolving to the international standard Open Systems Interconnect model (OSI), and industry maintenance through the Corporation for Open Systems (COS).

The second level addresses *transfer*, that is the ability to get a message from one part of an enterprise to another in a consistent way. The ability to deal with transfer of files and data is often termed interfacing, and is addressed by a very large family of specifications, most of them currently or proposed standards. Recently, a users' union, the Houston 30, was created to accelerate the introduction of connectivity and interfacing standards to the marketplace. The Open Software Foundation also supports many of these fundamental requirements.

At one time, many believed that connectivity and interfacing were sufficient to integrate islands of automation. It is now recognized that a third component is required, the most difficult of them all.

Integrated Enterprise Framework
This problem is a natural result of why the islands of automation were built in the first place, and is shown in the table as *integration*, differentiated from interfacing. Each island is built around a group of application programs which structure and manipulate data in a way that is best for that island, but probably not for another. Generally, this structured data is seen by users and applications as *models*. So we find financial models, management models, engineering models, process models,... the list is prodigious and growing daily as innovative application programs are addressed.

This problem of creating a framework for truly integrated enterprises is not yet totally solved, though major elements of the problem are well understood. The table shows that the problem is resolved into two major elements; a *common execution*

environment (CEE) and a *common application architecture* (CAA). The
CEE consists of the real platforms and operating services
supplied by vendors. Standard capabilities required by such an
environment for integrated systems are well understood and are
one deliverable under the Air Force EIF program.

More important is the ability for the CEE to be modeled
unambiguously by the CAA. This will allow various applications
to understand the different CEE's with which it is presented
and make adjustments accordingly. Key to this common CEE model
is a universal parsing of services based on logical services
and service subsumption. An example of such a CEE services
subsumption hierarchy is shown in Table 2.

```
LOGICAL BUS
  USER SERVER
  REPOSITORY SUPERSERVER
        CASE SERVER
             VERSION SERVER
        DATA SERVER
             INDEX SERVER
        KNOWLEDGE SERVER
             WINDOW SERVER
        NETWORK SERVER
             INTERBUS SERVER
                      TABLE 2.
```

Note that this parsing of the CEE emphasizes repository
services. The "bridge" between the CAA and the CEE is the
assembly and delivery to the user and/or application of
integrated models which are acclimatized to their environment.
The integration strategy therefore depends heavily on the
underlying technologies of how the models are integrated, and
what services the repository provides.

It is easy to predict with confidence what shape repository
services will take in the future. Both IBM and DEC have major
product offerings which provide a foundation, and next steps
for consolidated action may transpire through future
Information Resources Dictionary System (IRDS) standards;
repository work being performed by the Product Data Exchange

using STEP (PDES) standards activities; and the Object Request
Broker specification of the Object Management Group. As a
result, the CASE tools and methods component of the CAA
probably is in good shape. The primary challenge to a
successful CAA, and resulting repositories, is an improved
understanding of *integrated modeling*.

Integrated Models

Consider the case where a user is designing parts for an
aircraft engine, a case typical of enterprises. The user may
have one type of modeled information as an (artificial
intelligence) knowledge base consisting of a collection of
design rules....indicating various business, safety,
manufacturing types of things. The user would need another
type of model, a geometric model to tell what the shape of the
turbine blade (or whatever) was, sometimes called a 3-D model.
Also important is the parametric model of the fluidflow around
and through the turbine. It is called a parametric model
because it collects all of the parameters from the equations
used as the model basis. Spread sheets are a simple type of
parametric model.

These are fundamentally different types of models. Many
enterprises contain thousands of these models, discriminated by
different modeling methods and local control. Moreover, many
of them change frequently based on the perception of others'
models, themselves changing. Temporary "integration" at the
workstation level could conceivably solve one user's problem,
but in the general case, the models must be fused in a
repository. There have been a number of solutions of the non-
repository, "brute force", type tried to bring order to this
problem, but none have proved satisfactory. These are not
surveyed here.

Possible Approaches

It has already been indicated that at one time many people believed that interfacing alone was a sufficient answer to this problem. There still is a sizable community who feels this way, including many in the government. Moreover, there are many cooperative initiatives and existing products which are still based on this assumption. In terms of bonafide possible solutions to the problem itself, there are roughly four major approaches which can be taken. These can be indicated by analogy to the problem of differing spoken languages.

Even though there are thousands of languages spoken over the world, linguists have discovered that those actually used are only a very small portion of the languages which are theoretically possible. The reason the languages are limited to this small portion of possibilities has something to do with how the human brain is constructed. At a basic level, all humans' brains are wired up fundamentally in the same way. This means that at some level, French and, say, Japanese languages have the same basic skeleton.

Linguists call this *deep structure*, and the study of linguistics now focuses on discovering and refining our concept of what this deep structure is. Even with today's imperfect understanding, intelligence gathering agencies can extract deep structure from messages of different types and languages and integrate them into a deep structure skeleton. They can then get a "big picture", if there is one, from all of the component bits.

In the case of languages, as in the case of models, this underlying structure has two main elements, *syntax* and *semantics*. It is bending the definitions to simplify them thus, but semantics deals with the meaning of the components, and syntax deals with how those components are combined to convey meaning. This brings us to four possible approaches. Syntax and semantics appear to be closely related, at least in

everyday languages and models. So an approach to complex deep
structure could be made up of simple semantics if the syntax
was very complex, or vice versa.

The four possible approaches are: Complex Syntax/ Complex
Semantics; Complex Syntax/ Simple Semantics; Simple Syntax/
Simple Semantics; and Simple Syntax/ Complex Semantics. The
former two are termed *hypersemantic* because the deep structure is
the combination of all of the component structures. The latter
two are termed *metasemantic* because the deep structure is the
overlap of the components.

Complex Syntax, Complex Semantics
The most straightforward approach to deep structure for
integrated modeling is to expand both the syntax and the
semantics as required. This is informally known as the
"esperanto" approach, where the capabilities of the methodology
are derived from a detailed survey of the syntax and semantics
of all the "source" models. Critics point out the substantial
penalties for maintaining such high degrees of complexity,
maintaining that the problem has not been simplified at the
higher level, as one would expect.

Supporters of the approach emphasize the direct relationship
between requirements and capabilities in grammatic (ie
syntactic and semantic) space. On a single case basis, the
result is a sizing of the capability to the required robustness
of the integrated modeling requirements. However, as a
candidate for a global reference, the full spectrum of
esperanto complexity would have to be maintained. This is
clearly impossible.

This approach originated in the artificial intelligence
community. The best examples of research within this

methodology are PACIS[*], supported by the Air Force, and
HyperClass[**], supported by DARPA. PACIS supports the
hypersemantics in its grammatic design, and HyperClass does it
in a more traditional ad hoc fashion.

Complex Syntax, Simple Semantics

A more common approach to hypersemantic integration is to keep
the semantics as simple as possible. This pleases users,
especially if the semantic requirements for integration are
fewer than are required for the application proper. The
penalty is to increase the syntax. However, the tradeoff is
not equal. Complexity in syntax increases at a dramatically
greater rate as hypersemantic complexity increases.

The two best examples of this approach deal differently with
the problem. ROSE[***], sponsored by DARPA has the cleanest
solution. It localizes the added syntax in a model wrapper,
and avoids maintenance of all but the simplest constructs (such
as and/or constructs) in the global space. PTECH[****], which has
a CALS application sponsored by the Air Force does maintain
syntactic complexity globally. Much of this is successfully
hidden from the user, but most is not. Consequently, the
system may be difficult to scale up for enterprise-wide
hypersemantics.

Simple Syntax, Simple Semantics

The current state of the art employs this method. It is
inadequate for the metasemantic complexity requirements of

---

[*] Trademark of ONTEK Corp., 22951 Mill Creek Rd., Laguna
Hills, CA 92653.

[**] Trademark of Schlumberger, information from Center for
Integrated Systems, Stanford University, CA 94305.

[***] Trademark of Rensselaer Polytechnic Institute, Troy NY,
12180.

[****] Trademark of Associative Design Technology, 2 Westborough
Business Park, Westborough MA 01581.

enterprise integration. However, it is important for two reasons. The first is that it works well for what it does, the first criterion of endearment. The second is that a huge investment in initial, albeit low-level metasemantic models is currently being made. Any successful approach to integrated models must incorporate these heritage models. This approach is that employed by PDES and IRDS, financially sponsored in the government by CALS and the Air Force.

Simple Syntax, Complex Semantics

The reason why this fourth choice is not populated by any widespread examples is the immense difficulty of devising a simple syntax which can support a complex semantics. The problem is that successful syntax needs to be created according to laws or rules, and it is always easier to have a whole lot of rules than it is to have a few, simple rules which do the same job.

This is exactly the situation in other sciences, for example physics. Physicists look for the few, simple rules, in this case basic laws of the universe. The fewer and simpler the rules, the better as a skeleton. The laws then provide a basis (a syntax) for modeling the physical nature of the universe, using the semantic flesh of mathematics. So it is with the metasemantic modeling problem.

It is our opinion that this approach is the one required for successful enterprise integration. Its most attractive features include its natural evolution from existing efforts, both in supplier repository strategies, and in current standards research. A powerful technical advantage is that if a simple syntax can be maintained, the semantic complexity increases much more slowly than the complexity of the source models. As a result of the abstraction inherent in the approach, this guarantees that repositories can be simply and reliably maintained in real enterprises.

The challenge is to find basic laws of deep structure, or *metastructure*, to support a simple, yet robust abstract syntax. Original research performed by our group has indicated a promising approach to such metastructure as a concise set of grammatic primitives, with strictly formal mechanics.[1][2]

Conclusions

Competitive, integrated enterprises depend on integrated models and coordinated repository technologies. Successful strategies in this regard are dependent on sparse elegance in metasemantic deep structure. Recent study into the nature of the integrated enterprise problem has given us new insights into requirements for such services, and recent discoveries in metastructure primitives may provide a basis for future research.

References

1. Goranson, H. T., Deep Structure for Activity Modeling, Proceedings of the Second Annual Concurrent Engineering Symposium, Concurrent Engineering Research Center, Morgantown West Va., 1990.

2. Goranson, H. T., Deep Periodic Structure in Semantic Nets and Related Operating Grammars, in Computers and Mathematics with Applications, to appear, 1990.

The Selection and Prospect of CAD/CAM System for Diesel Engine Design and Manufacturing

GU ZE-TONG and HU GANG

Shanghai Marine Diesel Engine Research Institute
Shanghai, China

1. FOREWORD

It is highly necessary to use the CAD/CAM system in machine design and manufacturing because it will help the designer to modify the design method and thus promote productivity. According to traditional design method, the first step is the projective design; the second step is the detail design; then the last step - where the technology of machine components are considered - the manufacturing design. Furthermore, the prototypes of a machine are built, tested, and refined until the technical assignment is satisfied. To execute this design method, a long time is needed and it is too expensive to build the prototype of the machine and to carry on the testing process. Now the design - refine cycles are carried on in the CAD system partly to replace the pure prototype - testing cycles that can save vast amounts of money, and will reduce the period of design and manufacturing.

According to John Crouse's statistics in America, the CAD system is used for the engineering drawings only, and its efficiency is three times the manual drawing. If the interactive feature of the CAD system, i.e., the interactive feature between graphics with analysis system and between graphics with technical condition database is completely utilized, the design efficiency will be increased more than 10 times and the quality of products will be upgraded. Because the computer is good at working analysis and calculation, we can calculate in detail: the combustion of a diesel engine in a cylinder; the matching of a diesel engine with a turbo-charger; and the high-cycle and low-cycle fatigue. Of course, for the analysis of reliability of engine components, the application of the CAD system is more important. In addition, from the aspect of manufacturing, the application of the CAPP, CAM technique will significantly reduce the period of preparation of the technology file and the period of manufacturing. The efficiency of these periods can be raised about 5 times. The numerical control card can be made directly by the computer (including the cutting tools, selection of cutting allowance and cutting order diagram). The path of rough and precise cutting can be defined and the medium of numerical control data can be produced. Therefore, the automation of manufacturing of components can be enhanced and the reliability of components can be warranted also, because it will always avoid the error of workers and thus promote the product quality definitely.

2.   FEATURE REQUIREMENTS OF CAD/CAM SYSTEM

For diesel engine design and manufacturing, the following special requirements should be provided:

2.1 Apart from the individual component, the components in a diesel engine are constructed by a 3-dimensional solid body. For example, the piston, cylinder head, body, crankshaft, and so on are wholly constructed by a solid body, so the 3-dimensional solid modeling software must be provided.

2.2 Relating to these components, the 3-dimensional isoparametric element must exist in finite element analysis software package.

2.3 In a diesel engine and gas turbine, there are thermal stress problems as well as contact problems. These problems are rare in some finite element analysis package.

2.4 In addition, in power machine design the fluid dynamics calculating programs are also needed for computation of the velocity field and concentrations

2.5 In diesel engine manufacturing, the precise numerical control machine tools and fixtures are needed, so the best accuracy of CAM software is required.

Due to the above mentioned items, the CAD/CAM system of diesel design and manufacturing must have the following features:

1) To guarantee the accuracy of solid modeling, the CRT of the work station must have high resolution and with graphic accelerator and graphic processor. It is better to have the workstation with a Z-buffer.

2) The system must provide 3-dimensional solid modeling software and finite element analysis software. The system must also be capable of establishing a standard components library and tolerance - clearance library.

3) The system ought to have parametric design ability; it will satisfy the design of a series of machine products.

4) The system's software must provide the CAM feature and then the tool path of working cycles can be produced automatically. It also can process the technical information and produce the numerical control data for numerical control machining.

3. THE PRINCIPLE OF SELECTION OF A CAD/CAM ·
SYSTEM FOR DIESEL ENGINE DESIGN

3.1 Reliability

The reliability of the CAD/CAM system is very important. After installation, the system can be used at once. The selected workstations must be supported by different software systems, and CAD/CAM software must be established on a unified engineering database. The graphics from the CAD system can be transferred to the CAM system conveniently and perfectly.

The CAD/CAM company must have good credit, and the service support must be very strong.

3.2 Advancement

The speed of the CPU in the workstation must be calculated quickly and the workstations must have a graphic accelerator and graphic processor. It is better for the Risc-Unix system to be used as an operating system.

The software system must be the best CAD/CAM system which is rich in excellent features, such as the 3-dimensional solid modeling software, curved surface producer, finite element analysis package, 3-5 axis surface cutting, cavity cutting, and tool path calculation ability for many types of machine tools with numerical control.

3.3 Utility

The selected workstation must have large internal and external storage because the operating software, CAD/CAM software and graphics need several storage units. As a result, the hard disk storage must be over 500 MB. The database must be able to operate the graphic data and other data files. In addition, the system must be provided with the following two features: these are graphic management and bill of materials feature. It can produce the drawings which satisfy the engineering drawing requirements.

3.4 Economics

The system which satisfies the above mentioned feature requirements is a perfect system; its price/performance ratio must be fine.

3.5 Network ability

The system communication and network ability must be strong and all materials in each node can be shared.

3.6 System Opening

The system must be open for users, and they can develop the application software by the simple tools including in the system, i.e., it must be possible to add secondary developments. The system should also have the graphics interface for users to develop their own application programs.

3.7 Compatibility

The selected CAD system must be compatible with old type workstations sold by this company and also compatible with the new versions of the software.

4. THE CONFIGURATION OF THE CAD/CAM SYSTEM

We suggest that the graphic workstations with resolution of 1024X1280, internal memory 16MB-20MB, and the hard disk with 500-5000MB are suitable. The calculating speed of the CPU must be over 10MIPs, the speed of vector processing 100000/sec. It is better to have the system with a Z-Buffer.

The workstation must have a digitizer menu plate, and dial button box.

For the graphic input and output device, we suggest that the AO size digitizer or high precision scanner be adopted and the plate type or the roll type plotter be selected for output of the engineering drawings. If we need to output the numerical control data, a puncher or portable floppy must be included. Other peripheral devices are about the same as computers in general.

In the network of the CAD system we highlight that a IBM PC/AT micro computer with Autocad software must be connected. The aim is: 1) to communicate programs and, 2) to help the workstation to output the 2-D mechanical drawings and to punch the numerical control data.

5. CONCLUSION

For diesel engine design and manufacturing we must buy a turn-key system because it can be used immediately. Based on this foundation, we develop the CAD/CAM system especially for the diesel engine detailed design, i.e., we link the diesel engine performance program to the CAD/CAM graphics system and database. In aspect of CAPP and CAM, we develop the technology and numerical control technique of typical components of a diesel engine. Furthermore, we want to develop the diesel engine expert system which includes two parts: 1) the logical system, 2) the experiences library. Using the expert system we can carry on the project design of a diesel engine on line.

The selection of the CAD/CAM system must be in accord with the point of view of "development" and "up-to-date". In the research institute, it is more advantageous to take a CAD/CAM system in network with different types of workstations because it can transfer the graphics of any version easily.

In addition, we must take care of the orientation of development of new techniques in the world and continually enhance the software of the CAD/CAM system.

A Model for the Factory of the Future for Industrialized Housing

AHMAD K. ELSHENNAWY, MICHAEL A. MULLENS, WILLIAM W. SWART and
SUBRATO CHANDRA

Department of Industrial Engineering and Management Systems
University of Central Florida
Orlando, FL

ABSTRACT

The objective of this paper is to explore opportunities to improve quality, affordability and energy efficiency of industrialized housing in the United States. The paper presents current research efforts at the University of Central Florida to develop a conceptual model for building homes in the 21st century. The model is a fully-automated facility that employs a number of advanced information and manufacturing technologies. This paper describes the 21st century manufacturing facility, its layout and operation, the automation building blocks that are required and likely impacts.

INTRODUCTION

The American homebuilding industry is divided into two general segments:

o   Production builders - those homebuilders who stick-build a house (largely from scratch) on the owner's lot

o   Industrialized housing manufacturers - those home builders who build a house in a factory or on-site using major building components produced in a factory. Housing manufacturers' products range from complete homes (HUD Code/mobile homes), to 3-dimensional modules (rooms or partial rooms) to major two-dimensional components (walls, floors, roofs, etc.)

The American housing market is split roughly 50-50 between production builders and manufacturers (Brown, et al, 1989). American housing manufacturers currently use technologies and methods which are similar to those used by production builders. They have taken traditional low technology construction techniques and moved them inside a factory. Using this manufacturing strategy, manufacturers have been able to gain some competitive advantages including volume purchasing, no weather related delays, a captive work force, standard methods, etc. (Nutt-Powell, 1985).

Manufacturers have been unwilling to stray far from production builder technologies and methods. World class manufacturing strategies such as Total Quality, Continuous Flow Manufacturing, Flexible Manufacturing, Concurrent Engineering, Factory Automation and Computer Integrated Manufacturing have largely been ignored. There are a number of important reasons underlying this reluctance to accept more progressive manufacturing strategies:

o The industry is highly fragmented. It is composed primarily of a large number of small contractors. Most contractors do not operate with economics of scale sufficient to absorb significant R & D and capital costs.

o Manufacturers must compete against production builders, many of whom have little or no overhead. Incurring R & D costs or additional capital costs may further limit manufacturers pricing flexibility relative to that of production builders (Beighle, 1990).

o Industry sales are highly cyclical and driven largely by forces beyond the manufacturers' control. During peaks, demand cannot be satisfied. During slumps, manufacturers struggle to survive. An increase in overhead may make it even harder for manufacturers to survive a slump (Lucas, 1990).

o There is a lack of standardization in building codes. Codes vary from state to state and even county to county (Branson, 1990).

o Government support of R & D programs which is needed to foster an environment conducive to technological innovation has lagged (OTA, 1986).

Clearly, a problem exists. Compared to other industries, manufactured housing makes only marginal use of technological advances. There is virtually no industry funded R & D efforts and government supported R & D has lagged. To complicate the problem, foreign competitors, primarily Japan and Sweden, have developed highly advanced manufacturing processes (Kando, 1988; Mckellar, 1985). The housing industry in these countries receive considerable government R & D support. Although currently not price competitive, these potential competitors wait in the wings. However, where there is adversity, there is also a challenge. There are also significant opportunities for American housing manufacturers.

o To lower manufacturing costs
o To improve housing quality
o To increase energy efficiency
o To increase profit margins
o To increase market share
o To provide more affordable housing

The objective of this paper is to explore opportunities to improve quality, affordability and energy efficiency of industrialized housing in the United States. The paper presents current research efforts at the University of Central Florida to develop a conceptual model for building homes in the 21st century. The model is a fully-automated facility that employs a number of advanced information and manufacturing technologies. This paper describes the 21st century manufacturing facility, its layout and operation, the automation building blocks that are required and likely impacts.

THE 21st CENTURY FACILITY

The following is a concept of an automated manufacturing facility for building houses in the 21st century. The facility employs a number of advanced information and manufacturing technologies.

Factory Layout

o   The factory has a flow line layout. Subassemblies are manufactured in workstations located on either side of the factory and transported to the centrally located main assembly line using a computer-controlled overhead conveyor system. Subassembly workstations include:

- Floor systems
- Roof systems
- Exterior wall panels
- Interior wall panels

Installation of electric wiring as well as mechanical and plumbing fixtures is done at the appropriate workstations. A maintenance shop and a computer control room are located at the beginning of the main assembly line while engineering and administrative offices will be located at the end of the line.

o   In addition to the automated material handling system used for transporting panels and trusses between the sub-assembly workstations and the assembly line, the following equipment is employed in the factory:

- computer-controlled cutting, sawing, and drilling machines.
- variety of robots to tend machines and to perform other tasks such as nailing, gluing, interior and exterior finishing, and pick-and-place operations.
- an automated ultrasonic welding machine to be used for joining and assembling different components when non-traditional building materials (e.g. plastics and plastic reinforced structures) are to be used.

Factory Operations

o   Based on sales forecasts and customer/market demands, an order is placed to manufacture ten varieties of homes of different materials (traditional vs. non-traditional building material), floor sizes, finishes, and customized preferences. Production volumes are assumed to be different for each home type.

o   The computer-aided design and drafting system will generate the required architectural as well as structural drawings of each home type with the assistance of an artificial intelligence module or an expert system advisor for home design.

o   The computer software will take the CAD drawings and automatically generate a bill of material (BOM) for each home type. The bill of material is a complete listing of all parts and materials required for a finished home. Bill of materials are then input to the MRP II (Manufacturing Resource Planning) system. The MRP II system generates reports required for the purchase of necessary materials and tools.

o Increasing the effectiveness of operations will be achieved by a reliable inventory control system which depends heavily on a long-range commitment between producer and suppliers (vendor-vendee relationship). To achieve this objective, the factory will employ the Just-In-Time (JIT) inventory system. JIT insures that the required material is delivered at the right time at the right location.

o Automated process plans for each home type are generated using a computer-aided process planning (CAPP) system. Process plans show details of how the home is to be manufactured. Such plans include routing slips which are used to schedule the movement of material into assigned workstations, instruction sets which direct the tools through the motions required to perform the different operations, and operation sheets which detail the sequence of events for each workstation. Process plans are transmitted electronically to a computer terminal which assigns operations to the different NC machines, robots, and other equipment at different workstations.

o Material will be delivered to the factory in batches. Factory floor data collection systems will be available using bar code scanning technologies. A bar code is placed on an arriving batch of material. The code includes information about material and size for each home type. Upon arrival, the code is scanned to identify the material and accordingly instruct the machines, robots, and other equipment to set the appropriate process plans for that specific home type.

o Cutting workstations employ NC cutting/sawing machines and presses. Once the different pieces are cut as required, they are placed on a template or jig located on top of a conveyor. A pick-and-place robot is employed for this purpose. The conveyor will then move this arrangement towards the press which automatically detects the arrival of the panel and performs the assembly. After assembly is complete, the panel or truss moves to the other end of the press where two robots place insulation (when necessary), place drywall or plywood cover, and fasten the panel.

o The assembly process described above using the press is not necessary in case non-traditional or plastic materials are to be used. For such materials, the cut panel takes a different path towards an ultrasonic welding machine for inserting a joining strip that will connect the panel to the floor and roof systems in the main assembly line.

o The panel or truss will now move to a stacker where it will be adjusted vertically in a position where the overhead conveyor will take over.

o At this point, a decision must be made:

- If the customer desires fabricated panels which will be assembled on site, the overhead conveyor system will deliver the panels to the shipping area where they are to be transported to the construction site, or

- If the customer desires factory assembly of the home components, the overhead conveyor system will deliver panels to the main assembly line.

o   The overhead conveyor will move the assembled roof or wall to the assembly line to be installed on the floor which will be completed before the arrival of other components.

o   Robots are employed at the main production line for installing walls on the floor. Roofs are installed after the wall panels.

o   Another crew of robots will now move in to perform interior and exterior finishing operations.

o   The unit or component of the home is now complete. It will then move on a rail system and be pulled outside the plant where it will be loaded onto a truck for delivery to its destination.

o   It is recommended that carpeting and appliances be installed on site. This appears to be economically feasible and facilitates the interior finishing operations performed by robots inside the unit.

AUTOMATION BUILDING BLOCKS

The application of automation to industrialized housing manufacturing requires the following automation building blocks:

o   Computer-aided design (CAD)

o   Computer-aided process planning (CAPP)

o   Manufacturing resource planning(MRP II)

o   Operations control system

o   Computer aided manufacturing

o   Automated material handling

o   Safety requirements

Each of these steps is briefly explained in the following sections.

Computer-aided design (CAD)

The CAD system will be driven by customer orders. The system might be integrated to the extent that it becomes a computer-aided-sales/design system, providing design options and costs directly to the potential customer. CAD features will include expert system support of the design process with respect to design feasibility, cost and energy efficiency, generation of architectural and structural drawings and development of a bill of material which can be passed to the Manufacturing Resource Planning (MRP II) system.

<u>Computer-aided process planning (CAPP)</u>

The CAPP system will be driven by input from the CAD system. CAPP features will include the determination of material, process, machine and tooling requirements and the development of process plans which detail how the house will be made. Process plans will include material routings and operations sheets detailing operations at each workstation.

<u>Manufacturing Resource Planning (MRP II)</u>

The MRP II system will be driven by input from CAPP system. MRP II features will include development of material procurement schedules and manufacturing schedules to meet customer commitments and better utilize manufacturing resources.

<u>Operations control system</u>

The operations control system will be driven by CAPP and MRP II systems. The operations control system will be responsible for directing all operations on the manufacturing floor. This will include direct control of automated processing equipment, robots and automated material handling equipment. Where automation is not applied, the system will provide indirect control to the operator through computer terminals (portable or stationary) or printed action documents. The system will also be responsible for maintaining real time location and status control for all materials and work orders on the manufacturing floor.

<u>Computer-aided manufacturing (CAM)</u>

The CAM system will be driven by the operations control system. Modern computer numerical control (CNC) machines and robots will perform manufacturing operations where technologically feasible and cost effective. Potential applications include: cutting/sawing, drilling, assembly (nailing, fastening, gluing, welding), finishing, and machine loading/unloading (stacking/destacking).

<u>Automated material handling</u>

The automated material handling system will be driven by the operations control system. The system will store, move and stage materials between manufacturing workstations. Several key design features will facilitate material handling on the manufacturing floor.

o   A Just-In-Time (JIT) based manufacturing strategy will minimize the materials that are moving on the floor.

o   The manufacturing layout will support the JIT strategy and will minimize material movements.

Material handling will be automated where technologically feasible and cost effective. Potential applications include automated electrified monorail, power & free, or automated guided vehicle (AGV) for panel handling and powered conveyor or AGV for small parts handling.

Safety

Several safety issues are of concern and require consideration when designing the automated housing facility. Such issues include:

o   Dust generated during operation - Ventilation systems can be used or equipment will be designed with dust control filters which minimize the dust produced in the plant.

o   Movement of large structures such as roof and floor systems and wall panels - Since automation is to be employed, the risk of having workers in the work area is reduced.

IMPACT

Development and use of the 21st century manufacturing facility can be expected to have considerable impact, both positive and negative, on American housing manufacturers.

Key positive impacts include:

o   Quality improvements - driven by the consistency of the automated CAD, CAPP and CAM subsystems.

o   Enhanced energy efficiency - driven by the same factors.

o   Shorter lead times - resulting from the integration of the various subsystems under control of the operations control system.

o   Reduced labor costs - driven by automating design, processing and material handling functions.

o   Reduced inventory levels - driven by enhanced control provided by MRP and operations control system.

o   Greater flexibility in manufacturing - efficiently accommodating design variations over a limited range.

Potential negative impacts include:

o   High capitalization and a reduced capability to cut costs during market downturns.

o   Loss in larger scale flexibility associated with automation. Automation, whether computer systems, processing equipment or material handling equipment, is typically very powerful and flexible within its specified design range. The design range is typically bounded by limitations in both functionality and capacity. When a new product or process design exceeds this design range, the automation frequently cannot accommodate it and must be retrofitted or scrapped. Perhaps a more detrimental consequence is that the new design may be scrapped because it can't be efficiently accommodated by existing automation. Typically, the higher the level of automation, the more narrow is the effective design range. This is typically an economic decision based on the increased cost of providing more flexible automation, particularly, when the increased flexibility is not needed initially.

CONCLUSION

This report presents one vision of the future of industrialized housing manufacturing in the 21st century. Currently there are no such technologies in use within the American homebuilding industry. However, these technologies do exist in other manufacturing industries as well as foreign housing industries. It is believed that these technologies are technically feasible and can be adopted by the industry. Factory automation can help achieve the customer demand for a high quality home at an affordable price. In contrast, traditional methods have no major avenue for cost reduction while maintaining quality.

Other industrialized nations have implemented these technologies into their building industries . It is inevitable that these nations will ultimately target the U.S. housing market as a prime area for expansion. When this happens, U.S. manufacturers will be forced to find ways of remaining competitive.

Investment costs will be a major issue in any step towards higher automation, and this evolution will require a joint effort of industry, government, and academia. However, if the industry does not move in this direction the consumer will turn to other sources for their homes as they did with the automotive industry.

Automation technologies provide the capabilities needed for U.S. manufacturers to compete in both the domestic and global marketplace. In addition, the automation of industrialized housing manufacturing will achieve several objectives of national as well as international concerns. Such objectives include increased productivity, improved quality, low cost and increased energy efficiency.

REFERENCES

Beighle, J. Wayne, 1990. "Summary Statement," Steering Committee Meeting, Energy Efficient Industrialized Housing Research Program, Center for Housing Innovation, University of Oregon, Eugene, OR, (July).

Branson, Timothy R., 1990. Potential Methodologies and Concepts for Improving Industrialized Housing Manufacture, Internal Research Project Report, University of Central Florida, Orlando, FL.

Brody, Herb, 1987. "CAD meets CAM," High Technology, (May), 12-18.

Brown, G.Z., et al, 1989. Energy efficient Industrialized Housing Research Program: Summary of FY 1989 Research Activities. Center for Housing Innovation, University of Oregon and Florida Solar Energy Center, a Research Institute of the University of Central Florida, Orlando, FL (February).

Kando, Paul, F., 1988. When the Best Costs Less: An Economic Comparison of the Swedish Factory Crafted House Construction System and Conventional Homebuilding, Center for the House, Washington, D.C.

Lucas, Regina, 1990. "Kaptan Building Systems Expands Into Steel Modular Frame Construction," Automated Builder, (June).

McKellar, James, 1985. Industrialized Housing: The Japanese Experience. Alberta Municipal Affairs, Edmonton, Alberta, (Dec.).

Nutt-Powell, Thomas E., 1985. "The House That Machines Built," Technology Review, (December).

_____, 1986. Technology, Trade, and the U.S. Residential Construction Industry, Special Report, Office of Technology Assessment, Congress of the United States, Washington, D.C., (September).

Enabling Automation Technologies for an Automated Mail Facility of the Future

JAY LEE and GARY HERRING

Office of Advanced Technology
United States Postal Service
Washington, D.C.

Summary

The primary goal and obligation of the United States Postal Service (USPS) is to provide its customers, people, and organizations with high-quality products and services. By the year 2000, the volume of mail in the United States is expected to exceed 250 billion pieces per year. To maintain quality service into the 21st century, the Postal Service must identify, develop, and adopt new technologies if it is to keep pace with demand. The Office of Advanced Technology has the responsibility to conduct research to meet the future needs of the USPS.

Future postal technology will use fully automated machines to feed, transport, stack, and sort a wide variety of mail pieces. Furthermore, these machines will be integrated to become a fully automated facility in the manner of the factory of the future.

To realize such a vision in the next century, the USPS is counting on private industry, universities, government agencies, and international organizations to develop the enabling technologies, some of which are presented below.

Process and System Control Technology

A programmable logic controller (PLC) has been the traditional controller for machines and manufacturing processing applications. However, in recent years, the trend has been to link PLCs to microcomputer-based products. These links allow data gathering and control to be executed by the PLCs, while real-time data processing is performed by personal computers (PC). Thus, the functions of PC and PLCs have overlapped, and their capabilities have been expanded to include communication networking, more powerful instruction sets, math capability, and analog input-output. Integrated PLCs combine a PLC and computer on the same backplane designed for vertical or hierarchical integration of plant-floor PLCs to facility-level information systems (MIS).

Under this concept, a modular system for manufacturing applications (MSMA) is considered as an alternative to the traditional use of PLCs. The MSMA is a multi-tasking DOS application for control of material handling. For future automated material handling, a distributed hierarchical control system will be required to provide a control and distribution system for work in progress. Such a system will be needed to manage work center queues, and provide rapid storage and handling of mail trays under direct computer control, and real-time control of tray location.

A major requirement for future USPS applications is to establish a common hardware and software shell structure that facilitates incremental improvements in automation and robotics technology. Guidelines must be provided for industry to develop and market a diverse line of control system components which will be interchangeable and usable on many different vendors' control system platforms. The key element of the future enabling technology is the system architecture for the future fully automated facility. The USPS endorses the basic concept of hierarchical structure as developed by the National Institute of Standards and Technology (NIST). The concept simplifies process design by vertically partitioning the control system according to a task's complexity, and horizontally partitioning the control system by function.

Simulation

Simulation is a technique that has been employed extensively to solve problems. The process consists of using a computer to construct a model to represent a real system. In the postal environment, engineers and managers use simulation to plan for new facilities, to determine transportation routes, or to schedule workloads.

A major trend is to simplify the programming language so that more and more people, not just programmers, can learn to use it and can build their own models. To accomplish this, the line-by-line coding approach will be replaced by graphic symbols, and the use of pulldown menus.

Another trend for future development is animation. Instead of providing numbers, the future simulation model will provide dynamic pictures showing movements of men or materials through various operations.

For the USPS, the simulation output will be used as input for decision support systems of the future. Future postal management will be able to examine 'what if' operational decisions on simulation models. Part of the input to simulation may be derived from the expert systems that will suggest rules for consideration. Simulation will also be a part of all computer-aided design (CAD) and computer-aided process planning (CAPP) programs.

Graphic Workstation and Operator Interface

Workstations are the fastest-growing technology in the computer industry. Today's workstations are far superior to PCs and are replacing minicomputers in many applications. Ideally, a workstation should provide the following functions: monitoring, control, diagnostics, identification, decision, execution, and verification. Workstations should perform as a client-server; when an output is requested, it should be sent directly to the workstation, or server, and displayed in the appropriate window.

For the future, to enhance user requirements and to facilitate internetworking, six levels of operator interface will need to be developed. The workstation for the future for the man-machine interface of material handling systems will probably have a variety of levels: level 1, operational; level 2, supervisory; level 3, informational; level 4, diagnostic; and level 5, maintenance.

The future needs of workstations for the USPS will be mostly at the facility level, where engineers will use graphic interfaces to design facilities and to arrange the processing equipment. In addition, the workstation is expected to interface with other electronic media, such as voice and video image, to distribute information.

Database

The heart of a computer system is its database. It is the final depository of all the data processing effort. A major trend in database technology is to get data closer to the user, i.e., to use a distributed system rather than a central system.

The problem facing the enterprise as whole is that the proliferation of distributed systems may result in different database management systems. The use of so-called relational models and the structure query language (SQL) for data access and retrieval will simplify the interface problems. If the proper software is used, these distributed databases will appear as if they were a single information source. A user will be able to get the information he or she needs without having to know where the data are stored. Furthermore, such software will run under different types of network protocol, media, and operating systems.

Within the future postal environment, different functions, such as mail processing or transportation management, will each have their own database to meet their own requirements. However, the USPS needs a database administration system in order to access, update, and maintain both menu-driven and ad-hoc access to all reporting systems.

Communication

The decentralization of data processing and the proliferation of PCs are expected to continue. However, more and more of these computers are expected to be inter-connected into networks. To function in such an environment, the various computers must talk the same language, which is called a protocol. The protocol defines the rules, format, timing, and error-detection schemes.

Computer network protocols are complex. To keep the complexity to a manageable level, protocols are divided into several levels called layers. Each layer provides different and specific functions. The open system interconnection (OSI) is the proposed international standard using seven layers. Different manufacturers implement open systems differently to achieve product differentiation. However, all will behave the same in order to communicate with other open systems.

The challenge for the future in networking is the establishment, acceptance, and adherence of open system standards. At present, proprietary and incompatible systems are still common. Even MAP is not yet fully compatible with OSI.

The USPS supports the government effort represented by the National Institute of Standards and Technology (NIST), which promotes open system architecture. The USPS expects to obtain its future hardware and software from multiple vendors. For these heterogeneous systems to communicate and form a network, the need for a common shell structure cannot be overemphasized. The goal is to achieve full connectivity so that any user on any terminal can sign on to any application at any location in the network.

The USPS will be a large user of communication systems. Hopefully, its need for open architecture will hasten the standardization process, which will benefit both users and manufacturers.

Material Handling System

A material handling system must be fully integrated into the total automated facility. At present, a typical postal facility consists of "islands of automation." Material handling is the key that will link these islands into a total integrated system. However, in the future, an automated facility will be designed from the top down, and not patched together from the bottom up. The following descriptions cover mostly enabling technologies from a mechanical or equipment point of view. The control and information architecture forms the complementary part of the physical components.

Automated Guided Vehicle (AGV). The early applications of AGVs were limited to warehouse operations for towing trains of carts. In recent years, the trend has been toward auto loading in a fully automated facility. The new AGVs have the ability to pick up a load and deliver it anywhere in a plant without human intervention. In smaller production systems, miniload AGVs have been developed to handle small items.

The use of on-board computers and remote communication links greatly enhanced the capabilities of the AGVs. With advanced controls, the AGVs can be tracked, dispatched, and manipulated from remote locations. The AGVs carry not only products but also information, which can be transmitted electronically to the workstations. Another development includes the use of telemetry guidance instead of imbedded wires.

In the future, AGVs will be more intelligent and able to perform more functions, such as automatic positioning and manipulation of products for automatic retrieval and placement. If needed, an AGV will become an robot cart that can interface with workstations and storage devices without human intervention. It will also have an advanced guidance system that can be programmed to guide the path of the AGVs, and can be changed rapidly.

The USPS will need intelligent AGVs to perform a wide variety of work, such as transporting containers, mail bags, and trays from trucks and processing machines. These new generations of AGVs must be easy to maintain, rugged, and reliable.

Conveyor Systems. Conveyors are the mainstay of material handling in automated facilities. This trend will continue. Conveyors are comparatively low in cost, easy to install, and flexible. They can also be used for temporary storage and surges. In recent years, specialized conveyors have been developed for customized application, especially for the auto industry. For material distribution, conveyors may carry tilting trays and function as sortation devices.

One special development is the inverted power-and-free conveyor that provides positive control for each carrier and conserves floor space. Loads can be stopped and started individually, which is ideal for manufacturing cells.

The future trend in conveyor technology will stress several key concepts. The system must be modular and operate at higher speeds. It should be easy to maintain and easy to install and more reliable.

The USPS is expected to use a considerable number of conveyors to handle trays in future facilities. Most of these will be mounted overhead to conserve floor space. The desirable features of these conveyors are: light weight, quiet running, easy to install and maintain, high speed, and low failure rate. Down time must be minimized and mean time to repair should be less than 15 minutes.

Automated Electric Monorail. In this system, the carriers are not linked in a continuous chain, but are autonomous vehicles. They can also operate in an inverted position, carrying loads through curves, switches, and different elevations. The system is especially suited for applications where carrier and inventory identification is critical, and precision positioning or robot interface is required. By the use of individual carriers, the load in each carrier can be tracked and controlled at all times.

For the postal environment, a monorail system is being developed to link islands of automation and to complement conventional conveyors and other transportation devices. This system is shown in Figures 1 and 2. The monorail system consists of multiple loops of rails and can be deployed in a modular manner. The carrier is designed to have the capability to run at different speeds to meet different load demands.

For the future, the USPS believes that a monorail system will be a major element in an intelligent material handling system. For such a system to be effective, the control technology must be developed to be modular and flexible. Each future monorail carrier will be a smart cart with on-board system control and remote communication capabilities.

Automatic Storage and Retrieval Systems. Storage and retrieval systems have seen many changes in past years. They began primarily as unit loading devices for large pallets. The major trend is the shift from large load to mini- and micro- stackers. These smaller units are more suited for factory than for warehouse environments. More and more, these units are used for work-in-process buffers at the point of use to meet just-in-time and other processing requirements. The current technology for AS/RS is well developed, with sophisticated controls to provide random access to thousands of stock-keeping units (SKUs). Some of the systems have the capability to fill orders remotely.

At present, the USPS uses carts and containers for staging. The system is space consuming and requires manual handling. But the cost per SKU is low. For the future, the Postal Service will need many AS/RSs to stage millions of pieces of mail pieces per day on a cost-effective basis.

The USPS is exploring the feasibility of AS/RS systems for staging mail pieces between processes. Both a centralized and distributed storage approach are planned. The distributed system has a higher system reliability, and provides storage at the point of use, but the capacity is limited.

For the future, the USPS needs higher speeds for the pick and place function. The ultimate goal of the USPS is to sort mail to the individual delivery point in a single pass. The AS/RS is expected to play a part in this.

Automatic Identification. The USPS is currently using several different identification technologies. The POSTNET code has been used on letters for over 10 years. Bar codes of different designs have been used to identify trays, containers, and pallets. Other technologies under investigation include radio frequency (RF) and infrared (IR) technology.

A number of exotic coding technologies are being developed by industry; these include dye-polymer, holographic scanning, magnetic coding, etc. Many research laboratories are developing smart cards. By the year 2000, a number of these technologies will need to mature and be perfected for postal applications.

The future trend will include the increasing use of electronic data interchange and work sharing with business mailers. A system will be needed to transfer identification of mail pieces, trays, pallets, and other containers. The code tagging system will employ advanced processing technologies to eliminate redundant and labor-intensive handling.

The USPS will need to standardize a limited number of identification systems. The emphasis will be to use industrial standard codes and to avoid proprietary products and coding schemes.

Some Optimization Problems of Scheduling in a Flexible Manufacturing System

TOMASZ AMBROZIAK

Technical University of Warsaw, Poland

1. AN ACTIVITY GRAPH

A specified number of technological processes are assumed needed to achieve a certain goal. Every process consists of elementary stringently specified activities. The set of activities which add up to a process is a partially ordered set. Every activity will be represented as a graph arc. An activity or activities is an event. Every event is represented as a graph vertex.

With this notation a set of partially ordered activities is represented as a graph G, or $G=(\mathbb{X}, \mathbb{M}, P)$ where: $\mathbb{X}$ is a set of ordinal numbers of vertices (events) in G; $\mathbb{M}$ is a set of ordinal numbers of arcs (activities) in G; P is a certain three-positioned predicate. It is assumed that $\mathbb{X}=(x(i): i \in \mathbb{I}$, $\mathbb{I}=(1,\ldots,i, \ldots,I)$, $\mathbb{M}=(m(k): k \in \mathbb{K}$, $\mathbb{K}=(1,\ldots,k,\ldots,K)$,or the cardinalities of the sets $\mathbb{X}$ and $\mathbb{M}$ are known.

The predicate P maps a Cartesian product $\mathbb{X} \times \mathbb{M} \times \mathbb{X}$ into the set $(0,1)$, or P: $\mathbb{X} \times \mathbb{M} \times \mathbb{X} \longrightarrow (0,1)$. An arbitrary three-tuple $(x(i),m(k),x(j)) \in \mathbb{X} \times \mathbb{M} \times \mathbb{X}$ such that $P(x(i),m(k),x(j))=1$ is interpreted in the following way: the arc $m(k)$ connects the vertex $x(i)$ with the vertex $x(j)$. If $P(x(i),m(k),x(j))=0$ the arc $m(k)$ is said not to connect $x(i)$ and $x(j)$.

2. QUALITATIVE AND QUANTITATIVE DESCRIPTION OF ACTIVITIES

Facilities are assumed to be needed to perform an activity. A set $\mathbb{S}$ of facility types needed in performance of activities is assumed to be specified,or $\mathbb{S}=(s: s=1,\ldots,\mathbb{S})$. In addition, resources are needed to carry out activities. A set $\mathbb{P}$ of resource kinds is assumed needed, or $\mathbb{P}=(p: p=1,\ldots,P)$.

Assume that a mapping r of the Cartesian product $\mathbb{K} \times \mathbb{S}$ into the set $\langle 0,1 \rangle$ is specified, or r: $\mathbb{K} \times \mathbb{S} \longrightarrow \langle 0,1 \rangle$. It is true that $r(k,s)=1$ iff the s-th facility type is employed in Performing the k-th activity, otherwise $r(k,s)=1$. Consequently, for every $k \in \mathbb{K}$ the vector $r(k)$ of components $r(k,s)$ can be determined, or $r(k)=\langle r(k,1),..,r(k,s),..,r(k,S) \rangle$. Assume that the mapping a of the Cartesian product $\mathbb{K} \times \mathbb{S}$ into the set $\mathcal{R}^+$ is specified, or a: $\mathbb{K} \times \mathbb{S} \longrightarrow \mathcal{R}^+$. The value $a(k,s) \in \mathcal{R}^+$ determines the productivity of the s-th facility type employed in the k-th activity. Consequently, for every $k \in \mathbb{K}$ a vector $a(k)$ can be determined of components $a(k,s)$, or $a(k)=\langle a(k,1), \ldots, a(k,s), \ldots, a(k,S) \rangle$.

Assume also that the mapping z of the Cartesian product $\mathbb{K} \times \mathbb{P}$ into the set $\langle 0,1 \rangle$ is specified, or z: $\mathbb{K} \times \mathbb{P} \longrightarrow \langle 0,1 \rangle$. It is true that $z(k,p)=1$ iff the p-th resource is utilized in the k-th activity, otherwise $z(k,p)=0$.

Assum e that the mapping n of the Cartesian product $\mathbb{P} \times \mathbb{S}$ into the set $\langle 0,1 \rangle$ is specified, or n: $\mathbb{P} \times \mathbb{S} \longrightarrow \langle 0,1 \rangle$. It is true that $n(p,s)=1$ iff, utilization of the s-th facility type entails utilization of the p-th resource, otherwise $n(p,s)=0$.

Assume that the mapping b of the Cartesian product $\mathbb{K} \times \mathbb{P} \times \mathbb{S}$ into the set $\mathcal{R}^+$ is specified, or b: $\mathbb{K} \times \mathbb{P} \times \mathbb{S} \longrightarrow \mathcal{R}^+$. The value $b(k,p,s) \in \mathcal{R}^+$ describes the consumption norm of the p-th resource when the s-th facility type for carrying out the k-th activity. For the $k \in \mathbb{K}$ and $p \in \mathbb{P}$ a vector $b(k,p)$ with components $b(k,p,s)$ is determined, or $b(k,p)=\langle b(k,p,1), \ldots, b(k,p,s), \ldots, b(k,p,S) \rangle$.

Qualitative description of the set of activities is assumed to be reducible to determining for every $k \in \mathbb{K}$ the values of $a(k)$ and $b(k,p)$. Qualitative description of the set of activities is supplemented with quantitative description.

Assume that the mapping Q is specified which moves elements of set $\mathbb{K}$ into the set $\mathcal{R}^+$, or Q: $\mathbb{K} \longrightarrow \mathcal{R}^+$. The quantity $Q(k) \in \mathcal{R}^+$ stands for the size of the k-th activity. Determining $Q(k)$ for every $k \in \mathbb{K}$ is quantitative definition of the set of activities. Let $\mathbb{Q}$ be a set of elements $Q(k)$, or $\mathbb{Q}=\langle Q(k): k \in \mathbb{K} \rangle$.

3. SCHEDULE OF A SET OF ACTIVITIES

Let $\mathbb{T}$ denote a set of times (intervals of a certain length), or $\mathbb{T}=\{t: \ 0 \leq t < \infty\}$. It is assumed that $\mathbb{T}(k)$ such that $\mathbb{T}(k) \subseteq \mathbb{T}$ is a set of times at which the k-th activity is performed.

Let $t'(k)$ denote the time at which the k-th activity starts. It is obvious that $t'(k)=\min\{t: \ t \in \mathbb{T}(k)\}$. Let $tt'(k)$ denote a set of elements $t'(k)$, or $tt'=\{t'(k): \ k \in \mathbb{K}\}$.

The mapping u is assumed specified of the Cartesian product $\mathbb{K} \times \mathbb{S}$ into the set $\mathcal{R}^+$, or u: $\mathbb{K} \times \mathbb{S} \ ---------> \ \mathcal{R}^+$. The quantity $u(k,s) \in \mathcal{R}^+$ defines the amount of facilites of the s-th type utilized in performance of the k-th activity. For every $k \in \mathbb{K}$ a vector $u(k)$ of components $u(k,s)$ is determined, or $u(k)=<u(k,1),\dots,u(k,s), \ \dots,u(k,S)>$. Let $\mathbb{U}$ be a set of elements $u(k)$, or $\mathbb{U}=\{u(k): \ k \in \mathbb{K}\}$.

The schedule of the k-th activity is said to be known if the numerical values are known of elements in the pair $(t'(k),u(k))$. The schedule of the set of activities is said to be known if the numerical values of elements in the pair $(tt',\mathbb{U})$ are.

4. CHARACTERISTICS OF ACTIVITIES AS A FUNCTION OF A SCHEDULE

The mapping w of the Cartesian product $\mathbb{K} \times \mathbb{T} \times \mathbb{S}$ into the $\mathcal{R}^+$ is assumed specified, or w: $\mathbb{K} \times \mathbb{T} \times \mathbb{S} \ -------> \ \mathcal{R}^+$. The quantity $w(k,s,t) \in \mathcal{R}^+$ denotes the number of facilities of the s-th type are utilized at the time t in performance of the k-th activity. For every $k \in \mathbb{K}$ and $t \in \mathbb{T}$ the vector $w(k,t)$ is determined of components $w(k,s,t)$, or $w(k,t)=<w(k,t,1), \dots,w(k,t,s), \ \dots,w(k,t,S)>$. Let $\mathbb{W}$ be a set of elements $w(k,t,s)$, or $\mathbb{W}=\{w(k,t,s): \ k \in \mathbb{K}, s \in \mathbb{S}, \ t \in \mathbb{T}\}$ and $\mathbb{A}$, a set of elements $a(k,s)$, or $\mathbb{A}=\{a(k,s): \ k \in \mathbb{K}, \ s \in \mathbb{S}\}$.

The mapping f is assumed known that moves the Cartesian product $\mathbb{A} \times \mathbb{W}$ into the set $\mathbb{Q}$, or f: $\mathbb{A} \times \mathbb{W}-----> \ \mathbb{Q}$. The quantity $f(a(k,s),w(k,s,t))=Q(k)$ relates the activity size and the number of facilities employed in its implementation.

In further discussion f is assumed to take the form

$$f: \sum_{s \in \mathbb{R}(k)} \int_{t'(k)}^{t''(k)} a(k,s)w(k,s,t)dt = Q(k) \quad \text{where:} \quad t''(k) = \max\{t \quad t \in \mathbb{T}(k)\}$$

is the time of activity and $\mathbb{R}(k) = \{s: \ r(k,s) = 1, \ s \in \mathbb{S}\}$.

The dependence $w(k,s,t)$ can be specified in different ways; here it is assumed that

$$w(k,s,t) = \begin{cases} u(k,s), & t \in [t'(k), \ t''(k)] \\ 0, & t \notin [t'(k), \ t''(k)] \end{cases}$$

In the light of the above, f may be said to take the form

$$f: \sum_{s \in \mathbb{R}(k)} \int_{t'(k)}^{t''(k)} a(k,s)u(k,s)dt = Q(k)$$

With $\tau(k)$ being the time taken by the activity assume that

$$\tau(k) = t''(k) - t'(k) \quad \text{or} \quad \tau(k) = Q(k) / \sum_{s \in \mathbb{R}(k)} a(k,s)u(k,s).$$

From the latter it follows that $w(k,s,t) = w(k,s,t,t'(k),t''(k))$

or $w(k,s,t) = w(k,s,t,t'(k),u(k))$.

The mapping λ is assumed specified of the Cartesian product $\mathbb{K} \times \mathbb{P} \times \mathbb{T}$ into the set $\mathcal{R}^+$, or $\lambda: \mathbb{K} \times \mathbb{P} \times \mathbb{T} \dashrightarrow \mathcal{R}^+$. The quantity $\lambda(k,p,t) \in \mathcal{R}^+$ denotes the rate of utilizing the p-th resource in performance of the k-th activity at time t. Let $\mathbb{L}$ and $\mathbb{B}$ be sets of elements $\lambda(k,p,t)$ and $b(k,p,s)$, respectively, or $\mathbb{L} = \{\lambda(k,p,t): \ k \in \mathbb{K}, \ p \in \mathbb{P}, \ t \in \mathbb{T}\}$, $\mathbb{B} = \{b(k,p,s): \ k \in \mathbb{K}, \ p \in \mathbb{P}, \ t \in \mathbb{T}\}$.

The mapping ψ is assumed specified of the Cartesian product $\mathbb{B} \times \mathbb{W}$ into the set $\mathbb{L}$, or $\psi: \mathbb{B} \times \mathbb{W} \dashrightarrow \mathbb{L}$. The quantity $\psi(b(k,p,s),w(k,s,t)) = \lambda(k,p,t)$ specifies the relation between the resource utilitization rate and the number of facilities used in performance of the activity.

In futher discussion ψ is assumed to have the form

$$\psi: \sum_{s \in \mathbb{R}(k) \cap \mathbb{N}(p)} b(k,p,s)w(k,s,t) = \lambda(k,p,t) \quad \text{where:} \quad \mathbb{N}(p) = \{s: \ n(s,p) = 1, s \in \mathbb{S}\}.$$

Because $w(k,s,t) = w(k,s,t,t'(k),u(k))$ it is true that $\lambda(k,p,t) =$

$$= \sum_{s \in \mathbb{R}(k) \cap \mathbb{N}(p)} b(k,p,s)w(k,s,t,t'(k),u(k)).$$

Consequently, $\lambda(k,p,t) = \lambda(k,p,t,t'(k),u(k))$.

The other Characteristics of the activities one may introduce in similar way.   Certain Proposition follow.

PROPOSITION 1. If- $w(k,s,t)$ is stepwise, or equal to $u(k,s)$
to $t \in [t'(k), t''(k)]$ and to zero for $t \notin [t'(k), t''(k)]$; -

$-\lambda(k,p,t,t'(k),u(k)) = \sum\limits_{s \in R(k) \cap N(p)} b(k,p,s)w(k,s,t); - \quad \tau(k) = t''(k) - t'(k)$

then $V(k,p) = V(k,p,u(k))$

Indeed,

$$V(k,p) = \int\limits_{t'(k)}^{t''(k)} \sum\limits_{s \in R(k) \cap N(p)} b(k,p,s)u(k,s)dt = \sum\limits_{s \in R(k) \cap N(p)} b(k,p,s)u(k,s)\tau(k) =$$

$$= Q(k) \frac{\sum\limits_{s \in R(k) \cap N(p)} b(k,p,s)u(k,s)}{\sum\limits_{s \in R(k) \cap N(p)} a(k,s)u(k,s)}$$

From the latter dependence it follows that $V(k,p) = V(k,p,u(k))$
The similar Propositions may be proved in a similar way.

5. CHARACTERISTICS OF A SET OF ACTIVITIES AS A FUNCTION OF A SCHEDULE

The amount of facilities of the s-th type utilized for
performance of a set of activities at time t is described
by the formula $\sum\limits_{k \in K} w(k,s,t,t'(k),u(k)) = w(s,t,tt',U)$.
The labor consumption of the s-th facility type utilized

in performance of a set activities is given by the formula

$\sum\limits_{k \in K} A(k,s,u(k)) = A(s,U)$. The similar Characteristics may be
defined in a similar way.

6. SOME OPTIMIZATION PROBLEMS IN SCHEDULING A SET OF ACTIVITIES

Problem 1. Determine the numerical values of elements in the
pair (tt', U) so that a criterion of the form

$$\sum\limits_{s=1}^{S} \int\limits_{T_0}^{T^*} [w(s,t,tt',U) - y(s)]^2 \, dt$$

take on a minimal value under the following constrains:

    1. $\min\limits_{k \in K}(t'(k)) = T_0; \quad \max\limits_{k \in K}(t''(k)) = T^*$

2. $t''(k) = t'(k) + \dfrac{Q(k)}{\displaystyle\sum_{s=1}^{S} a(k,s)u(k,s)}$, $\forall\ k \in K$

where $y(s) = A(s,U)/(T^* - T_0)$; $T^*,\ T_0$ are specified.

Problem 2. Determine the numerical values of elements in the pair (tt', U) so that a criterion of the form

$$\sum_{s=1}^{S} \int_{T_0}^{T^*} [w(s,t,tt',U)]^2 dt$$

take an a minimal value under the constrains:

1. $\min_{k \in K}(t'(k)) = T_0$; $\qquad u(k,s) \le u^*(k,s)$, $\forall\ k \in K$, $\forall\ s \in S$

2. $t''(k) = t'(k) + \dfrac{Q(k)}{\displaystyle\sum_{s=1}^{S} a(k,s)u(k,s)}$, $\forall\ k \in K$

The above optimization problems are nonlinear and are hard to solve by directed methods. An approximate solution parallel algorithm is proposed in which the specified graph is replaced by another which is a sequence of subgraphs of the original graph. The computing load of this algorithm is analyzed.

Some Methods of Modeling for Computer Integrated Workshop

V. N. KALACHEV and YE. N. KHOBOTOV

International Research Institute for Management Sciences
Moscow, USSR

Summary

This paper considers problems and methods of pre-design modeling of computer integrated manufacturing (CIM) which consist of several flexible manufacturing systems (FMS).

Problems and a model for pre-design CIM

Let us take up a fairly general case of pre-design modeling for a workshop which consists of several FMS.

The workshop consists of automatic modules, transportation, stores and buffers, and a control system. As a rule, all machining facilities are incorporated in the FMSs which add up to a workshop and automatic transportation, stores, and buffers may be included either in some FMSs or utilized for transportation of parts and ingots between them.

Pre-design modeling of an automatic workshop amount to choice of the number and types of FMS machining facilities, a transportation system for serving these facilities and supply of the FMS with parts and ingots, to determining the buffer and store sizes, to choice of fixtures and of products from a list that could be profitably manufactured in the workshop.

The workshop structure and the above variables should be chosen so that, with specified constraints on the price of the facilities, maximize its production, extend the output potential and reduce the part manufacturing time.

Let us take up assumptions under which the model is built. L types of products are manufactured in the works. For every product the number of its parts $\tilde{L}_i$ (i=1,...,L) is known.

On every part in any product the designer knows:

-the facilities with which all products and parts may be machined;
-the route of machining every product and part (a unique route is assumed to exist for every part and product);
-the time of machining every part in any facility along the route and the cost of its machining along the route;
-the average size of the lot of every type of product;
-the mean frequency at which every lot of products is included into the workshop schedule.

The workshop is also assumed to include FMSs for machining base members, prismatic parts, and bodies of rotation and for assembly jobs. As noted above, automatic transportaition is also included.

Under these assumption the model for choice of the workshop structure is built of models for the FMSs [1-3] but the FMSs should be compatible as far as the part machining is concerned and the transportation and store must be capable of keeping and timely of part and ingots to all work stations. If this is not the case, balanced operation of the shop areas woud be out of the question as different orders arrive varying in lots of products of every type.

The model is linear and its functional and constraints are

$$\sum_{l=1}^{\tilde{M}_o} \hat{D}_l + \tilde{D} + \overset{v}{D} \le D, \tag{1}$$

$$\sum_{i=1}^{L} \sum_{k=1}^{\overset{v}{L}_i^l} x_i \{ \tilde{n}_i^{lk} t_{ij}^{lk} + \tau_{ij}^{lk} \} + \overset{v}{t}_j^l \le y_j^l V_j^l , \quad j=1,\dots,M_1, \tag{2}$$

$$y_j^l - w_{ij}^{lk} t_{ij}^{lk} \ge 0 , \quad j=1,\dots,M_1, \ k \in I^l, \ l=1,\dots,\tilde{M}_o, \tag{3}$$

$$w_p^{lk} - w_{ip-1}^{lk} + \tilde{w}_{ip}^{lk} \ge 0, \quad k \in I^l, \ l=1,\dots,\tilde{M}_o, \ p \in J_i^{lk}, \tag{4}$$

$$p_{\underset{im}{\sim}}^{lm} \tilde{w}_{ik}^{im} - w_{ik}^{lm} + \overset{v}{w}_{ik}^{lk} \ge 0, \quad l \in J^l, \ \tilde{m} \in I^l, \ m \in I^l, \tag{5}$$

$$\sum_{j=1}^{M_1+1} d_j^l y_j^l \le \hat{D}_l, \quad l=1,\dots,\tilde{M}_o, \tag{6}$$

$$J=\max\{\alpha_1^0 \sum_{l=1}^{L} c_i x_i n_i - \sum_{l=1}^{\tilde{M}} \{\alpha_2^1 \sum_{j=1}^{M_1+1} d_j^1 y_j^1 + \alpha_3^1 \sum_{i\in I} \sum_{p\in J_i} \sum_{k=1}^{\overset{v}{L_i^1}} \tilde{w}_{ip}^{lk} + \qquad (7)$$

$$+ \alpha_4^1 \sum_{j=1}^{M_1+1} b_j^1 y_j^1 + \alpha_5^1 \sum_{i\in I} \sum_{k\in J_i} \sum_{m=1}^{\overset{v}{L_i^1}} \overset{v}{w}_{ik}^{lm} \}.$$

Here $\tilde{M}_0$ denotes the number of FMSs in the workshop; $\hat{D}_1$, the price of facilities in the l-th FMS; $\tilde{D}$, the price of the transportation system; $\overset{v}{D}$, the price automatic stores; D, the maximal admissible price of the workshop facilities; $\overset{v}{L_i^1}$, the number of part types which are machined in the l-th FMS and are used in manufacture of the i-th product; x_i, integral variabes such as 0 and 1 ($x_i=1$ if the i-th product is included into the workshop schedule and 0 if otherwise); $\tilde{n}_i^{lk}$, the number of parts of the k-th kind that are machined in the l-th FMS and utilized in manufacture of the i-th product; t_{ij}^{lk}, the time of machining the k-th part in the j-th facility of the l-th FMS; τ_{ij}^{lk}, the time taken by adjusting the j-th facility for machining the k-th part which is utilized in manufacture of the i-th product; $\overset{v}{t_j^1}$, the idle time of the j-th facility in the l-th FMS which is determined by simulation [2]; y_j^1 ($j=1,...M_1$), the number of machining facilities of the j-th type in the l-th FMS; v_j^1, the time allowed the j-th facility for execution of the machining schedule; M_1, the number of machining facility types that can be incorporated into the l-th FMS; w_{ij}^{lk}, the productivity of the j-th facility of the l-th FMS in machining parts of the k-th type that are used in manufacture of the i-th products; w_{ip}^{lk} and w_{ip-1}^{ik} in the constraints (4) are the productivity of facilities in the l-th FMS engaged in machining k-th parts in related jobs; I^1, the set of part types that are used in manufacture of the i-th products and for which the related jobs must be made compatible in the l-th FMS; J_j^{lk}, an ordered set of facility types that are used in the l-th FMS for machining parts of the k-th type ($k\in I^1$); $\tilde{W}_{ip}^{lk}$, auxiliary positive variables ($\tilde{W}_{ip}^{lk} > 0$); $p_{\tilde{im}}^{lm}$, the number of parts the m-th type which are used in manufacture of the m-th product; $\tilde{W}_{ik}^{lm}$, auxiliary variables ($\tilde{W}_{ik}^{lm} > 0$); d_j^1, the price of a unit of the j-th facility in the l-th FMS; y_{M+1}^l , the number of robocars in the l-th FMS; and a_n^l ($l=1,\ldots,\tilde{M}$, $n=1,\ldots,5$) are weights.

The constraint (1) determines the maximum cost of the workshop designed.

The constraints (2) relate the time needed for machining in the j-th facility of the l-th FMS of the chosen list of parts, idle time of the j-th facility, $t_j^{\overset{v}{l}}$ (the left- hand side of the inequality (2)) with the time available for utilizing the j-th facility (the righthand side). By varying parameters $t_j^{\overset{v}{l}}$ ($j=1,\ldots,M_1$; $l=1,\ldots,\tilde{M}_o$) in a certain way the optimization model (1) – (7) is made compatible with simulation models [2] which lead to optimal schedules of machining the chosen list of parts in every FMS and determine the idle time of the machining facilites.

The constraints (3) and (4) enable matching the products from the list.

The constraints (5) make it possible to match the productivity of related FMSs, or those that are upstream and downstream of a given FMS.

The constraint (6) determines the admissible price of facilities in the workshop.

The optimizing functional (7) maximizes the profit made by selling the products, minimizes the facility prices and service costs, and minimizes the mismatch ($\tilde{W}_{ip}^{lk}$ and $W_{ik}^{\overset{v}{lm}}$) between the facility and FMS productivities in related machining jobs.

In addition to the above general constraints, the model also includes those which define the number of pallets, sizes of stores and buffers, and the number of transportation facilities. Because they are numerous and cumbersome, these constraints are left out. In order to determine the configuration of an automatic workshop in such a model with constant $t_j^{\overset{v}{l}}$ ($l=1,\ldots,\tilde{M}_o$, $j=1,\ldots,M_i$), a linear programming problem is solved.

It is important that general constraints in this model are not too numerous and so decomposition algorithms may help divide the original problem into numerous subproblems by solving which the structure and composition of FMSs are determined. Consequently, the software of modeling these FMS and knowledge and expertise of the designers may be essential.

The software where the model is implemented includes an intellectual interface which facilitates the input of fairly complicated data and interaction with the modeling system. As a result, a non-mathematic or programm user can interactively try various options of the workshop. An expert system SPEIS is a good tool in choosing the most desirable option and in making a decision that modeling must be terminated.

CONCLUSIONS

Computing experiments with the model showed its firly high efficiency and utility in CAD of workshops which consist of several FMSs.

REFERENCES

[1] Khobotov, Ye.N., Design principles of FMS pre-design modeling. Abstracts of the VI International Conference on Flexible Manufacturing Systems. Moscow: MNIIPU, 1989.

[2] Khobotov, Ye.N., On one approach to development of methods of pre-design modeling for certain types of flexible manufacturing systems. In: The State-of-the-Art and Development of Flexible Manufacturing Systems. Moscow: MNIIPU-MTSTI, 1989.

[3] Kalachov, V.N. and Khobotov Ye.N., Models of choice of the optimal structure for some types of flexible manufacturing systems. In: The State-of-the-Art and Development of Flexible Manufacturing Systems. Moscow: MNIIPU-MTSNTI, 1986.

Combined Procedures for Simulation of Manufacturing Systems

S. N. DWIVEDI
West Virginia University, Morgantown, WV, USA

YE. N. KHOBOTOV
International Research Institutre for Management Sciences, Moscow, USSR

Summary

The paper discusses methods for pre-design simulation of flexible manufacturing systems (FMS) and computer integrated manufacturing (CIM). Unlike conventional simulation methods used for this purpose, the proposed methods successively employ models of two kinds, one optimization and the other simulation. Both kinds of models work together.

Introduction

The manufacturing system pre-design methods which will be described below make it possible to choose the type and number of machine tools, the buffer and storage spaces, the transportation system for serving the machine tools, and the most profitable list of products for the workshop or factory. The structure of manufacturing systems is chosen so that with specified constraints on the price of the equipment and the floor area the system throughput and profit are maximized, the manufacturing potential is expanded, and the product manufacturing time is reduced.

The proposed methods which combine optimization and simulation models significantly reduce the modeling time, improve the modeling perfomance, and solve problems for which other methods are either ineffectual or inexistent.

The FMS structure will first be represented as a straightforward linear model in which every part is assumed to be machined along a single possible route.

The linear model

The constraints imposed on the resources of the i-th tool are [1,2]

$$\sum_{i=1}^{L} x_i f_i (n_i t_{ij} + \tau_{ij}) + \hat{t}_j \le y_j V_j, \quad j=1,\ldots,M, \tag{1}$$

where L is the number of part types which add up to the list of FMS products, x_i are integer variables such as 0,1 ($x_i=1$ if the i-th part is in the list and 0 if otherwise), f_i is the rate at which the i-th lot of parts emerges over the scheduled time span T^*, n_i is the size of the i-th type, t_{ij} is the machining time of the i-th part in the j-th facility, τ_{ij} is the time for rearrangement of the j-th facility for manufacture of the i-th parts, $\hat{t}_j$ is the idle time of the j-th facility while awaiting parts to be machined ($\hat{t}_j$ is computed by simulation methods and is used for matching the operation of simulation and optimization models), y_j is the number of j-th facilities which must be incorporated into the system ($y_j=0, 1, 2, \ldots$), V_j is the time slack of the j-th type of tools during T^* ((T^*-V_j) is the idle time of the j-th tool in scheduled repair and maintenance), and M is the number of machining equipment types that add up to an FMS.
If the j-th facility is not engaged in machining the i-th part, then $t_{ij}=0$ and $\tau_{ij}=0$.
When estimates must be obtained of the number of required tools and of the most profitable schedule, the constraint (1) is represented in the form [3]

$$\sum_{i=1}^{L} x_i f_i (n_i t_{ij} + \tau_{ij}) \le y_j V_j,$$

where $\hat{t}_j$ is assumed equal to $(1 - \mu_j)y_j V_j$, or the idle time is assumed to be part of the time slack of the j-th facility. In this case simulation methods are not needed in developing an optimal machining schedule and the parameters $\hat{t}_j$ ($j=1,\ldots,M$) need not be determined. With $\mu_j=1$ the upper estimate is obtained of the FMS manufacturing potential when no facility has to idle because of the machining sequence.

As noted above, in FMS design the throughputs of all machining areas have to be matched. The throughput of the j-th group of facilities in machining the i-th parts W_{ij} is dictated by the constraints

$$y_j - W_{ij}t_{ij} \geq 0, \quad j=1,\ldots,M, \; i=1,\ldots,L, \tag{2}$$

where y_j is the number of parts that can be simultaneously machined in the j-th group of facilities. To avoid bottlenecks of parts anywhere along the line, the throughput of the succeeding job must exceed that of the preceding job. Therefore another constraint in the model is

$$W_{ip} - W_{ip-1} \geq 0, \quad p \in J_i, \; i=1,\ldots,L, \tag{3}$$

where W_{ip} and W_{ip-1} denote the throughputs of facilities of the j-th type in machining the i-th parts in the preceding and succeeding jobs and J_i is the ranked set of facilities used in manufacture of the i-th parts.

All inter-job movements of parts must be over within time $\overline{T}$ which is specified by the designer. For this reason the number of transportation facilities must be sufficient for these movements,

$$2 (\tilde{t} + \tilde{\mu}) \sum_{i=1}^{M} y_j \leq y_{M+1}\overline{T}, \tag{4}$$

$$(\tilde{t} + \tilde{\mu}) \leq \overline{T},$$

where $\tilde{t}$ denotes the maximal inter-job transportation time, $\tilde{\tau}$ is the time of loading and unloading the pallets with parts into the buffer or table of the machining facilities.

The profit of manufacturing must exceed $\tilde{D}$ which is specified by the designer or customer while the cost of the FMS equipment must never exceed D. For this reason the following constraints are introduced in the model

$$\sum_{i=1}^{L} c_i x_i f_i n_i \geq \tilde{D}, \tag{5}$$

$$\sum_{i=1}^{M+2} d_j y_j \leq D, \tag{6}$$

where c_i denotes the price of the i-th part and d_j is the price of a unit of the j-th type of facilities.

The required number of pallet, y_{M+2}, is determined from the equality

$$\sum_{j=1}^{M} \aleph_j \, y_j + K = y_{M+2},\tag{7}$$

where K denotes the number of areas in the loading station and $\aleph_j$ is the number of areas on loading tables of the j-th type of facilities.

In this model all parts with their fixtures are assumed to be taken in by pans of the same type.

The functional to be optimized in this model is

$$J = \{ a_1 \sum_{i=1}^{L} c_i n_i f_i x_i - a_2 \sum_{j=1}^{M+2} d_j y_j - a_3 \sum_{i=1}^{L} \sum_{p \in J_i} (W_{ip} - \tag{8}$$

$$- W_{ip-1}) - a_4 \sum_{j=1}^{M+2} b_j y_j \},$$

where a_i (i=1,4) are weights and b_j is the cost of servicing a unit of the j-th type of facilities (j=$\overline{1, \ M+2}$) during the scheduled period T^*.

With $\hat{t}_j$ (j=$\overline{1,M}$) constant, the run of the model (1)-(8) for estimating the FMS configuration reduce to solution of a linear programming problem with some varibles being integers. The overriding fact in development of algorithms for solution of this problem, however, is the price of the machining center which is many times that of machining and so an accurate integer solution can be obtained only for variables y while for remaning variables approximate solution will do. Besides,

non-integer values of x_i (i=$\overline{1,L}$) can be, rather unexpectedly, interpreted as a fraction of the lot of parts ($0 \le x_i \le 1$, i=$\overline{1,L}$) that are the desired products. The more x_i are nonzero, the more types of parts can be machined in the system, which is at a premium when the market changes fast.

Consider an heuristic procedure which makes it possible to match the optimization model (1)-(8) and simulation methods. By means of these methods a suboptimal schedule is developed for the chosen list of products. At the first itertion of this matching procedure the optimization problem (1)-(8) is solved with some values of $\hat{t}_j$ (j=$\overline{1, M}$). Then for the list which has been obtained from (1)-(8) an optimal sequence is obtained for machining every lot. Once the schedule is available, the downtime $\bar{t}_j$ (j=$\overline{1,M}$) is determined for every group of facilities. The sum total is taken of the downtime for every facility in the group from the start of FMS operation to the end of machining the last part from the chosen list. The downtime of the j-th group is the sum total of those of all y_j facilities in the group. At the next step the values of $\hat{t}_j^{k+1}$ are calculated

$$\hat{t}_j^{k+1} = \hat{t}_j^k + h_k(\bar{t}_j^k - \hat{t}_j^k)., \quad j=1,\ldots,M,$$

and included into the constraints (1); the model (1)-(8) is recalculated with new values of $\hat{t}_j^{k+1}$.

The superscript k denotes here the ordinal number of iteration in coordinating the optimization and simulation models and h_k is the

$$h_k: \quad ||\hat{t}_j^{k+1} - \bar{t}_j^k|| \leq \mu_1 ||\hat{t}_j^k - \bar{t}_j^k||,$$

where μ_1 is a constant value $0 < \mu_1 < 1$.

Reiteration of this process usually leads to a good approximate solution of the problem.

In this model all the variables, x, y and W are nonnegative. The solution should yield the values of x and y, the remaining variables being auxiliary.

The problem (1)-(8) is essentially a multicriterial optimization problem; the weights may be changes with customer's requirements.

In particular, pre-design models of FMS intended for bodies of rotation are obtained in this way.

In the cases where every part can be machined in several various sequences non-linear models yield the most economical sequence for every scheduled part.

Conclusions

Pre-design FMS simulation systems where optimization and simulation models are used in succession will significantly save the design time, improve the performance of the resultant FMS, and yield structures, optimal by one or several criteria. Besides, unlike simulation methods, the optimization models will lead fast enough to an approximate solution if it exists or report, on certain cues, the absence of a solution for the actual initial data and parameters. Consequently, the above procedures will result in CAD systems for FMS and CIM.

References

[1] Khobotov Ye.N. On one approach to development of pre-design modeling methods for flexible manufacturing systems of certain kinds. In: The State-of-the-Art and Development of Flexible Manufacturing Systems. Moscow: MTsNTI and IRIMS, 1989 (in Russian).

[2] Khobotov Ye.N. Design principles of FMS pre-design modeling Abstracts of the VI International Conference on Flexible Manufacturing Systems. Moscow: IRIMS, 1989.

[3] Kalachov V.N. and Khobotov Ye.N. Models of choice of the optimal structure for some types of flexible manufacturing systems. In: The State-of-the-Art and Development of Flexible Manufacturing Systems. Moscow: MTsNTI and IRIMS, 1986 (in Russian).

Expert Systems in CIM

V.M. PONOMARYOV, V.V. IVANISTCHEV, A.A. LESKIN, and N.N. LYASHENKO

U.S.S.R. Academy of Sciences
Leningrad Institute for Informatics and Automation
Leningrad 199178, U.S.S.R.

The necessity to use decision-making expert systems arises at various stages of CIM (marketing, production planning, research, design, manufacturing control, technological processes control, checking and testing). So the problem of the general approach to the control of data acquisition and analysis processes, and inference rules design has acquired great significance. The problem was dealt with in a number of works which were carried out in the Leningrad Institute for Informatics and Automation of the USSR Academy of Sciences (LIIAS).

It can be stated generally that the main task may be reduced to data transformation into a form that would be most convenient for decision making or, to the available information co-ordination with the means using it. Data acquisition and subsequent transformation may be represented then as the stages of various application expert systems automated design. In is supposed that the data being acquired may include experimental information as well as data obtained by means of simulation. So models design automation aimed at getting models of required consistency has also become very important.

Acquisition of a data base concerning an object, and diagnostic and control decision-making are based upon the studied object statistical models use. To construct a model it is necessary to have a sufficiently representative experimental data sample on hand. Diagnostics and control problems may be solved at early stages of collecting information on the object, classic sequential analysis techniques originating from Bayes approach being used for the purpose because as a rule the required volume depends upon unknown parameters for most practical situations ([1], ch. 1,4).

If the required data volume is rather large, the problems of the object behaviour prognosis precision and stability, as well as those of control decision-making are attached with primary importance. In this case an appropriate analysis strategy depends upon an adequate object model availability.

If the have a trustworthy analytical description or an simulator at our disposal, then the problem of the optimum or an acceptable decision is reduced to some regressive models parameters estimation. Otherwise inductive inference procedures seem to be to the point. These are able to generate decisions outside the model, using minimum suppositions as to the parameters links pattern. Besides it is possible to construct an analytical description by inductive inference using fast analytical descriptors (regressive model selection).

So we have get the analysis scheme presented at Fig.1.

As the version "a" is known well enough, let us come into some more details of the version "b". In many situations when the integrated manufacturing is dealt with (an object design, technological processes control robot control etc.) it is possible to use a highly formalized object description and to construct an analytical model, e.g. a Petri net [2]. At early stages of the integrated manufacturing (marketing, ecological estimations production planning etc.) the weakly structured, initial information prevails. The interactive automated simulation technique and instrumental programming nets [3] were designed for such applications providing experts in a given object area with the possibility of creating adequate simulators using a special ideographical language. The initial model structure represented as a block-diagram is transformed with the aid of these means into a operators' net; an example of such a net is shown in Fig.2. This net elements and quantitative parameters are put into the instrumental system SAPFIR, which composes the needed set of simulation and reports generating programs.

In the version "c" [4] the inductive inference procedure is used for quick construction an analytical model, describing the links between the initial information and the condition of the studied object parameters set being analysed belonging to a certain class.

The mathematical interpretation of the problem is given below. If a condition $F(x_0)$ describes the belonging of an object to a certain class, and $A_1(x_1)$, ..., $A_n(x_n)$ are some initial descriptors, then assignment of a pair (Q,Level), were Q - statistics of the relation measure between $F(x_0)$ and some expression G containing variables $A_j(x_j)$ and Level - present threshold numbers, formalizes the criterion which makes terms "the best classification rule" and "an allowed rule" sensible. The best rule maximizes Q, while an allowed G satisfies the inequality $Q(F,G) >$ Level and thus defines the empirical interpretation of such statements as "$F(x_0)$ is equivalent to $G(A,(x_1),...)^n$".

Inductive inference basic software algorithmic structure includes the following modules:
1. Relation criteria definition.
2. Inference rules synthesis (for a criterion defined).
3. Inference acceleration computing schemes construction.
4. Derivation of the formulas maximizing the present relation measure with a fixed goal attribute.
5. Acquisition of the empirical quantifiers via relation statistics and confidence probabilities.

6. Acquisition of the empirically true formulas lists via high performance algorithms.
7. Generation of all given formula consequences.
8. Recovery of all the formula empirically true precursors.
9. Acquisition of "empirical axioms", i.e. empirically true formulas, which it is impossible to infer from other true formulas.
10. Generated lists stability estimation for random data base disturbances.

Software design for cylindrical articles shape deviations from standard detection and correction in real time is an example of application of the described approach for a complex object analysis.

The information sought for is an equation the "x-y" coordinates, which describes a part cross-section. The relation between the values directly measured p_1, p_2 and p_3 and the equation desired established by means of inductive inference.

The analysis sequence includes the following steps:
1. The object description.
2. The object simulation.
3. Data base acquisition based on the simulator.
4. Data base analysis with the aid of inductive inference.
5. The analytical descriptor application for the classes selected.
6. The descriptions constructed application for monitoring and control regressive problems.
7. Synthesis of test and control programs for the design of real-time hardware-implemented modules.

References

1. Leman A., Statistical Hypotheses Checking

98

2. Leskin A.A., Maltsev P.A., Spiridonov A.M. Petri Nets for
   Simulation and Control. Leningrad, Nauka, 1989.
3. Icvanisthev V.V. Flow Systems Simulation Automation.
   Leningrad, Nauka, 1986.
4. Lyashenko N.N. Inductive Inference Methods and Algorithms.
   Leningrad, LIIAS, 1989.

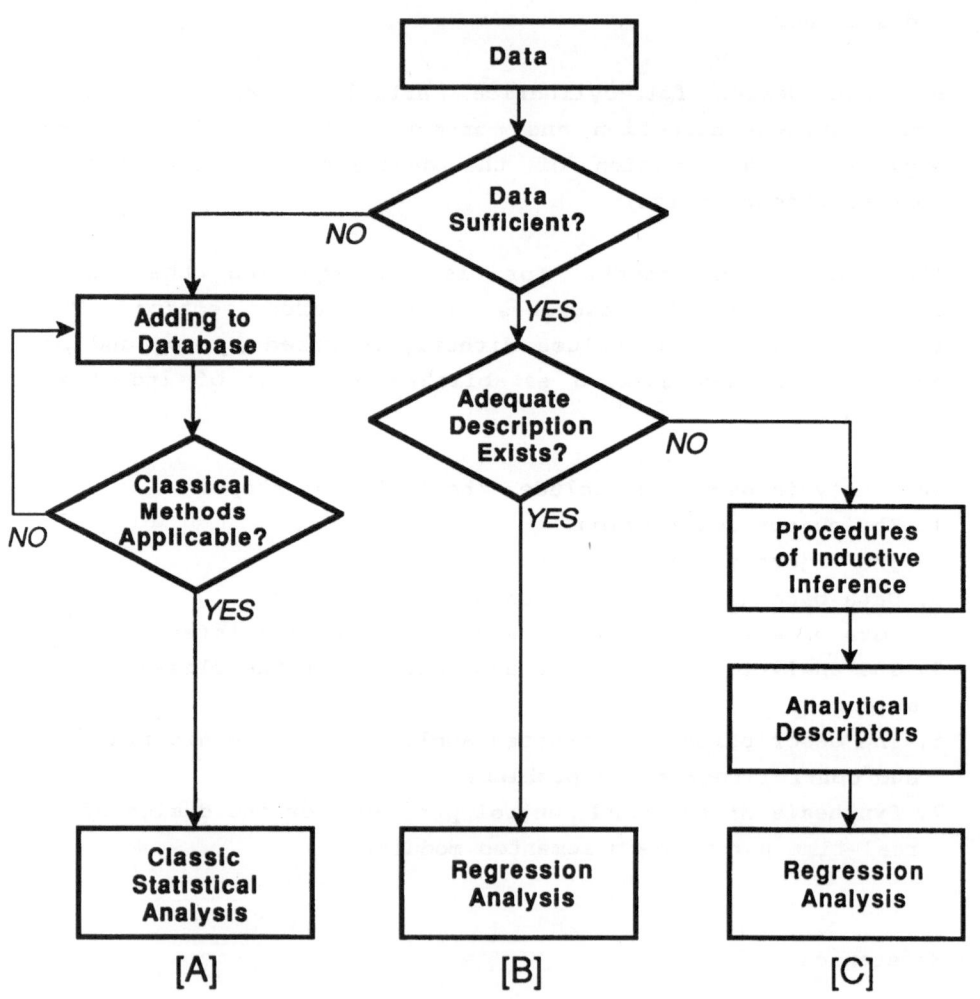

Fig. 1. General Data Analysis Algorithm

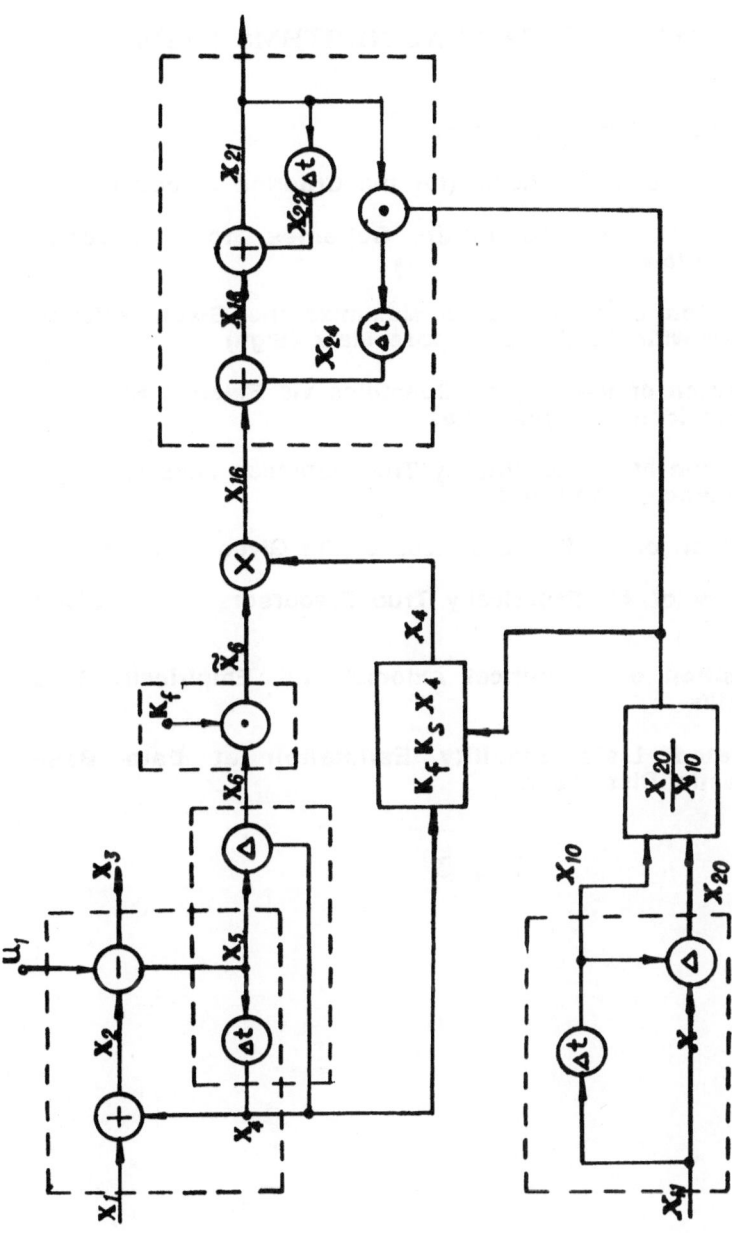

Fig. 2 IDEOGRAPHIC REPRESENTATION OF AN OPERATORS' NET

INDUCTIVE INFERENCE ALGORITHMS BASIS

1. Selection of Relation Criteria

2. Inference Rules Synthesis (for the Criterion Selected)

3. Acquisition of Computer Schemes for Inference Acceleration

4. Acquisition of Formulas to Maximize the Given Relation Measure with the Fixed Purpose Mark (Sign)

5. Acquisition of the Empiric Quantities via Relation Statistics and Confidence Probabilities

6. Acquisition of the Empirically True Formulas Lists via High-Performance Algorithms

7. Generation of All Consequences of the Given Formula

8. Recovery of All Empirically True Precursors of the Given Formula

9. Acquisition of "Empirical Axioms", i.e. Empirically True Formulas

10. Generated Lists Stability Estimation at Data Base Stochastic Disturbances

Fig. 3

REAL-TIME DETECTION AND CORRECTION OF CYLINDRICAL SHAPE DEVIATION FROM STANDARD

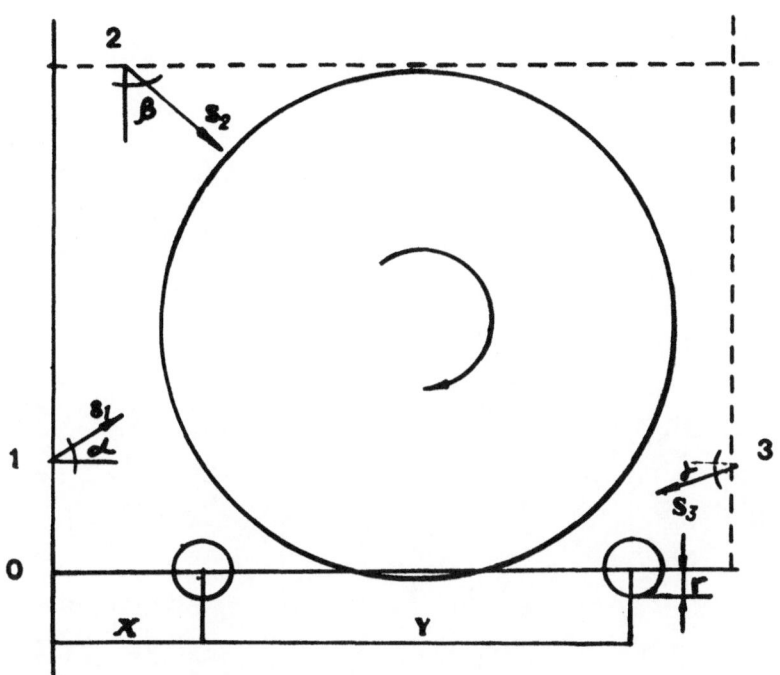

1. Object Description

2. Creation of the Object

3. Object Simulation

4. Data Base Acquisition via Object Simulation

5. Analytical Description Application

6. Application of the Acquired Descriptions to Solve Monitoring and Control Regressive Tasks.

7. Synthesis of the Test and Control Programs to Create Real-Time Hardware Modules.

Fig. 4

Chapter II

Production Planning

Introduction

Diverse aspects of production planning and related issues are addressed by the papers in this chapter. A taxonomy of event-driven AI-based expert systems is presented in the first paper. It focuses especially on the synergy between the use of such systems and the issues of deriving parallel algorithms for real-time applications. The second paper compares the new lot sizing policy methodology termed DLIC (Dynamic Least Incremental Cost) for variable demand pattern with the well known policies using a variety of demand patterns. The problem of real-time scheduling is addressed by the third paper. Specifically, the potential application of discrete event computer simulation to aid system control in real-time decision-making is analyzed.

The job-shop scheduling problem from an algorithmic point of view is the focus of the fourth paper. The decomposition approach is analyzed, and a method of making the resulting sub-problems independent, and thus amenable to separate resolution, is presented. The fifth paper evaluates the impact of sophisticated scheduling techniques, real-time monitoring systems, and manufacturing plant automation on job-shop performance. The last paper presents a software package for generating time standards for turned components, including both machining time and non-machining time (handling, setup, idle time, tool change, etc.). The non-productive time determined by the use of this package serves as a basis for evaluation of computer-generated process plans.

A Taxonomy on Event-Driven Production Systems

S. S. IYENGAR, NITIN S. NAIK and RAJENDRA SHRIVASTAVA

Robotics Research Laboratory
Department of Computer Science
Louisiana State University
Baton Rouge, LA

ABSTRACT:
Intelligent Computing Systems require an integration of the traditional Expert system technologies with real-time response and control capabilities. The purpose of this paper is to exploit the synergism between the use of an event-driven AI-based expert system and the issues in deriving parallel algorithms for various applications. The two issues that will be considered in this proposal are: (i) the effect of variations in the dynamics of the environment on overall system stability, and (ii) the structure and properties of event-driven parallel algorithms for some real time applications.

1. INTRODUCTION

Today, the intelligence analysis and automation community is being swept by broad, pervasive technological demands. The area of Artificial Intelligence has been fast developing but has to overcome the hurdles of real-time environments. The qualities of a Real-time Artificial Intelligence system is its capability to provide intelligent, knowledge-based responses to real-time, unforeseen occurrences. Research in this area is still in its infancy and there has been a fair amount of activity in two of the most important practical implications of this research - in the areas of navigation and process control. The popularity of expert systems in this discipline is due to their capabilities of symbolic computation and knowledge representation along with their ease-of-implementation and practicality and this has pushed them to the forefront of current research in intelligent, real-time systems.

Traditionally, Expert Systems have been successfully applied to solve the problems of diagnosis, design and classification in environments whose characteristics are known a-priori. Such application areas have boosted the popularity of expert systems in a wide variety of significant disciplines. The application of existing expert system technologies to problems in real-time environments has, however, revealed certain basic inadequacies, thus requiring the development of methodologies which combine the features of knowledge-based expert systems with the ability to respond to real-time changes. The design of such real-time expert systems depends to a large extent on the capability of the underlying expert system tool to cope with real-time data. This makes the development of such tools and techniques an important focus of research in the effective design of real-time expert systems.

1.1   PRODUCTION SYSTEMS AND PROBLEM SOLVING STRATEGIES

Production Systems are the basis of many rule-based Expert Systems. Rule-based systems like expert and production systems are computer programs that emulate the search behavior of human experts in solving a problem. These are not procedural programs, but data driven programs, that are sensitive to current data and future modifications on the data. This data is also termed as the problem state; any changes in the state are carried out by the *production systems* operators like modification of some existing data or the creation of new data. The flow of program execution is controlled by the order in which the states are generated by the data in the *working memory*. For a broader treatment on expert systems see [10].

Production systems are capable of representing knowledge, expertise, and problem state. They also have a set of state evaluation functions and operators for modifying the state that makes *production systems* simpler than other models of conventional computation, for certain class of problems. *Production Systems* have been widely used in implementing expert systems like the new DENDRAL [3] system for inferring chemical structures from spectrometry data, MYCIN [4] system for diagnosis and therapy of bacterial infections of the blood, PROSPECTOR [5] for mineral exploration, the R1 [9] system for configuring DEC VAX computer systems and HEARSAY-III [6] production system for understanding speech.

Production systems are defined as consisting of a collection of *production rules*, representing the knowledge and forms the *production memory*, a collection of data in the form of *elements*, that describe the current status of the system which is defined to be the *working memory and inference engine* which provides the decision making mechanism or expertise to the program. Each *production rule* has two parts. The LHS consists of assertions that have to be satisfied by the *elements* currently in the *working memory* and the RHS consists of actions that need to be performed and may involve some modification to the data or providing conclusions to the user. The search phase in production systems, that evaluates the data with respect to its knowledge and then the system performs the RHS actions of the selected rule, in what has traditionally been called the *recognize-act or match-act cycle*.

Real-time production systems like R1 and HEARSAY-III are computationally intensive in the *recognize or Match* phase. Some have an extremely large rule base, forming the *production memory* as in the case of R1 production system which has to be consider all the possible configurations for the VAX computer; whereas others may have an extremely large *working memory* as in the case of HEARSAY-III production system for understanding speech. For HEARSAY-III program data in the *working memory* represents the entire spectrum of sound frequencies with different modulations that may have been generated by the speech pattern that is being evaluated by the production system.

Depending on the type of problem that *production systems* solve, they are classified into either *goal directed* forward chaining systems or *data directed* backward chaining cognitive systems. Robotics applications, path routing and navigation systems are *goal directed* systems where the system tries to achieve a primary goal by modifying states in a sequence that generates the goal state. MYCIN, R1 and DENDRAL are *data directed* cognitive systems that evaluate the data and try to identify a sequence of problem states that would generate a state corresponding to the current data.

The basis for production systems is the *MATCH-ACT* or the *recognize-act cycle* [8] where the production system will recognize all the *production rules* that have a successful *MATCH* by iteratively evaluating the *production rules* over the *working memory*. The *act* phase is the action to be executed by the production rule that has been selected by the *conflict resolution* phase within the *inference engine* of a production system. This cycle is repeated till no production rules are true or the required results have been obtained (see Figure 1).

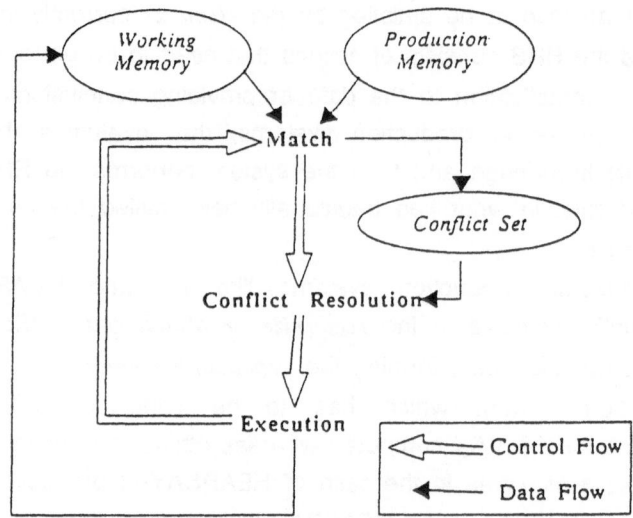

Figure 1: Traditional Rule-Based Expert System

Production systems are query based programs that are similar to large data-base application programs; but unlike the data-base programs systems are expected to evaluate the programs' initial response to the query and proceed to determine the correct sequence of queries that need to be answered if the original query does have an answer. Production systems have reasoning capabilities in the *recognize-act* phase and may have multiple sub queries that are active at any given instance. Data-base programs have to be provided one query at a time to perform the *MATCH* phase only and cannot spawn sub queries that are active concurrently.

There are two approaches that have traditionally been implemented as problem solving strategies in production systems; they are *Forward Chaining and Backward Chaining.* Any production system may use a combination of the two strategies or use either one exclusively, depending on the application at hand. The deduction or reasoning amongst both problem solving strategies may be based either on *Deterministic or Probabilistic* principles at any given *production rule.* For more details see [16].

Traditional expert systems are not intended to serve as real-time control devices, and have several significant drawbacks related to this:

　*　*Inadequate infrastructure*:　lacks infrastructure for collecting and

representing external sensor and actuator data; only internal states and events are represented.

* *Slow Speed*: cannot guarantee a real-time response to a critical external event.

* *No interruptability*: control structure or production cycle of traditional system is inherently sequential and synchronized; system cannot be interrupted and resume previous activity in a graceful manner. This limits the ability to provide event-driven response in a dynamic, dense threat environment in real-time, since during an external emergency a new set of responses cannot be concurrently initiated to respond to the new situation. For a broader treatment on this see [1].

1.2   ORGANIZATION OF THIS PAPER

This paper is organized as follows: Section 2 details the architecture and operation of an Asynchronous Production System. Section 3 presents the implementation issues involved with Asynchronous Production Systems in an MIMD architecture. We conclude with section 4 which draws some conclusions from previous sections.

2.   ASYNCHRONOUS PRODUCTION SYSTEMS (APS)

Conventional *production systems* are data driven programs that execute the *recognize act* cycle till completion where the data is modified only by the RHS actions of *production rules*. APS are a version of *production systems* that enable external events to modify the data in *working memory*. This ability of APS to recognize the occurrence of external events facilitated the development of real-time expert systems which retain the convenience of traditional rule based system's knowledge base representation. Unlike conventional systems, the contents of *working memory* are not the only information that is available to the APS during the *recognize-act* cycle.

A situation is asynchronous if multiple events are occurring at different times and at different rates for a given time period of interest. Asynchronous production systems use the knowledge base representation of traditional production systems but incorporate a new data structure for representing external memory and it has a high precedence over the data in working memory at execution time. The need for recognizing external events is apparent for applications in the fields of Autonomous Robots, Process Control, Medical Monitoring, Navigation in Unknown terrain, etc. All these programs require a rule base for representing knowledge and also should be able to respond to a real-time unknown or unexpected external event. The corresponding data memory definitions are described in Figure 2 for a robot (HERMIES IIB) attending a fire emergency operation.

```
/* WORKING MEMORY DATA DEFINITIONS */

    (goal ^status ^type ^start ^destination)
    (robot ^position)
    (exception ^status ^type ^priority)

/* EXTERNAL MEMORY DATA DEFINITIONS */

    (fire-transducer ^smoke ^temp)
    (sonar-infra-red ^distance ^height ^width)
    (sonar-acoustic ^distance ^height ^width)
    (sonar-ultra-sonic ^distance ^height ^width)
```

Figure 2: A set of DATA elements for HERMIES IIB

The working memory data is used for the navigational aspect of the program which computes the shortest path from an initial point *goal ^start A* to the final destination *goal ^destination B* using intermediate positions X for the robot as *robot ^position X*. Any modification to this data has to originate as an action of the APS itself and this represents the conventional part of the system.

The external memory data is used for the asynchronous exception handling aspect of the program which reacts immediately to the stimuli from any external event that HERMIES-IIB is capable of detecting. Figure 3 illustrates the conceptual layout of an Asynchronous Production System.

An APS contains rule-based inference engine capable of dynamic and rapid interactions with its environment. In the context of remote rendezvous and docking, the tasks of physical device control, sensor data interpretation, and scheduling the use of the limited communication and computational resources at the remote site are very time-critical and also highly complex. The two traditional approaches to these types of problems, both have significant drawbacks: control-theoretic algorithms are too restrictive to deal with the large-range of possible anomalies during real-time

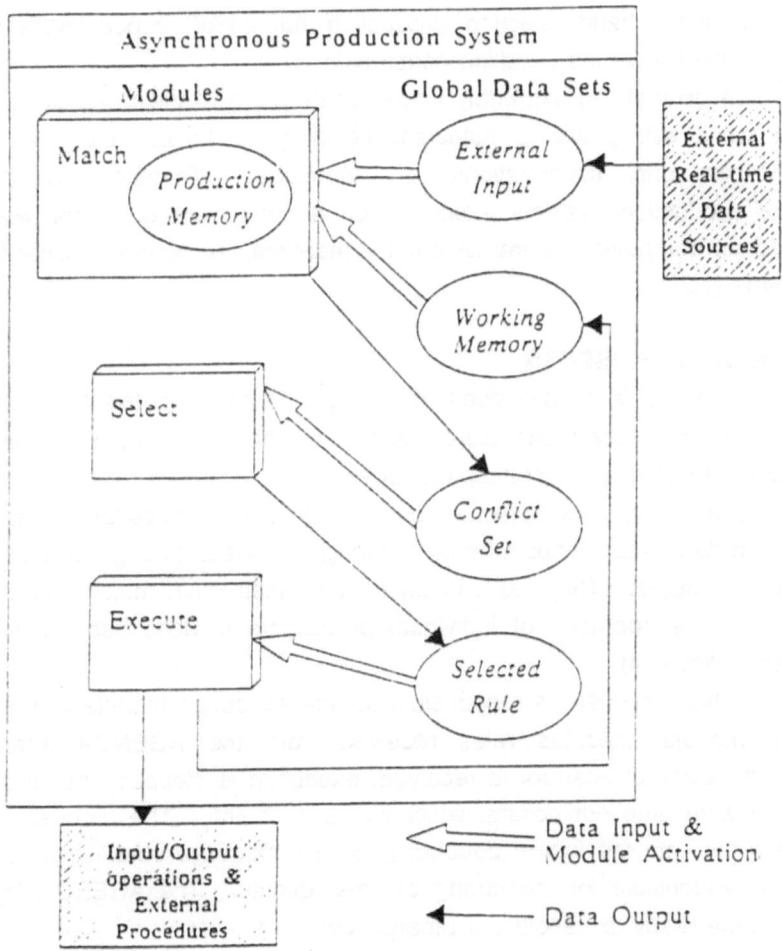

Figure 3: Asynchronous Production System Architecture

control; on the other hand, existing production rule-based expert system models do not meet real-time deadline constraints.

In order to met the extremely time-critical nature of the tasks at the remote site while still providing sufficient flexibility to handle the many possible classes and combinations of anomalous real-time events, asynchronous production systems executing on a concurrent computer will be advocated as computational framework for real-time, event driven expert control systems [1,2].

3. IMPLEMENTATION ISSUES

In this section is a description of an implementation of the APS for CLIPS on a multiprocessor architecture. A typical application scenario has been described to illustrate the conceptual ideas.

The computational architecture consists of three processors, each performing a distinct task independently sharing the data through shared memory data structures. They work in an multiple instruction multiple data (MIMD) mode. The operation of individual processors is described briefly below (shown in figure 4):

Processor 1: This processor is responsible for the execution module of the APS. This module executes rules received from the AGENDA after matching. If an external interrupt is received, execution is stopped after the current rule or after time out occurs, whichever occurs first. The processor then asserts a fact into the factlist depending on the message obtained from the processor responsible for monitoring external events. The AGENDA is cleared and made ready to handle the emergency.

A typical sequence of activities for processor 1 is as follows:
1. performs all required initializations.
2. loads the file containing rules provided by the user.
3. sets up the mechanisms necessary to receive interrupts from other processors.
4. resets the CLIPS environment.
5. initiates a run command to execute rules from AGENDA.

During the execution of rules from the AGENDA, the system may encounter assertion of facts on the RHS of the rules. These assertion of facts result in (a) inclusion of the fact in the factlist, and (b) subsequent matching of the fact with the rules. After the addition of the fact to the factlist, processor 1 sends a message to processor 2 which is responsible

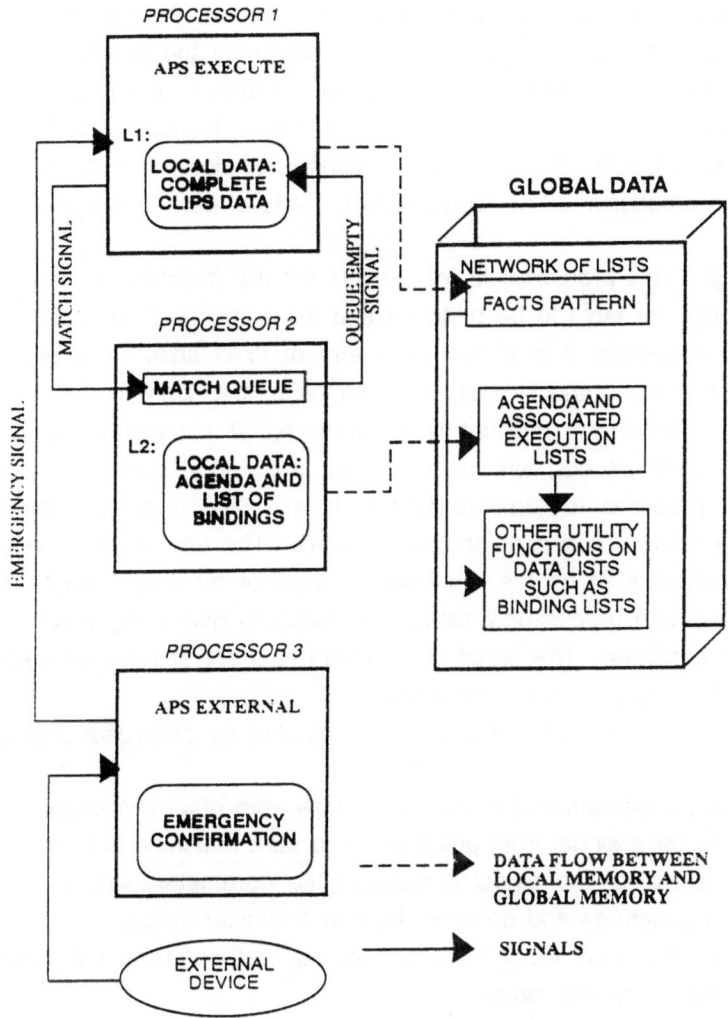

Figure 4: Implementation of APS on a Multiprocessor Architecture

for the matching activity. Also, processor 1 transfers the parameters and data structures required for matching into a queue in the processor 2 data area. Processor 1 continues performing the remaining actions in the RHS until the rule is completed. All other activities, besides matching, are performed by processor 1. Processor 1 stops execution when the AGENDA is empty and processor 2 has no matches to perform from its queue.

Processor 2: This processor is responsible for the matching activity of the APS. It stores the facts to be matched and the associated database details provided by processor 1 in a circular queue of fixed size. The processor then performs the matching of the fact with the data in the queue sequentially. When an interrupt is encountered, it completes the current matching or is timed out, whichever occurs first. The matching queue is cleared and prepared for emergency matching. The emergency matching requests are queued and executed sequentially. The end of the emergency is either taken care of by the rule-based design or an explicit signal to all the processors and processor 2 clears the matching queue again for normal execution to continue. The result of matching is a set of rules deposited in the AGENDA for execution by processor 1.

A typical sequence of activities performed by processor 2 is given below:

1. sets up mechanisms to receive signals from other processors.
2. waits for a signal from processor 1 to start matching activity.
3. on arrival of the signal it receives the parameters and associated data structures and deposits them in a circular queue.
4. takes the next entry from the queue and performs the matching operation for that entry.
5. when matching is over it repeats step 4.
6. when an interrupt signalling an external emergency is received it stops or completes the current matching in progress and clears the queue for emergency matching.
7. when the emergency is over the queue is cleared and the processor is ready for normal execution.
8. if the queue is empty then it waits for the signal from processor 1 and repeats step 4.

Processor 3: This processor monitors for external events and evaluates the event to determine whether an emergency exists or not. Also, the processor is an interface between the system and the transducers used to monitor the external world. It calibrates and validates the data received from the transducers. The data compared to datalimits determines whether an emergency situation exists or not. If an emergency situation exists then a signal is sent to processor 1 indicating the emergency and type of emergency. The type facilitates the assertion of a fact corresponding to the emergency as different emergency situations call for different handling procedures and hence different fact assertions. Processor 1 asserts the fact corresponding to the emergency type as stated earlier.

A typical sequence of activities for processor 3 is given below:

1. sets up mechanisms for receiving interrupts from external sensors whenever there is new data.
2. accumulates the data in its local data area.
3. evaluates and validates the data from the transducers.
4. calibrates the data using a predetermined calibration procedure for each transducer.
5. checks for emergency situations using predefined data limits.
6. sends an interrupt signal to processor 1 in case of emergency and a Type parameter depending on the transducer causing the emergency.
7. continues monitoring of external devices.

Let us consider an application of the above scheme.

Problem: Assume a security robot is monitoring a closed room by navigating continuously in a rectangular path close to the walls of the room. The environment is completely known to the Robot. An emergency situation occurs when any door is open or when it encounters any unknown obstacle in the room. The emergency is handled by the Robot navigating to a known alarm switch location, out of its normal path of navigation and activating the alarm. Subsequently, it resumes its normal routine of navigating and monitoring along the rectangular path around the room.

A rule base will be necessary to co-ordinate the activities of the Robot. A typical initial activity rule is as follows:

(DEFRULE startrule
        (initial-fact) =>

                (initialize ROBOT parm)
                (estimate cur_position)

(assert (status normal))
)

(DEFRULE normrule
   (status normal) =>
      (goto nearest point in the rectangular path)
      (start rectangular navigation)
      (assert (navigation rectangular))
)

(DEFRULE navigation
   (navigation rectangular) =>
      (move in straight line)
      (detect edge)
      (if edge occurs (assert (edge arrived))
)

and so on.

An emergency handling rule may be applied as follows:
(DEFRULE emergenrule
   (status emergency) =>
      (stop ROBOT)
      (estimate cur_position)
      (plan path to alarm)
      (assert (navigation alarm))
)

   Our scheme starts with processor 1 loading the above rule base file into the environment. The trace of the NORMAL activity yields the following:

1.   Processor 1 loads rule file into the CLIPS environment.
2.   Processor 1 resets CLIPS. The reset process asserts initial-fact into CLIPS. This gives rise to a match and so Processor 1 sends signals to processor 2 as soon it encounters a match.
3.   Processor 1 activates the run command. At this time the AGENDA is empty and the match queue has an entry. The run command waits for a rule to enter the AGENDA. At the same time, Processor 2 performs the matching operation for initial-fact and loads the rule into the AGENDA.
4.   Processor 1 starts with the new rule in AGENDA. It executes all the RHS actions sequentially. When an assert is encountered, processor 1 signals processor 2 and continues execution.

During the above operation in processors 1 and 2, processor 3 has been independently monitoring the external devices.

Consider the situation of processor 1 executing the rule named NAVIGATION, and an emergency occurs. The processor must start the emergency handling procedures according to the rules for the emergency situation. To activate theses emergency rules, processor 1 has to assert EMERGENRULE. Control is transferred to these rules and execution is continued. In the above scheme, it is assumed that the user or programmer will specify the course of action to be taken during and after the emergency. The action taken is not common to all tasks as they are dependent on the task performed by the CLIPS program. For example, in the above setup processor 1 must abandon NAVIGATION and start EMERGENRULE. If processor 1 was executing RHS action (navigate in a straight_line) when interrupted by the emergency signal, then the robot would execute the RHS action of planning a path to the alarm switch and navigate along the path to reach the switch. If the programmer has not specified an operation to be executed after the emergency then the robot will remain at the switch. An alternative to this predicament is to save the original state and resume with this state after the emergency routine is completed. The programmer may just initiate the NAVIGATE rule again after the emergency is handled but this will cause the robot along a different path starting at the switch. Thus the programmer must specify a rule for "after emergency" conditions. The robot will be in the right state to continue its normal routine after the execution of the emergency rule. A sample rule is given below:

```
(DEFRULE after_emergency_rule
         (status emergency_over) =>
                          (plan path to the nearest point on the
                          rectangular path from the alarm switch)
                          (goto nearest point)
                          (start rectangular navigation)
                          (assert (status normal))   )
```

The above rule puts the Robot on the right path after the emergency from the alarm switch location and then resume its routine navigation. Thus the Robot is back into its normal activity.

Hence the above scheme is best suited to emergency based environments as compared to the state recovery scheme.

4. CONCLUSIONS

Expert systems are well suited to solve the complex real-time control problems due to their structured approach. Expert systems provide a structured problem solving methodology by separating program, control and data. Separation of control and data enables a program to respond quickly to dynamic changes in either data or control. Real-time control, unlike the conventional applications of expert systems, involves the manipulation of physical devices with large operational time constants. Thus, the execution of actions on the right hand side of rules can take as long a time as the computationally expensive match phase of the production cycle. Real-time control systems require dynamic, rapid, and guaranteed responses to changes in the working environment. If response time needed is less than the period of a production cycle, it should be possible to interrupt the time consuming phases of the production cycle.

Asynchronous Production Systems (APS) methodology described in this paper has the distinction of external memory elements and working memory elements (in the production memory), and concurrency for Match, Select and Executive phases (of the production cycle). External memory elements provide a natural way to express variables representing external events. Values of these variables may be modified from external functions. Concurrency of the phases allow asynchronous processing of data at the earliest response time without waiting for other phases to be completed. Concurrent processing allows interrupts to intervene the execution of any phase of the production cycle. Thus, Asynchronous Production System exploits the advantages of expert system technology to meet the requirements on real-time control.

ACKNOWLEDGEMENTS

The authors would like to thank Shiva Subramaniam and friends at RRL, LSU for their assistance in the research work.

REFERENCES

[1]   Iyengar, S.S., A. Sabharwal, F.G. Pin and C.R. Wesbin, "Asynchronous Production Systems for Control of an Autonomous Mobile Robot in Real Time Environments," To appear in *Applied Artificial Intelligence Journal (Dec. 90)*

[2]   Sabharwal A.,S.S. Iyengar, C.R. Weisbin and F.G. Pin, "Asynchronous Production Systems," *Journal of Knowledge Based Systems*, Vol. 2, No. 2, p.p. 117-127, June 1989.

[3]   "The Dendral Project," *Artificial Intelligence*, Vol II, No. 2, pp.5-24, 1978.

[4]   Shortliffe, E.H. "Computer Based Medical Consultations: MYCIN," *Elsevier*, New York 1976.

[5]   Duda, R., J.G. Gasching and P.E. Hart, "Model Design in the Prospector Consultant System for Mineral Exploration," in *Expert Systems in the Microelectronic Age*, pp.153-167, Edinburgh University Press, 1980.

[6]   Erman, L., P. London and S. Fickars, "The Design and an Example Use of HEARSAY-III," *Proceeding of the Seventh International Joint Conference on Artificial Intelligence*, Vol. 1, pp. 409-415, 1981.

[7]   Newell, A., "Production Systems: Models of Control Structures," in *Visual Information Processing*, pp. 463-526, Academic Press, 1973.

[8]   Rich, E., "Artificial Intelligence," McGraw-Hill Series, 1985.

[9]   McDermott, J., "R1: A rule-based configurer of computer systems," *Technical Report*, CMU-CS-80-119, 1980. Department of Computer Science, Carnegie-Mellon University.

[10]   Shrivastava, R., "Parallelization of Goal Driven Production Systems on Hypercube machines in C Environment," Ph.D. Thesis, 1990. Department of Computer Science, Louisiana State University.

An Improved Lot Sizing Policy for Variable Demand

M. D. SREEKUMAR
HMT Ltd., Kalamassery
Cochin, India

C. ESWARA REDDY and O.V. KRISHNAIAH CHETTY
Department of Mechanical Engineering
Indian Institute of Technology
Madras, India

Summary

The methodology and usefulness of a new lot sizing policy for variable demand pattern termed Dynamic Least Incremental cost policy (DLIC) is describe. It is compared with the well known policies using a variety of demand patterns. The superiority of DLIC has been well established considering different performance criteria. It performs equally well like WWA when EOQ/$\bar{D}$ and CV are less than or equal to 1.5 and 1.17 respectively.

Introduction

A variety of lot sizing policies are available for decision making considering variable demand pattern [1,3-5,7,8]. Compared to economic order quantity (EOQ), Economic Order Quantity Rounded (EOQR) determines the order quantity as the sum of the requirements of certain periods which is close to EOQ value. In Periodic Reorder System (PROS) [3] orders are placed at fixed time intervals. Heuristics are used to develop new policies, which will allow both the lot size and the time between orders to vary. The aim is to include more periods of lean demand and increase the time between two orders and lot sizes are dynamic. Least unit cost (LUC) and least total cost (LTC) least period cost (LPC) [7], part period balancing (PPB), Freeland and Colley heuristic (FCH) and Groff's heuristic (GH) [4] policies perform very well in typical situations. Solutions can always be improved by adjustments; look ahead/look back policy [1]. Wagner-Whitin algorithm (WWA) provides best solution for static demand [8]. It is sensitive to the number of periods and solution has inbuilt `nervousness', thus limited to dynamic situations. This paper describes a computationally simple lot sizing policy called

'Dynamic Least Incremental Cost policy, DLIC'. This methodology is found superior to the existing ones and is second only to WWA.

Dynamic Least Incremental Cost Policy (DLIC)

Basic aim of any lot sizing strategy is to determine the schedule of replenishment orders which minimizes the sum of the ordering cost and inventory carrying cost. DLIC attempts in "least incremental cost" to arrive at optimal solution. Ordering of components during periods of lean demand is avoided and they are clubbed with periods of high demand. Assumptions common to all such cases are made in the present case also [4].

Methodology

Initial solution for any stage is arrived at, by comparing the cumulative carrying cost (CCC) of alternate feasible solution with ordering cost (OC). It is then refined by combining the requirements of next period, if the demand is lean, with the previously planned order. Therefore the ordering for a higher demand will be shifted by one period ahead. This is done only if the adjustment can result in net savings. This methodology is found to yield a better solution. The computations involved are explained below considering a typical demand pattern, Table 1.

Step 1: Consider all the alternate solutions possible at the beginning of the current period, i.e., issuing an order for the requirements of the 1st period only; 1st and 2nd periods; 1st, 2nd and 3rd ....periods, etc. In each case compute the cumulative carrying cost. Choose the lot size as the sum of the requirements of all the periods (say t periods) excluding that of the period ((t + 1)th) which caused the cumulative carrying cost to be greater than OC. This is the initial solution at this stage. Table 2 shows the computations.

Step 2: Calculate the carrying cost, CC1 of the next period's, [i.e. (t + 1)th period] requirement, if the same is combined with the previously planned order. In the present case, CC1=6x3x2=36/-

Step 3: Is CC1 less than or equal to OC? If 'yes' go to step 4, otherwise go to step 7. In this case, the answer is 'yes'.

Step 4: Calculate the carrying cost CC2 of the (t + 2)th period's requirement if the same is ordered in (t + 1)th period. In the present problem CC2 = 20 x 1 x 2 = Rs.40/-

Step 5:Is CC1 < CC2? If yes go to step 6, else 7. Here it is YES
Step 6: Include the requirements of the (t + 1)th period also in
the initial solution of this stage. Increment t by one and go to
step 2 for further improvement of the solution.

In the current problem 4th period reqirement will be included
after 1st looping and that of 5th period in the 2nd looping.
Therefore solution to first stage = 85 + 6 + 20 = 111 units.
Step 7: Accept the initial solution of this stage as the optimal
solution at this stage.
Step 8: Repeat steps 1 to 8 for next stage till all the requirements are met. Table 3 shows the results for data in Table 1.

Performance of the proposed strategy
The performance of the new policy has been checked with 30
numerical data sets some of them generated by the authors to
represent a variety of datas and others available in the literature [1-8]. The performance of EOQ, EOQR, PROS, LUC, LTC, LPC,
PPB, PCH, GH, WWA and DLIC policies are compared.

1. Performance ranking: Ranking based on total cost of different
strategies for each of the demand pattern is calculated and the
frequency of the ranks is computed, Table 4 The DLIC policy
generated an equally good solution as WWA in 18 out of 30 sets
and stands second in 9 cases.
2. Performance index:If weightages of 1,2,...etc, are attached to
the 1st, 2nd,...ranks respectively, total score of each model can
be computed. Ratio of this total score to that of WWA, identified
as performance index, PI, is taken as a measure of effectiveness
of each policy. Table 5 shows the PI of different policies with
respect to WWA, showing better position occupied by DLIC.
3. Total cost: Many industries, having a multiproduct environment, prefer the sum of the total cost of different demand data
to the frequency of their performance ranking for selecting a
lot sizing policy. Table 6 shows the sum of the optimal costs of
all the demand pattern for each of the policies considered. DLIC
is superior compared to all others except WWA. The increase in
cost of DLIC for the entire set of data is only 0.8% over WWA.
This result is of significance. Policies namely LPC and GH with

1.8% increase occupy the next position. The EOQ model is seen to be the one with maximum cost.

4. Computer Processing Time: Industries considered computer time also as vital in choosing a lot sizing strategy.Table 6 also shows the total time (in milli units) for processing 30 data sets by an ICIM-6004 computer. DLIC requires the same time as compared to many policies. EOQ and PROS require lesser time while WWA needs more than 50% extra time compared to DLIC.

From the above it is inferred that DLIC stood 2nd only to WWA in its performance. Its validity is to be further investigated analytically. Kaimann [6] suggested an experimental frame work for comparing the performance of lot sizing policies. Berry [2] modified it considering coefficient of variation CV as proposed by Kaimann while the cost ratio OC/CC was replaced with the ratio EOQ to the average demand $\overline{D}$ (to represent the degree of mismatch between integral multiply of product demand). Six data sets with values of CV - 0 to 3.32 and EOQ/$\overline{D}$ = 0.75 to 3 were used wherein DLIC yielded same results as WWA when EOQ/$\overline{D}$ and CV were less than or equal to 1.5 and 1.17 respectively. DLIC yielded same result as WWA when CV becoms very large (typical value 3.32).

Conclusion

The DLIC, a computationally simple and efficient algorithm to calculate the lot sizes for a variable demand is presented. Its superiority over others is established in a static planning horizons with a variety of examples. DLIC policy consistantly resulted in better solutions whereas others varied widely in their performance depending on the demand pattern.Considering the performance criteria DLIC stood next to WWA. However, DLIC should be preferred for the following reasons. It is simpler and requires 30% lesser computer time than WWA. It gives equally good solution when EOQ/$\overline{D}$ and CV of the demand pattern less than or equal to 1.5 and 1.17 respectively. In general, the probability of getting the best solution is 60% and in no case the total cost of DLIC solution will be higher by 10% of the best solution provided by WWA. DLIC can take care of dynamic situations also. These factors favour the proposed strategy.

TABLE 1 - A typical demand pattern
Ordering cost:200/- per order Carrying cost:2/- per piece/period
----
Period	1	2	3	4	5	6	7	8	9	10	11	12
Demand	10	60	15	6	20	95	10	45	8	40	170	25

TABLE 2 - Illustration of the First step of DLIC Computation

Sl.	Alternate solutions	Lot size	ordering cost(OC)	Cumulative carrying cost(CCC)	Is CCC<OC	Remarks
1.	Period 1	10	200	0	No	Continue
2.	Period 1&2	70	200	120	No	Continue
3.	Period 1,2&3	85	200	180	No	Continue
4.	Period 1,2,3&4	91	200	216	Yes	Stop

Result - Initial solution is 85 units which is the sum of the
          requirements of the first 3 periods.

TABLE 3 - Performance of DLIC policy

    No. of orders = 4   Ordering Cost = Rs 800.00
    Carrying Cost = Rs 674.00   Total Cost   = Rs 1474.00
    Orders are placed in first, sixth, tenth and eleventh period.
    Inventory carried forward are 101,41,26,20,0,63,53,8,0,0,25
    and 0 from 1st to 12th periods respectively.

TABLE 4 - Frequency distribution of performance ranking

Rank	Lot sizing Policy										
	EOQ	EOQR	PROS	LUC	LTC	LPC	PPB	FCH	GH	WWA	DLIC
1	-	6	5	8	5	15	16	14	14	30	18
2	3	8	6	5	7	8	7	3	9	-	9
3	-	6	8	6	10	3	3	6	5	-	1
4	6	3	6	3	5	3	3	-	1	-	2
5	3	5	3	5	2	-	-	4	1	-	-
6	11	1	1	2	1	1	-	3	-	-	-
7	4	1	1	-	-	-	1	-	-	-	-
8	2	-	-	1	-	-	-	-	-	-	-
9	1	-	-	-	-	-	-	-	-	-	-
10	-	-	-	-	-	-	-	-	-	-	-

TABLE 5 - Performance Index (PI) of different lot sizing
strategies (PI for WWA is taken as 1)

Policy	EOQ	EOQR	PROS	LUC	LTC	LPC	PPB	FCH	GH	DLIC
PI	5.47	3.00	3.10	3.10	2.83	1.93	1.93	2.53	1.87	1.57
Rank	8	6	7	7	5	3	3	4	2	1

TABLE 6 - Total cost comparison of lot sizing strategies

Policy-->	EOQ	EOQR	PROS	LUC	LTC	LPC	PPB	FCH	GH	WWA	DLIC
Total cost in 1000s	64.5	44.8	45.1	47.3	43.8	41.6	41.8	42.3	41.6	40.8	41.2
Time	17	19	17	19	19	19	19	20	19	30	19
Rank	10	7	8	9	6	3	4	5	3	1	2

References
1. Aucamp, D.C., 'A variable demand lot sizing procedure and a comparison with various well known strategies', Int. Jl. Prodn. and Inv. Mgt., 2nd Quarter, (1985), 1-20.
2. Berry, W.L., 'Lot sizing procedures for requirements planning system: A frame work for analysis', Production and Inventory Management, 13, (2), (1972), 19-34.
3. Buffa, E.S., 'Modern production/operations management 7/E'., Wiley Eastern Limited, New Delhi, (1985).
4. Freeland, J.R. and Colley, J.L., 'A simple heuristic method for the lot sizing in atime phased re-order system', Int. Jl. Prodn. and Inv. Mgt., 23, (1), (1982), 14-22.
5. Groff, C, 'A lot sizing rule for time phased component demand', Int. Jl. Prodn. and Inv. Mgt., 20(1), (1979).47-53.
6. Kaimann, R.A, 'EOQ vs dynamic programming - which one to use for inv. ordering', Int.Jl.Prodn. & Inv.Mgt., (1969).
7. Silver, E.A. and Meal, H.C., 'A heuristic for....replenishment', Int.J. Prodn. and Inv. Mgt., 14, (2), (1973),64-74.
8. Wagner, H.M. and Whitin, T., 'Dynamic version of economic lot-size model', Management Science, 5, (1), (1958), 89-96.

Simulation for Real-Time Control: Advantages, Potential Pitfalls, Opportunities

C. M. HARMONOSKY

Department of Industrial and Management Systems Engineering
Pennsylvania State University
University Park, PA

Abstract

Real-time scheduling and control has always been a desirable goal in manufacturing. With more Computer Integrated Manufacturing Systems evolving, new opportunities exist to pursue this problem. This paper discusses the potential application of discrete event computer simulation as a real-time decision-making tool to aid system control, based on experience attempting application and discussion with industrial experts in the area.

Introduction

In the manufacturing environment, the terms Computer Integrated Manufacturing (CIM) Systems and Flexible Manufacturing Systems (FMS) have become more commonplace over the last ten to fifteen years. Due to the dynamic nature of these systems, intelligent real-time scheduling and control has always been a desirable goal. The hypothesis is that real-time adjustment of scheduling or sequencing rules (e.g. SPT, EDD) attempting to adapt to current system conditions, such as machine breakdowns or parts shortages, may lead to better long-term system performance. Ideally, a system scheduling or control decision made in real-time should have a positive impact upon long-term system performance. With increased shop floor level computing power and more emphasis placed on networked computer communications, there has been increased industrial and academic interest in real-time scheduling and control.

Tools are needed to aid in real-time control decision assessment, which have a capability to look ahead into future system conditions. The application of simulation techniques to this problem seems very natural, since simulations have proven to assist the manufacturing community in initial system design evaluation (Harmonosky and Robohn 1990, Grant, et.al. 1988, Erickson, et.al. 1987, Sadowski 1985). More specifically, using discrete event simulation languages, which have been specially designed for application in environments such as manufacturing and are accepted by the manufacturing community, seems logical. Although this is conceptually an easy match, an attempt at actual application to a physical system highlights many important potential pitfalls that accompany the advantages and opportunities.

This paper discusses the application of discrete event simulation for real-time production control. The information is based upon experience working with an existing discrete event simulation language being applied as a real-time decision tool in a laboratory setting, as well as interchanges with other industrial experts working in this area.

Background--Simulation Interface for Real-Time Control

The scenario for using a discrete event simulation language as a real-time decision making tool is illustrated in Figure 1. A computer simulation of a CIM system is linked with the actual physical system, enabling the simulation logic to be controlled by the actual system communication signals, dictating start and stop of robot movement, equipment processing, and cart movement. The simulation will be effectively monitoring the system, always reflecting the current system status. While monitoring, graphical

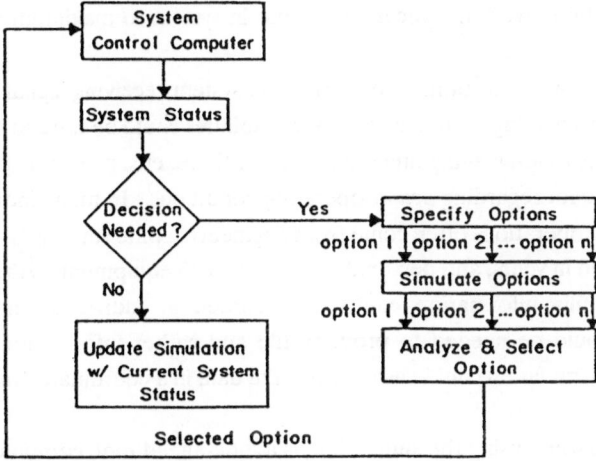

Figure 1. Interfacing simulation with physical system.

animation capabilities of a simulation language can be invaluable, allowing the user to effectively "view" the physical system status without being on the factory floor. Then, when a system production control decision is needed, the starting condition for the simulation is the actual system status, with no time lost for initializing the simulation. For each different control decision option, a simulation run may be executed for some period of time, and the future impact upon the system due to different decisions may be evaluated by analyzing simulation statistical results (Harmonosky and Barrick 1988). In this mode, the simulation model is used as a real-time production control tool with look-ahead system assessment capabilities.

Advantages

One major advantage of using a commercially available discrete event simulation language is the growing familiarity with simulation in the manufacturing community. There is great acceptance using simulation for "what if" analysis of long-term system performance measures when designing a new system or modifying an existing facility. Therefore, when used as a look-ahead mechanism in the real-time control environment, the long-term performance measure information will be understood and accepted by the user, increasing the probability that the decision-making tool will actually be used. Also, because simulation languages have many internal statistical analysis capabilities, it allows easy specification of most performance measures actually used in industry, ranging from average values for throughput, work-in-process, and time in system to maximum values for queue lengths or waiting times.

Because the simulation is constantly monitoring the system receiving signals from actual operations, another advantage is that statistical capabilities could be used to obtain records of actual processing times, waiting times, changeover times, etc., providing power-ful and reliable historical data regarding *actual* operating conditions. In many industrial settings, time estimates for operations often come from engineering time standards, which may not have been updated in years, and personal experience with equipment. Although in a CIM environment this information could be recorded at individual machine computers, extra code would be needed to properly file and gather information from different computers. The simulation could easily record the data in a coordinated manner on one computer.

Another advantage when using the simulation as a look-ahead tool, computer run times of models can be very fast. For example, obtaining a run time of 8 hours in simulated time in less than 5 minutes is quite common. Although run time is affected by the number of jobs moving through the system, it is still considered a very rapid tool, with run times decreasing as personal computers become more powerful.

Potential Pitfalls

Although the advantages listed in the previous section indicate a great potential for simulation applied to real-time control, there are potential pitfalls that become evident when trying to actually implement the methodology. Issues include data retrieval, system status saving and recovery, decision alternative knowledge, and response time.

When retrieving data from a physical system, compatibility issues arise. The base computer language of the simulation (e.g. FORTRAN) may differ from the control language used by the system computer control hierarchy (e.g. C). Also, the flagging messages used by the system computers relaying start and stop of activities will have to be translated into values needed by the simulation variables to communicate the appropriate action. In both cases, post-processing of data may be necessary, which may not be an easy task. Another data issue concerns where in the system computer hierarchy does the data exist at a level of detail necessary to update the simulation. For example, the simulation may be at the detail of individual machines in a workstation and require data when parts begin and end processing and when parts request another machine. However, the central cell control computer may not receive information regarding intra-work station activity--it only gets data when the part is ready to leave the entire work station. Therefore, the simulation could not receive data only from the central system computer. Decisions must be made concerning the appropriate level of detail for the simulation or how to obtain data form different sources in the hierarchy.

Associated with data retrieval is the capability to save data describing the system state at a decision point. Because the simulation will need to be run with several analysis alternatives using the same initial system state, system state data must be recalled for each alternative. Most simulation languages offer a "save and restore" option; however, it may not only save system state information, such as number and types of parts in queues, but may also restore control logic, such as queue priority rules. For example, if alternatives include testing different queue priorities (a very common alternative), external user code must be written to circumvent the built-in save and restore function. Consequently, another pitfall is the need for a priori knowledge of likely decision alternatives. This allows several compiled extensions of the basic model to exist, so appropriate save and restore capabilities can be established.

Once a decision has been made, the simulation must be brought back to on-line monitoring mode. While the simulation was off-line in decision-making mode, the physical system continued operation, and now, the simulation must be brought back to current system status. The biggest potential problem is mobil material handling devices in continuous motion. However, possible solutions to the decision recovery problem include (1) making a copy of the simulation at the decision point, having 1 simulation running as system monitor and 1 running as decision-maker, or (2) receive system signals during decision-making mode into a file then download to the simulation in a batch. Option (1) depends upon parallel computer capability and option (2) will not work if too many signals were received and updating becomes time intensive.

An issue of a statistical nature regards how long to make the look-ahead time horizon, which has a direct effect on the execution time of the model (Wu and Wysk 1989). The longer a valuative simulation runs, the longer it takes to make scheduling and control decisions, affecting the degree of truly "real-time" control. The tradeoff is having the look-ahead horizon long enough to produce valid statistics on performance measures versus having a rapid control decision (Harmonosky and Robohn 1990).

Opportunities

Despite the potential pitfalls of this technique, many opportunities for application still exist. An opportunity for application is to consider using a more basic computer language (e.g FORTRAN, C) to create the simulation. Although the queue structures, file manipulation, event calendar maintenance, and statistical calculation capabilities inherent in simulation languages would have to be coded and animation would be lost, interfacing with the physical system's computer control hierarchy may be more easily facilitated. The technique also allows for cause and effect monitoring, recording the cause of a decision point, the action taken, and the future result. A detailed cause-effect mapping is an opportunity for a strong scientific foundation for future knowledge base systems to automate future scheduling and control at decision points.

Another opportunity exists if a "quasi-real-time" mode could be employed. In an environment that does not require a decision in a matter of minutes, it may be appropriate to have a simulation of the system available, but not interfaced to the system in real-time. For example, in a parallel manufacturing press operation, a catastrophic press failure could mean several days downtime. Therefore, a re-scheduling decision made within an hour is workable. When the failure occurs, a visual inspection of the system could record status values needed to initialize a system simulation, and a program could be developed for the simulation to quickly read this information and be initialized. Then, the look-ahead alternative analysis may proceed as previously described. In other words, definition of "real-time" (i.e. how quickly a decision is needed) for a given environment opens opportunities for applying variations on the basic technique which circumvent many of the interfacing and data retrieval pitfalls.

ACKNOWLEDGEMENT

This material is based upon work supported by the National Science Foundation under Grant No. DDM-8909760. The Government has certain right in this material.

References

1. Erickson, C., Vandenberge, A., Miles, T., 1987, "Simulation, Animation, and Shop-Floor Control", Proceedings of the 1987 Winter Simulation Conference (1987) 649-653.

2. Grant, F.H., Nof, S.Y., and MacFarland, D.G., "Adaptive/Predictive Scheduling in Real-Time," Advances in Manufacturing Systems Integration and Processes: 15th Conference on Production Research and Technology (1988) 277-280.

3. Harmonosky, C.M. and Barrick, D.C., "Simulation in a CIM Environment: Structure for Analysis and Real-time Control," Proceedings of the 1988 Winter Simulation Conference (1988) 704-711.

4. Harmonosky, C.M. and Robohn, S.F., "Real-Time Scheduling in Computer Integrated Manufacturing: A Review of Recent Research," Working Paper No. 90-108, Department of Industrial and Management Systems Engineering, Pennsylvania State University (1990).

5. Sadowski, R.P., "Improving Automated Systems Scheduling" CIM Review, Vol. 2, No. 1 (1985) 10-13.

6. Wu, S.D. and Wysk, R.A., "An Application of Discrete-Event Simulation to On-Line Control and Scheduling in Flexible Manufacturing," International Journal of Production Research, Vol. 27, No. 9 (1989) 1603-1623.

Decomposition Approach for the Job-Shop Scheduling Problem

H.D. LEMONIAS and Z. BINDER

Laboratoire d'Automatique de Grenoble
Institut National Polytechnique de Grenoble
St. Martin d'Heres, France

Abstract: The job-shop scheduling problem is known to be NP-hard in the strong sense. Existing algorithms are thus not operational when the number of jobs and/or machines increases. To simplify the resolution of large-scale job-shop problems the decomposition methods have been investigated. However, whenever remaining links exist between the production sub-systems (the most common case in practice), then the decomposition method becomes debatable. The purpose of this paper is to show that we can replace the remaining sub-system links by fictitious entities introduced into sub-problems. At this moment we can resolve separately the new independent sub-problems.

1 INTRODUCTION

1.1 Definition of problems

The classical job-shop scheduling problem is defined as follows: n jobs have to be processed on m machines. The processing of a job on a machine is called an operation. For each job a processing order for its operations is also given. Then for the operations there are three kinds of constraints to respect.

No preemptions constraints:	Operations do not accept preemptions.
Machine capacity constraints:	A machine can execute at most one operation at a time.
Processing order constraints:	If i,j are consecutive operations in the processing order of a job then j cannot be started before i is finished.

The problem finally consists in finding a schedule which respects all constraints and minimizes makespan (for the definition of the job-shop scheduling problem see also [1]).

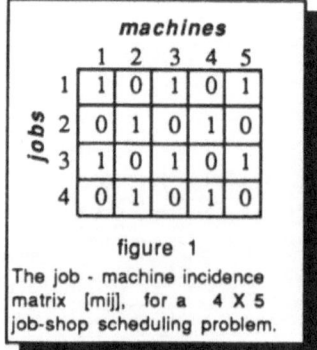

figure 1

The job - machine incidence matrix [mij], for a 4 X 5 job-shop scheduling problem.

figure 2

After rearranging rows and columns of [mij] two sub--problems P1, P2 appear.

1.2 Decomposition principle

Suppose P to be a job-shop scheduling problem and let us construct the associated job-machine incidence matrix $[m_{ij}]$. So $m_{ij}=1$ if there exists an operation of the ith job that requires the jth machine; $m_{ij}=0$ otherwise (see for example figure 1).

Remark The information that matrix $[m_{ij}]$ furnishes is simply the list of machines that every job needs in order to be processed. Thus by permuting columns and \ or rows of $[m_{ij}]$ we obtain an equivalent job-machine incidence matrix for the job-shop P.

Taking the previous remark into account we can use a group technology algorithm (see for example [3],[4]) in order to transform the initial incidence matrix $[m_{ij}]$ of P into a more structured (possibly block-diagonal) form. Thus when a block-diagonal form is found for $[m_{ij}]$ then we can decompose the original problem P into independent sub-problems $P_1, P_2,...$ (see figure 2).

1.3 Analysis of decomposition interest

To illustrate the usefulness of decomposition let us consider the CARLIER-PINSON's branch and bound algorithm for the job-shop scheduling problem (see [2]). The complexity function of this algorithm is $c(k)=2^k$ where the instance size k is defined as the number of disjunctive arcs in the associated disjunctive graph. So suppose that a job-shop P whose size is k is decomposed into r independent sub-problems $P_1, P_2,..., P_r$ whose sizes are $k_1, k_2,..., k_r$ respectively $(k=k_1+k_2+...+k_r)$. It is clear that the time requirements that the algorithm in [2] needs to resolve P is proportional to $c(k)= 2^k= 2^{k_1+k_2+...+k_r} = 2^{k_1}.2^{k_2}. .... .2^{k_r}$. However the time requirements the same algorithm needs to resolve $P_1, P_2,..., P_r$ independently is analogue to $c(k_1)+c(k_2)+ ... +c(k_r)= 2^{k_1}+2^{k_2}+..+2^{k_r}$.

Thus $c(k_1)+c(k_2)+ ... +c(k_r) << c(k)$ and so by the decomposition the computing speed of a job-shop algorithm increases considerably.

2 THE IMPERFECT DECOMPOSITION : A more realistic case

It's frequent today that in real production systems many machines are used to carry out a large number of jobs. But a job needs in general a restricted number of machines. So the associated job-machine incidence matrix $[m_{ij}]$ may contain few no zero elements. Thus it is natural to try the decomposition method. However it is not always possible to transform $[m_{ij}]$ in a block diagonal matrix. What we can hope for the general case is to obtain $[m_{ij}]$ almost block-diagonal (let us call this case **imperfect decomposition**). That means that some non zero elements are left outside the diagonal blocks (see figure 3).

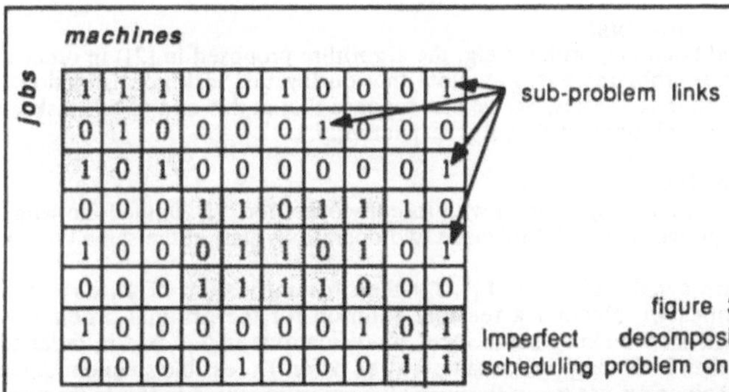

figure 3

Imperfect decomposition of a job-shop scheduling problem on three sub-problems.

In the following we mean by **sub-problems link** any non zero element of $[m_{ij}]$ which remains outside the diagonal blocks, the associated operation will be called **link operation** .

In the case of an imperfect decomposition we feel that we must resolve the sub-problems independently but because of the remaining sub-problems links this is impossible. Clearly if we forget these links and resolve the sub-problems independently, then their solutions risk to be incompatible with respect to the original problem.

134

The purpose of this paper is to show that we can replace the remaining sub-problem links by fictitious entities introduced into these sub-problems and then resolve them independently. To describe our results, some preliminary elements are indispensable.
So suppose P be a job-shop scheduling problem we are interested in resolving.

(1) Throughout this paper T^0 represents the best known solution for P. This solution can be found using a fast heuristic method.

(2) Clearly in the following we are interested in better than T^0 solutions of P (a solution T is better than T^0 iff makespan(T)<makespan(T^0)). But in [5] we show how to define r_i, d_i **release date** and **tail** respectively associated with every operation i, in such a way that in any T better than T^0 solution of P then every operation i starts after the time r_i and finishes before d_i. Denoting by p_i the **processing time** of operation i we can also define $m_i = (d_i - r_i) - p_i$ the **marge** of this operation.

(3) We define finally as **strong decomposition** a decomposition which contains one link operation at most per job (or per row of $[m_{ij}]$).

In the following we start by presenting the application of our method to the case of strong decomposition with exactly two sub-problems; we call this case **2-strong decomposition**.

2.1 A basic result the 2-STRONG DECOMPOSITION
The following three tasks (I), (II), (III) constitute the resolution of this case.

(I) unmatching task
Clearly whenever at least one link operation exists after the application of the group technology method on $[m_{ij}]$ then the two sub-problems P_1, P_2 cannot be resolved independently. So this task consists in uncoupling the two sub-problems by using the technique of figures 4.a, 4.b . These two figures reflect what must be done for one link operation. We apply this technique to every link operation; note so that for any link operation two new fictitious machines and five new fictitious operations are added. For the rest of the paper P_1^*, P_2^* denotes the new independent sub-problems we obtain at the end of this task.

(II) Resolution of P_1^*, P_2^* task
Here we use a branch and bound algorithm (e.g. the algorithm proposed in [2]) in order to resolve independently sub-problems P_1^*, P_2^*. Note that during this task no distinction is made between real and fictitious entities. So we can suppose that at this end of this task we dispose of T_1^*, T_2^* optimal solutions of P_1^*, P_2^* respectively.

(III) Synchronization task
Let us call **pair** the couple of a link operation i with its copy operation c(i). In the following a pair is said to be **synchronized** in T_1^*, T_2^* iff these two operations start at the same time in T_1^*, T_2^*.
Obviously if all pairs are synchronized in T_1^*, T_2^* then by unification of T_1^*, T_2^* and erasing all fictitious entities we obtain T a feasible solution for the original problem P. Otherwise this task consists in making some local modifications in T_1^*, T_2^* in order to synchronize all pairs and then in finding (in the same way as before) a feasible solution for the original problem. The technique we use to synchronize all pairs is presented in [5]. In the same paper we show that the solution T we obtain for the original problem by synchronization is obtained in a polynomial time. Furthermore if T^* is an optimal solution of P then T verifies:

$$\text{makespan}(T) - \text{makespan}(T^*) \leq \sum_{i \in L} m_i \qquad (1)$$

where L=link operations set and m_i=marge of operation i.

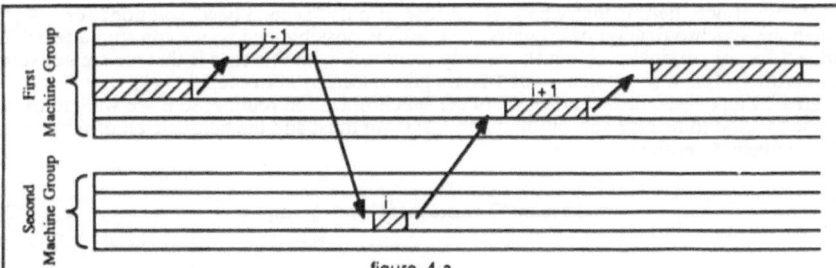

figure 4.a

This figure reflects all information about a job A with operations ....,i-1,i,i+1,... . The processing time of each operation of A is proportional to the associated rectangle length. Moreover the required machine is defined by the zone in which the rectangle is placed. The processing order of operations is defined as well. Obviously i is a link operation.

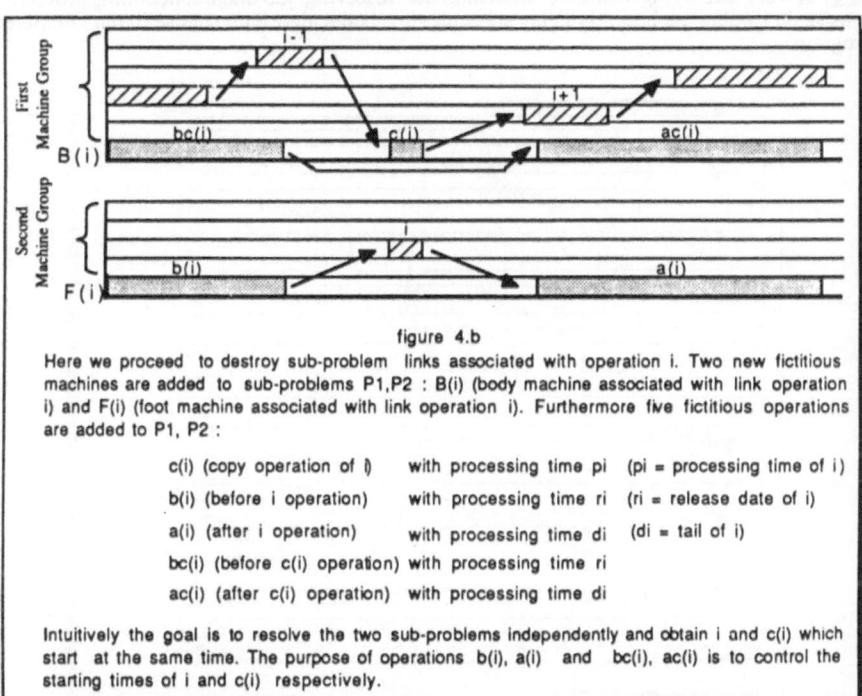

figure 4.b

Here we proceed to destroy sub-problem links associated with operation i. Two new fictitious machines are added to sub-problems P1,P2 : B(i) (body machine associated with link operation i) and F(i) (foot machine associated with link operation i). Furthermore five fictitious operations are added to P1, P2 :

c(i) (copy operation of i)   with processing time pi   (pi = processing time of i)

b(i) (before i operation)   with processing time ri   (ri = release date of i)

a(i) (after i operation)   with processing time di   (di = tail of i)

bc(i) (before c(i) operation) with processing time ri

ac(i) (after c(i) operation) with processing time di

Intuitively the goal is to resolve the two sub-problems independently and obtain i and c(i) which start at the same time. The purpose of operations b(i), a(i) and bc(i), ac(i) is to control the starting times of i and c(i) respectively.

2.2 Some extensions
2.2.1 Using heuristic algorithms to resolve P_1^*, P_2^*

A disadvantage of the above method (tasks (I), (II), (III)) is that we need optimal solutions for the independent sub-problems P_1^*, P_2^*. The reason is that when P_1^*, P_2^* are large size then the exact algorithms (e.g. branch and bound algorithms), are unable to resolve these in a reasonable CPU time.

136

What we must note here is that when we dispose of only "second best" solutions for P_1^*, P_2^* then the synchronization of all pairs (task (c)) is possible and so we can find a feasible solution T for the original problem P; the only drawback with this case is that we are not sure that T verifies (1) and so the quality of T is not guaranteed by above relation (1).

The "error" Σm_i of our method arises from the synchronization routine; so if we want to use a heuristic method to resolve sub-problems P_1^*, P_2^* we must add the heuristic algorithm error. At this moment T verifies:

$$\text{makespan}(T)-\text{makespan}(T^*) \leq \sum_{i \in L} m_i + e \qquad (2)$$

where: $e=\max\{\text{makespan}(T_1)-\text{makespan}(T_1^*),\text{makespan}(T_2)-\text{makespan}(T_2^*)\}$,
T_1, T_2 = our heuristic solutions for P_1^*, P_2^* respectively,
and T_1^*, T_2^* = optimal solutions for P_1^*, P_2^* respectively.

note: A very satisfying heuristic algorithm for resolving job-shop scheduling problems consists in finding the best non-delay schedule; this is the algorithm we use in our computer program.

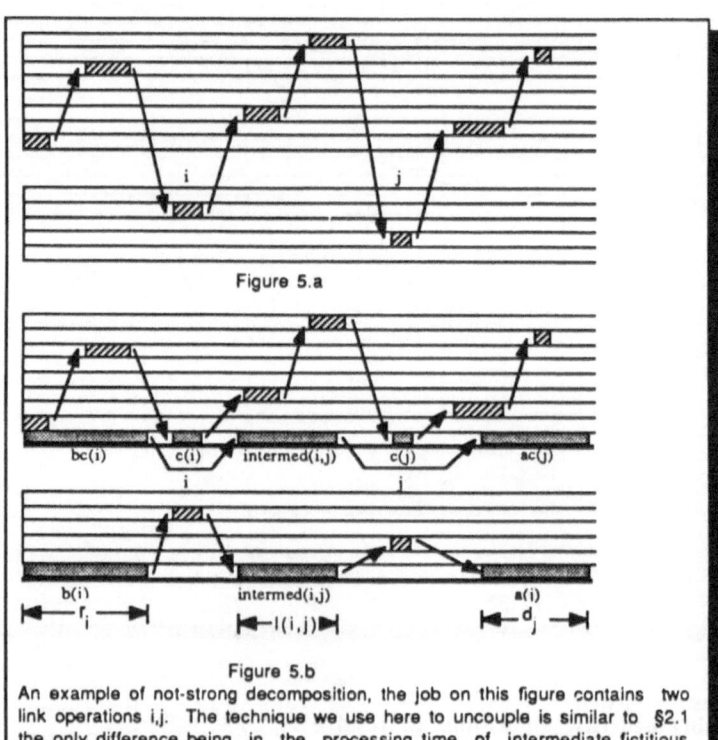

Figure 5.a

bc(i)   c(i)   intermed(i,j)   c(j)   ac(j)
i   j

b(i)   intermed(i,j)   a(i)
$\leftarrow r_i \rightarrow$ $\leftarrow l(i,j) \rightarrow$ $\leftarrow d_j \rightarrow$

Figure 5.b
An example of not-strong decomposition, the job on this figure contains two link operations i,j. The technique we use here to uncouple is similar to §2.1 the only difference being in the processing time of intermediate fictitious operations: the processing time of the fictitious operation which separates the two link operations i,j is l(i,j). (remember that r_i, d_j are the release time and the tail of operation i)

2.2.2 The 2-general decomposition

Up to this point we have discussed the "simplified" case of strong decomposition. To reach the 2-general decomposition we need some complementary elements:

In the following we need the notion of disjunctive graph associated with a job-shop P (for definition see also [1]).

It is well known (see [5]) that when we dispose of a solution T^0 of P then we can find a set D of imposed disjunctive arcs in any better than T^0 solution of P. We suppose now C to be the conjuctive arc set for P and G to be the graph formed by arcs of D and C. Let also l(i,j) be the length of a longest path under i and i on G (the length of a path is the sum of processing times for the operations this path contains).

To resolve the 2-general decomposition we will use the three tasks of § 2.1. At the unmatching task we proceed as in figures 5.a , 5.b to create the independant sub-problems P_1^*, P_2^*. The second task consists in resolving P_1^*, P_2^* independently; thus we find two heuristic solutions T_1, T_2. The last task consists to synchronize all pairs in T_1, T_2 but it is sufficient here to apply the synchronisation routine of §2.1.

We must note here that in the case of 2-general decomposition no performance bound is proved, so we can deal with this case but the final solution quality is not guaranteed.

Prospect: Using the 2-general decomposition technique a possible future of the above method is the general decomposition.

- The above theoretical results have been evaluated in a SUN work station.

-This research was supported in part by the French Ministry of Research.

REFERENCE

[1]   K. R. BAKER. Introduction to sequencing and scheduling, John Wiley (1974).

[2]   J. CARLIER and E. PINSON. An algorithm for solving the job-shop problem. Management science vol.35, no2 February 1989.

[3]   N. DRIDI, J. M. PROTH. Ordonnancement des tâches une methode basée sur la technologie de groupe. 2e Conf. Internationale SYSTEMES DE PRODUCTION, 6-10 Avril 1987.

[4]   A. KUSIAK and W. S. SHOW. Decomposition of manufacturing systems. IEEE journal of robotics and automation, vol. 4, no. 5 October 1988.

[5]   H. LEMONIAS and Z. BINDER. Résolution approchée d'un problème de job-shop fortement décomposé (algorithme, complexité, performance). Colloque international sur les methodes de blocs seriation et applications. STRASBOURG 3-5 Avril 1990.

Evaluation of the Impact of Plant and Production Management Automation on Job-Shop Manufacturing Performances

M. PESSINA, A. POZZETTI and A. SIANESI

Department of Economics and Production
Politecnico di Milano
Milano, Italy

Summary

This paper summarizes the most important results of a research program - recently carried out at the Department of Economics and Production of the Politecnico di Milano - concerning the evaluation of competitive advantages produced by the automation of manufacturing plant and of production management information system.

1. Introduction

This paper would like to provide a triple contribution:
1) evaluation of how it is possible to improve job-shop performances by means of "management" tools (i.e. sophisticated scheduling techniques);
2) evaluation of which is the impact of a pervasive utilization of information technologies (utilization of a real-time monitoring system, that allows the implementation of more and more sophisticated scheduling techniques);
3) evaluation of which is the further advantage obtainable by manufacturing plant automation (that allows to remarkably reduce negative effects due to processing time variability, set-up times and influence of waiting and handling times).

In this paper we summarize the wide research program structure and, consequently, we present the major results obtained.

2. Research program structure

In this research we tested, by means of computer simulation, six different experimental cases corresponding to different automation levels.
The simulation language used is SIMAN; we completed the program with FORTRAN subroutines for the implementation of the scheduling rules.
The test cases are the following.
- The first experimental case (EC1) consists of a job-shop that processes sequences of jobs using RND, NDD, BL2 rules (which will be mentioned in the following).
- The second experimental case (EC2) consists of a job-shop with a batch monitoring system (daily feed-back about system status), that allows the use of rules able to effectively

manage production, thanks to the knowledge of system status (BL3, LOA).
- The third experimental case (EC3) consists of a real-time monitored job-shop.
- In the fourth experimental case (EC4) we analyzed the effects of reduced processing time variance and reduced set-up times.
- The fifth experimental case (EC5) consists of a job-shop with a material handling system so advanced as to allow an unitary-size lot management.
- The sixth experimental case (EC6) consists of a flexible manufacturing system characterized by a real-time control system, by deterministic processing times, by a zero value of set-up times and by a material handling system which allows an unitary-size lot management.

We tested these cases with the following system loading rules:
- RND - random loading sequence;
- NDD - nearest due date [1];
- BL1 - balancing set of "already processed jobs" [4]; this rule loads one lot of the job type "i" that presents the maximum value of the expression:

$$I(i)=N(i)/Ntot - np(i)/nptot$$

where:
$N(i)$ is the total number of lots of the job type i to be produced,
$Ntot$ is the total number of lots to be produced,
$np(i)$ is the number of already processed lots of the job type i,
$nptot$ is the total number of already processed lots;
- BL2 - balancing set of "not yet released jobs" [4]; this rule loads one lot of the job type "i" that presents the minimum value of the expression:

$$I(i)=N(i)/Ntot - nnr(i)/nnrtot$$

where:
$N(i)$ is the total number of lots of the job type i to be produced,
$Ntot$ is the total number of lots to be produced,
$nnr(i)$ is the number of not yet released lots of the job type i,
$nnrtot$ is the total number of not yet released lots;
- BL3 - balancing work-load [3]; this rule loads one lot of the job type "i" with the minimum value of the expression:

$$I(i)=\Sigma_n \left[Cs_n/max_n\{Cs_n\} - F_n(i)/F^*(i) \right]^2$$

where:
Cs_n is the total work-load on machine n due to all the jobs to be produced,
$F_n(i)$ is the work-load on machine n following the loading of one lot of the job type i into the system,
$F^*(i)$ is the present work-load in the system - following the introduction of one lot of the job type i - on the machine for which $Cs_n=max$;
- LOA - Bechte algorithm [2]: its objective is to maintain balancing of work-load of the whole system by controlling the work-loads of single work-centers.

For all test cases NDD rule was used for job dispatching on the machines.

For a detailed discussion about the contents of the above mentioned rules we suggest to refer to the bibliography.

3. Results

In this chapter we would try to summarize the major results obtained by the application of the above mentioned rules to the six different experimental cases proposed.

3.1. System monitoring impact

In this section we compare a traditional job-shop (EC1) with a monitored one (EC2); the introduction of a monitoring system (even a batch one), enabling the use of more sophisticated scheduling rules based on work-load balancing (BL3, LOA), allows a great reduction in mean lead-time (time elapsed between the arrive of an order and its completion) (fig. 1); in particular Bechte algorithm performs a 50% mean lead-time reduction (fig.1), a 7% makespan reduction and a 5% net utilization increase. This first result shows that, to increase job-shop performances, it is at least necessary to avoid situations of unbalanced work-loads; with the adoption of a load-balancing technique, the job flow is more regular so to turn mean lead-time and makespan to lower values.
The adoption of a real-time monitoring system (EC3) further allows a reduction of WIP, lead-time, makespan and flow-time (time elapsed between the loading into the system of a job and its completion), because any rule can rely on an immediate knowledge of the global system status.
However we would notify that, in the particular simulated system ("make to order" production of lots characterized by large size and very long processing times), this reduction is not as meaningful as the reduction obtained in proceeding from EC1 to EC2.
An interesting result is mean flow-time (fig. 2) and makespan (fig. 3) lower values obtained by BL3 and LOA rules, due to the knowledge in real-time of the work-center work-loads; this allows to maintain balancing of work-loads (fig. 4) and to reduce WIP (fig. 5).

3.2. Plant automation impact

In this section we describe the advantages obtained by a plant automation. The first reachable advantage is the repealing of processing time variance and set-up times. In this way it is possible to consider processing times as "deterministic" and zero set-up times.
The results of the research concerning this experimental case (EC4) show an increase of work-center utilization (due to the zero set-up times) and a makespan reduction (fig. 3). But the most significant result is the mean flow-time reduction (fig. 2).
For the above mentioned loading rules aiming the load-balancing, this result can be explained by their inner

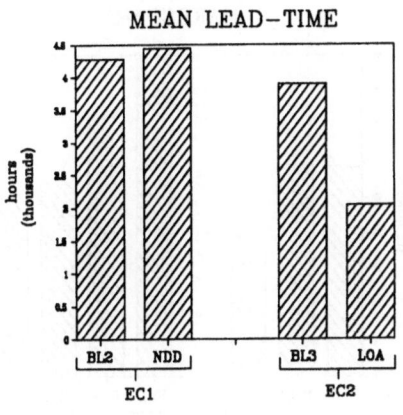

Fig. 1

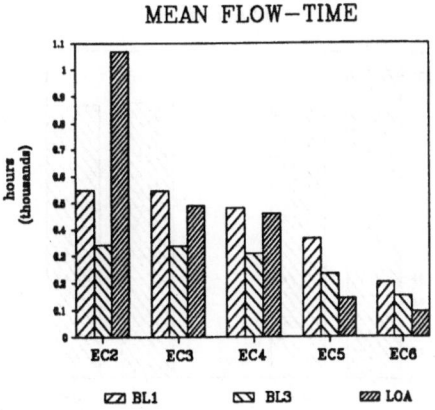

Fig. 2

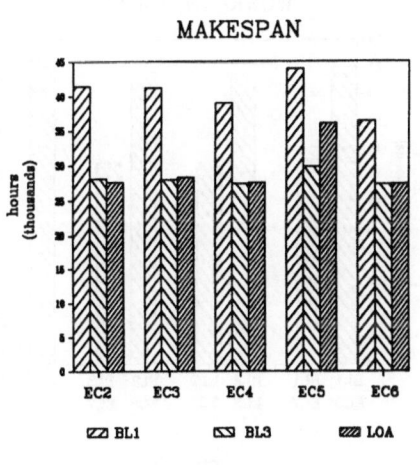

Fig. 3

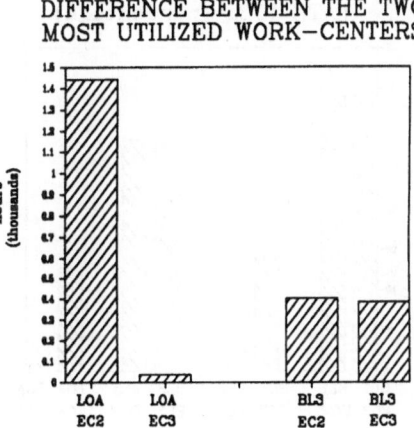

Fig. 4

142

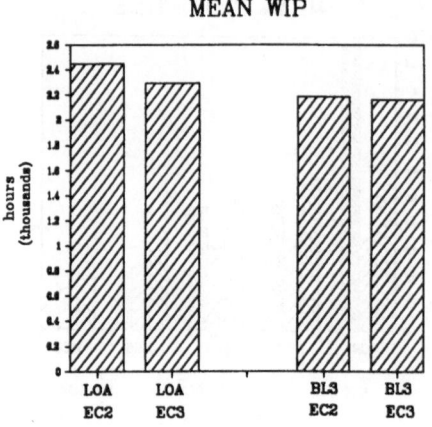

Fig. 5

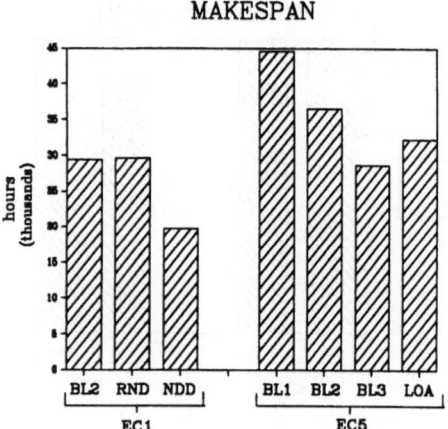

Fig. 6

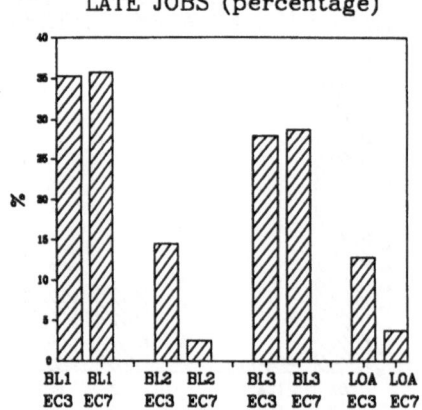

Fig. 7

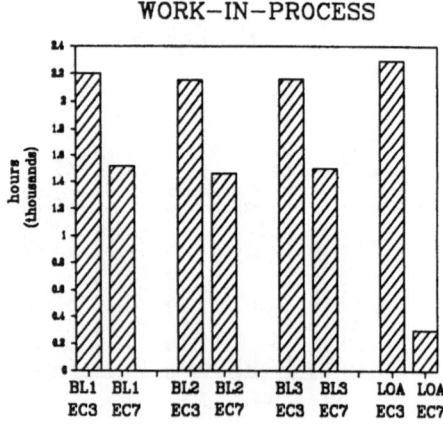

Fig. 8

structure; in fact these rules do not consider set-up times (it is quite impossible to foresee the real processing sequence at any work-center) and so every set-up reduction carries out performance increase.

3.3. Material handling system impact

In this section we briefly present the results of the fifth experimental case (EC5) concerning the evaluation of a real-time monitored job-shop that processes unitary-size lots with negligible handling times, but without a plant automation allowing zero set-up times and deterministic processing times. The main result (fig. 6) is that this solution is very dangerous; this means that it is quite impossible to process unitary-size lots, achieving good performances, simply adopting sophisticated management rules not supported by a suitable plant automation level.

3.4. Flexible automation impact

In this last section are presented the results of the simulation of a flexible manufacturing system characterized by:
- an integrated material and tool handling system that allows an unitary-size lot management;
- an automation level that allows to produce with deterministic processing times and zero set-up times;
- a real-time monitoring system;
- a wide set of sophisticated control rules.

The research program had the objective to investigate which performance increase is achievable by every management rule; however the global result shows better performance in terms of flow-time (fig. 2), makespan (fig. 3), number of late jobs (fig. 7) and work-in-process (fig. 8). Also these experiments show that the best rules are those aiming the load-balancing (LOA and BL3).

References

1. Baker, K.R.: "Introduction to Sequencing & Scheduling", John Wiley & Sons, 1974.

2. Bechte, W.: "Theory and Practice of Load Oriented Manufacturing Control", International Journal of Production Research, Vol. 26, No. 3.

3. Garetti, M.; Pozzetti, A.; Bareggi, A.: "On-line Loading and Dispatching in Flexible Manufacturing Systems", paper accepted for publication in International Journal of Production Research.

4. Perona, M.; Sianesi, A.; Spotti, D.; Turco, F.: "Group Technology e Programmazione Operativa per un Flow-Shop di Assemblaggio Piastre Elettroniche", Convegno Nazionale ANIPLA, Torino, November 1988.

Role of Non-Productive Time in the Evaluation of Computer Generated Process Plans

N.K. MEHTA, P.C. PANDEY and A.V.S.R.K. PRASAD

Department of Mechanical and Industrial Engineering
University of Roorkee
Roorkeee, India

Summary

A software has been developed for generating time standards for turned components which include machining as well as non machining time elements such as handling, set up, tear down, idle time between operations, tool change time etc. The software is based on COFORM system of description of part geometry and also takes into account type of lathe. The non productive time determined with the help of this software alongwith the machining time serves as the basis for evaluation of computer generated process plans.

Introduction

Computer Aided Process Planning (CAPP) has emerged as an efficient interface between CAD and CAM. CAPP techniques are classified into two categories: variant [1] and generative [2]. In an earlier work [3] the authors reported on a generative process planning system for turned components in which the process plans were evaluated on the basis of machining time and the number of tool changes. The present work is an extension of [3] and describes a method of evaluation of process plans on the basis of manufacturing time, i.e., machining time plus non productive time. In recent years a number of computer packages have been introduced to generate time standards for calculation of non productive time [1,4,5], using standard data stored in computer files in the form of elemental time and mathematical formulae. In the present work a computer program has been developed in FORTRAN-IV for computer generation of time standards for turned components, based on COFORM system [2].

Elements of Non Productive Time

Non productive time consists of the following components: batch set up time and tear down time; loading/unloading time; job set up time; miscellaneous handling time during and between operations; and idle approach & retrackting time.

Batch setup time includes the time necessary to prepare the machine for production, such as NC programming, mounting work holding devices, fixtures, tools etc. On the contrary tear down time represents the time taken for removal of work holding devices, fixtures, tools etc., so that the machine is ready for starting the next set up. From the batch set up and tear down time, the set up and tear down time per component is calculated depending on batch size.

The loading/unloading time for rotational components depends on the work holding devices, viz., chuck or between centres, whereas the job set up time is operation dependent. It involves the time required for changing the tool, moving it to a predetermined position called home position and subsequent movement to the starting point of cut for the current feature. Thus, job set up time is the sum of tool changing time and tool homing and start up travel time.

The miscellaneous handling time during operations is also operation dependent and includes elements of non productive time such as time to start spindle/machine, time to change speed, time to stop machine, time to reverse spindle to back up tap in case of threading operation, etc. The idle approach and retracting times represent non productive times between operations. These include time to retract the tool from finish point of previous operation, time to move the tool to the start point of present operation if no tool change is required. If a tool change is required then the times involved are the same as explained in job set up time.

Machining time elements

The machining time elements are those elements which are performed by the machine. They start and end when the tool touches and leaves the component respectively. The machining time depends on machining process, tool and component materials, geometry of the component and speed and feed assigned for the particular operation.

Algorithm for determining manufacturing time

A software has been developed to determine the manufacturing time for different sequences of operation of the feasible process plans. From among these, the process plan(s) with least manufacturing time is selected. The software is modular in structure and is interactive. The modules have been developed in FORTRAN-IV on MicroVax II computer. The various modules are described below.

The batch set up and tear down times are calculated with the help of data taken from Ref.[6], depending upon the type of lathe, type of holding device and number of tools used. The interactive input, for this module is shown below.

Choose type of lathe: Enter E for engine lathe and T for turret lathe.

Choose type of work holding device: Enter 1 for chuck and 2 for bicenter.

Choose type of chuck used: Enter 1 for 3-jaw chuck and 2 for 4-jaw chuck.

Type of true up tool: Enter 1 for chalk, 2 for surface gauge and 3 for indicator.

Indicate part weight: Weight should be in grams only.

Indicate batch size:

Indicate attachment used for taper cut: Enter 1 for taper attachment and 2 for compound rest.

Indicate distance bet. tote box and holding device: The unit should be in centimeters.

Type of tool used for thread cutting: Enter 1 for single point thread cut tool, 2 for die and 3 for tap.

Indicate speed change mechanism: Enter 1 for lever and 2 for belt/back gear.

Indicate no. of levers required to change speed:

Indicate length of rough blank (mm):

Indicate diameter of rough blank (mm):

Indicate home position coordinates (mm): All coordinates should be given with respect to the right hand free end of blank. Enter X-coordinate first and Y-coordinate next.

The module for calculating loading/unloading time takes the input from the previous module. The data for this module is taken from Ref.[6,7]. The machining time for each operation depends on the cutting speed, feed, machining process and geometry of component and can be calculated with the help of standard formulae. Having determined the machining time, the tool changing time for the given operation is determined as the ratio of the product of machining time and unit tool changing time to the tool life.

The homing time, start up travel time, idle approach time and retrackting time can be determined only after specifying the actual start and finish points of each cut and the home position. The latter is a convenient fixed point, therefore, its specification presents no difficulty. For determining the coordinates of the start and end point of each cut a module has been incorporated in the software. The COFORM system of part geometry description used in the earlier work of the authors [3] has 27 attributes. To this the following four attributes are added.

Attributes 28 and 29: Start point X and Y coordinates of feature on finished component
Attributes 30 and 31: End point X and Y coordinates of feature on finished component.

All these coordinates are taken from the left free end of component with origin at the point of intersection of left free end surface and centre line of component. The attributes are tabulated below for a sample component (Fig.1).

For U cuts, form turning and chamfering, the X coordinates of actual start and finish points are taken from the previous module. To determine the Y coordinate another attribute is defined as follows.

Attribute 32: Height of material present above and below a feature but above the axis of component.

There is provision in the software for a module to check whether the component needs reversing at any stage during

148

Feature No.	Values of Attributes			
	28	29	30	31
1	0	25	30	25
2	30	25	30	20
3	30	20	55	20
4	55	20	55	15
5	55	15	80	15

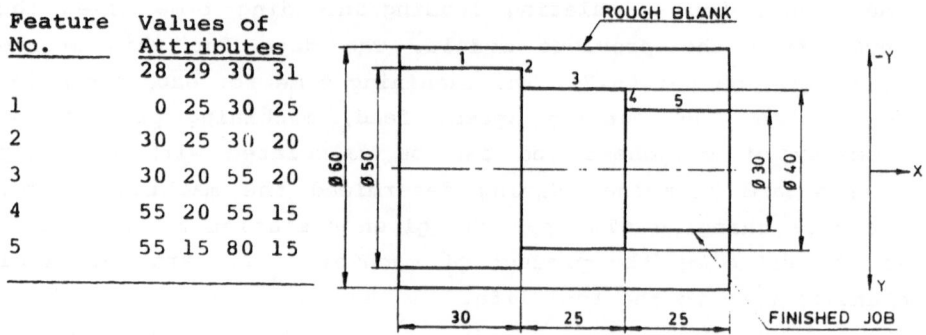

Fig.1: Sample component and assigning of attributes 28 to 31.

machining of the features in a given sequence. The reversal time and the other miscellaneous handling times during and between operations are taken from time standard data [6,7].

The complete algorithm for determination of manufacturing time is shown in the form of a flow chart (Fig.2).

References

[1] Groover, M.P., and Zimmer, E.W., Computer aided design and manufacturing, Prentice Hall of India, 1986.

[2] Chang, T.C., and Wysk, R.A., An introduction to the automated process planning, Prentice Hall, 1985.

[3] Mehta, N.K., Pandey, P.C. and Jain, P.K., A computer aided graphic generative process planning system for turned components, Proc. Conf. CAD, CAM, Robotic and Factories of the Future,Vol II,Dec.19-22,1989,IIT,Delhi,pp 194-205.

[4] Weaver, R.F., Kollmar, J.T. and Boepple Jr., E.A., Developing standards by computers, Ind. Engg.,Jan.1978.

[5] McNeely, R.A., and Malstrom, E.M., Computer generates process routings, Industrial Engg., July 1977.

[6] Hadden, A.A. and Genger, V.K., Handbook of standard time data. Thames and Hudson Book Co., London,1960.

[7] Nordhoff, Machine shop estimating, McGraw Hill,1960.

149

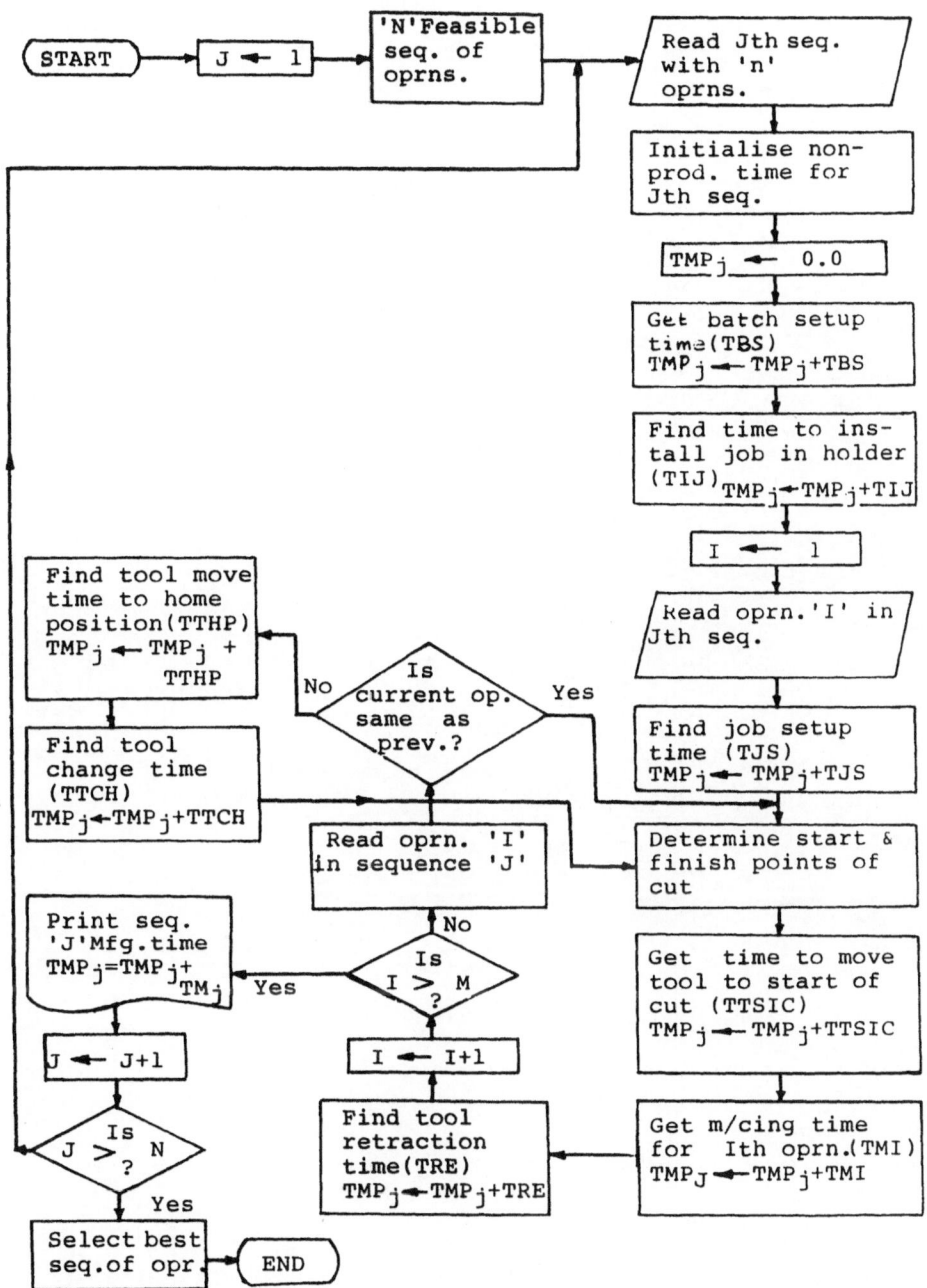

Fig.2: Flow chart for determination of manufacturing time.

Chapter III

Process Technology

Introduction

The planning and performance evaluation of various manufacturing processes constitute the theme of the third chapter. The first paper discusses research efforts aimed at the development of interfaces for the smooth transition of information from automated design to the Computer-Aided Process Planning (CAPP) system and then directly to the automated shop floor equipment. The next paper reports on the application of nonlinear goal programming to electrical discharge machining of ceramic composites, implementing a modified search technique to obtain optimal operating conditions. The third paper presents basic concepts of expert systems and then illustrates them with the prototype expert system aiding the designer of a metal-forming process of axisymmetric bars with simple extrusion. The fourth paper presents a CAPP system for generating optimal process plans for robotic assembly which uses a feature data base and a connectivity data base to generate feasible assembly sequences, which results in robot arm trajectory plans. The automated assembly of printed circuit boards is considered in the fifth paper. The author provides an analysis of the combined effects of angular and linear errors on successful part registration during assembly.

The sixth paper is concerned with the problems of evaluation of new manufacturing technologies such as material handling systems like Automated Guided Vehicles (AGVs). The authors propose a new evaluation framework based on theoretical underpinnings and drawn from decision support systems and logistics theories. The authors of the next paper propose the so-called "index method" to solve the machine loading problem, i.e., assigning the jobs to work centers, and show it to be superior to the "first come - first served" method.

The last paper presents a thorough theoretical derivation of an optimal parallel algorithm for channel assignment in the routing problem in design of integrated circuits and printed circuit boards.

Computer Managed Process Planning for Cylindrical Parts

AMY J. KNUTILLA and BOBBY C. PARK

U.S. Army Missile Command
Redstone Arsenal, AL

Abstract

Process planning can be an integral part of the manufacturing cycle, and Computer Aided Process Planning (CAPP) has emerged as a key requirement in a Computer Integrated Manufacturing (CIM) environment. At the U. S. Army Missile Command (MICOM) in the Manufacturing Technology Division of the Research Development and Engineering Center, a CAPP system has been developed to meet this requirement. Current research is being performed to develop the necessary interfaces for the smooth transition of information from automated design to the process planning system and then to create code to run the automated shop floor equipment directly from the process plan output. This paper will discuss this research effort.

Computer Managed Process Planning

Process planning is the activity which translates the part design and its specifications from the engineering drawing into the sequence of manufacturing operations that will convert raw material into the finished part. It has been estimated that as much as 75% of manufacturing costs are predetermined at this point based on how effectively a company uses its materials, labor, and machines. [1] In most companies, process planning is accomplished by highly skilled manufacturing engineers whose level of expertise varies and is subject to disappearing as these engineers retire or move. Surveys indicate that the

number of skilled process planners has declined in many industrial countries. [2] Consequently, process planning has become
a viable candidate for automation.

Computer Aided Process Planning (CAPP) systems can standardize
and optimize this technology while reducing the time and cost
for process planning. CAPP systems provide manufacturing
engineers with standard process planning tools and information
by building and storing machine tool and part data and using
logic built into the software to generate the process plans.
Not only do CAPP packages optimize process plans, but they also
can serve as training tools for process planners. CAPP is an
integral component of a Computer Integrated Manufacturing (CIM)
strategy, yet to achieve the optimal benefit of CAPP, it is
necessary to integrate CAPP with Computer Aided Design (CAD) and
Computer Aided Manufacturing (CAM).

Computer Managed Process Planning (CMPP) is CAPP software that
was developed for the MICOM's Manufacturing Technology Division
in the 1980's by United Technologies Research Center. It
contains over 50,000 lines of FORTRAN 77 code in fourteen main
programs, using seven data bases. CMPP is a robust CAPP system
addressing precision machined cylindrical parts characterized by
expensive materials, tight tolerances, and complex manufacturing
processes. A major feature of CMPP is that it is manufacturer-
independent. A local manufacturing data base, containing both
the planning logic and machine data of the manufacturer, drives

the system. The software was originally developed to run on IBM
and Sperry mainframe computers and has since been modified to
also run on DEC VAX computers.

[3] CMPP is an interactive, generative process planning system.
The generative approach uses decision logic, formulae, algorithms, and geometry-based data to determine the plan. Unlike
the variant approach to process planning, which recalls, identifies and retrieves existing plans for similar parts and
requires the engineer to make necessary modifications, generative process planning develops a specific plan for a specific
part without depending on the user to input decisions. The user
has the option of reviewing some or all system decisions interactively, and can suspend and resume the planning sessions as
required.

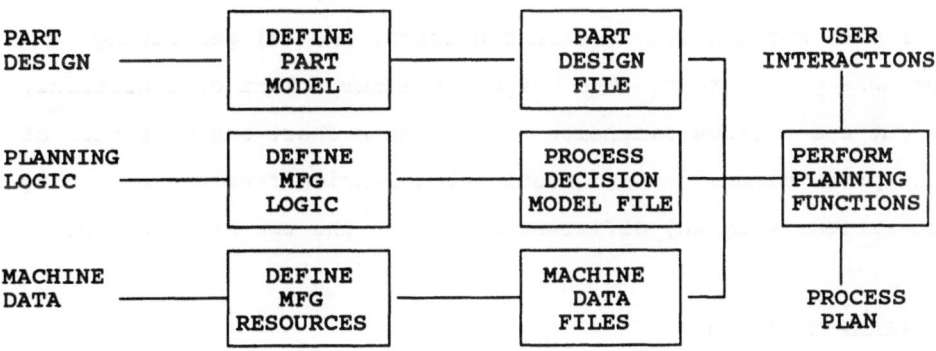

Fig. 1. CMPP System

CMPP is made up of four basic subsystems which form a data base that can be customized for any machine shop. (Fig. 1.) The first subsystem defines the part model based on the completed part design, specification input, and raw material description. The user has the option of entering the part data interactively or in batch mode. The second subsystem defines the manufacturing logic based on user input of such items as the order and purpose of the operations. The third subsystem, for defining manufacturing resources, establishes information about shop facilities and shop rules and procedures. The final subsystem creates and displays the process plan.

The process planning technical functions include four steps: (1) generating a summary of operations; (2) selecting dimensioning reference surfaces, clamping surfaces, and locating surfaces for each cut in each operation; (3) determining and analyzing machining dimensions, tolerances, and stock removals for each surface cut in each operation; and (4) generating process plan output. The output includes a list of operations, a cut and balance dimension chart, and a chart and/or sketch of blueprint dimensions and resultant dimension from the machining operations with any differences between the two highlighted.

Current Research

The full benefits of CAPP as an key component of a Computer Integrated Manufacturing (CIM) facility can be realized with its integration with Computer Aided Design (CAD), Computer Aided

Manufacturing (CAM), and Manufacturing Resource Planning (MRP II). Although interfaces between specific CAPP packages to specific CAD or CNC systems exist, research and development of this technology is not generally widespread. Defining the part for CMPP can be very time consuming, cumbersome, and redundant considering that most design information is available in a CAD file. CAD and CAM interfaces to an earlier version of CMPP have been developed at an aerospace firm as proprietary developments. [5] MICOM's Manufacturing Technology (MT) Division is developing these interfaces to be available with CMPP. Current research goals include: (1) developing a generic interface from any CAD package to any CAPP package; (2) developing a CMPP Post Processor for input to CNC equipment; (3) establishing methods for remote access from machine shops to CMPP; and (3) improving the customization capabilities of CMPP.

Specifically, Louisiana State University (LSU) is under contract to develop the interface from CAD to CMPP and to develop user-friendly improvements to the software. Auburn Research Center is developing software and methods to provide remote access capability which permits small machine shops to generate a process plan via a Personal computer/modem link to CMPP resident on a main frame, and use the output to automatically generate CNC code for the machine shop's equipment.

CAD/CMPP Interface

The CAD/CMPP Integration effort requires taking the digital data
generated by a CAD drawing and translating this into the digital
input format required by CMPP. For example, primitive design
features such as lines, arcs, and circles of a CAD drawing must
be converted into diameters, faces, and holes for CAPP processing. This poses a formidable problem due to the differences in
syntax of data and the variability in CAD data storage techniques. In order to meet the requirement for generic integration
capability, an interface has been developed which adheres to the
Initial Graphics Exchange Specification (IGES) since over 30
major CAD vendors support IGES to some degree. [4] To demonstrate the interface, LSU is using AutoCAD for the CAD input and
CMPP for the CAPP output.

The CAD interface requires a standard, user-defined template
which is based on Department of Defense specifications. The
information in the template is defined as technological data or
organizational data required for CMPP input and converted to
Initial Graphics Exchange Specification (IGES) neutral format
using the AutoCAD IGES preprocessor. This information is passed
to an IGES postprocessor for CMPP which was developed by LSU.
This component will process four types of drawing information:
template, geometry, dimensions/annotations, and notes. The
template and note information will be processed directly into a
CMPP input file, and the geometry and dimension information will
be further processed by a program called Feature Recognition and

Identification for Process Planning (FRIPP). The processing of dimensions and annotations continues to be the most difficult task due to the variability in IGES interpretation amoung CAD systems. This may improve as application standards become more defined.

The methodology of the FRIPP processor is based on the concept that a feature satisfies some generic, geometric properties and therefore can be recognized without resorting to heuristics. [7] The methodology decomposes feature recognition into the following six functions: 1) preprocessing the geometric data; 2) categorizing and marking edges; 3) extracting feature subgraphs; 4) classifying into simple and complex features; 5) organizing complex feature hierarchy; and 6) identifying features. The FRIPP postprocessor then generates the CMPP Audit Trail which is used by CMPP to generate the process plan. (Figure 2.)

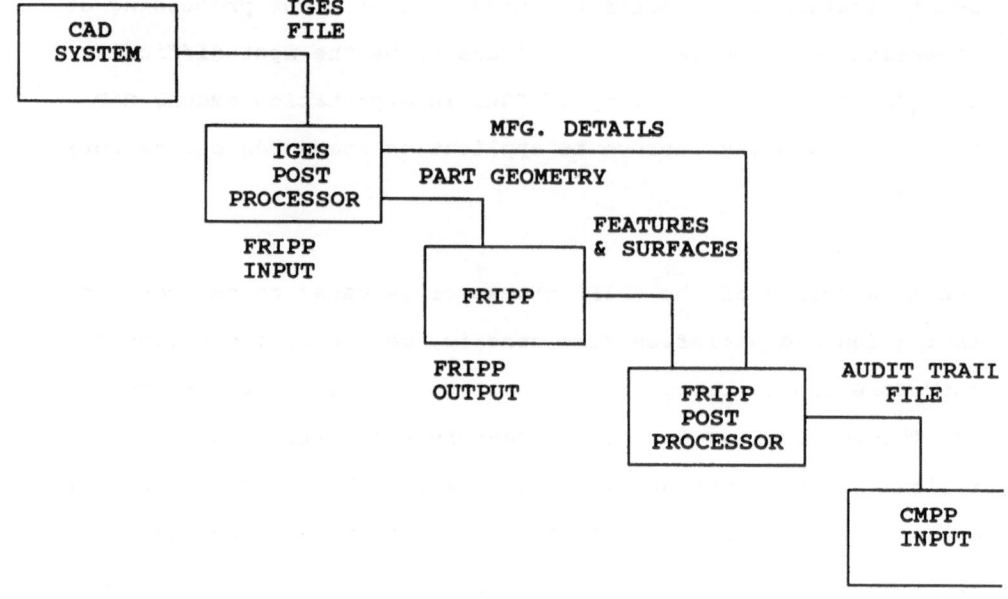

Fig. 2.   CAD/CMPP Interface

The AutoCAD to CMPP translation has been successfully demonstrated at LSU.   Currently, this interface is capable of automating over 50% of the design information input to CMPP, significantly reducing the process time.   However, it is necessary to
supplement the input with manual data entry.   Research is con
continuing and estimated completion is December 1990.

CAPP/CAM Interface

The Cut and Balance Dimension Chart of the CMPP output furnishes
the machinist with the information necessary to order the
machining operations to make the part.   This information is the
data   used for the CAPP postprocessor which Auburn University
developed to generate computer aided manufacturing programs.

[8] The CAM post processor is made up of four major parts: 1) a program called CNCIN to rearrange the data from the Cut and Balance Dimension Chart to reflect the cutting sequence; 2) the main program called CNCPLAN which reads the rearranged data file and generates a temporary output file with each line of this file requesting a machine operation or function; 3) the Computer Numerical Control (CNC) codes generator called CODE which generates a contrasting table of cutting functions and CNC command codes; and 4) the output format program called CNOUT for generation of the final CNC program. (Fig. 3.) The CODE module can be modified to represent different CNC controller codes.

Auburn is working with The Bevill Center for Advanced Manu-facturing Technology in Gadsden, Alabama to demonstrate CAPP/CAM interface capability. Part design data is manually entered at the remote shop and submitted via a modem and telephone line to CMPP residing on the mainframe computer at Auburn. A process plan is generated and the output downloaded back to the PC at the "machine shop" at the Bevill Center. This output is processed by the CAPP/CAM interface program which generates machine code for the EMCO lathe that produces the part.

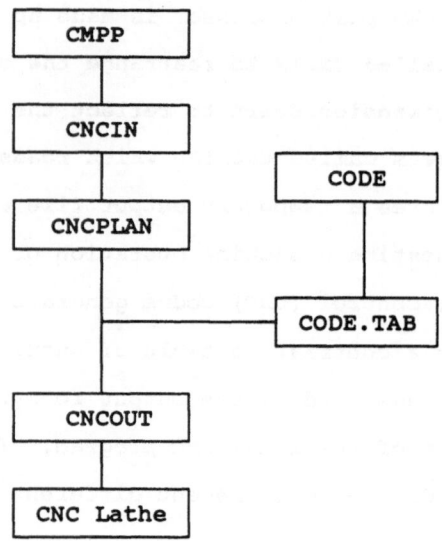

Fig. 3. CAM Interface

Benefits of CAD/CMPP/CAM Integration Research

With the integration of CMPP with automated design and shop
facilities, true savings can be achieved as process plans are
standardized and productivity is improved due to time and labor
savings and the increased capability to handle surges in the
production cycle. Because CMPP generates new process plans for
new items, companies which are subject to surges in major new
products such as defense items can significantly reduce planning
labor and lead times.

CMPP also provides for a reduction of scrap and rework in the
shop which is difficult to quantify. The savings result from
CMPP's powerful tolerancing algorithm which produces a process

plan that defines realistic cuts so that quality parts can be produced from the first cut. This reduces the chances for machining errors resulting in fewer scrapped parts or rework.

Because less time is spent producing a process plan, the plan can be reprocessed repeatedly to optimize the use of material or to specify necessary engineering changes. This is a difficult task for the manual process planner.

Not only is CMPP a process planner, but it can also be used to anticipate lead times required for tool and jig delivery. CMPP stores and catalogs shop machine information and can be used to order tools necessary to produce a part.

Conclusions

CMPP as a stand-alone CAPP system can provide significant savings when replacing a manual process planning function. However, its real benefits cannot be realized until it is integrated with a firm's automated design and shop floor facilities. A CIM environment is not achieved until digital output from one source is automatically transferred to the next computerized tool rather than manually entered. The research being conducted at MICOM is an attempt to provide generic interfaces from any CAD system, to any CAPP system, and then to any shop floor facility. CMPP is a CAPP system which was developed to be customized to any machine shop which produces high precision cylindrical parts. The CAD interface is being developed to utilize the commonly accepted IGES standards so that any compliant CAD system can participate in this environment. The integration to the shop floor equipment is being developed so that all shops with their variety of cutting machines can utilize the interface provided they have PC's and remote access to a mainframe which can run CMPP.

The current research does not completely show smooth integration, yet feasibility has been demonstrated. Future research efforts will include developing interfaces to master scheduling programs and providing for smoother implementation of an integrated CAD/CAPP/CAM system.

References

1. Granville, C.; Computer Aided Process Planning: Last Piece of the CIM Puzzle. Computer Aided Engineering, August, 1989, pp. 46-50.

2. Alting, L., Zhang, H.; Computer Aided Process Planning: the state-of-the-art survey. International Journal of Production Research, 1989, Vol. 27, No. 4, pp. 553-585.

3. Chen, J., Hsieh, J., Shy, F., Kao, P., Lin,. S., Beckett, R., Madsen, N. Mechanical Engineering Department, Auburn University. Progress Report: Adaptation of Computer Managed Process Planning Code for Monitor and Control by a Personal Computer. Contract No. DAAH01-88-C-0561. September 30, 1989.

4. Graves, G.R., Parks, C.P.. Louisiana State University. Interim Report: Knowledge-Based System for CAD/CMPP Communication. Contract No. DAAH01-88-C-0503. September 30, 1989.

5. Gray, S.; Cox, M.V.; Weston, S.; Williams, J.E.: Final Report. Delivery Order No. 001 of Contract #DAAH01-90-D-0074, U.S. Army Missile Command, Redstone Arsenal, AL: July 1990.

6. Austin, B.L.; 1986, Computer Aided Process Planning for
   Machined Cylindrical Metal Parts. Autofact '86, Detroit,
   Michigan, November 12-14.

7. Sahay, A., Graves, G.R., Parks, C.M., Mann Jr., L..
   Louisiana State University. A Methodology for Recognizing
   Features in Two-Dimensional Cylindrical Part Designs. 1989.

8. Kao, C., Shy, L., Hsieh, S., Chen, C., Madsen, N., Beckett,
   R.. Auburn University. Post Processor for CAPP. CSME
   Mechanical Engineering Forum. 1990.

An Application of Non-Linear Goal Programming in Electrodischarge Machining of Composite Material

M. RAMULU, H.-W. SEE and D. H. WANG

Department of Mechanical Engineering
University of Washington
Seattle, WA

SUMMARY

This paper reports on the investigation that was carried out on the application of Non-linear Goal Programming to an R-C circuit based Electrical Discharge Machining of ceramic composites. The model was developed from the EDM machining experimental data on ceramic composites. This process utilizes the modified Hookes and Jeeves search technique to obtain the optimal operating conditions.

INTRODUCTION

The emergence of advanced ceramics in various high technology applications has increased the need for the development of an efficient machining method applicable to ceramic materials. Silicon carbide SiC based ceramic composite materials are found to have higher fracture toughness, strength, and stiffness than the base SiC ceramics with good thermal conductivity [1]. When conventional machining methods are used, the presence of the silicon carbide leads to very rapid tool wear, frequent expensive tool changes, and excessive time required to complete the job. Moreover, it is very difficult to produce a quality hole or slot due to the brittle nature of the materials. Recently, the authors studied the electric discharge machining (EDM) method as a viable manufacturing process to produce quality slots and holes in electrically conductive composites [2-5]. In recent years various methods were suggested to optimize the R-C circuit based EDM process [6]. However, the goal programming approach proved to be effective in optimizing multiple objectives [7-10]. The technique is also efficient in dealing with conflicting objectives by ranking of the constraints according to their importance. In order to optimize the machinability of Titanium diboride TiB_2/SiC ceramic material by EDM process, and to establish EDM as an effective and efficient machine tool to machine ceramic composites, one must optimize the material removal rate, electrode wear rate, surface roughness, and the quality of the hole which is expressed in terms of taper angle. Therefore, optimization of EDM process involves multiple objectives and the selection of EDM operating conditions will be crucial to satisfy all of the above-mentioned objectives.

In this paper we utilized our hole drilling experimental results of ceramic composites, TiB_2/SiC to establish the objective functions and the constraints. A nonlinear goal programming model is developed to optimize the EDM operating conditions in machining ceramic composite materials.

MATHEMATICAL MODEL

The advanced ceramic material machined data used in this study was ceramic particulate composite 20 percent volume TiB_2/SiC, and is a two phase particulate composite material. Tools made of Poco Graphite (EDM-3), brass (Cu, Zn) and copper (Cu, Fe) were used as electrodes for drilling. All the experiments were conducted on ram type Hansvedt model SE-380. The 120V R-C circuit generator provided various current and frequency conditions. A summary of experimental conditions are given in Table 1. Further details can be found in References 4 and 5. Based on the data material removal rate (MRR) and electrode wear rate (EWR) surface roughness R_a and the taper $(Tan\alpha)$ relations were established for different tool materials and are given in Table 2.

TABLE 1. EDM Experimental Conditions

Work Material:	20% Vol. TiB_2/SiC Composite
Tool Materials:	Graphite (Poco), Copper (Cu, Fe), Brass (Cu, Zn)
Tool Polarity:	Negative
Dielectric fluid:	Cutzol EDM 220-30
Open circuit Voltage:	120 Volts
Circuit resistance:	Variable between 16 and 333.3 ohms
Circuit Capacitance:	Variable between 1 and 160 µF
Current:	Variable between 1/3 and 6 amps

NON-LINEAR GOAL PROGRAMMING PROBLEM

The technique used in optimizing the EDM process is the iterative Nonlinear Goal Programming (NLGP) method. Each sub-problem is then solved iteratively for the optimal solution, starting with the highest priority level. A general goal programming problem statement is:

Find $X=(X_1,X_2,\ldots\ldots,X_n)$ so as to
Minimize $a=[a_1(d^-,d^+),a_2(d^-,d^+),\ldots\ldots,a_i(d^-,d^+)]$
Subjected to $g_i(X)+d_i^- -d_i^+ = C_i,\ i=1,\ldots\ldots,m$
$f_i(X)+d_{m+i}^- -d_{m+i}^+ = b_i,\ i=1,\ldots\ldots,k$
$d_i^-,d_i^+ \geq 0,\ d_i^-.d_i^+ = 0$ for all i

The vector X is the set of all decision variables of the process. The aim is to obtain the optimal solution of X. $a_i(d^-,d^+)$ is the achievement function corresponding to the i-th priority level. The achievement function indicates how completely the constraints are attained at each priority level. Achievement functions, which are a function of the deviational variables, may consist of overachieved (d^+), underachieved (d^-) or both deviational variables from each priority level. The $g_i(X)$'s are the absolute constraints of

Table 2. Ceramic Composite (TiB$_2$/SiC) Machinability Relations in EDM Process

Machinability Parameters	Graphite Electrode	Copper Electrode	Brass Electrode
Material Removal Rate, MRR	$0.76413R^{0.0798} C^{0.5348}$	$0.78391R^{0.17764} C^{0.461}$	$0.20296R^{0.7813} C^{0.0975}$
Electrode Wear Rate, EWR	$0.05323R^{-0.3091} C^{0.2909}$	$0.26972R^{0.514} C^{0.2953}$	$0.55616R^{1.1737} C^{0.0574}$
Surface Roughness, R$_a$	$1.9521 \times 10^8 C^2 - 0.0214 \times 10^6 C$	$9.5565 \times 10^7 C^2 + 0.0393 \times 10^6 C + 2.3606$	$-7.784 \times 10^8 C^2 + 0.14545 \times 10^6 C + 2.3553$
Tapered Angle, Tan(α)	$1.4758 \times 10^7 C^2 - 2.376 \times 10^3 C$	$1.3923 \times 10^6 C^2 - 9.746 C + 0.0227$	$-1.3366 \times 10^7 C^2 + 2.3663 \times 10^3 C + 9.011 \times 10^{-3}$
Resistance	78.8 ohms $\leq R \leq 333.33$ ohms	16.25 ohms $\leq R \leq 333.33$ ohms	$26.15 \leq R \leq 333.33$
Capacitance	$1\mu F \leq C \leq 160\mu F$	$1\mu F \leq C \leq 160\mu F$	$1\mu F \leq C \leq 160\mu F$

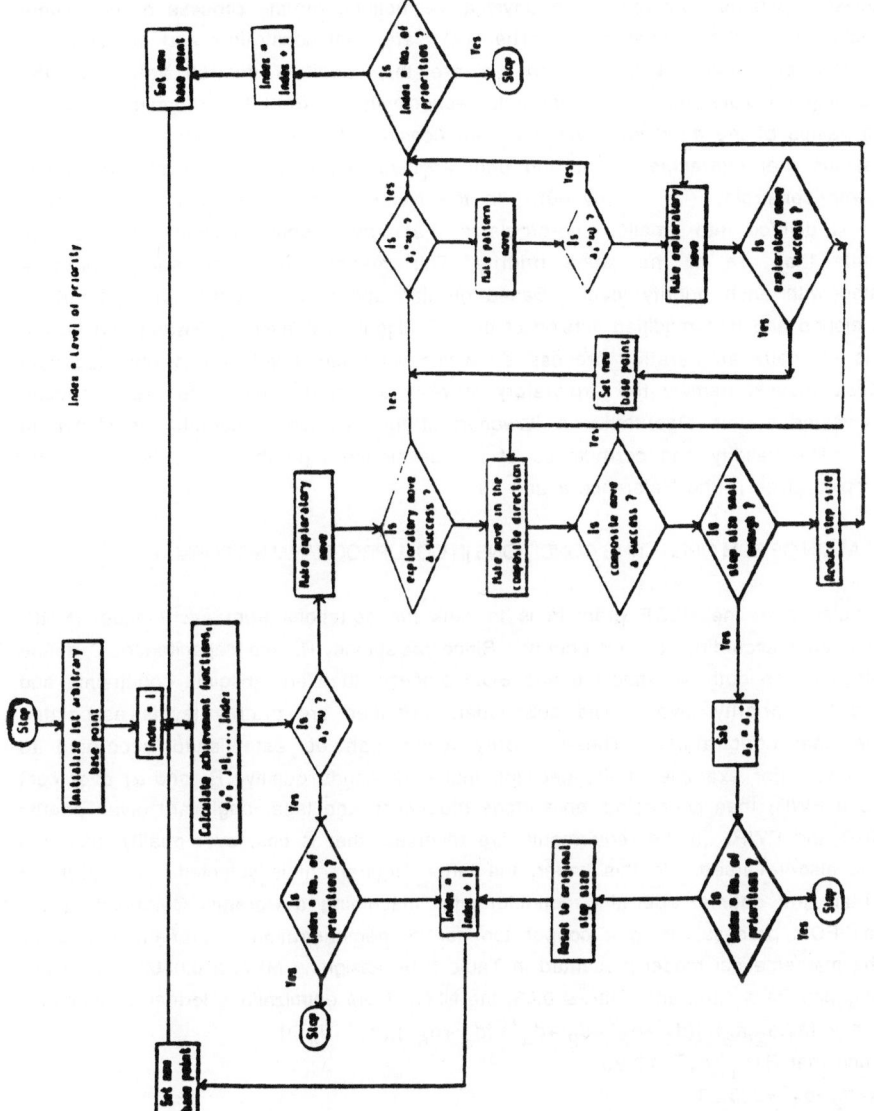

Figure 1. Flowchart of NLGP algorithm for EDM process.

the process. Absolute constraints are physical restrictions on the process or equipment and must be given the highest priority. The $f_i(X)$'s are goal constraints and are specified by the decision maker. The b_i's and c_i's are the specified desired values of the corresponding absolute and goal constraints respectively. In Goal Programming, only the absolute values of the deviational variables are considered, (i.e., $d_i^-, d_i^+ \geq 0$). Furthermore, it is obvious that overachievement and underachievement cannot occur simultaneously for a given constraint, (i.e., $d_i^- \cdot d_i^+ = 0$). In the NLGP problem, the goal programming program is divided into smaller sub-problems. Each sub-problem consists of a set of constraints that are of the same priority. This means that each sub-problem is associated with each priority level. Based on this approach an optimization algorithm was developed and is a modified version of the GP algorithm derived by Hwang and Masud [8]. It is basically an iterative process by which an optimal solution is obtained with three basic moves, namely the exploratory move, the pattern move, and the composite move, involved in this algorithm. A flowchart of the developed algorithm is shown in Figure 1. The validity and correctness of the developed algorithm is verified by using the examples given in the Reference 8 and 10.

FORMULATION OF EDM OPERATING CONDITIONS IN GOAL PROGRAMMING FORMAT

The formulation of the NLGP problem is to rank the equations from each model (MRR, EWR, R_a, etc.) according to their priority. Since resistance, R, and capacitance, C define the restrictions on both the machine and EDM process, they are absolute constraints and have the first priority level. The subsequent priorities are priority levels associated with the goal constraints. These priority levels can be established according to requirements. For example, if the decision maker demands quality (R_a and α) over cost (MRR, and EWR) then constraints on surface roughness and taper angle are given priority over MRR and EWR. If the requirments are reversed, that is cost over quality, then the priority is also reversed. In this paper, the latter requirement is selected. The problem is to obtain the optimal operating conditions for machining a Ceramic Composite (SiC-TiB2) by EDM process, using a copper tool for its demonstration is presented here by using the mathematical model presented in Table 2 by assigning MRR ≥ 0.0003g/min; EWR ≤ 0.0005g/min; Ra $\leq 4\mu$m; and Tan$\alpha \leq 0.03$, the NLGP EDM optmizations format becomes:

$$\text{Min } a = (a_1, a_2, a_3) = [(d_1^- + d_2^+ + d_3^- + d_4^+), (d_5^- + d_6^+), (d_7^+ + d_8^+)]$$

Such that $R + d_1^- - d_1^+ = 16.25$

$R + d_2^- - d_2^+ = 333.33$

$C + d_3^- - d_3^+ = 0.000001$

$C + d_4^- - d_4^+ = 0.00016$

$0.78391R^{-0.17764}C^{0.461} + d_5^- - d_5^+ = 0.0003$

$0.26972R^{-0.514}C^{0.2953} + d_6^- - d_6^+ = 0.0005$

$9.5565 \times 10^7 C^2 + 0.0393 \times 10^6 C + 2.3606 + d_7^- - d_7^+ = 4$

$1.3923 \times 10^6 C^2 - 9.746C + 0.0227 + d_8^- - d_8^+ = 0.03$

$d_i^-, d_i^+ \geq 0, \ d_i^- \cdot d_i^+ = 0$ for all i

Similar formulations can be made by using other two electrode tools with the corresponding models. The optimization problem is now solved by using the developed NLGP algorithm.

RESULTS AND DISCUSSION

The decision variables x_1, x_2 are taken for the operating conditions, resistance, R, and capacitance, C, respectively. Taking the starting base point as x_1 = 100, x_2 = 0.00001 with the correspnding step sizes as 5 and 0.000001, the solution of the achievement function vector $[a_1, a_2, a_3]$ is obtained as:

$$x_1^* = 73.91, \quad x_2^* = 0.000001, \text{ and } a^* = [0,0,0].$$

The solution indicates that based on the prescribed constraints, R=73.91ohms and C=1μF are the optimal process parameters and operating at this setting, the machine will be running at an optimal level. a^* also indicates that all the constraints are fully satisfied since its entries are all zero. Using the same set of initial and step size values and the specified values of the goals, the remaining tool materials were computed and the results are summarized in Table 3.

Table 3. Optimal Values of R-C Circuit Based EDM Operating Conditions and Values of Objectives for Different Electrode Materials

Electrode Material	Resis-tance $x_1^* = R, ohm$	Capaci-tance $x_2^* = C, \mu F$	objective function $a^* = [a_1, a_2, a_3]$	MRR g/min	EWR g/min	Surface Roughness R_a, μm	Taper Angle Tan α, radians
Copper	73.91	1	[0,0,0]	0.0006	0.0004	2.34	0.0226
Graphite	81.72	12	[0,0,0]	0.0013	0.0005	3.99	0.0248
Brass	333.13	160	[0,0.0005,1.7]	0.0009	0.001	5.70	0.0454

By comparing the results for all the three electrode tool materials, it is obvious that brass is the least efficient of the three. This can be noted by observing the non-zero value of a^* for brass and a zero value for copper and graphite tools. Therefore using either the graphite or copper tool to machine ceramic composite by R-C based EDM gives the same efficiency according to the prescribed goals. It is then up to the decision maker to choose the appropriate tool material. The results obtained in this optimization investigation are consistent with our previous slot cutting results [6] and based on the machinability and the cost of the material, one might choose copper over graphite.

CONCLUSION

The approach to optimizing of Electrical Discharge Machining by using Non-Linear Goal Programming was investigated in this paper. An advantage of using the NLGP approach is

172

that no weights need to be figured out to stress the relative importance of each constraint. Only a ranking of the constraints according to their importance is required. Based on the optimal results, copper tools were found to give superior performance to brass or graphite electrodes.

REFERENCES

1. L.M. Sheppard, "Machining of Advanced Ceramics," *Advanced Materials & Processes*, .132, 6 (1987) 40-48.

2. M. Ramulu, "EDM Sinker Cutting of a Ceramic Particulate Composite SiC-TiB$_2$," *Advanced Ceramics Materials*, 3, 2 (1988) 324-327.

3. M. Ramulu and M. Taya, "EDM Machinability of SiC$_w$/Al Composite," *Journal of Material Science*, 24 (1989) 1103-1108.

4. M. Ramulu, H.W. See, and D.H. Wang, "Machining of Ceramic Composite TiB$_2$/SiC by Spark Erosion," *Manufacturing Review* (in press).

5. Hung-Wah See, "Optimization of Electrical Discharge Machining," MS Thesis, June 1988, University of Washington.

6. S.K. Mukherjee and M.N. Pal, "Optimization of R-C Based Electrodischarge Machining Process : A Multicriteria Approach," *Proceedings of 12th AIMTDR Conference*, IIT Delhi (1986) 508-511.

7. S.M. Lee, Goal Programming for Decision Analysis, Auerback Publ., Philadelphia (1972).

8. Ching-Lai Hwang and Abu Syed Md. Masud, "Multiple Objective Decision Making - Method and Applications," *Lecture Notes in Economics and Mathematical Systems*, 164, Springer Verlag, Berlin Heidelberg, New York (1979).

9. R.M. Sundaram, "An Application of goal programming in Metal Cutting," *Internal Journal of Production Research*, 16, 5 (1978) 375-382.

10. B. Satyanarayana, P.N. Rao and N.K. Tewari, "Application of Non-Linear Goal Programming Technique In Metal Cutting," *Proceedings 12th AIMTDR Conference*, IIT Delhi (1986) 483-486.

An Expert System for Metalforming

K. HANS RAJ and V. M. KUMAR

Department of Mechanical Engineering
Dayalbagh Educational Institute
Dayalbagh, Agra, India

Abstract

Expert system technology as an area of artificial intelligence is coming to the field of metalforming processes. A number of expert systems have been developed or are under development. An intelligent knowledge based system that has user friendliness with tools to implement CAD is attempted to help the engineer in industry and research. An example of forming of axisymmetric bars with simple extrusion is used to illustrate new possibilities. The software package described in this paper is applicable on personal computers and is provided with comprehensive functions. The principle is that the designer will interrogate the knowledge base for the production of a particular component starting with material selection going upto design. The paper consists of two parts. A brief discussion of the basis of expert systems and their concepts is given in the first part. The second part illustrates the prototype expert sytem developed to aid the designer.

Introduction

Expert systems are problem solving programs which model human expertise and apply logical reasoning to the knowledge base in solving problems. They are regarded as a means of recording and accessing human expertise in a particular field. Expert systems have been developed and applied successfully in the field of medical diagnosis (e.g., MYCIN, INTERNIST, CADUCEUS). In the field of metalforming there is a growing interest for the possibilities that the technology of knowledge based systems could bring into the domain. The metalforming techniques can greatly reduce material waste, ensure good surface finish and tolerances in the manufacture of engineering components. For their full potential to be realised the assistance of experienced engineer is essential. Expert system could be of great value in this application by making the knowledge of human expert more easily accessible and widely available. In the following section we will briefly summarize the concepts of expert systems. Further a prototype expert system for cold extrusion with its salient features will be described.

Concepts of Expert Systems:

Expert systems are usually applied to narrow specialised fields to make high level expertise available and accessible to many users at any time in a consistent quality unlike the human expert who cannot always provide the same advise in the same situation. The basic elements of an expert system are the knowledge base containing the domain of knowledge and the inference engine which solves the problem by interpreting the domain knowledge. The user interacts with the system through a User Interface(fig.1) and gets access to the Knowledge Base via the Inference Engine. The Data Acquisition Module enables the knowledge engineer to furnish additional knowledge. It is also possible to interact with External Databases or conventional programs through interfaces taking the help of Explanation Module.

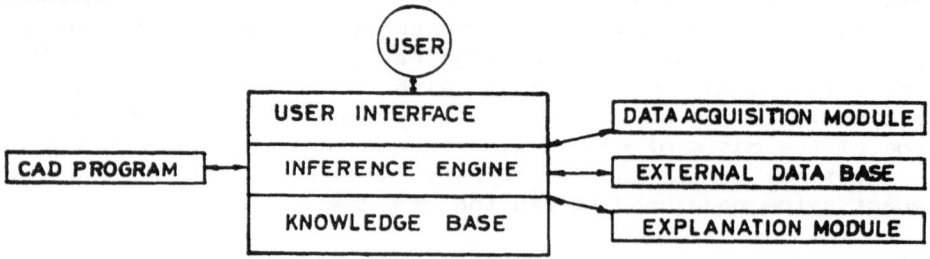

FIGURE - 1.

Knowledge representation plays an important role in developing an expert system. To get an efficient reasoning process the level of representation of knowledge should conform to the real world knowledge closely. Knowledge can be classified in many different ways depending on what it represents (e.g., knowledge based on experience and heuristics, knowledge based on first principles and general theories, metaknowledge). In general, knowledge representation can roughly be divided into two main categories: declarative and procedural. In declarative mode, knowledge is represented as a static collection of facts which will require a set of general procedures to manipulate them. Procedural mode represents knowledge as procedures. Advantages of declarative representation are convenience of adding new facts to the knowledge base and that facts only need to be stored once whereas the procedural representation is convenient for heuristic knowledge, for describing how to do things and for representing complex logic such as probabilistic reasoning.

To represent more complex knowledge structures, a number of schemes have been developed: e.g.,(i) Semantic nets which can describe both events and objects. Information is represented as sets of nodes connected to each other by arcs. Nodes represent objects and the arcs represent relationships between the nodes. (ii) Frames which consist of a collection of slots that describe aspects of the objects. Procedural information can also be associated with a slot and related frames can be grouped together to form a frame system.

Procedural information is often represented by means of rules. Rules consist of a set of premises (IF clauses) and one or more conclusions (THEN clauses). In the present system a combination of knowledge representation schemes was used. The reasoning process can propagate in two main search directions, viz., forward and backward: Forward reasoning starts from the initial states and propogates the reasoning towards the goal states. On the contrary, backward reasoning starts from the goal states and reasons backward, trying to find a solution that matches the initial state.

Prototype Expert System for Cold Extrusion:

The knowledge base is provided with information on number of components and various materials with which these components are to be manufactured based on operating conditions and design requirements. The moment the expert system is started it gives the root menu as shown below.

```
┌─────────────────────────────┐
│  1. SYSTEM INFORMATION      │
│  2. MATERIAL SELECTION      │
│  3. LUBRICANT SELECTION     │
│  4. EXTRUSION               │
│  5. CAD OF DIES             │
│  6. EXIT                    │
└─────────────────────────────┘
```

The information module explains the capabilities and the working mode. As the material selection module is evoked it asks certain preliminary questions. There is provision to select either ferrous or nonferrous materials for the job to be forged. Information about a number of commonly used parts is stored in the knowledge base (such as gears, piston rods, shafts, bars etc.). Based on operating conditions, surface hardness and strength requirements of the components a suitable set of metals along with their codes (DIN/AISI) and properties like yield strength, hardness number, ultimate strength are made available to the user and the system.

Once the material is selected and the component information is drawn from the database directly, the CAD Module is called in and the process parameters to produce the components such as punch pressure, punch force, tolerances are recommended. The punches are designed and material information of both punches and dies is made available to the user.

Rule base development:

A majority of extruded products are axially symmetric to ensure axial loading on the tool. The products are classified into two main categories: solids and hollows. The solids are further sub-classified into plain and stepped shafts and similar components. The divisions in hollows are made according to the shape of the steps, whether the inside diameter is blind or through hole. Based on this information rules are formed to decide a particular component's formability.

Rules are also formed relating to tool life, extrusion pressures and loads and the properties of cold extruded parts. Owing to the heavy expense of tool replacement a decision relating to the tool life must be made at the outset. Once the CAD programme gives away punch pressures ,loads and die design the aforementioned rules are checked to verify whether the designed tool can give the required number of extrusions . If any of these rules fail the designer will be given an option to change certain parameters and the CAD module is reexecuted.

For an expert system used on micro computers the principal limitation is memory space. Hence the program is made in modules and as per the requirement these modules are executed. To illustrate some of these features, a sample consultation with the expert system is incorporated.

```
                    CONSULTATION
                MATERIAL SELECTION SYSTEM
Do you want to select a non-ferrous material ? (y/n)    y
Select from the following groups
1. Aluminium and its alloys
2. Copper and its alloys
3. Magnesium and its alloys
```

SPECIFY THE APPLICATION REQUIREMENTS

TO MANUFACTURE :   electrical_parts.
OPERATING TEMPERATURE :   room_temperature.
SURFACE FINISH REQUIREMENT :   high.
LOADING CONDITION (light,medium,heavy) : light.

THE RECOMMENDED MATERIAL IS DIN Al 99.5,5F10
                             AISI AA 1050 - H14
        DETAILS

MECHANICAL PROPERTIES:-
S_Y = 20 MPa
S_u = 70 MPa
e% = 20 %
HB = 20   PRESS <SPACE BAR>
                                     TO CONTINUE.

PROCESS INFORMATION

Guidance values for possible workpiece dimensions and
reductions of area for AA 1050 - H14
The max. dia. of billet should be within the limits of
300mm to 500mm, Area reduction limit (A_0/A_1) = 50

Select the process you wish to perform
1. SOLID FORWARD EXTRUSION
2. HOLLOW EXTRUSION

SOLID FORWARD EXTRUSION

1. SINGLE REDUCTION   2. MULTIPLE REDUCTIONS
DIAMETER OF BILLET   = 325   mm.
PRODUCT DIAMETER   = 300   mm.
SINGLE GO LENGTH   = 650   mm.
PUNCH PRESSURE   = 505.2307 M Pa.
DESIGN DETAILS OF PUNCH:
d2 = 300.000
d3 = 300.500
d4 = 420.000
h2 = 650.000
h3 = 300.500
h4 = 210.000

Extrusion

Punch for solid forward
extrusion.

press <SPACE BAR>

```
RECOMMENDED MATERIALS FOR DIES

PRESSURE PLATES: AISI 4840; HB 270-330
PUNCH:           AISI S1; HRC 56-58
SHRINK RINGS:    AISI 4840; HB 270-380
DIE INSERT:      AISI D2; HRC 60-62
COUNTER PUNCH:   AISI M2; HRC 62-64
EJECTOR:         AISI S2; HRC 56-58
LANOLIN LUBRICANT(u=0.05) RECOMMENDED
FOR LIGHT EXTRUSION AT ROOM TEMPERATURE.
```

Conclusions:

A knowledge based system approach has been developed to design solid and hollow rotationally symmetric parts using extrusion process. As this expert system is written in PROLOG it is easy to add and delete rules. The basic concepts of Expert Systems and their utility as expert advisers to metalforming process is shown by developing a prototype expert system to a narrow field of extrusion.

References:

1. Carl Townsend: Mastering expert systems with Turbo Prolog, Howard W. Sams & Co.(1987).
2. Kurt Lange: Handbook of metal forming, McGraw Hill Book Company (1975).
3. Bariani,P., Knight,W.A.: Computer-aided cold forging process design:A knowledge-based system approach to forming sequence generation, Annals of CIRP, Vol.37,1 (1988).
4. Brucker, M., Keller, D., Reissner,J., Computer-aided drawing of profiles from round and square bar', Annals of the CIRP, Vol.37,1 (1988).
5. Mackerle, J.,Orsborn, K., Expert systems for finite element analysis and design optimization - A review, J. of Engg. Comput., Vol 5, (1988).
6. Rowe, G.W., An intelligent knowledge-based system to provide design and manufacturing data for forging, J. of Computer -Aided Engineering, (1987).

Optimal Process Planning for Robotic Assembly Operations

SHYANGLIN LEE and HSU-PIN WANG

Department of Industrial Engineering
University of Iowa
Iowa City, IA

Summary

A systematic approach for generating optimal process plans for robotic assembly is presented. A CAPP system is proposed which uses a feature database and a connectivity database to generate connectivity graphs and calculate the final position for every part forming a product. It also identifies sub-assemblies, generates feasible assembly sequences and finally performs trajectory planning for the robot arm to carry out the assembly.

Introduction

Computer-Aided Process Planning (CAPP), the function utilizing computers in process planning, was first introduced by Niebel in 1965 [1]. Since then, it has received much attention from both academic researchers and industrial participants. Several reviews show that the most existing CAPP systems are dedicated to machining process planning, only few of them are designed for assembly planning [2,3,4]. This reveals the need for more assembly planning research work.

The basic requirement for assembly process planning is to determine the sequence in which parts are brought together to the final product. This task may not be very difficult for the human process planner, because of human intelligence. Yung [5] developed a CAPP system which is capable of generating assembly sequence for manual assembly.

In recent years, industrial robots have been used in a variety of manufacturing tasks for the flexibility the robots provide and to reduce the risk of human operators to get exposed to hazardous materials. One of the major applications of industrial robot is assembly. When a robot is used to perform the assembly task, more information is required in the proces plan than that for manual assembly. Information ought to be provided to support robot arm trajectory planning in order to carry out the assembly. The trajectory plan would instruct the robot where to place a part, in what orientation and through what path. Therefore, two more classes of information become essential. They are the final position and orientation of each component part with respect to the global coordinate system, and the trajectory paths through which the robot assembles the parts without colliding into any object or obstacle.

Deshmukh developed a system which generates the final position and orientation of each part which goes to a product based on the feature and connectivity information of the parts [6]. Shpitalni, *et. al.,* developed a system for robotic assembly, which deals with straight-line insertion operations along the principal axes [7]. Chang and Wee reported a system for robotic assembly process planning using a knowledge base approach [8].

In real world robot application, the environment (manufacturing setup) is much more complicated than it is assumed in the the previous studies. Moreover, it would be desirable to be able to optimize the assembly trajectory. Therefore, the authors propose an approach for planning robot assembly path and hope that the approach will extend the useability of previous studies.

The objectives and the approach are presented first. An example follows. Some future research work is then discussed. Since this is an ongoing research, this article covers only the first half of the study.

Objectives

The primary objective of this work is to develop a feature-based, generative CAPP system for robot assembly. This system is is capable of doing the following tasks:

- identifying the spatial relationship among the parts and sub-assemblies,

- recognizing the subassemblies,

- generating all feasible operation sequences,

- planning the trajectories, and

- identifying the one which will consume the least amount of time and energy.

Methodology

The proposed CAPP system is feature based in that all component parts of a product are described by the basic features. A feature is a collection of geometric entities to form a geometric shape and/or for a function, e.g. cylinder, ring, keyway, thread, groove, etc. All parameters associated with the features are stored in a feature database. The connectivity relationship, such as fit, contact, etc. is available in a spatial relationship database. The information as to how two or more parts are positioned next to each other is derived from the spatial relationship. Figure 1 shows an example of features and connectivity relationships.

```
Part name - Container
  Feature - Cylindrical hole, cylinder, Internal thread
  Cylinder:
      dia - 30       (diameter - varies with different shapes)
      len - 10       (length)
      tol - 0.050    (tolerance)
(X,Y,Z) - (0,0,8) (location of the coordinate system)
(a,b,c) - (0,0,0) (orientation of the coordination system)
      .
      .
      .
```

(a) Feature Representation

```
Hori-IN(Cap(Cylinder,30),Container(Cylinder-Hole,30))
```

(b) Connectivity Relation

Figure 1: Feature Representation and Connectivity Relation

Based upon those two databases, a connectivity graph can be generated, as well as the position and orientation of each component part with respect to a global coordinate frame. The connectivity graph is nothing but a undirectional graph whose nodes represent parts of a product and arcs between nodes indicates the connection condition of respective parts. two nodes indicates these two parts are connected to each other. Figure 2 shows a connectivity graph of a screw fastening a cap to a container.

The connectivity graph forms the basis on which sub-assemblies are identified. For grouping parts into sub-assemblies, each of the graph nodes is removed to test if the connectivity graph can be divided into independent, disjoint sub-graphs. If so, each sub-graph represents a sub-assembly.

A base part is required to start an assembly. As the name implies, a base part provides the base for an assembly. Generally, a base part is chosen based on the following criteria:

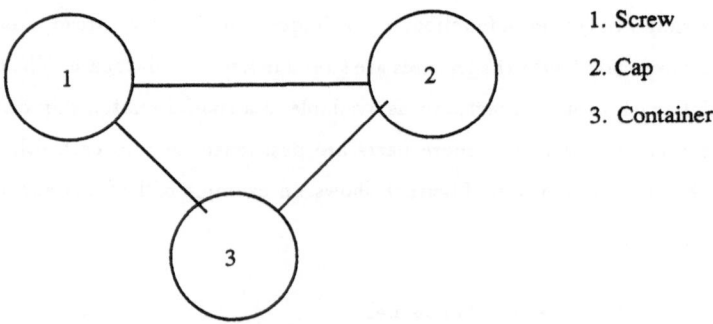

Figure 2: Connectivity Graph

(1) the most stable one (one with the largest solid feature). (2) the heaviest one. After the base part is chosen, any other part can be brought to it as long as it does not violate the precedence constraints. Therefore, it is common to have more than one assembly sequence. The nature of having more than one assembly sequence matches the required flexibility of using robots to perform the tasks.

At this stage, the assembly sequence(s) and the position and orientation of each part at its final assembled situation are obtained. The following tasks include generating the required trajectory for the robot to carry out the desired operations, comparing the optimality of different trajectories, and identifying the optimal one.

Most assembly work reqiures a revolute robot. This class of robots has most flexibility in terms of posing and accessing capabilities. The cost of gaining such flexibility is the non-linear characteristics of robot trajectory planning and a highly coupled robot structure. Finding a collision-free, optimal assembly path is not easy. Two approaches are proposed:

- finding a set of collision-free paths which leads the robot to move parts from initial positions to final assembled position and developing a control scheme which optimizes our interests along that particular set of paths.

- digitizing the entire work space such that a finite number of paths can be found in that space and searching for the optimal one.

Example

A water pump, taken from [6], is used as an example in this presentation. It is composed of 19 parts (Please refer [6] for the pictorial view of the water pump.). The feature representation of a package ring is shown in Figure 3.

```
Part name - package ring
  Feature - Cylinder, cylindrical hole
  Cylinder:
        dia - 35
        len - 3
        tol - 0.030
  (X,Y,Z) - (0,0,0)
  (a,b,c) - (0,0,0)
Cylindrical hole:
        dia - 20
        len - 3
        tol - 0.030
  (X,Y,Z) - (0,0,0)
  (a,b,c) _ (0,0,0)
```

Figure 3: Feature Representation of a Package Ring

The connection of a bolt to the flange is described in Figure 4.

```
Vertical_IN(bolt(Ext_thread,4),Flange(Int_thread,5))
```

Figure 4: Connection of a Bolt to the Flange

Seven sub-assemblies are identified and three sequences are found to assemble the water pump.

Conclusions

The work of developing a CAPP system for robot assembly has been discussed along with an example which shows the feasibility of the proposed methodology. This system can be used off-line for robotic assembly operations.

Future Work

The remaining work of this research includes developing

- a geometric representation scheme with which the assembly environment (robot arm, parts, obstacles) can be described

- a collision detection scheme which will work in the dynamic assembly environment

Reference

1. Nieble, B.W. *"Mechanized Process Selection for Planning New Designs"* ASME paper No.737

2. Alting, L. and Zhang, H. *"Computer Aided Process Planning: The State-of-the-art Survey"* IJPR, Vol.27 No.4 1989, pp 533-585

3. Ham, I. and Lu, S .C-Y. *"Computer-Aided Process Planning: The Present and the Future"* CIRP Annals, Vol.37 No.2 1988, pp 591-601

4. Eversheim, W. and Schulz, J. *"Process Planning for Assembling"* CIRP Annals, Vol.34 1985, pp 607-613

5. Yung, J.P. *"Automated Process Planning for Mechanical Parts Assembly"* Unpublished M.S. Thesis, Department of I.E., SUNY-Buffalo, 1988

6. Deshmukh, A. *"Knowledge Processing for Assembly Process Planning"* Unpublished M.S. Thesis, Department of I.E. SUNY-Buffalo, 1989

7. Shpitalni, M., Elber, G., and Lenz, E. *"Automatic Assembly of Three-Dimensional Structures via Connectivity Graphs"* CIRP Annals, Vol.38 1989, pp.25-28

8. Chang, K-H. and Wee, W.G. *"A knowledge-based Planning System for Mechanical Assembly Using Robots"* IEEE Expert, Vol.3 1988, pp.10-30

Effect of Angular Errors in Part Registration for PC Board Assembly

T. RADHAKRISHNAN

Mechanical Engineering Department
Villanova University
Villanova, PA

ABSTRACT
Automated printed circuit (pc) board assembly is a significant area today, affected by various critical factors. The effects of different assembly process parameter values and their variations (tolerances or errors), on the successful assembly of a part, needs to be properly understood to design an effective and efficient assembly system. Part registration before final assembly is an important phase in the assembly process. This paper analyzes the combined effect of angular and linear errors in the system on successful part registration.

INTRODUCTION
PC board assembly is ever-increasing in automation, along with the associated complexities of newer parts and smaller part sizes. A number of systems currently exist [1,2] that are dedicated (high-speed, low-flexibility) or general-purpose (low-speed, high-flexibility). However, a system with both high-speed and high-flexibility is very desirable to increase productivity. However, the difficulties increase even more in this case, in achieving a high rate of successful assembly.

The high-speed, high-flexibility pc board assembly process is schematically shown in Fig. 1. Parts (components) are fetched from the storage area, holding a variety of parts, by a high-speed shuttle. Pinned components are considered in this study, since they still represent a major portion of all pc board components. The shuttle places the component on a registration plate, having a grid of holes made according to the type of pc boards to be handled. Only parts that are properly registered (pins properly inserted in the holes) are then picked up by the assembler head and taken for final assembly. A mis-registered part needs to be identified and replaced with a properly registered part; otherwise, it will lead to mis-assembly or part jamming at the pc board. In any case, improper part registration calls for an interruption and delay in the production process for part replacement. More the occurence of improper registration, more the chances are for expensive delay and repair required.

The part registration process is therefore an important phase in the assembly of a component. Critical to proper registration are the errors due to the variations in the parameter values present in the system. Earlier studies [3,4] focussed on linear errors in the system. This study includes the effects of angular errors, in addition. Such errors can arise from any angular errors present in the position of the registration plate and any errors in the orientations of the part and the pins (including pin bending).

Both round and rectangular cross sections, which are common, are considered for the pins in this study.

DESCRIPTION OF THE ERRORS

The following are the sources of angular errors included in this study:

a) Error in the YZ plane (Fig. 2) during part pick up by the shuttle (or robotic) gripper, due to the to-and-fro angular motion required for picking the part from the feeder on an inclined axis. This is due to the tolerance on the gripper mechanism motion. Pin axis moves from O'P' to OP_1.

b) Error in the YZ plane, due to bending of the pin (Fig. 2). Pin axis moves from OP_1 to OP.

c) Error in the XY plane due to the positioning of the registration plate (Fig. 3). The plate is assumed to have rotated about an axis parallel to Z axis and passing through the plate center. The hole center shifts from H' to H.

d) Error due to the twisting of a pin about its axis; significant in the case of pins with rectangular cross sections (Fig. 4).

The Z axis is assumed perpendicular to the XY plane. No other angular errors are considered to be significant. It is assumed that the directions of these errors is such that their effects add up at least in one direction (along Y axis, in this case). Combined with the total linear errors discussed in the earlier study [3], these angular errors present a limiting (extreme) case for part registration. If the part can be registered even with all these errors, successful part registration can be achieved all the time. This can be done by compensating for the errors with a suitable size for the holes on the registration plate. Alternatively, errors can be reduced (controlled) wherever economically feasible.

In order to quantitatively analyze the effect of the various errors, mathematical relationships are required, linking the various relevant parameters and error quantities of the registration process. In order to develop these relationships, referring to Figures 1-4, let

D_p = maximum diameter of a round pin
D_a , D_b = maximum length and width of a rectangular pin
D_r = minimum required diameter of a hole on the registration plate (= diameter of the upper portion of the hole if the hole is made up of two sections, with a larger upper section to facilitate pin insertion, tapering to a smaller lower section)
l_{cp} = distance between the part center (of the longest part to be considered) and the bottom center of the farthest pin on it (length CP',

l_p = Fig. 2), in the YZ plane
maximum length of the pin (length OP, Fig. 2) to be considered

l_r = distance between the center of the registration plate and the center of the farthest hole on it (in which a pin may be inserted), in the XY plane (length C'H', Fig. 3)

β = angle between CO' and CP' in Fig. 2

α = total angular error (in the YZ plane) due to the to-and-fro pickup action of the part gripper (Fig. 2)

ω = angle of bend in YZ plane for the pin (Fig. 2)

γ = angle between C'H' and Y axis in Fig. 3

ϵ = angular error in the XY plane, in the position of the registration plate (Fig. 3)

d'_x , d'_y = total linear errors of the pin center relative to the hole center, along the X and Y directions, respectively, from various sources described in [3].

Angles γ and β can be obtained, given the configurations of the registration plate and the parts to be considered. Also, the maximum values of quantities such as D_p, D_a, D_b indicate their nominal values plus their maximum tolerances. The additive method of compounding the various errors will be used in this study. The mathematical relationships developed can be modified to incorporate the use of statistical compounding methods, if needed, as discussed in [4].

ANALYSIS OF THE EFFECT OF ERRORS
From Fig. 2, the Y-direction shifts of the bottom of the pin centerline, due to angular errors, are

$$d_{ya} = l_{cp} \{\cos(\beta-\alpha) - \cos\beta\} \qquad \text{--- (1)}$$
$$d_{yb} = l_p (\tan\omega) (\cos\alpha) \qquad \text{--- (2)}$$

From Fig. 3, the X- and Y-direction shifts, due to angular errors, of the center of the farthest hole on the registration plate are

$$d_{xc} = l_r \{\cos\gamma - \cos(\gamma+\epsilon)\} \qquad \text{---- (3)}$$
$$d_{yc} = l_r \{\sin(\gamma+\epsilon) - \sin\gamma\} \qquad \text{---- (4)}$$

Fig. 3 also shows the limiting case of successful registration (pin insertion) for round pins, with all these errors compounded. J represents the point of contact between the pin and the hole, in this extreme case where pin insertion is just achieved. Due to the bend in the pin axis, the projection of the bottom surface of the pin on the XY plane will be an ellipse as shown, with center P (shifted from center P' denoting the pin position with linear errors only). The major and minor axes of the ellipse will be 2a = D_p and 2b = $D_p \cos(\alpha+\omega)$. H is the center of the hole, shifted from H' denoting the case with no angular errors. d'_x and d'_y denote the X and Y distances between the pin and hole centers, due to the effect of linear errors alone. With the additional effect of angular errors, the total X and Y distances, between the final pin and hole centers P and H, are given by

$$d_x = d'_x - d_{xc} \qquad\qquad \text{--- (5)}$$
$$d_y = d'_y + d_{ya} + d_{yb} + d_{yc} \qquad\qquad \text{--- (6)}$$

The point of contact between the pin and the hole will lie along a quarter segment of the elliptical pin configuration. Hence, the limits for the required hole diameter will be between $2R_1$ and $2R_2$, where $R_1 = k+b$ and $R_2 = k+a$, with $k = (d_x^2 + d_y^2)^{\frac{1}{2}}$. The exact value of the required hole diameter is difficult to solve from the equations representing the circular hole and the elliptical pin, since the point of contact is unknown. For small angles α and ω, the difference between $2R_1$ and $2R_2$ is small and the upper bound, $2R_2$, can therefore be taken as a safe approximation of the required hole diameter. Hence, for round pins, the minimum required hole diameter for the registration plate is given by

$$D_r = 2\{R_p + (d_x^2 + d_y^2)^{\frac{1}{2}}\} \qquad\qquad \text{--- (7)}$$

For the case of the rectangular pin, Fig. 4 shows the twist of the pin about its axis by an angle ρ in the X'Y' plane, which is inclined from the XY plane by the angle $\alpha+\omega$. The projection of the bottom surface of the pin on the XY plane is also shown in Fig. 4, with K being the point of contact between the pin and the hole in the limiting case of registration. d_x and d_y represent the total relative displacement between the pin and hole centers, as before.

Here, the distances in the X and Y directions, between points H and K, are given by

$$d_{xr} = d_{ab} \sin(\rho+\psi) \qquad\qquad \text{--- (8)}$$
$$d_{yr} = d_{ab} \cos(\rho+\psi) \cos(\alpha+\omega) \qquad\qquad \text{--- (9)}$$

where d_{ab} and ψ are as shown in Fig. 4 and given by

$$d_{ab} = \{(D_a/2)^2 + (D_b/2)^2\}^{\frac{1}{2}} \text{ and } \psi = \tan^{-1}(D_b/D_a).$$

Hence, the minimum required hole diameter for the registration plate, in the case of rectangular pins, is given by

$$D_r = 2\{(d_x+d_{xr})^2 + (d_y+d_{yr})^2\}^{\frac{1}{2}} \qquad\qquad \text{--- (10)}$$

In a given situation involving both round and rectangular pins, the larger of the values given by Equations (7) and (10) for D_r should be used.

EXAMPLE APPLICATION
To illustrate the use of the mathematical models, the following practical values were used, for both cases of round and rectangular pins:

$$d'_x = 0.533 \text{ mm } (0.021"), \quad d'_y = 0.203 \text{ mm } (0.008")$$
$$l_{cp} = 12.7 \text{ mm } (0.5"), \quad l_p = 6.35 \text{ mm } (0.25")$$

$$l_r = 35.56 \text{ mm } (1.4"), \; D_p = 0.457\text{mm } (0.018")$$
$$D_a = 0.533 \text{ mm } (0.021"), \; D_b = 0.381 \text{ mm } (0.015")$$
$$\beta = 26.57°, \; \gamma = 45°, \; \alpha = \rho = 0.5°$$

With these values, the required minimum hole diameter was calculated as a function of angular errors ϵ and ω, for both types of pins. Fig. 5 shows the results graphically, for a range of ϵ and ω values. It can be seen that the hole diameter values increases non-linearly as a function of both angles. For the chosen set of parameter values, the required hole size is greater for the rectangular pin than for the round pin.

The equations may be rewritten to express any other parameter (or tolerance) as a function of the others, for successful part registration. Hence, for a given situation, the optimum set of parameter and tolerance (error) values can be chosen. Such an optimum set of values will be based upon the economical feasibility of controlling the different types of errors. It will also indicate those errors to be controlled outside the process (such as tolerance specifications for the component vendors) and those to be controlled during the process (such as the error in the part gripper motion).

CONCLUSIONS
This study provides a quantitative means of analyzing the effects of various system parameters and related tolerances (errors) on the successful registration of pinned components before final assembly. It can be used to generate a feasible set of optimum values for the parameters and tolerances required to specify a high-speed, flexible pc board assembly system. Also, the relative effects of different errors can be quantitatively studied, for proper control of their values to achieve successful part registration.

REFERENCES
1.   J. Garin and T. Stiles, 'Robotic Assembly for Printed Circuit Boards', SME/Robots 11 Conference Proceedings, April 1987, pp. 4.33-4.47.

2.   H.R. Stillwell, 'An Automatic Printed Circuit Board Assembly Factory', NEPCON-East Conference Proceedings, June 1985, pp. 357-360.

3.   T. Radhakrishnan and S.M. Hegde, 'An Analysis of the Errors Involved in Component Pick Up and Registration for Automated Printed Circuit Board Assembly', paper no. D-453, accepted for publication in the ASME/Journal of Engineering for Industry.

4.   T. Radhakrishnan, 'An Analysis of the Critical Parameters Involved in Automated PC Board Assembly', CAD/CAM, Robotics and Factories of the Future Conference Proceedings, vol. 1, December 1989, pp. 775-783.

190

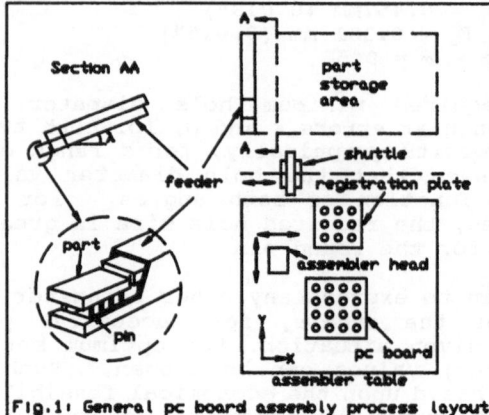

Fig.1: General pc board assembly process layout

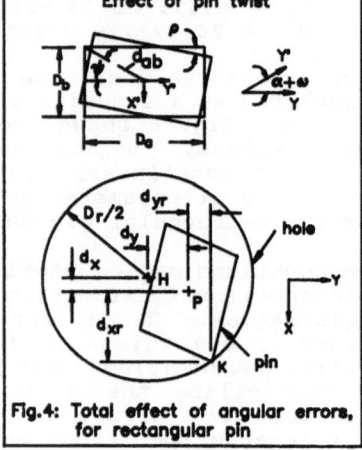

Fig.4: Total effect of angular errors, for rectangular pin

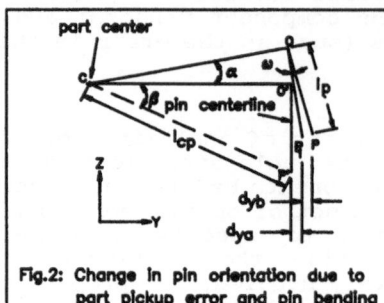

Fig.2: Change in pin orientation due to part pickup error and pin bending

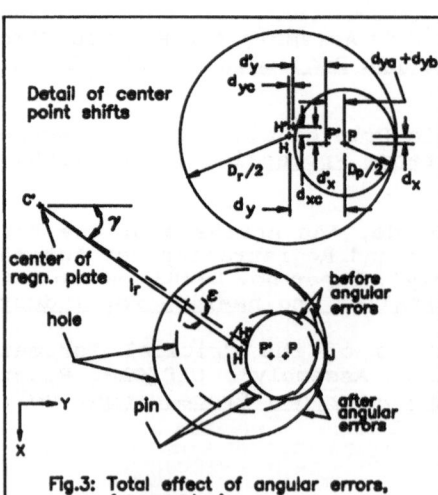

Fig.3: Total effect of angular errors, for round pin

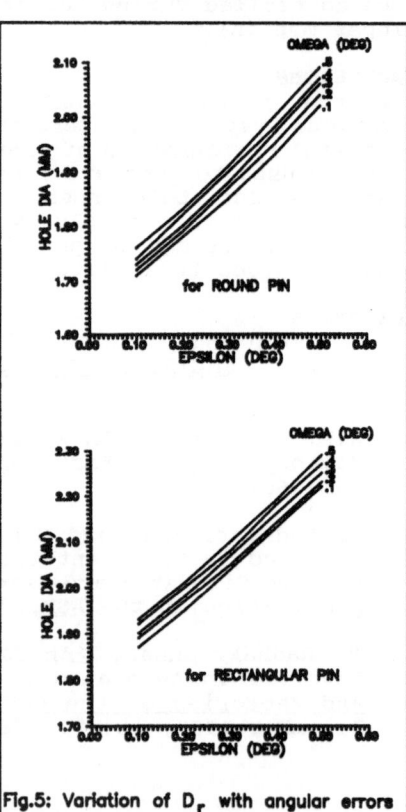

Fig.5: Variation of D_r with angular errors

An Evaluation Framework for AGVS Within FMS

P. F. RIEL
Genie industriel
Ecole Polytechnique de Montreal
Montreal, Canada

M. S. JONES
School of Business Administration
Winthrop College
Rock Hill, SC

SUMMARY

Manufacturing can be seen as the design, evaluation, implementation and control of production systems. This article concentrates on the evaluation aspect as it proposes theoretical underpinnings drawn from the design of decision support systems (DSS) on the one hand, and the theory of logistics on the other hand. An evaluation framework is presented for new manufacturing technologies, in particular, for material handling systems such as AGVS, typically found within FMS. The role of material handling systems within FMS is essentially one of physical integration. Investments in these systems are essentially investments in flexibility. Therefore, this aspect is particularly emphasized in the proposed framework which emphasizes the interdisciplinary nature of the evaluation process and the fact that flexibilities pertaining to MHS may be in a trade off relationship. The methodology involved in this evaluation process includes multiattribute models such as the Analytic Hierarchy Process and the Displaced Ideal Model and suggests how these could be integrated with technological and systems aspects of AGVS.

INTRODUCTION

Much as been written on the problem of evaluating new manufacturing technologies, in particular, flexible manufacturing systems or FMS [1-5]. The problem of evaluating this type of investment seems fundamentally different from assessing a conventional replacement of machine or production line, since an FMS has far reaching implications pervading several aspects of the organization. The problem is essentially an *ill-structured* one. An ill-structured problem can be defined in several ways; one definition that grasps especially well its essence is stated by Saaty and Kearns [6], and is the following: " Ill-structured problems are actually interdependent systems of problems in which multiple decisions makers in a pluralistic environment consider unlimited alternatives whose outcomes are either unknown or very uncertain." In other words, problems of that kind cannot be solved with traditional algorithmic techniques in a systematic way. Mathematical modeling, which has been used successfully in solving the usual well defined problems, must be used in conjunction with other kinds of modeling when problems become ill-structured. In certain cases, operations research methodologies are even useless, as only heuristics are able to solve these problems.

Therefore, if a particular problem cannot be structured in a systematic way (and is thus ill-structured), then the process by which formal mathematical models and *non-formal* models are combined must be designed. Guidelines must be determined in order to direct the design process itself and have theoretical underpinnings for the evaluation problem of new manufacturing technologies. These can be determined by examining design criteria of decision support systems and the analysis of logistics within organizations. The field of logistics looks, among other things, at the flow of information throughout a system as it is imbedded within the layers of the organizational structure. Along this line of thinking, the design of what are called by Hax and Candea, [7] hierarchical systems must be made with the following criteria in mind:

• "The ability to partition the overall problem into subproblems, or the decision process into modules which properly represent the various levels of decision making in the organization.

• The ability to *properly* aggregate and disaggregate the information through the various hierarchical levels.

•The ability to solve each of the subproblems identified by the partitioning procedure.

•The ability to have *linking mechanisms* among subproblems.

• The ability to evaluate the overall performance of the system, particularly with regards to issues of suboptimization introduced in the hierarchical design. "

Another set of design criteria can be used that are more practical and are also relevant to the design of an evaluation process for new manufacturing technologies. These are more in line with the mechanics of the decision process itself as they are related directly to decision support systems DSS (Sprague and Carlson 1985, [8]):

• "The ability to support hard, underspecified, or unstructured decisions as well as structured decisions and assist the decision maker to handle, novelty, time constraints, lack of knowledge, large search space, and need for nonquantifiable data.

• The ability to support the decision making at all levels of the organization and to integrate between levels (strategic, tactical and operational) when appropriate.

•The ability to support communication between decision makers, so that decisions that are interdependent as well as independent may be possible.

•The ability to support all phases of the decision making process and facilitate interaction between the phases.

•The ability to support a variety of decision-making processes but not dependent on any one. In other words, a good DSS is one that is process independent.

•The ability to be modified in response to changes in the user, the task, or the environment."

THE EVALUATION FRAMEWORK

In this section, a framework that integrates different aspects of the evaluation problem for an example of new manufacturing technology, AGVS, is presented. Material handling systems are recognized as an important component within FMS, especially when a computer integrated manufacturing environment is involved [9-10].

The framework is an attempt to make the overall evaluation process of these systems as *systematic* as possible by helping the decision maker in generating alternatives which are improvements over the preceding ones. In particular, it also attempts to structure, to the extent possible, the analysis of AGVS within FMS, since this type of MHS is used more in FMS [11]. This type of framework combines the analytical hierarchy framework (AHP) and the displaced ideal model (DIM) [12] for the interdisciplinary aspect of the evaluation process. It also emphasizes the multiattribute aspect of the evaluation process as several performance criteria are likely to be involved. The multiattribute aspect can occur at more than one level in the analysis. The framework presented here concentrates on trade-offs between flexibilities by looking at systems that might be performing well in one type of flexibility while be deficient in another and vice versa. This framework is also useful when opportunity costs, which are mainly intangibles, cannot be estimated by using flexibility indices as surrogates for these costs [13].

The analytic hierarchy process (AHP) and the the displaced ideal model (DIM) are systems analysis and multi-attribute methodologies (developed respectively by Saaty [6] and Zeleny [12]) which are well suited for a problem such as the evaluation of new manufacturing technologies. Both methodologies can be combined together in an overall evaluation process [13].

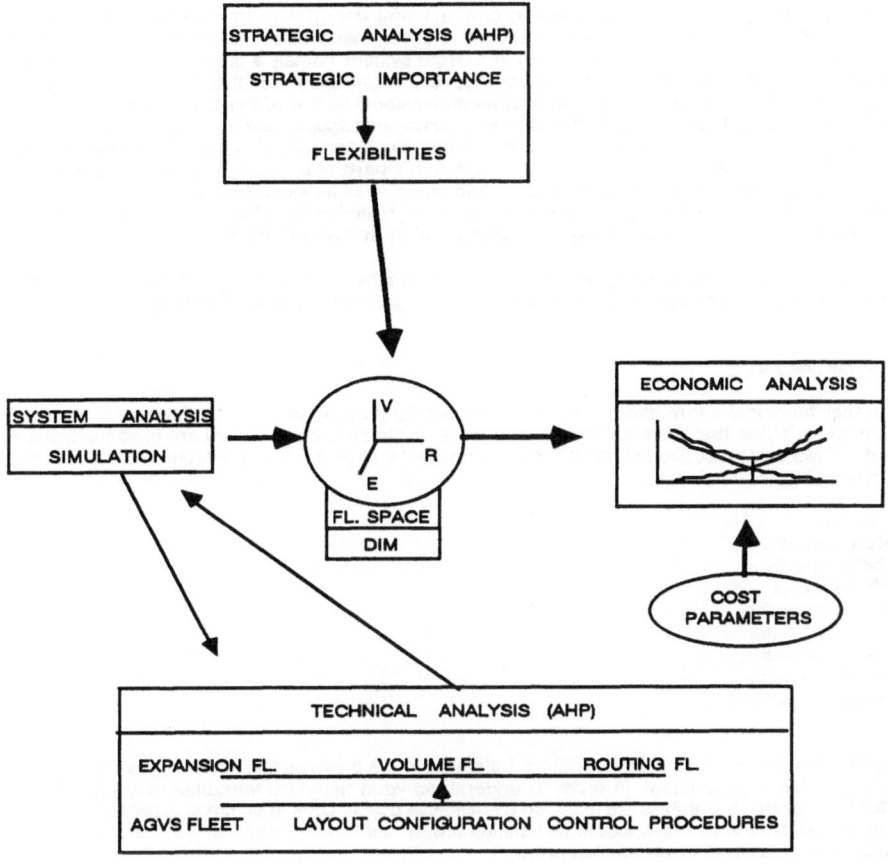

FIGURE 1. FRAMEWORK INVOLVING THE AHP AND THE DIM

This evaluation paradigm is illustrated in Figure 1. The decision maker is at the *center* of the overall evaluation process. The strategic analysis, looks at business objectives as they are linked to the flexibilities. The technical analysis looks at available alternatives and at alternatives that could be designed (AGVS and conveyors). It is also at the technical analysis that operational factors (pertaining to the alternatives involved) that affect these flexibilities are examined. The system analysis looks at the interaction of these operational variables and attempts to measure the flexibility of the MHS's. Finally, an economic assessment is done in order to balance the opportunity costs (known or assumed) with the other costs in order to minimize, to the extent possible, overall costs. The methodology of the proposed framework has been designed to integrate the various aspects of the evaluation process. This framework has also been designed for the case when flexibilities are in a trade-off relationship from a technical point of view. Figure 2 shows the techniques involved in this framework. The full implementation of the AHP is recommended for the strategic analysis, since this technique is well adapted to tackle qualitative variables and their interrelationships from a systemic standpoint. This technique is also recommended for the technical part of the evaluation process so that operational parameters pertaining to AGVS and conveyors as they affect flexibility can be assessed. The system analysis then looks at the quantitative relationships between these parameters; at this stage, math based techniques such as network models and simulation are used.

A technical analysis of material handling can be first undertaken in order to have a better understanding of the variables involved. A second analysis, made in parallel with the first one, looks at materials handling systems as imbedded in a larger system, namely a production system. These two analyses lead to the use of indices for routing, volume and expansion flexibilities (which would at this point constitute the components of a three dimensional vector of flexibility). A third analysis, more succinct, done from a more organizational or managerial aspect, attempts to indicate how each of these flexibilities may affect strategic goals of a firm. Such analysis also indicates which flexibility type is more important than the others with respect to a particular strategic goal. At this point it is possible to integrate these flexibility indices into one overall index of flexibility for a hypothetical material handling system. Finally, an economic analysis of materials handling systems is done with first costs, operating costs and opportunity costs expressed by that overall index.

A flow chart of the overall decision process is given in Figure 2. The sequence of steps necessary to reach the objectives underlined by these analyses are proposed to be the following:

Technical analysis

1. Gather technical information about material handling systems so as to determine the most important variables that influence their performance. Systems to be examined are the ones typically found in FMS and most likely to fulfill FMS requirements. In particular, the technical aspects that affect flexibility, for AGVS are :

- fleet size
- vehicle speed
- vehicle capacity
- buffer distance
- alternative routing
- dispatching rules
- spurs, cutbacks
- segment capacity
- segment directionality

2. Determine the most important variables (gleaned in the previous step) affecting each flexibility (routing, volume, expansion). In order to understand what technical variables may affect these flexibilities, specific hierarchies are used (on a qualitative basis). Look at common variables that affect flexibility. Determine to which extent flexibilities are in conflicting relationships. Determine those technical variables from which conflicts arise.

System analysis

3. Choose a particular layout (loop, ladder, open field) with hypothetical cells defined in terms of their cycle time, set up time, and part types.

4. Choose sequences of operations for a given set of parts.

5. Transform the layout into a graph and/or network where the nodes represent workstations and the arcs (directed to show sequence of operations) , the flows of materials. Use network flows or/and simulation for analyzing proposed AGVS/FMS.

6. By using the flexibility indices pertaining to the material handling systems discussed earlier, evaluate MHS alternatives.

Strategic analysis

7. Link strategic parameters of MHS to these indices.

8. Integrate the AHP and the displaced ideal approaches.

Empirical analysis of MHS and integration

9. Choose a set of material handling systems of a certain type. Define specific ranges for parameters such as speed, capacity etc, and start with a small set of MHS that fall within these ranges. Compute flexibility indices for each MHS alternative. Convert these into a common closeness scale and position each MHS in a *flexibility space.*

10. Compare MHS alternatives in terms of their flexibilities and examine trade offs between them.

11. Build an overall index of flexibility using the integrated framework.

Economic analysis

12. Examine possible trade-offs between first costs, operating costs and opportunity costs (the value of flexibility) with the overall flexibility index.

Recommendations

13. At this point in the process, the decision maker should have a fairly good idea of available alternatives, and the one that seems the most desirable.

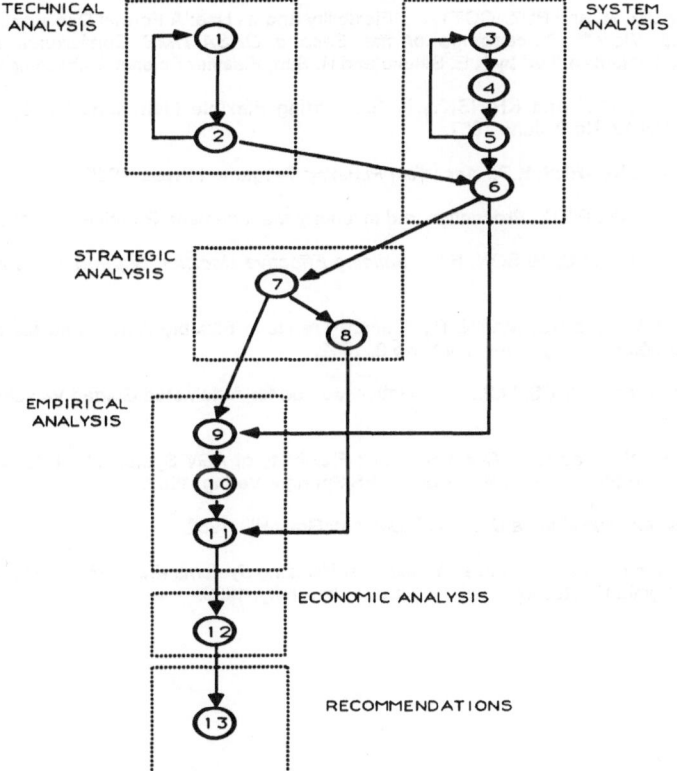

FIGURE 2. FLOW CHART OF DECISION FRAMEWORK WITH THE AHP AND DIM

CONCLUSION

This paper dealt with the evaluation of new manufacturing technologies as it proposed a working framework for AGVS within FMS. The interdisciplinary aspect of this problem was emphasized attempting to grasp its complexity. Other points not discussed here are the measurement of flexibility as it is becoming increasingly important among performance criteria for manufacturing systems, alternative material handling equipment such as conveyors, AS/RS, and cranes which also exhibit some form of flexibility, and the measurement of intangibles such as opportunity costs. These finer issues should constitute promising research opportunities.

REFERENCES

1. HAYES,R. and WHEELWRIGHT S., *Restoring our Competitive Edge*, John Wiley and Sons, New York, 1984.

2. HUTCHINSON, G.K. and HOLLAND, J.R., "The Economic Value of Flexibility." *Journal of Manufacturing Systems*, Vol.1, No.2, 1982.

3. KAPLAN, R.S., "Must CIM be justified by faith alone." *Harvard Business Review*, March-April, 1986.

4. MANDELBAUM, M. and BUZACOTT, J., "Flexibility and its Use: A Formal Decision Process and Manufacturing View." *Proceedings of the Second ORSA/TIMS Conference on Flexible Manufacturing Systems* edited by K.E. Stecke and R. Suri, Elsevier Science Publishings, 1986.

5. MILTENBURG, G.J. and KRINSKY, I., "Evaluating Flexible Manufacturing Systems." *IIE Transactions*, Vol.19, No 2. June 1987.

6.SAATY, T.L. and KEARNS, K.P., *Analytical Planning*, Pergamon Press, 1985.

7. HAX,A.C. and CANDEA,D., Production and Inventory Management, Prentice Hall, 1984.

8. SPRAGUE, R.H. and CARLSON, E.D., *Building Effective Decision Support Systems*, Prentice-Hall, 1982.

9. EVERSHEIM, W. and HERMANN, P., "Recent Trends in Flexible Automated Manufacturing." *Journal of Manufacturing Systems*, Vol.1, No.2, 1982.

10. SHELTON, D. and JONES, M.S., "A Selection Method for Automated Guided Vehicles." *Material Flow*, Vol.4, 1987.

11. KARLSSON, S., "Capacity, Availability and Flexibility of AGV Systems." *Automated Guided Vehicle Systems*, Edited by Prof. R.H. Hollier, IFS Springer Verlag, 1987.

12. ZELENY, M. *Multiple Criteria Decision Making*, McGraw Hill, 1982.

13. RIEL,P.F., *An Evaluation Process for Material Handling Systems within FMS*, Unpublished Ph.D. dissertation, Virginia Polytechnic Institute & State University, 1989.

Computer Aided Machine Loading Technique

USHIR SHAH, SAUMIL TRIVEDI, KETAN SHAH and P.B. POPAT

Mechanical Engineering Department
L.D. College of Engineering
Ahmedabad, India

Introduction

Detailed day-to-day planning of operations involves basically assigning different jobs to different facilities which is called loading. The main problem in machine loading i.e. assigning the jobs to work centres arises when two or more machines are capable of performing the same set of jobs but take different times to complete various job and have different total production capacities. Here appropriate criteria for assignment is minimum time to complete all jobs. In this paper "Index Method" is being used to solve this type of problem. Solution to problem is obtained using a computer. It has been assumed that out of given machines one machine is most efficient machine for all jobs. It is also observed by others that this method is superior to 'First come first serve' method of machine loading. It can be said that after making few modifications in the developed routine, it will be of great help to the manufacturing industries for their machine loading work.

Every order whether it is from a customer or from the assembly bench is required to be completed on the stipulated date. To ensure that the desired delivery date is met, each component entering the assembly should be made available on time. Scheduling function of Production Planning & Control makes it possible by determining starting and completion date of each of the operations listed on the process sheet.

Index Method :

A heuristic method of loading, which would yield better results than the simple and intuitive methods is the INDEX METHOD.

Before this method was invented "First come First Serve" method was used. In this method, when orders are received by the schedule clerk, jobs would be assigned to the machine which is best available at that time. But in such cases the backlog of work at the machines is usually very small and such methods is not very efficient or effective when there is a large number of jobs and machines.

In Index Method appropriate criteria for assignment is minimum time to complete all jobs. It is also assumed that out of given machines only one machine is most efficient machine for all the jobs and a job cannot be split.

Solution Algorithm :

1. Accumulate a backlog of jobs for the facility (work centre or machine) to be scheduled.
2. Arrange the jobs in a tabular form showing the production time for various job-machine combinations. The bottom row of the table shows the available machine time.
3. Develop the index numbers for all of the non-optimal combinations. This will be used as a guide in rearranging the assignment in order to secure the best feasible number. This is done with each job separately, by dividing with the number of hours for the best machine time to numbers of hour for non-optimal machines.

4. Best machine time is the lowest production time having the base Index of 1.0. Assign all of the jobs to the best facility regardless of the machine time availability.

5. Calculate required time by that machine to perform all the jobs.

6. If required time is more than available time on that machine, find the immediate next higher index number from whole table.

7. Remove the task from the optimal machine, reassign it to alternate machine with the smallest index number among all others, without exceedding the constraints on capacities.

8. As the orders are shifted from one machine to another, keep track of reduction and increment in various work loads.

9. Repeat steps 5,6,7,8 till all the jobs are assigned and best feasible solution is reached.

THIS IS DATA & OUTPUT TO THE GIVEN PROBLEM

JOB = 5   MC = 5

HOURS REQUIRED BY EACH MACHINE FOR EACH JOB

	MC1	MC2	MC3	MC4	MC5
JOB1	112.50	168.75	281.25	213.75	337.50
JOB2	80.00	120.00	200.00	180.00	150.00
JOB3	78.75	87.50	157.75	140.00	87.50
JOB4	120.00	210.00	270.00	285.00	285.00
JOB5	67.50	90.00	120.00	112.50	82.50
AVHR=	150	300	400	200	250

INDEX NUMBERS

	MC1	MC2	MC3	MC4	MC5
JOB1	1.0000	1.5000	2.5000	1.9000	3.0000
JOB2	1.0000	1.5000	2.5000	2.2500	1.8750
JOB3	1.0000	1.1111	2.0032	1.7778	1.1111
JOB4	1.0000	1.7500	2.2500	2.3750	2.3750
JOB5	1.0000	1.3333	1.7778	1.6667	1.2222
REQHR=458.7500		.0000	.0000	.0000	.0000

ASSIGNMENT OF EACH JOB OF MACHINE

	MC1	MC2	MC3	MC4	MC5
JOB1	.000	168.750	.000	.000	.000
JOB2	.000	.000	.000	.000	150.000
JOB3	.000	87.500	.000	.000	.000
JOB4	120.000	.000	.000	.000	.000
JOB5	.000	.000	.000	.000	82.500

FINAL TOTAL REQUIRED HOURS OF MACHINE

MC1	MC2	MC3	MC4	MC5
120.0000	256.2500	.0000	.0000	232.5000

Result & Discussion :

Using the solution algorithm a computer program was prepared in FORTRAN Language. Program was tested with number of loading problems. Fig. 1 shows the data & output of one such problem.

Results of given problem shows that JOB-4 should be assigned to Machine-1, JOB-1 and JOB-3, should be assigned to Machine-2, JOB-2 and JOB-5 should be assigned to Machine-5, Machine-3 and Machine-4 are not loaded. This can be seen in the output of the problem.

If we observe the table for index numbers we can find that there are two or more machines which have same index numbers. But this particular program is such that it will take the machine which is earlier in the table and shift would be done with that machine. Due to this limitation total hours required to complete all the jobs, sometimes, may not be optimal.

Conclusion :

This technique is capable of providing a systematic procedure for arranging the work tasks in a better fashion than is normally achieved with the "First Come first Serve" basis. By adopting index method, saving in machine hours is realized.

REFERENCE BOOKS

1. Principles and Design of Production Control System
    By Scheele, Westerman, Wimmert.

2. Elements of CAD/CAM
    By John Willey and Sons.

3. Production Management and Planning
    BY S.N. Chary

4. Production Control A Quantitative Approach
    By Biegel.

An Optimal Parallel Algorithm for Channel-Assignment

STEPHEN OLARIU, JAMES L. SCHWING and JINGYUAN ZHANG

Department of Computer Science
Old Dominion University
Norfolk, VA

Abstract

The channel-assignment problem is central to the integrated circuit fabrication process. Given a two-sided printed circuit board, the problem is to make n pairs of components electrically equivalent. The connections are made using two vertical runs along with a horizontal one. Each horizontal run lies in a channel. The problem is to minimize the total number of channels used. We propose a cost-optimal parallel algorithm to solve this problem, running in $O(\log n)$ time and using $O(n)$ processors in the EREW-PRAM model of computation.

Index Terms: VLSI, CAD, intervals, parallel algorithms, EREW-PRAM, optimal algorithms.

1. Introduction and Terminology

A classical problem in computer aided design is the routing of carriers such as printed circuit boards, and IC chips. In spite of the fact that more and more CAD packages for layout are available nowadays, the circuit layout problem is far from solved [9-12].

The layout problem that we are interested in can be specified as follows: we

---

† This work was supported by NASA under grant NCC1-99

are given a two-sided printed circuit board featuring horizontal lines called *channels* on one side, and vertical lines on the other. We are also given 2n components along with an interconnection pattern specified in the form of n pairs of components. The interconnections are to be realized with horizontal and vertical wire runs according to specifications. The basic constraint is that two pairs can use the same channel only if their connections do not conflict with one another. The problem of interest here is referred to as the *channel-assignment* problem and asks for a layout that minimizes the total number of channels.

A number of solutions for the channel-assignment problem have been proposed in the literature. In particular, Gupta et al [6] have proposed an $O(n\log n)$ sequential algorithm along with a proof that the channel-assignment problem has a lower bound of $\Omega(n\log n)$. Later, Dekel and Sahni [4] have proposed a parallel algorithm to solve the same problem: their algorithm runs in $O(\log n)$ time using $O(\frac{n^2}{\log n})$ processors being, therefore, *suboptimal*.

The purpose of this note is to propose a cost-optimal parallel algorithm to solve the channel-assignment problem. To anticipate, our algorithm runs in $O(\log n)$ time using $O(n)$ processors in the EREW-PRAM model of computation.

To specify our result, we shall find it convenient to state the channel-assignment problem in terms of intervals. In this context, an interval $I_i = [a_i, b_i]$ describes the positions of a pair of components. Now the collection of pairs of components can be represented by a family $I = \{I_i = [a_i, b_i] \mid a_i \leq b_i, 1 \leq i \leq n\}$ of intervals on the *real* line.

As it turns out, families of intervals are very useful when it comes to modeling a vast array of practical situations involving time dependencies or other restrictions that are linear in nature. Applications include such areas as archaeology, biology, psychology, management, engineering, VLSI design, circuit routing, file organization, scheduling, transportation, and many others [1,4,6,9]. It comes as no surprise, therefore, that interval families have been studied intensely from both the theoretical and algorithmic point of view [1,5-7].

The advent of highly parallel computing capabilities has motivated an active search for large classes of computational problems that can be solved fast in parallel. We assume the Parallel Random Access Machine model (PRAM, for short) which consists of autonomous processors, each having access to a common memory. At each step, every processor performs the same instruction, with a number of processors masked out. In a Concurrent Read Exclusive Write PRAM (CREW-PRAM) model, several processors may simultaneously read the same memory location, but exclusive access is used for writing. In the Exclusive Read Exclusive Write PRAM (EREW-PRAM) model, a memory location cannot be simultaneously accessed by more than one processor. The interested reader is referred to [13] for a competent discussion on the PRAM family. It is easy to see that the more restrictive EREW-PRAM is the *weakest* member of the PRAM family. Additionally, several authors argue that EREW-PRAM is the only model that is reasonably close to real machines, making it of a particular practical interest. With this observation in mind, we shall adopt the EREW-PRAM as our model of computation.

2. The Algorithm

Consider a family $I = \{I_i = [a_i, b_i] \mid a_i \leq b_i, 1 \leq i \leq n\}$ of intervals on the real line; a *color assignment* for the family I is a partition of I into nonempty, disjoint subsets $C_1, C_2, ..., C_k$ such that within every set C_i the intervals are pairwise non-overlapping. Intuitively, it is clear that if we assign the same color to all the intervals belonging to some C_i, then what results is a coloring of I such that no overlapping intervals share the same color.

It is immediate [4,6] that an optimal solution to the *channel-assignment* problem amounts to a coloring of I with the *least* number of colors.

To make this note self-contained we shall reproduce the details of the optimal sequential coloring algorithm in [6]. The idea of the algorithm is simple: the intervals are colored sequentially from left to right. As soon as an interval has ended, its color is released (pushed on a stack) and can be reused for the first interval that starts after the current one ended: in other words, we can reuse the same channel for both pairs of components. The details follow.

Sequential-algorithm Optimal-Coloring($\{I_i\}$); (see [6])
{Input: a family $I = \{I_i = [a_i, b_i] \mid 1 \le i \le n\}$ of intervals on the real line;
 Output: an optimal coloring for I.}
Step 1. let $c(1)$, $c(2)$, ..., $c(2n)$ be the left and right-endpoints
         sorted in ascending order [†];
Step 2. **for** $i \leftarrow 1$ **to** $2n$ **do**
            if $c(i) = a_k$ **then begin**
               avail $\leftarrow$ pop();
               color(I_k) $\leftarrow$ avail;
            end
          else {assume $c(i)=b_{k'}$}
               push(color($I_{k'}$));
Step 3. return(color);

Before we discuss our parallel algorithm to solve the channel-assignment problem, we need to review a number of techniques that are instrumental in the efficient implementation of our result. We assume that a linked list containing n elements is stored in an unordered array. The following computational problems can be solved optimally in parallel:

(1) given a linked list, transmit a value to all the elements of the list;

(2) given a linked list of n real numbers $x_1, x_2, \ldots, x_n$, and an associative binary operation o on $x_1, x_2, \ldots, x_n$, compute all the *prefixes* of the form x_1, x_1 o x_2, ...,

---

[†] if $c(i)=c(i+1)=...$, we assume that left-endpoints precede right-endpoints

$x_1 \circ x_2 \circ ... \circ x_n$.

Problems (1)-(2) can be solved in O(log n) time using $O(\dfrac{n}{\log n})$ processors in the EREW-PRAM model of computation. See [3] and [7] for more information and detailed algorithms.

Our parallel algorithm follows the idea of the sequential algorithm of Gupta et al [6]. There are, however, a number of differences worth noting. First, we need to introduce some terminology. Specifically, let $c(1)$, $c(2)$, ..., $c(2n)$ stand for the sorted sequence of the 2n endpoints in the family I of intervals[†]. Assign to $c(1)$ the weight $w(1) = 0$, and for all i, $(2 \leq i \leq 2n)$ assign to $c(i)$ the weight $w(i)$ defined as follows:

$$w(i) = \begin{cases} 1, \text{ if both } c(i-1) \text{ and } c(i) \text{ are left–endpoints} \\ -1, \text{ if both } c(i-1) \text{ and } c(i) \text{ are right–endpoints} \\ 0, \text{ otherwise} \end{cases} \qquad (3)$$

Next, we perform prefix sum on the sequence of weights assigned above; let $e(1)$, $e(2)$, ..., $e(2n)$ be the result (i.e. $e(i) = w(1) + w(2) + ... + w(i)$).

Call two intervals $I_i = [a_i,b_i]$ and $I_j = [a_j,b_j]$ *related* whenever the following conditions are satisfied:

$$b_i < a_j, \ e(a_j) = e(b_i) \text{ and} \qquad (4)$$
for all endpoints u with $b_i < u < a_j$, $e(u)$ is distinct from $e(a_j)$ and $e(b_i)$.

The following simple result follows directly from the definition of related intervals. More precisely we have

Lemma 0. Let I_{i_1}, I_{i_2}, ..., I_{i_k} be a sequence of intervals such that every pair of consecutive intervals are related. Then no intervals in this sequence overlap.

Proof. Clearly, by (4) together with the assumption that every pair of consective intervals are related we can write: $a_{i_1} \leq b_{i_1} < a_{i_2} \leq b_{i_2} < ... < a_{i_k} \leq b_{i_k}$.

[†] if $c(i)=c(i+1)=...$, we assume that left-endpoints precede right-endpoints

Consequently, all the intervals in the sequence must be non-overlapping. □

In fact, it is easy to confirm that the sequential algorithm of Gupta et al assigns the same color to a sequence $I_{i_1}, I_{i_2}, ..., I_{i_k}$ of intervals in the family I such that for all t ($1 \leq t \leq k$), I_{i_t} and $I_{i_{t+1}}$ are related. Sequentially, finding related intervals is done elegantly by using a stack. In parallel, however, emulating a stack is rather inefficient. Instead, our approach is based on the weighting scheme devised in (3).

The following result and its consequences are at the heart of our parallel algorithm for the channel-assignment problem.

Lemma 1. Consider the subsequence $c(i_1), c(i_2), ..., c(i_r)$ obtained by scanning $c(1), c(2), ..., c(2n)$ from left to right and retaining every i_t with $e(c(i_t)) = k$ for some fixed $k \leq$ max{ $e(i) \mid 1 \leq i \leq 2n$ }. Then $c(i_t)$ is a left-endpoint or a right-endpoint depending on whether t is odd or even.

Proof. Let $k \leq$ max{ $e(i) \mid 1 \leq i \leq 2n$ } be arbitrary, but fixed. Obviously, $c(i_1)$ must be a left-endpoint. We only need show that no two consecutive endpoints in the subsequence are left-endpoint or right-endpoints. Suppose to the contrary that both $c(i_p)$ and $c(i_{p+1})$ are left-endpoints. But now every $c(j)$ with $i_p < j < i_{p+1}$ must have $e(c(j)) > k$. Since, by assumption, $c(i_{p+1})$ is a left-endpoint, we have $e(c(i_{p+1})) > k$, a contradiction. The case where both $c(i_p)$ and $c(i_{p+1})$ are right-endpoints is similar. □

Lemma 1 implies the following simple result.

Observation 1. Let $c(i_1), c(i_2), ..., c(i_r)$ be as in the statement of Lemma 1. All intervals $I_k, I_{k'}$ with $c(i_p) = b_k$ and $c(i_{p+1}) = a_{k'}$ are related.

[The justification follows directly from Lemma 1 and (4), combined, and is left to the reader.]

In addition, Lemma 1 motivates us to define a linear order $<$ on the set of ordered pairs $(e(i), c(i))$ ($1 \leq i \leq 2n$) such that

$(e(i),c(i)) \prec (e(j),c(j))$ whenever $e(i) < e(j)$ or $e(i) = e(j)$ and $c(i) < c(j)$.

For further reference we write

$$m = \max \{ e(i) \mid 1 \le i \le 2n \}. \tag{5}$$

Lemma 2. The minimum number of colors in a coloring of the family I of intervals is $m + 1$.

Proof. We only need prove that the largest number of pairwise overlapping intervals in I is $m + 1$. For this purpose, let i stand for the smallest subscript for which $e(i) = m$. Clearly, $c(i)$ must be the left-endpoint of some interval in I.

Since $e(i)=m$, (3) implies that there are m more left-endpoints than right-endpoints among $c(1)$, $c(2)$, ..., $c(i-1)$. The corresponding intervals must end to the right of $c(i)$. Consequently, the number of intervals overlapping at $c(i)$ is $m+1$ (including the interval whose left-endpoint is $c(i)$).

To see that no more than $m+1$ intervals are pairwise overlapping, consider a counterexample $c(j)$. Specifically, $m' > m+1$ intervals are overlapping at $c(j)$. But now, (3) implies that $e(j) > m$, contradicting (5). $\Box$

Now consider the sequence $\{(e(i),c(i)) \mid 1 \le i \le 2n \}$ sorted by $\prec$. It will have the form:

$$(e(i_1),c(i_1)), (e(i_2),c(i_2)), ..., (e(i_{2n}),c(i_{2n})). \tag{6}$$

It is convenient to visualize (6) in the following way:

$$(0,c_1^0), ..., (0,c_{n_0}^0), (1,c_1^1),...,(1,c_{n_1}^1), ...,(m,c_1^m),...,(m,c_{n_m}^m). \tag{6'}$$

In the remainder of the note, we shall use either (6) or (6') interchangeably.

Observation 2. For all j, $(0 \le j \le m)$ n_j is even. Furthermore, for all even values of t $(t < n_j)$, the intervals whose endpoints are c_t^j and c_{t+1}^j are related.

[To see this, apply Lemma 1 and Observation 1 for all j, $(0 \le j \le m)$.]

Now the algorithm proceeds as follows, for all $0 \le j \le m$ perform:

- mark the intervals having c_t^j as its left-endpoint;

- set link(I_t) ← nil for $b_t = c_{n_j}^j$;

- for all even $t < n_j$, write $c_t^j = b_k$ and $c_{t+1}^j = a_{k'}$, and set link(I_k) ← $I_{k'}$.

As we are about to point out, what results is a set of linked lists with the first element of every list marked. Now using (1), broadcast j to all the intervals belonging to the list whose first element has c_j^j as its left-endpoint. When this is done, every interval knows its own color (this will be proved later).

The details of this procedure are spelled out as follows.

Procedure Parallel-Optimal-Coloring($\{I_i\}$);

{Input: a family $I = \{I_i \mid 1 \le i \le n\}$ of intervals on the real line;
Output: an optimal coloring for I.}
0.   **begin**
1.   let $c(1)$, $c(2)$, ..., $c(2n)$ be the left and right-endpoints sorted in ascending order[†];
2.   assign weights $w(1)$, $w(2)$, ..., $w(2n)$ as in (3);
3.   perform prefix sum on the sequence $w(1)$, $w(2)$, ..., $w(2n)$
          and let the result be $e(1)$, $e(2)$, ..., $e(2n)$ (i.e. $e(i) = \sum_{j=1}^{i} w(i)$);
4.   let $(e(i_1),c(i_1))$, $(e(i_2),c(i_2))$, ..., $(e(i_{2n}),c(i_{2n}))$ be sorted as in (6);
5.   mark the interval having $c(i_1)$ as its left-endpoint;
6.   set link(I_t) ← nil for $b_t = c(i_{2n})$;
7.   **for** all even j $(2 \le j < 2n)$ **do in parallel**
8.   **if** $e(i_j) \ne e(i_{j+1})$ **then begin**
9.   mark the interval having $c(i_{j+1})$ as its left-endpoint;
10.   set link(I_k) ← nil where $c(i_j) = b_k$
12.   **end**
13.   **else**
14.   set link(I_k) ← $I_{k'}$ where $c(i_j) = b_k$ and $c(i_{j+1}) = a_{k'}$;
15.   let I_{k_0}, I_{k_1}, ..., I_{k_m} be the marked intervals obtained in lines 5 and 9;
16.   **for** i = 0 to m **do in parallel**
17.   transmit i to all the intervals on the linked list starting at I_{k_i}
18.   **end**;

The correctness of our approach relies on a number of intermediate results that we present next.

Lemma 3. Every interval in I belongs to exactly one of the lists created in lines 5-14.

Proof. We need prove that every interval is either marked or else it is pointed to

[†] if $c(i) = c(i+1) = ...$, we assume that left-endpoints precede right-endpoints

by another interval, and that every interval points to another interval or to nil.

Consider an arbitrary interval I_p of I. It is either marked in lines 5 or 9 or else we find some interval I_q and a subscript k such that in the sequence $(e(i_1),c(i_1))$, $(e(i_2),c(i_2))$, ..., $(e(i_{2n}),c(i_{2n}))$, $b_q = c(i_k)$ and $a_p = c(i_{k+1})$. Since I_p is not marked, $e(i_k) = e(i_{k+1})$. Consequently, line 14 guarantees that $\text{link}(I_q) \leftarrow I_p$.

Similarly, either $\text{link}(I_p)$ is set to nil in lines 6 or 10, or else we find an interval I_r and a subscript t such that in the sequence $(e(i_1),c(i_1))$, $(e(i_2),c(i_2))$, ..., $(e(i_{2n}),c(i_{2n}))$, $b_p = c(i_t)$ and $a_r = c(i_{t+1})$. Again, line 14 guarantees that $\text{link}(I_p) \leftarrow I_r$. $\square$

The following result justifies the fact that all the intervals belonging to the same list can receive the same color in a coloring of the family I.

Observation 3. All the intervals belonging to the same linked list constructed in lines 5-14 are pairwise non-overlapping.

[Follows immediately from Lemma 0, Lemma 1, and Observation 1 combined.]

Observation 4. The number of linked lists obtained after performing lines 5-14 is exactly m+1.

[To see that this is the case, note that that the number of lists created in lines 5-14 is exactly the number of marked intervals, which is m+1.]

Theorem 1. Procedure Parallel-Optimal-Coloring correctly colors the family I of intervals with a minimum number of colors in O(log n) time using O(n) processors in the EREW-PRAM model of computation.

Proof. The correctness follows immediately from Lemma 2, Observation 3, and Observation 4, combined. We note that since no read or write conflicts occur, the computation can be carried out in the EREW-PRAM model of computation.

To argue about the complexity, note that line 1 takes O(log n) time and O(n) processors if the sorting algorithm of Cole [2] is used. Line 2 takes O(1) time and o(n) processors. By (2), line 3 runs in O(log n) time using $O(\frac{n}{\log n})$ processors.

By using the sorting algorithm of Cole, again, line 4 takes $O(\log n)$ time and $O(n)$ processors. Lines 5-14 run in constant time if $O(n)$ processors are used. Finally, by (1), lines 16-17 run in $O(\log n)$ time using $O(n)$ processors. Altogether, therefore, the entire algorithm runs in $O(\log n)$ time and uses $O(n)$ processors, as claimed. $\square$

To illustrate our algorithm consider the example in Figure 1.

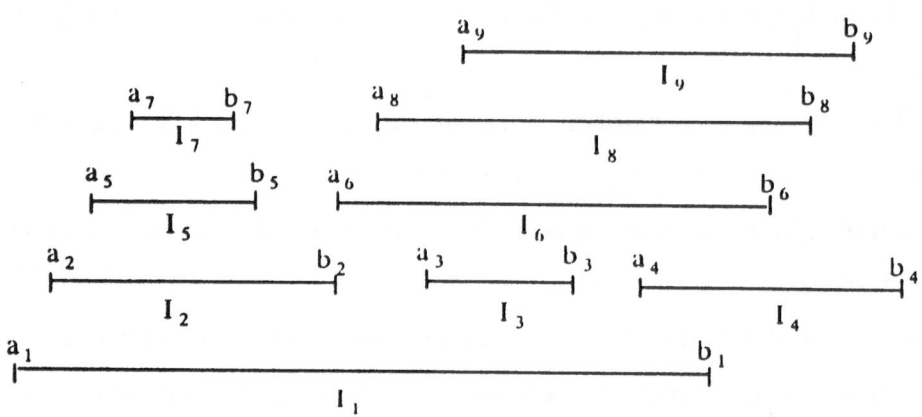

- Figure 1 -

*	1	2	3	4	5	6	7	8	9	10	11	12	13	14	15	16	17	18
c's	a_1	a_2	a_5	a_7	b_7	b_5	a_6	b_2	a_8	a_3	a_9	b_3	a_4	b_1	b_6	b_8	b_9	b_4
w's	0	1	1	1	0	-1	0	0	0	1	1	0	0	0	-1	-1	-1	-1
e's	0	1	2	3	3	2	2	2	2	2	4	4	4	4	3	2	1	0

e's	0	0	1	1	2	2	2	2	2	2	3	3	3	3	4	4	4	4
c's	a_1	b_4	a_2	b_9	a_5	b_5	a_6	b_2	a_8	b_8	a_7	b_7	a_3	b_6	a_9	b_3	a_4	b_1

I's	I_1^*	I_2^*	I_3	I_4	I_5^*	I_6	I_7^*	I_8	I_9^*
link	nil	I_8	I_4	nil	I_6	nil	I_3	nil	nil

Now the lists and the colors assigned are as follows:

color 0: $I_1 \rightarrow$ nil

color 1: $I_2 \rightarrow I_8 \rightarrow$ nil

color 2: $I_5 \rightarrow I_6 \rightarrow$ nil

color 3: $I_7 \rightarrow I_3 \rightarrow I_4 \rightarrow$ nil

color 4: $I_9 \rightarrow$ nil

References

1. A. A. Bertossi and M. A. Bonuccelli, Some parallel algorithms on interval graphs, *Discrete Applied Mathematics* (1987) 101-111.

2. R. Cole, Parallel Merge Sort, *SIAM Journal on Computing*, 17 (1988) 770-785.

3. R. Cole and U. Vishkin, Approximate parallel scheduling. Part I: The basic technique with applications to optimal parallel list ranking in logarithmic time, *SIAM Journal on Computing*, 17, (1988) 128-142.

4. E. Dekel and S. Sahni, Parallel Scheduling Algorithms, *Operations Research*, 31 (1983) 24-49.

5. P. C. Gilmore and A. J. Hoffman, A characterization of comparability graphs and interval graphs, *Canad. J. of Math.* 16, (1964) 539-548.

6. U. I. Gupta, D. T. Lee, and J. Y. T. Leung, An Optimal Solution for the Channel Assignment Problem, *IEEE Trans. Comput.* 28 (1979) 807-810.

7. U. I. Gupta, D. T. Lee, and J. Y. T. Leung, Efficient algorithms for interval graphs and circular-arc graphs, *Networks*, Vol. 12 (1982) 459-467.

8. C. P. Kruskal, L. Rudolph, and M. Snir, Efficient parallel algorithms for graph problems, *Algorithmica*, 5 (1990) 43-64.

9. B. T. Preas and M. J. Lorenzetti, Physical Design Automation of VLSI Systems, Benjamin/Cummings, Menlo Park, California, 1988.

10. R. Raghavan and S. Sahni, Single row routing, *IEEE Trans. on Computers*, C-32 (1983), 209-220.

11. Z. Syed, A. El Gamal, and M. A Breuer, On routing for custom integrated circuits, *Proc. 19th Design Automation Conference*, 1982, 887-893.

12. S. Tsukiyama, E. S. Kuh, and I. Shirakawa, An algorithm for single-row routing with prescribed street congestions, *IEEE Trans. on Circuits and Systems*, CAS-27, (1980) 765-771.

13. U. Vishkin, Synchronous parallel Computation - a Survey, TR 71, Dept. of Computer Science, Courant Institute, NYU, 1983.

Chapter IV

Product Engineering

Introduction

Papers grouped in this chapter all present various solutions and systems of product design and engineering. The first paper describes the work in automating the design process with an artificial intelligence technique called Case-Based Reasoning, as applied to process planning, product redesign, and design testing. The system of computer-aided design of mechanical clutches is presented in the second paper, whereas the third paper presents an expert system (using the VP-Expert shell) for design of precise ball and roller bearings. The fourth paper discusses a two-stage design framework for design of spur gears. The two stages proposed are preliminary design and design optimization, and the paper focuses especially on the optimization stage. Then, the fifth paper presents an approach to computer-aided design of underground structures.

The sixth paper proposes a mathematical model of the computer-aided design of technical systems, based on notions of increasing sequence of macromodels, block frontal algorithms in the hypermatrix algebra, and a database approach. The next paper focuses on solid modeling, namely on using curves and surfaces with tension control to build complex surface models with geometric discontinuities.

The next four papers address issues of Finite Element Analysis (FEA). The eighth paper reviews the design optimization algorithms integrated with FEA programs, used in optimization for weight, shape, and structural response, the ninth proposes the direct tessellation approach to automatic generation of FEA meshes from CAD geometry models, while the tenth one proposes a 3-D isoparametric mesh generation method. The last paper in this group presents an effective FEA strategy to model complex mechanical systems, using the practical example of Stiller-Smith engine design.

The twelfth paper presents a simplified design approach for selection of suspension systems of cross-country trailers and trucks, aimed at minimization of shock levels.

Design Using Case-Based Reasoning

COSTAS TSATSOULIS

Center for Computer-Aided Systems Engineering
Department of Electrical and Computer Engineering
The University of Kansas
Lawrence, KS

Summary

Design is one of the most knowledge- and expertise-intensive processes in product development. Expert designers are individuals with deep understanding of the domain, of the physical laws governing it, of the manufacturing operations available, of the industrial setups, and so forth. A good designer does not produce only a functional design, but also one that is producible, maintainable, testable, and marketable. This paper will describe our work in automating the design process for the manufacturing domain. We are using an innovative technique in Artificial Intelligence called Case-Based Reasoning. This paper will briefly describe Case-Based Reasoning, and will concentrate on our work in process planning, product redesign for criteria other than functionality, and product design testing. We describe a total of three systems that we have developed and which are parts of a large project in Intelligent, Computer-Aided Design (ICAD).

1. Case-Based Reasoning

There are two broad research targets in Artificial Intelligence (AI). The first tries to understand the nature of intelligence and human thought. It examines cognitive behavior and looks for principles that play general descriptive and explanatory roles. The second seeks to create intelligent artifacts and to develop a technology of intelligence. Case-Based Reasoning is an AI paradigm that addresses both research agendas. Case-Based Reasoning is based on psychological theories of human cognition. It is also a new methodological approach in AI that provides the foundations for a new technology of intelligent systems.

Avoiding the details of the theory Case-Based Reasoning, we can describe it as based on the intuitive notion that human expertise is not based on rules or other formalized structures, but on experiences. Human experts differ from novices in their ability to relate problems to previous ones, to reason based on analogies between current and old problems, and to use solutions from old experiences. The process of reasoning using experiences or cases can be described by the following steps [3,4]:

1. Retrieve: Given a new problem, retrieve a similar past case from memory. The past case contains the prior solution.
2. Modify: The old solution is modified to conform to the new situation, resulting in a proposed solution.

3. Test: The proposed solution is tested for successful solution of the current problem.
4. Learn: If the solution fails, explain the failure and learn it to avoid repeating it. If possible, repair the failure, generate a new proposed solution and return to step **3**. If the solution succeeds, incorporate it into the case memory as a successful solution, and stop.

Case-Based Reasoning systems have been proposed as an alternative to traditional rule-based expert systems. Expert systems have been used in the last ten years to develop many intelligent systems for real-life applications. The basic unit of an expert system has been an *IF-THEN* rule. Several hundred rules might be required to handle a typical diagnostic or repair task; design tasks tend to require thousands of rules, as, for example, in R1, an expert system for the design of VAX computers that uses over 10,000 rules [6]. Building and using rule-based expert systems became very popular; they are easy to develop as a prototype, and very powerful in their results. As experience with rule-based expert systems increased, so did the understanding and awareness of some basic short-comings of the rule-based paradigm.

The first problem is *knowledge-acquisition*. To build an expert system a knowledge engineer had to interview human experts (informants) and to try to elicit appropriate knowledge in the form of rules. This knowledge was difficult to uncover. The human expert could not make a list of hundreds of rules he or she used, because the expert most often did not use rules in problem solving. The process of knowledge acquisition became very tedious and tiring for both the knowledge engineer and the expert. Often it resulted in incorrect rules, since the expert was forced to express his or her expertise in a format (rules) that did not truly reflect the way he or she thought. Knowledge acquisition became known as a bottleneck in the construction of expert systems [5].

The second problem is that expert systems lack memory. For example, when a medical diagnosis expert system was presented with a patient, it might use hundreds or even thousands of rules to reach a diagnosis. When presented with exactly the same case again, it would have to re-use the same set of rules to reach the exact same conclusion. This lack of memory led to computational inefficiencies.

The third problem is related to robustness. If an expert system was given a problem that did not match its rules fully, it would be incapable of producing an answer. The knowledge of an expert system is limited to its rules, and if no rules are applicable the system has no alternatives.

The behavior of human experts in no way matches that of a rule-based expert system. The central feature of expertise is *experience*. An expert is someone who has vast knowledge in a specialized domain, who has witnessed numerous cases in a domain, and has generalized this experience to apply it to new situations. When confronted with a problem an expert is reminded of previous, similar problems and their respective resolutions. Even rules that the expert may be using are rooted in actual experiences, which have been distilled into a general formula of action. Thus, the basic unit of knowledge for an expert is not a *rule* but a *case*. Consequently, it is easy to acquire expert knowledge if the knowledge engineer asks for cases and experiences, rather than for rules. Experts can easily articulate their experiences and the knowledge acquisition bottleneck is solved in Case-Based Reasoning.

Second, human experts remember their experiences. Similarly, Case-Based Reasoning systems are memory systems that remember all problem solving instances. Case-Based Reasoning systems keep adding to their experiences and knowledge and can grow as a more advanced reasoning system. Case-Based Reasoning systems learn from experience and add to their knowledge both successful cases and failures, so that they can avoid repeating the same mistakes.

Third, human experts can reason by analogy. Even if the expert is faced with a problem for which there is no exact similar experience, the expert is still capable of coming up with approximate or probable solutions that need to be tried out for correctness. By reasoning from analogy with past cases Case-Based Reasoning systems can address new and unknown problems, and are not constrained to solving only a limited set of known problems.

In conclusion, the technology of Case-Based Reasoning promises to offer new, improved methodological approaches to the creation of intelligent systems. Case-Based Reasoning provides the developer of intelligent systems with ease in knowledge acquisition, learning capabilities that will allow the system to evolve and improve from experience, and robustness in its reasoning and problem solving process.

2. Using Case-Based Reasoning in Design Problems

Our research in the intelligent automation of design has led us to the belief that Case-Based Reasoning is an excellent candidate for reasoning in design problem solving. In the last five years we have developed Case-Based Reasoners that deal with various facets of the design process. In this paper we will present three systems: TOLTEC, a case-based process planning system; REINRED, a case-based redesign system; and ASP, a case-based tester and simulator of designs. TOLTEC and REINRED are applied to part manufacturing and design while ASP is applied to the testing and analysis of communications equipment.

2.1. TOLTEC - A Case-Based Process Planning System

TOLTEC is a case-based system for process planning [7,8]. Process planning is the selection and sequencing of manufacturing operations to produce a given work piece. Due to the importance of process planning in the overall production process, there has been a significant research effort to create systems that can assist in process planning.

TOLTEC contains cases of solved process planning problems on various levels of abstraction. Using a weighted partial matching algorithm guided by constraints it selects an old process plan that is best applicable to the currently studied machine part. The process plan is on a very high level of abstraction. For example, for a single feature, the plan might be: (make-hole, thread, finish). Next, each general operation is selected and is expanded by re-evoking TOLTEC. For example, the hole making operation might be expanded as (counterbore, drill with a HSS tool). The process continues until TOLTEC creates a process plan consisting of only primitive (non-expandable) manufacturing operations.

During the process plan creation TOLTEC does not simply retrieve an old plan and use it as an expansion of an abstract operation. An old case will only rarely match a current problem perfectly. For example, we might know how to drill a hole, but have no previous case of counterboring. TOLTEC contains domain knowledge that allows it to *adapt* old cases to fit current problem descriptions. TOLTEC also contains cases of old failures. These cases describe situations in old process plans that resulted into some type of manufacturing error: tool breakage, costly production, tolerances not met, etc. Whenever an operation is planned TOLTEC checks to see whether it matches any of its old failure cases, and if so, it changes the process plan to avoid the error. To allow early detection of manufacturing failures we added cases that described failures due to poor designs and suggested redesign solutions. This way we allowed the manufacturing engineer to have some input in the design.

TOLTEC was an early project that showed the power of case-based solutions to manufacturing. The manufacturing engineers who used it seemed to find it very helpful. Although no benchmarks were attempted, the overall impression was that TOLTEC functioned on an expert level in process planning. The most positive feedback we received was for TOLTEC's ability to detect faulty or non-manufacturable designs and to provide suggestions for correcting them. This feedback has led us to the development of a case-based system that performed pure redesign for manufacturability.

2.2. REINRED - A Case-Based System for Redesign for Manufacturability

When we say *redesign for manufacturability* we mean correcting the design of work pieces in a way that allows optimal cost in their manufacturing. Rather than waiting for the manufacturing engineers to catch non-manufacturable designs, we would like the design engineers to produce manufacturable designs from the very start. This would minimize the time lag between receiving initial specifications for a part and its actual production. Of course, any speed-up in production time will result in substantial production increase and cost decrease. More importantly, designs that are manufacturable make better use of the manufacturing setup of a company, thus decreasing idle times, result in less tool ware and failure, decrease accidents, and allow specialized equipment and personnel to be used when truly needed, thus increasing productivity. One important realization is that small redesign changes can result in large improvements in the ease and cost of manufacture. Another important realization is that designing for manufacturability is a very knowledge-intensive process. Design engineers do not want to be burdened with the added concern of manufacturing the part they are designing; achieving functionality is difficult enough.

REINRED is a case-based system that can redesign parts for manufacturability. Its storage and use of cases is very similar to TOLTEC's, with the obvious difference that REINRED's knowledge consists of cases of redesigns taken from various manufacturing handbooks and from interviews with experts. Each case contains information about the original design, parts of it that were not easy to manufacture, an explanation of the reason a design feature was "bad" in terms of manufacturability, and the ways by which this feature was redesigned to make the part better manufacturable.

REINRED repairs all features independently and then must check whether the re-design requests conflict with each other and with other features of the design. Consider, for example, two holes on the same surface of a machine part. Let us also assume that both holes are of non-standard radii and a redesign request is generated for each one, to change the radius to a standard one. REINRED might discover that by changing the radii the two holes are now too close together and that the material between them is too thin. The question is how to resolve such conflicts and to make sure that the resulting redesign is feasible.

One solution to the problem of conflict resolution would be to backtrack, undo a redesign request and attempt to identify one that does not create a conflict. This method is computationally inefficient and cannot guarantee to produce no future conflicts. Another solution would be to completely eliminate the conflict problem by describing a design as a unified entity rather than a collection of features. While this might be possible in small, constrained domains, in manufacturing such a solution is not feasible due to the large number of possible combinations of features. The third possible solution would be not to resolve the conflicts, but to compromise between them; this is the technique employed by REINRED.

REINRED prioritizes features and redesign requests. The priorities are symbolic and belong to the ordered set (A-MUST, VERY-IMPORTANT, IMPORTANT, LESS-IMPORTANT, A-WISH). The priority of a feature can be computed from the functional requirements of this feature in relationship with the functionality of the whole part. For example, if a hole is used to secure a screw, the threads in this hole are prioritized as A-MUST. The functionality of a feature can be deduced from the explanations attached to old redesign requests. Prioritizing redesign requests is based on the priority of the analogous redesign requests in the old cases.

REINRED is still a prototypical system and is being constantly improved. In its current stage it has redesigned some complex parts successfully and has managed to resolve redesign conflicts by prioritizing features. Still, it has some problems: some of the compromises it makes may contradict functional requirements of the part, while other times the system can find no compromise, although some is possible. The shortcomings of our current compromise solution can be attributed to REINRED's inability to take into account the complete functionality of a part. Modification of redesign requests is very much dependent on the functionality and design specifications of a machine part and as of yet we have no good way of representing them. The current technique employed by REINRED has managed to produce successful, compromised redesign requests to a large number of parts, and additions to its compromise system keep improving its performance. REINRED is also the first system that deals with the problem of simultaneous adaptations of many cases that might be conflicting and that need to have these conflicts resolved or compromised.

2.3 ASP - A Case-Based System for Design Testing

Both TOLTEC and REINRED have no ability to independently and automatically judge the results of their work and must rely on a human expert to critique their reasoning. While this does not affect the reasoning ability of these particular systems, it is a serious shortcoming of any complete, intelligent design system. As a response we create ASP, a case-based system that assists in the simulation and testing of designs.

ASP is a case-based system that generates simulation runs for the testing of communications systems designs. With ASP we were forced to step away from the machine part domain because the performance of communication systems designs can be easier simulated than that of part design. We used the communication systems analysis engine COEDS [1] to interface with ASP and produce analysis plans.

The inputs to ASP are the description of a communications system, environmental information (noise environment, global position, etc.), and the analysis goals, that is what kind of information should be the final result of the analysis run. ASP then uses Case-Based Reasoning techniques to generate an *analysis plan*. An analysis plan represents the set of actions that must be taken to realize the analysis goals specified. The actions are analyses that must be sequenced, combined and run. After the analysis plan is generated it is evaluated and stored in a Case Memory. If the analysis plan fails it is identified as an error and stored as a failure. If it succeeds, it is executed and the results are presented to the engineer.

Due to the complexity of the domain of analysis and testing of communication systems designs ASP introduced the novel technique of using pieces of cases - rather than the whole case - during reasoning [2]. ASP has made major contributions to case-based reasoning theory, as well as solve many analysis problems.

3. A Common Thread

The three systems described in the previous pages, together with other we are currently developing, are all small pieces of a very large project in Intelligent Design. TOLTEC started as a process planning system and introduced us to the reasoning abilities of case-based reasoning systems. The feedback we received about TOLTEC showed us that most designers were interested in avoiding design failures. Thus we created REINRED which redesigns for manufacturability. Finally, ASP showed us how we should develop a system that will allow the automated analysis, simulation and evaluation of designs in progress.

We have adopted Case-Based Reasoning as the primary Artificial Intelligence methodology we are using (although other systems we are developing use other techniques, too), since we have found it to be best suited to the complex and non-formalized domain of expert design. Our work in Case-Based Reasoning has resulted in some innovative intelligent design systems, and in some major advances in the theory of case-based reasoning.

4. References

[1] Alexander, P., P. Magis, J. Holtzman and S. Roy, "A Methodology for Interoperability Analysis", *MILCOM-89,* 1989.
[2] Alexander, P., C. Tsatsoulis, J. Holtzman and G. Minden, "Case-Based Planning for Simulation", *IEE Conf. on Expert Planning Systems,* 1990.
[3] Hammond, K.J., *Case-Based Planning,* Boston: Academic Press, 1989.
[4] Riesbeck, C.K. and R.C. Schank, *Inside Case-Based Reasoning,* Hillsdale, NJ: Lawrence Erlbaum Associates, 1989.
[5] Hayes-Roth, F., D.A. Waterman and D.B. Lenat (Eds)., *Building Expert Systems,* Reading, Mass.: Addison-Wesley, 1983.
[6] van de Brug, A., J. Bachant and J. McDermott, "The Taming of R1", *IEEE Expert,* vol. 1, no. 3, 1986.
[7] Tsatsoulis, C, "Case-Based Planning in Manufacturing", *AAAI Spring Symposium,* 1989.
[8] Tsatsoulis, C. and R.L. Kashyap, "A Case-Based System for Process Planning", *J. of Robotics and Computer-Integrated Manufacturing,* vol. 5, no. 3/4, 557-570, 1988.

An Interactive Programming System for Design of Mechanical Clutches

B. SATYANARAYANA, K.V. MOHAN and M. MALLIKHARJUNA RAO

Department of Mechanical Engineering
College of Engineering
Andhra University
Visakhapatnam, India

ABSTRACT

This paper presents computer aided design of mechanical clutches. The design is made possible interactively using conventional design practices and also making use of optimization technique. Necessary graphic routines have also been developed in the proposed system.

1. INTRODUCTION

The present day trends in the field of machine design is the development of Interactive Software in CAD. This paper presents such an user friendly system for designing mechanical clutches. The development of this system is based on modular approach and provides the following options:

1. Create Design	4. Display Graphics
2. Modify Design	5. Plot Figure
3. Output Results	6. Exit

The proposed programming system has two phases. The first phase is an interactive session where the system prompts the user a series of logical questions under one of the above chosen options. It also provides a list of suggestive answers wherever applicable. In the second phase, the system interpret the responses and manipulates the data or computes the results which can be presented in the required form. Mechanical clutches, whether a disc clutch or a cone clutch or a centrifugal clutch with the strength criteria of uniform wear/uniform pressure, can be designed making use of empirical formulae or by using optimization techniques. The designs created earlier i.e., existing designs can also be modified based on the changes in environmental and/or load conditions.

2. SYSTEM MODULES
The proposed system has five modules as discussed below.

Create Module: Create module incorporates two approaches for the design of clutches. One is the formal design which is the default option and the other is through optimization technique. This module prompts user about the input details such as the torque to be transmitted, working conditions and material specifications etc.

Result Module: The design data created by earlier module is stored in a file. The present module invokes the file and display results in user convenient format.

Graphic Module: This module has provision to read the file having the data created by the first module. It then converts this data into co-ordinate data for the clutch to display. The actual co-ordinates are then transformed such that the display of the figure is possible to view in user defined window.

Modify Module: This module facilitate the user to modify existing design data of any clutch. It displays three windows. The top left window displays the existing data while the right window displays the modified data. Messages to the user are prompted through the bottom window.

Plot Module: The algorithm of this module is similar to that of Graphic Module except that it actuates the plotter and uses the plotter pens to draw the clutch.

3. DESIGN THEORY

Formal Design Appraoch

The formal design approach is by using the empirical formulae established in design data hand books [1]. According the pressure distribution between the friction surfaces, the design is based either on uniform pressure condition or on uniform wear condition. The design procedure for the friction clutches is outlined below.

Design of Disk Clutch:

For uniform pressure condition

$$\text{Torque } T = (2/3) \, \Pi \, \mu \, p \, (r_o{}^3 - r_i{}^3)$$

$$\text{where } p = \text{axial pressure}$$

$$\mu = \text{coefficient of friction}$$

$$r_i \quad = \quad \text{inner radius}$$

$$r_o \quad = \quad \text{outer radius}$$

For uniform wear condition

$$T \quad = \quad \Pi \mu C (r_o^2 - r_i^2)$$

Where C = constant equal to p $(r_o + r_i)/2$

By assuming a proper ratio of the outer radius to inner radius, the torque due to uniform pressure condition and uniform wear condition can be made nearly equal.

Assuming r_o = 1.5 r_i we get

$$r_i \quad = \quad (0.6315\,T/ \Pi \mu p)^{1/3} \text{ for uniform pressure condition}$$

$$= \quad (0.8\,T/ \Pi \mu p)^{1/3} \text{ for uniform wear condition}$$

Then face width, w = $r_o - r_i$ = 0.5 r_i

Design of Cone Clutch:

For uniform pressure condition

$$T \quad = \quad (2/3)\,(\Pi \mu p/\text{Sin } \alpha)\,(r_o^3 - r_i^3) \text{ where } \alpha \text{ is the cone angle}$$

For uniform wear condition

$$T \quad = \quad \Pi \mu C (r_o^2 - r_i^2)/\text{Sin } \alpha$$

$$= \quad \Pi \mu p(r_o + r_i)\,(r_o^2 - r_i^2)/2\,\text{Sin } \alpha$$

If it is assumed that r_o = 1.1 r_i, then

$$r_i \quad = \quad (4.55\,T\,\text{Sin } \alpha\ / \Pi \mu p)^{1/3} \text{ for uniform pressure condition}$$

$$= \quad (4.525\,T\,\text{Sin } \alpha\ / \Pi \mu p)^{1/3} \text{ for uniform wear condition}$$

Face width, w = $(r_o - r_i)/\text{Sin } \alpha$ = $(0.1\,r_i)/\text{Sin } \alpha$

Optimal Design Appraoch

To allow maximum torque to be transmitted by a clutch for given environmental conditions and material combinations, optimization techniques should be adopted for suitable selection of outer and inner radii of the clutch. The constraints which limit the maximum torque to be transmitted by a friction clutch are two. One is the maximum surface temperature (which increases as the torque T reaches very high values) and the other is the response time (which when too small will tend to damage the clutch) [2]. The mathematical models for the optimum design of the clutches are presented below.

Problem Model for Disk Clutch:

Maximize $T = \Pi \mu p r_i (r_o^2 - r_i^2)$

Subject to

Axial force $F = 2 \Pi p r_i (r_o - r_i) \leqslant F'$

Response time $t_o = I_1 I_2 (\omega_1 - \omega_2)/(T(I_1 + I_2) < t_o'$

Surface temperature $\theta = (E_s/1C_s) + (7/45) \dfrac{(E_s T^2 1^{3i} R_s^2 C_s (I_1+I_2)^2)}{I_1^2 I_2^2 (\omega_1 - \omega_2)^2} \leqslant \theta'$

$(r_o)_L \leqslant r_o \leqslant (r_o)_H$ and $(r_i)_L \leqslant r_i \leqslant (r_i)_H$

where $(r)_L, (r)_H =$ lower & upper limits of radii

$\quad I_1, I_2 =$ Interitia on drive and driven ends

$\quad \omega_1, \omega_2 =$ angular velocities of drive & driven shafts

$F', t_o' \& \theta'$ are the maximum allowable values of $F, t_o, \& \theta$

Problem Model for Cone Clutch:

The mathematical programming model for optimum design of cone clutch is:

Maximize Torque $T = (2/3) \Pi \mu p r_i (r_o^2 - r_i^2)/\mathrm{Sin}\,\alpha$

Subject to

Axial force $F = \Pi p (r_o^2 - r_i^2) \leqslant F'$

Response time $t_o = I_1 I_2 (\omega_1 - \omega_2)/ (T(I_1 + I_2)) \quad t_o'$

Surface temperature $\theta = \dfrac{I_1 I_2 (\omega_1 - \omega_2)^2 \times \mathrm{Tan}\,\alpha \times 10^{-2}}{2(I_1+I_2) C_s \rho_s \times 0.5 \times 427} + \theta_1 \leqslant \theta'$

$(r_i)_L \leqslant r_i \leqslant (r_i)_H$ and $(r_o)_L \leqslant r_o \leqslant (r_o)_H$

Solution Technique

The problems formulated as above, are solved by using complex box method [3]. In this method the simplex with 4 or 5 vetices is chosen in the solution space. The value of the objective function is found at each of these simplex points and the worst point (the point at which the objective value is highest) is replaced by its reflection point.

$$X_r = \alpha \, X_c + (1 - \alpha) \, X_w$$

where X_c is the centroid and α is the reflection coefficient

The above procedure is continued till the simplex converges such that the deviation in the function value at the vertices is less than a specified small quantity. Then the best point where the objective function value is minimum, becomes the optimum solution.

4. COMPUTATIONAL EXPERIENCE

The proposed Interactive Programming System for Design of Mechanical Clutches (IPSDMC) provides six options to the user. By choosing **Create** option the user is allowed to design a clutch either through formal design approach or optimal design approach. The design data thus executed is stored in a file which can either be retrieved at later use or be accessed through other options. **Results** option allows the user to view the various dimensions of the clutch or to get a print copy. The user can modify the existing data of any clutch through **Modify** option. It exhibits the existing data in one window and changed data in another window as shown in Fig. 1-A and 1-B. **Display** option provides the graphic display of the clutch in user specified location as shown in Fig.2 and Fig. 3. Drawing of any clutch that is created or modified can be obtained on a plotter by selecting the option **Plot**. Quiting off the system is possible through **Exit** option.

Part of an interactive session between the user and the system is presented in the appendix.

5. CONCLUSIONS

The Computer Code is written in GWBASIC and can be implemented on any IBM or IBM compatible Personal Computers. The interactive nature of the proposed system made the user easy to make the clutch designs in less time with no erros. The optimization technique can not be applied for centrifugal clutches, since the mathematical models have to be established yet. The modification of the earlier designs can be possible.

REFERENCES

1. HAROLD A. ROTHBART, Mechanical Design and Systems Hand Book, McGraw-Hill Book Company, New York, 1964.

2. COLIN CARMICHAEL, Kent's Mechanical Engineers Hnad Book, Design and Production Volume, John Wiley & Sons, INC, New York, 1958.

3. S.S. RAO, Optimization - Theory and Applications, Wiley Eastern Limited, 1979.

230

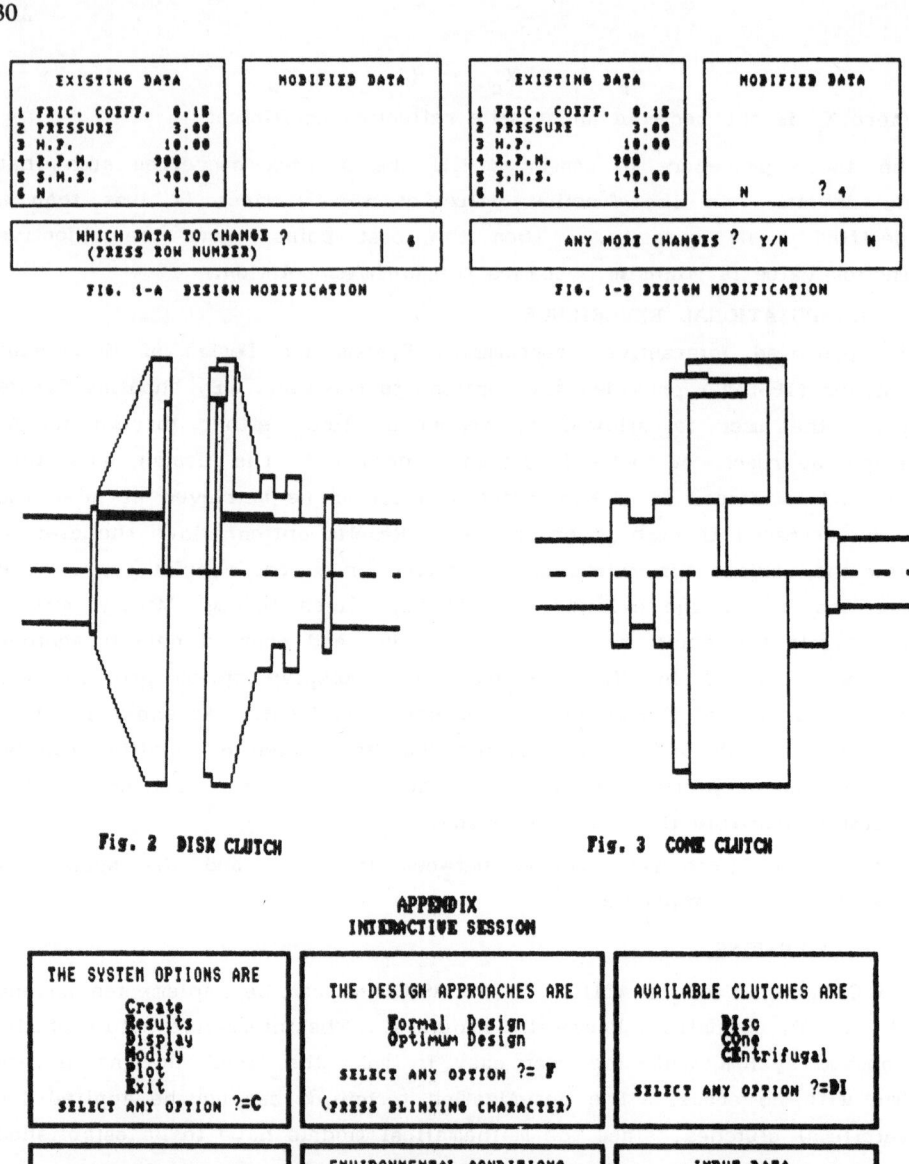

EXISTING DATA	MODIFIED DATA
1 FRIC. COEFF. 0.15	
2 PRESSURE 3.00	
3 H.P. 10.00	
4 R.P.M. 900	
5 S.M.S. 140.00	
6 N 1	

WHICH DATA TO CHANGE ?
(PRESS ROW NUMBER)   6

FIG. 1-A DESIGN MODIFICATION

EXISTING DATA	MODIFIED DATA
1 FRIC. COEFF. 0.15	
2 PRESSURE 3.00	
3 H.P. 10.00	
4 R.P.M. 900	
5 S.M.S. 140.00	
6 N 1	N ? 4

ANY MORE CHANGES ? Y/N   N

FIG. 1-B DESIGN MODIFICATION

Fig. 2 DISK CLUTCH

Fig. 3 CONE CLUTCH

APPENDIX
INTERACTIVE SESSION

THE SYSTEM OPTIONS ARE
Create
Results
Display
Modify
Plot
Exit
SELECT ANY OPTION ?=C

THE DESIGN APPROACHES ARE
Formal Design
Optimum Design
SELECT ANY OPTION ?= F
(PRESS BLINKING CHARACTER)

AVAILABLE CLUTCHES ARE
Disc
Cone
CEntrifugal
SELECT ANY OPTION ?=DI

DESIGN CRITERIA IS
Uniform Wear
Uniform Pressure
SELECT ANY OPTION ?= W

ENVIRONMENTAL CONDITIONS
NO	FACE1	FACE2	CONDITION
1	HSS	HSS	OILED
2	CI	CI	OILED
3	CI	CI	DRY
4	BRON	CI	OILED
5	ASB	CI	DRY
6	PMML	CI	DRY
SELECT ANY ROW ?= 3

INPUT DATA
POWER TRANSMITTED =? 10
(IN HP)
DRIVE SPEED(RPM) =? 900
ALLOWABLE SHEAR
STRESS(KG/CM^2) =? 140
NO. OF SURFACES =? 1

An Expert System for the Design and Selection of Ball Bearing Parameters

M. A. PATHAK and R. S. AHLUWALIA

Industrial Engineering Department
West Virginia University
Morgantown, WV

Abstract

Artificial Intelligence (AI) is an emerging technology. Research in AI is focused on developing computational approaches to intelligent behavior. The computer programs with which AI could be associated are primarily symbolic processes associated with complexity, ambiguity, indecisiveness, and uncertainty. One of these computer programs is referred to as Knowledge-based Expert System as it represents knowledge acquired from various experts in a particular field of interest to the user. The expert system emulates human behavior in solving problems thought to require experts for their solution by utilizing computer programs that incorporate experts' heuristic reasoning. In this paper, the application of Knowledge-based Expert System to aid the design of ball and roller bearings is discussed. The precision rolling-element bearings of twentieth century is a product of exacting technology and sophisticated science. A bearing supports radial and axial loads, at the same time allowing relative motion between two elements of a machine. Various requirements and steps in the design of ball and roller bearings are discussed. Equations are developed for the relevant design parameters and input into the expert system shell called VP-Expert. The expert system rules are also provided.

Introduction

Artificial intelligence (AI) has been applied to engineering problems, speech recognition and image analysis problems, medical consultation systems etc. Expert systems are considered a sub-group of AI. An expert system (ES) is a computer program that achieves high levels of performance on problems that normally require years of special education and training for human beings to solve. Expert systems us AI, problem-solving, decision-making and knowledge-representation techniques to combine human expert knowledge of a problem and methods of conceptualizing and reasoning about that problem.

An expert system consists of three components, the knowledge-base, the inference engine and the knowledge-acquisition shell. A knowledge-base represents facts and heuristic knowledge in the form of rules in the IF-THEN format. The inference engine concludes on a goal or data based on the knowledge described in the rules and a method of reasoning. There are two methods in practice today, namely, forward chaining and backward chaining. Forward chaining involves reasoning from data to goal while backward chaining finds data to conclude on a goal. As the name implies the knowledge-acquisition shell acquires the pertinent knowledge for the problem at hand from the user. In this paper a rule based expert system shell is utilized to demonstrate expert system application in the field of design of ball and roller bearings.

Ball & Roller Bearings

The purpose of a bearing is to support a load while permitting relative motion between two elements of a machine. The term rolling contact bearing refers to the wide variety of bearings that use spherical balls or some type of roller between the stationary and the moving elements. Eschmann [1] has given a complete treatise on the design of ball and roller bearings. Mott [2] discussed the general concepts in the design and applications of roller bearings. Matsumori [3] discussed the application of CBN abrasives in advanced ball bearing manufacture. Ashburn [4] provided a rare look at the manufacture of roller bearings at a bearing plant.

The most common type of bearing supports a rotating shaft, resisting purely radial loads or a combination of radial and axial loads. The essential parts of a ball bearing are inner and outer ring, the balls, and the separator. The inner ring is mounted on a shaft and has a groove in which the balls ride. The outer ring is usually the stationary part of the bearing and also contains a groove to guide and support the balls. The separator prevents the contact between the balls and thus reduces friction, wear and noise from the regions where sliding conditions would occur. There are three types of ball bearings, one that support only radial loads, only thrust loads, and a combination of both loads.. Here only those bearings are considered that support radial loads.

The design of a bearing requires consideration of the following, 1.) Characteristics of the bearing load, 2.) Relative motion between the bearing elements, 3.) Geometry of the bearing surfaces, 4.) Physical and chemical properties of the lubricant and bearing metals. The requirements that a bearing must satisfy are, 1.) Factor of safety, which depends on the bearing application, 2.) Bearing life, reliability and ambient conditions, 3.) Bearing precision, 4.) Power consumed in the bearing, 5.) Bearing installation and maintenance cost.

Following are the initial steps in the design of a ball-roller type bearings

1. Determine the axial force on the bearing from working condition.
2. Take radial and axial factors from the handbook, these factors are given below.

BEARING LOAD	RADIAL LOAD	AXIAL LOAD
Radial ball	1	0.85
Angular contact ball	0.5	0.4
Thrust ball	-	0.25
Cylindrical roller	0.5	-
Tapered and spherical roller (According to bearing type)	0.2	0.15
Deep groove	1	1
Filling notch	1.2	low
Double row	2.2	2.6
Self aligning	0.76	0.5

It should be noted that radial loads are forces at right angles to the axis of the shaft, such as the loads imposed by straight spur gears, drive chain or V-belts [5]. Thrust loads are parallel to the shaft axis e.g. turntable on vertical shafts, loads on crane hook.

3. Calculate bearing life (L) in million revolution of the bearing, the formula applied is as follows,

$$L = \text{(Life in hours} \times \text{r.p.m.} \times 60) / (10^6)$$

Following is a list of bearing-life recommendations for various classes of machinery [6].

MACHINERY CLASS	LIFE, kh
Instruments & Apparatus for infrequent use	<= 0.5
Aircraft Engine	0.5 -2
Machines for short operation	4 - 8
Machines for intermittent service	8 - 14
Machines for 8 hour service which are not fully utilized.	14 - 20
Machines for 8 - h service with full utilization	20 - 30
Machines for 24 hour service	50 - 60
Machines for 24 hour service with high reliability	100 - 200.

4. Take the service factor depending upon the nature of load. These service factors obviously depend on the type of machinery where these bearings are applied. For some basic types of machinery these service factors are,

TYPE OF APPLICATION	SERVICE FACTOR
Precision Gearing	1 - 1.1
Commercial Gearing	1.1 - 1.3
Application with poor bearing seals	1.2
Machinery with no impact	1 - 1.2
Machinery with light impact	1.2 - 1.5
Machinery with moderate impact	1.5 - 3

5. Determine the dynamic loading capacity (C).

$$C = [L/L_{10}]^{1/K} \; X \, (\, x \, . \, F_r \, + \, Y \, . \, F_a \,) \, K_s$$

Where,

L - required life of bearing in million revolution.

L_{10} - life of bearing for 90% survival at one million revolution.

F_r - Radial load on bearing.

F_a - axial load on bearing

K_s - Service factor.

x, Y - radial and axial factors.

K - exponent - 10/3 for roller bearing and 3 for ball bearing.

6. Select bearing that satisfies these requirements using manufacturer's catalogue for the required dynamic capacity and bore size.

7. Determine fits and tolerances for the shaft and housing. As there are tables available for the tolerances, statistical relations are developed for diametral and width tolerances from the relevant graphs and one of these graphs is shown in Figure 1. These linear or polynomial equations are then input into the expert system so that the tolerances can be computed.

8. If bearings are operated at high temperatures, employ correction factor to dynamic load bearing capacity.

9. Determine the type of lubrication.

These are the considerations that must be satisfied in order to design and select an appropriate bearing. When these considerations are put into the expert system in the form of IF-THEN rules, the expert system can suggest bearing parameters that will satisfy most of the bearing requirements and then the bearing can be chosen from the manufacturer's catalogue.

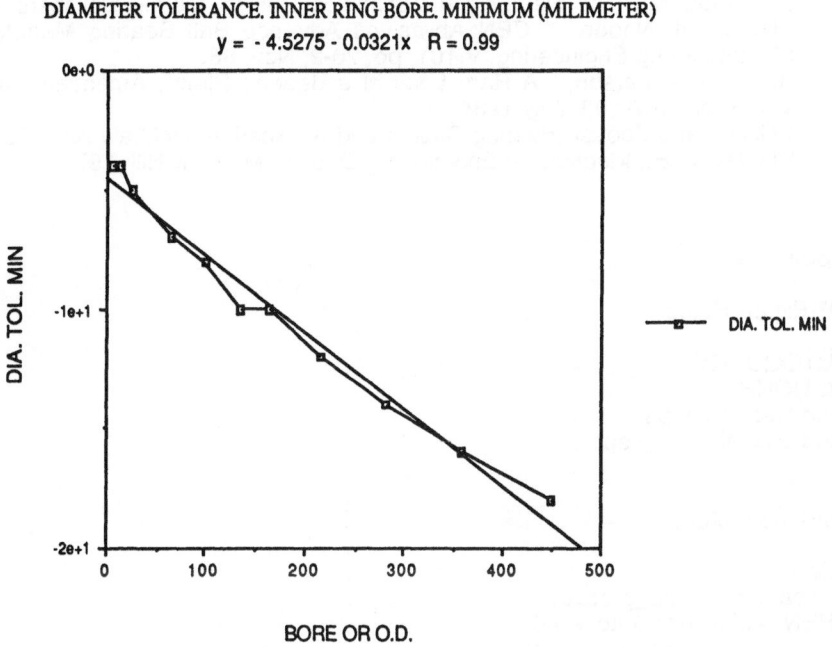

DIAMETER TOLERANCE. INNER RING BORE. MINIMUM (MILIMETER)

$y = -4.5275 - 0.0321x \quad R = 0.99$

Figure 1. Diameter Tolerance on Inner Ring Bore. (Minimum)

	Use Grease	Use Oil
Temperature	Below 200 - 250 F	Above 200 - 250 F
Speed Factor	Below 200k - 300k	Above 200k - 300k
Load	Low to moderate	High
Housing design	Simple	Complex
Long period, no attention	yes	no

236

Conclusion

The design steps and the design guidelines are converted into expert system rules (Appendix A) for ball and roller bearing design. This expert system when run by the user will provide him with the relevant bearing parameters and their values.

References

1. Eschmann, Hasbargen, and Weigend, Ball and Roller Bearings - Theory, Design and Application, Wiley Publishers, 1985.
2. R. L. Mott, Machine Elements in Mechanical Design, Merrill Publishers, 1985.
3. Matsumori, Noboru, " CBN Abrasives Advance Ball Bearing Manufacture", Manufacturing Engineering, V.101, pp. 70-2, Nov. 88.
4. Ashburn, Anderson, " A Rare Look at a Bearing Plant", American Machinist, V.128, pp. 102-103, July 1984.
5. Wilcock and Booser, Bearing Design and Application, McGraw Hill, 1957.
6. Shigley, J. E., Mechanical Engineering Design, McGraw Hill, 1977.

Appendix A

VP-Expert Rules

```
AUTOQUERY
ACTIONS
Find Bearing_type
Find Radial_load_cap
        .
        .
Find load_char;

rule 1
IF bearing = deep_groove
THEN radial_load_cap = 1.0
      thrust_load_cap = 1.0;
        .
        .
        .
rule 7
IF axial_load = (load)
THEN dyn_load_cap = ((l/l10)*(x*fr + Y*fa)*ks);
        .
        .
        .
rule 40
IF bore_dia = (outer_dia)
THEN diametral_tol = (-4.5275 - 0.0321* (outer_dia))
display " This is the minimum tolerance for inner ring bore diameter in tenths of an
inch with correlation coefficient = 0.99";
```

Computer-Aided Optimal Design of Gears

HUNGLIN WANG and HSU-PIN WANG

Department of Industrial Engineering
The University of Iowa
Iowa City, IA

Summary

This research is directed toward the development of a computer-based methodology for optimal design of spur gears. To achieve this, a two stage design framework including preliminary design and design optimization is proposed. In this paper, we focus on the design optimization stage.

Introduction

Gears are used in most types of machinery for power transmission. The design of a gear set is highly complicated involving the satisfaction of many constraints like bending strength, pitting resistance, and scuffing resistance, etc. Many approaches for gear design have been proposed. One approach is the use of optimization techniques [1,2,6,7,8,9,10,11].

Nomenclature

A : addendum constant

C : center distance (in)

E : modulus of elasticity (psi)

HP : horsepower (hp)

F : face width (in)

K_a, C_a : application factors

K_s, C_s : size factors

K_L, C_L : life factors

K_r, C_r : reliability factors

d : pitch diameter (in)

T_b : blank temperature (°F)

T_f : total flash temperature (°F)

D : dedendum constant

P_d : diametral pitch (1/in)

n_p : pinion speed (rpm)

W_t : transmitted load (lb)

ϕ : pressure angle (degree)

K_v, C_v : dynamic factors

K_m, C_m : load distribution factors

K_t, C_t : temperature factors

m_p : contact ratio

S' : relative surface roughness (μin)

ΔT : flash temperature rise (°F)

W_n : normal load (lb)

r_1, r_2 : pitch radius of the pinion and gear (in)

N_1, N_2 : number of teeth on the pinion and gear

δ : maximum deflection of gear teeth (in)

r_{a1}, r_{a2} : addendum circle radius of pinion and gear (in)

r_{b1}, r_{b2} : base circle radius of pinion and gear (in)

ρ_1, ρ_2 : density of pinion and gear (lbm/in³)

S_t : calculated bending stress at the root of tooth (psi)

J, I, G : strength, wear, and scoring geometry factors

S_{at}, S_{ac} : allowable bending and contact stress (psi)

λ : specific film thickness (microin)

h_{min} : minimum oil film thickness (microin)

σ_p, σ_g : average roughness for pinion and gear (μin)

L_1 : life of a pinion tooth in millions of rotation with 90 percent probability of survival

L_{2p} : life of a gear tooth in millions of rotation with 90 percent probability of survival

L_M : life of a meshing gear set in millions of rotation with 90 percent probability of survival

Design Variables and Parameters

In a design process, the selection of design variables and input parameters is important. In this spur gear design procedure, the following variables and input parameters are selected.

Input Parameters	Design Variables
operating horsepower	number of pinion teeth
gear ratio	face width
operating speed of pinion	diametral pitch
elastic modulus	pressure angle
Poisson ratio	
surface strength and sending strength limits for material	

Objective Functions

There are four design variables to be optimized in this model. They are the size and weight of the meshing gear set, the deflection of the teeth, and the life of the gear set. Each objective is discussed in detail as follows.

Minimization of gear size and weight

We usually like to design a spur gear set as small as possible because a smaller gear set would need less material and space to operate in [1,11]. The following two objective functions to this end are formulated.

$$MIN \ F(\pi r_1^2 \rho_1 + \pi r_2^2 \rho_2) \tag{1}$$

$$MIN \ (r_1 + r_2) \tag{2}$$

Minimization of gear tooth deflection

Deflection is generally ignored in a gear design process. Yet, when analysis needs to be done on the failure of a gear, we shall consider the deflection on teeth and it shall be minimized. According to [8], the objective function for minimizing deflection of teeth is shown below:

$$MIN \ \delta = \frac{1.512 \times 10^6 HPP_d}{(h1 - h2)^3 N_1 N_2 FE} ((\frac{h_1}{h_2} - 3)(\frac{h_1}{h_2} - 1) + 2\ln\frac{h_1}{h_2}) \tag{3}$$

where

$$h_1 = \frac{2}{P_d}(0.7854 - \tan\phi) \ , \ h_2 = \frac{2}{P_d}(1.25\tan\phi + 0.7854)$$

Maximization of the gear useful life

The useful life of a meshing spur gear set can be calculated as follows.

$$MAX \ L_M = \{N_1(\frac{1}{L_1})^e + N_2(\frac{1}{L_{2p}})^e \}^{-\frac{1}{e}} \tag{4}$$

where

$$L_1 = KW_n^{-4.3} F^{3.9} \sum \rho^{-5} l_1^{-0.4} \ , \ L_{2p} = (\frac{N_2}{N_1})(\frac{l_2}{l_1})^{-\frac{1}{e}} L_1$$

$\sum \rho$ is the curvature radius and l_1 is the involute profile arc length for the pinion, l_2 is the involute profile arc length for the gear.

Constraints

Constraint on bending strength

According to the AGMA standard, the bending stress of a standard spur gear tooth is given below:

$$S_t = \frac{W_t K_a}{K_v} \frac{P_d}{F} \frac{K_s K_m}{J}.$$

To avoid tooth breakage, the bending stress calculated shall be limited by the maximum allowable bending stress of the material which can be estimated from the following equation.

$$S_t \leq \frac{S_{at} K_L}{K_t K_R}. \tag{5}$$

Constraint on surface stress

The design equation for the contact stress is given as an AGMA standard:

$$S_c = C_p \sqrt{\frac{W_t C_a}{C_v} \frac{C_s}{dF} \frac{C_m C_f}{I}}.$$

The contact stress calculated from the above equation should be kept smaller than the allowable contact stress of the material. The contact stress can be calculated from the following equation.

$$S_c \leq S_{ac} [\frac{C_L C_H}{C_T C_R}]. \tag{6}$$

Constraint on number of teeth and face width

The following two constraints on the number of pinion teeth and face width are used [1,12,8].

$$N_1 \geq \frac{2.4 + 0.004 P_d}{\sin^2 \phi} \tag{7}$$

$$\frac{3\pi}{P_d} \leq F \leq \frac{5\pi}{P_d} \tag{8}$$

Also, N_1 and N_2 shall satisfy the ratio specified.

Constraint on scoring

Film thickness, flash temperature, and scoring criterion number are reported related to scoring failure [4,5,13]. First, the film thickness shall be within a certain limits.

$$1 \leq \lambda = \frac{h_{min}}{\sigma} \leq Upper \ bound \tag{9}$$

where

$$\sigma = \sqrt{\sigma_p^2 + \sigma_g^2}$$

Second, the flash temperature shall be within the allowable range. The flash temperature can be calculated as follows:

$$0.0 \leq T_f = T_b + \Delta T. \leq Allowable \ Temperature \tag{10}$$

where T_b is usually the oil inlet temperature and ΔT can be calculated as the temperature rise due to frictional heating for two bodies in contact.

Finally, the scoring criterion number shall be limited by the critical scoring criterion number. The value for scoring criterion number can be calculated in the following equation.

$$\text{scoring criterion number} = \left(\frac{W_t}{F}\right)^{0.75} \frac{n_p^{0.5}}{P_d^{0.25}} \leq Critical\ Scoring\ Number \qquad (11)$$

Constraint on contact ratio

For spur gear, the contact ratio should not exceed 2.0 and should be greater than 1.4. The contact ratio can be calculated as follows.

$$1.4 \leq m_p = \frac{(\sqrt{r_{a1}^2 - r_{b1}^2} + \sqrt{r_{a2}^2 - r_{b2}^2} - C\sin\phi)P_d}{\pi\cos\phi} \leq 2.0 \qquad (12)$$

Solution Algorithm

From the above mentioned objectives and constraints, we can see that gear design is essentially a multiple-objective, non-linear programming problem. This class of problems are difficult. What we propose here is a solution algorithm called Modified Iterative Weighted Tchcbycheff algorithm developed by Steuer [12]. At each iteration, the user will be provided a set of solutions by the algorithm. According to his/her preference, the user may select a most preferable solution. Using this solution, the program will focus on a smaller feasible solution region and provide the user with a set of feasible solutions. This process goes on until the designer is satisfied with the design.

Example

Using the same example in [1,2,8,11], we present our solution as shown below.

INPUT

Gear Ratio : 5:1	Transmitted power : 20 hp
Operating Speed : 1260 rpm	Pressure angle : 20 (degree)
Surface strength : 200 ksi	Bending strength : 60 ksi
Elastic modulus : 30 ×10⁶ (psi)	Poisson ratio : 0.25
Addendum constant : 1.0	Dedendum constant : 1.25

RESULT

Center Distance : 4.55 (in) Number of pinion teeth : 16

Diametral pitch : 10.44 (1/in) Face width : 1.08 (in)

Reference

1. Agrawal, R., and Kinzel, G. L., "The Optimum Design of Spur and Helical Gears Based on AGMA Strength and Wear Criteria," Computers in Engineering, Vol.2, 1987, pp. 377-385.

2. Carroll, R. K., and Johnson, G. E., "Optimal Design of Compact Spur Gear Sets," Journal of Mechanisms, Transmissions, and Automation in Design, Vol.106, Mar. 1984, pp. 95-101.

3. Deutschman, A. D., Michels, W. J., and Wilson, C. E., Machine Design - Theory and Practice, McGraw Hill, New York, 1975.

4. Drago, R. J., Fundamentals of Gear Design, Butterworth Publishers, MA, 1988.

5. Dudley, D. W., Handbook of Practical Gear Design, McGraw Hill, New York, 1984.

6. Lee, T. W., "Weight Minimization of a Speed Reducer," ASME Paper 77-DET-163, Sept. 1977.

7. Kamenatskaya, M. P., "Computer-Aided Design of Optimal Speed Gearbox Transmission Layouts," Machines and Tooling, Vol.46, No.9, 1975, pp. 11-15.

8. Onwubiko, C., "Spur Gear Design by Minimizing Teeth Deflection," Proceeding 1989 International Power Transmission Gearing Conference, Chicago, 1989, pp. 115-120.

9. Osman, M. O. M., Sankar, S, and Dukkipati, R. V., "Design Synthesis of a Multi-Speed Machine Tool Gear Transmission Using Multiparameter Optimization," ASME Journal of Mechanical Design, Vol.100, No.2, Apr. 1978, pp. 303-310.

10. Rao, S. S., and Eslampour, H. R., "Multistage Multiobjective Optimization of Gearboxes," ASME Journal of Mechanisms, Transmissions, and Automation in Design, Vol.108, 1986, pp. 461-468.

11. Savage, M., Coy, J. J., and Townsend, D. P., "Optimal Tooth Number for Compact Standard Spur Gear Sets," ASME Journal of Mechanical Design, Vol.104, 1982, pp. 749-758.

12. Steuer, R. E., Multiple Criteria Optimization : Theory, Computation, and Application, John Wiley & Sons, 1986.

CAD for Underground Structure

GU HANLIANG

Shanghai Underground Space Architectural Design Institute
Shanghai, China.

1. Introduction

A number of structural analysis and CAD system have been developed for structural design,but engineers only apply general purpose structure analysis program to analyze underground structure,which are inconvenient in using. The differences between underground structure and above ground structure are as follows:

1. Construction processes

To analyze the underground structure the construction process (including excavation) must be properly considered[1,2];in other words,we have to analyze a series of structures under its corresponding load, and accumulate the displacements and internal forces of the all step results.

2. Slab planar deformation and beam axial deformation

For above ground structure, all slab planar deformations and beam axial deformations are very small, and can be neglected [3,4],i.e. Rigid Membrane Assumption;almost all special programs or CAD systems for civil engineering employ the assumption in China;but,for underground structure, the internal forces of the exterior walls and/or retaining walls are sensitive to this deformations[2].

3. The change of the internal forces and displacements, which results from excavating soil or removing struting elements.

The effect of excavating manifests itself as two aspects. Firstly, it results in the unbalance of the soil pressure between the retaining wall's interior and exterior,and it can be considered as load acting on retaining wall;secondly,it causes part of the retaining wall to miss the elastic support, in the saturated clay, the interactive problem between structure and soil is considered as Winkler's elastic foundation model, so that it can be considered as the model change in the finite element analysis. For above ground structure, there is no excavating problem.

Removing struting elements also exists on above ground structure, without any significace. For underground structure,non struting is not equivalent to supporting struting elements, after doing somethig, then removing them [2].

For underground structure CAD system, it is necessary to consider water-proof,which manifest itself as reinforcing to control the crack of the exterior wall, bottom and top slab.

2. System design

The data flow chart of UCAD is shown in Fig.1.By data flow cutting, the system is divided into squential five programs,which are linked with a common database,that is a hierarchical file system and is made up with multilinks. The data access of the database is independent of the data physical structure, the databas e management routines supply logical interfaces for storing and/or loading.

244

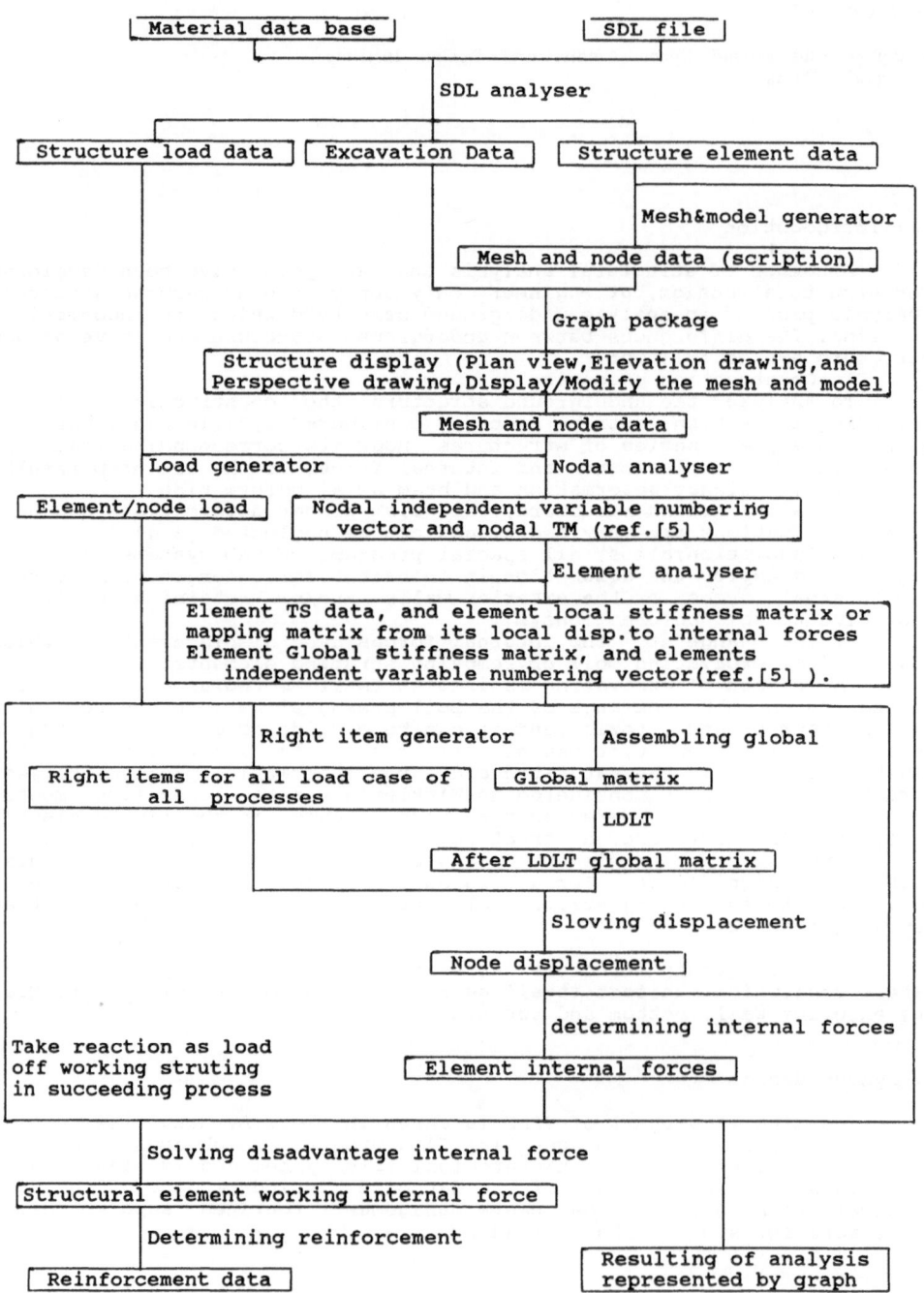

Fig.1 Data flow Chart

The first part of UCAD (named as UCAD1) deal with the SDL file.
In UCAD, the structure is defined as summarization of all on working structural elements, so far the structural design is sum of all structural elements design, that is determining the position or region, the section shape and its size, the processes of the element on working. After structural design, the load applied on the structure except for self_weight, should be declared. The grammar of the SDL is LL(1)[6], and the analyser and the interpreter may be constructed in a simple way.

The second part of UCAD(UCAD2) performs the pre-processing for FEM, which includes three parts.
1. Mesh & Model generation for FEM.
2. Graph display and modify the mesh and/or the model.
3. Load generation.

The paper[7] is a summary of a number of different mesh generation methods. But, these methods are not suitable for UCAD, There are two problems to be solved.
1. There are some nodes in the region or on the boundary of the structural element.(For example,in Fig.2)
2. The mesh can not be too dense.

The strategy adopted in UCAD is two passes dealing with and layer by layer generating the mesh.
The pass one is generating all conner node and end node.
The pass two is performed in three stages:
1. Finding out all nodes in the region or on the boundary.
2. Refining the boundary. If the structural element is of one dimension the mesh is generated.
3. Generating the mesh in the layer by layer way.
Now,the problems become how to extract a layer from the region and how to discretize the layer by finite element. The element standard size S is selected one-fiftieth of the largest size of the structure. A layer is of two borders (called as upper and lower respectively),the distance between upper and lower vary from 0.8*S to 1.2*S(in Fig.2).

In UCAD, Nodal displacement Quality [5] (NQ),is introduced:

$$NQ=MST*262144+NCO*4096+\sum_{i=1}^{5} ID_i \, 4^{6-i}$$

Where MST: The master node No. of the node.
       NCO: No. of the relative coordination system under which the NQ is defined.
       ID_i: The specific of the displacements of the i'th component of the node,and its meaning is declared in Tab. 1

Tab. 1

ID=0	ID=1	ID=2	ID=3
Kinetic	Independent	Dependent	Fixed

The modelling of the structural analysis consist of two aspects: that are selecting the element type (such as beam), and determining the all NQs. The formmer is done before mesh generation, and the relation of element type and structural element type is shown in Tab.2. During mesh generating, the closure of a node effective displacement subspace is calculated,derived from element's one concerned with the node.After mesh generation,the rigid membrane model is applied to all above ground floor slabs, and the master_servant relation is applied to all articulated joints.

Tab.2

Structural element type	Element type
Anchor and Pile	Spring
Column and beam above the bottom slab	Beam
Struting	Bar
Interior wall&above ground exterior wall	Membrane
Underground exterior wall,slab and staris	Bending plate
Beam under the bottom slab, and retaining wall except for diaphragm	Beam on Winkler's elastic foundation
Bottom slab and diaphragm	Bending plate on Winkler's elastic foundation

The third part of UCAD(UCAD3) performs nodal and element analysis, and generates all right items for the all loadcases of all processes.

The fourth part of UCAD(UCAD4) performs finite element analysis.
Process is defined as a stage on constructing or using. The whole construction is discreted to a number of processes (NPRC processes). The peace time using (process No.=NPRC+1) is necessary, and war time using (process No.=NPRC+2)is optional.The structures are different corresponding to different process.
UCAD4 may be described as follows
where MPRC=NPRC+1,or MPRC=NPRC+2
FOR IPRC=1 TO MPRC DO
   Calculate the closure of all nodal variable which are concerned with on working structural element.
   Solve the nodal variables (nodal displacements).
   Determine the internal forces of the on working elements.
   If IPRC <= NPRC then
    If IPRC>1 then accumulate the displacements and internal forces under dead loadcase with preceding process's one.
    Calculate the structure internal forces and disadvantage internal forcees, compare with past one, and save the most.
    Search all on working structual elements, if one is off working in succeeding process,take the reaction of the internal forces of the structural element under the dead loadcase as load for succeeding process.
   ELSEIF IPRC=NPRC+1 then
    Calculate the structure internal forces and disadvantage internal forces under all load combination.
   ELSEIF IPRC=NPRC+2 then
    Calculate the structure internal forces.
   ENDIF
ENDDO

The fifth part of UCAD (UCAD5) performs the post-processing, which consists of graph representation of displacements and/or internal forces, graph editting utility,and reinforcement determination.

As an CAD system, the UCAD is well considered for developement. A material database is developed,which supply with a logical interface for UCAD system to load the physical & mechanical property,prices,cost ratio and so on, by means of the building material name and its type model. To draw working drawing, the UCAD supply a logical interface, which allows other programs to get the detail reinforcement information of a element, which includes shape of structural element, its section shape and size, as well as main reinforcements and hoops on all concerned position,which is required by the code.

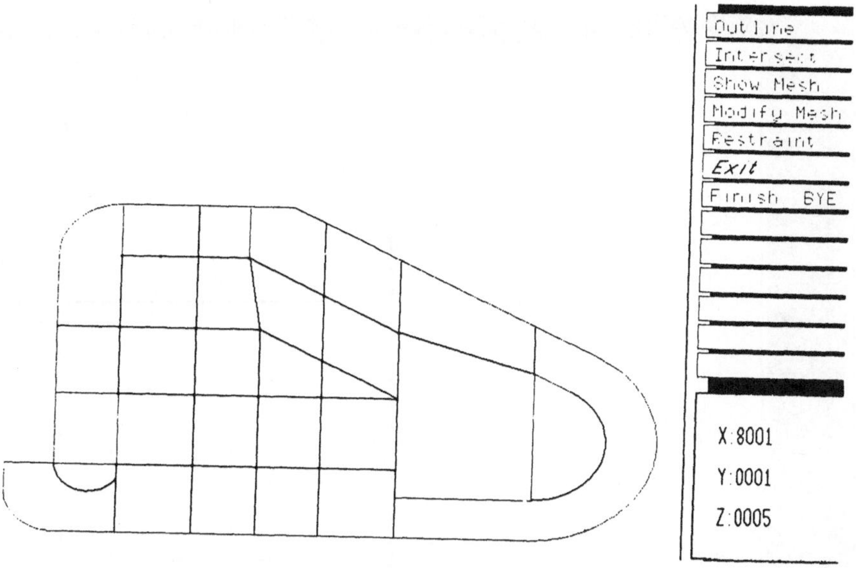

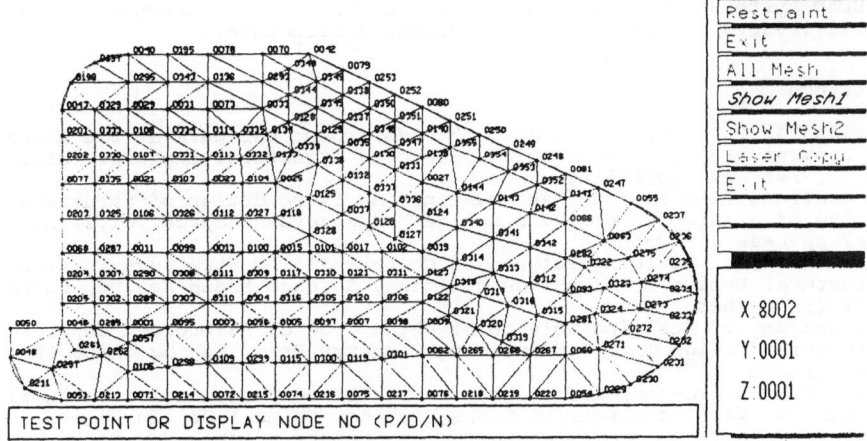

The menu items shown beside figure a):
Outline
Intersect
Show Mesh
Modify Mesh
Restraint
Exit
Finish BYE

X:8001
Y:0001
Z:0005

a)   The bottom planview

The menu items shown beside figure b):
Outline
Intersect
Show Mesh
Modify Mesh
Restraint
Exit
All Mesh
Show Mesh1
Show Mesh2
Laser Copy
Exit

X:8002
Y:0001
Z:0001

TEST POINT OR DISPLAY NODE NO (P/D/N)

b)   The bottom mesh
No. of node < 82 generated in the pass one
No. of node <276 generated in the pass two bonudary refining
The upper border of the first layer is consist of
following nodes (before refined)
48,262,105,109,115,119,62,265,266,267,60,271,231

Fig 2. Structural element and its mesh

248

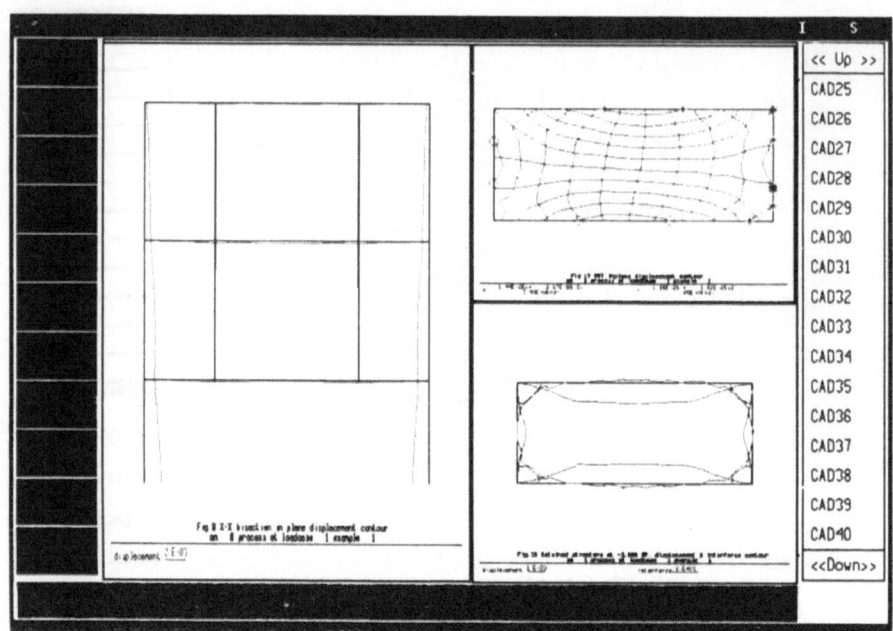

Fig.3 UCAD results of a structure

3. Remark

UCAD system solves the problem of the underground structure satisfactorily. Fig.3 is results of an undergound structure.

4. Reference

[1] Burland,J. B. ; Hancock,R. J. R. " Underground Car park at the house
    of Commons, London: Geotechnical Aspects" Structural Engineer Vol.55
    No. 2, Feb. 1977 pp81-100
[2] Gu Hanliang;Ling Xiaojun;Wan Quanlin " An Introudction of structural
    Design of a large Underground Garage " Underground Space Vol.27,No.2
    1987 pp30-36 (in Chinese)
[3] The Academy of Building Science of China,Building Structure Institute
    Structural Design of High-rise Building, Science Press Bejing, China
    1985 (in Chinese)
[4] The Academy of Building Sciencc of China Code of Reinforced Concrete
    Structural Design TJ10-74, Building Industry Press,Bejing,China 1974
    (in Chinese)
[5] Zhong,W.X. Micro-computer Program design of computational structural
    Mechanics Hydrolic electrics Press, Bejing, China 1986 (in Chinese)
[6] Pyster, Arthur, B. Compiler Design and Construction New York, Van
    Nostrand Reinhol, 1980
[7] Yuen,M.M.F;Tan,S.T.;Sze,W.S. "Current status and Trends in Automatic
    Mesh Generation for Geometric Modelling system" Hong Kong Engineer
    Vol.15, No.11, Nov 1987, pp27-33

A Microcomputer Aided Design of Technical Systems

W. PRZYBYLO

FEMA ENGINEERING Ltd.
Cracow, Poland

The main ideas of the microcomputer aided design of technical systems have been presented. A mathematical model of the problem has been formulated. The author's own approach to the problem - including the concept of an increasing sequence of macromodels, block frontal algorithms in the hypermatrix algebra and a database approach to the CAD/CAM systems - has been pointed out. The challenge of computational mechanics as well as this of expert systems have been called. The reliability of engineering design using computers and a validation of Finite Element systems have been cited. Some remarks on international associations for CAD/CAM have been included.

1. INTRODUCTION

In the paper the main ideas concerning the microcomputer aided design of technical systems have been briefly presented. A basic mathematical model of the problem has been used in order to describe the author's own approach to the problem - including the concept of an increasing sequence of macromodels, a block frontal algorithms in the hypermatrix algebra and a database approach to the CAD/CAM systems. The challenge of computational mechanics as well as this of expert systems have been called. The reliability of engineering design using computer software and a validation of Finite Element systems have been cited and discussed. Some remarks on international associations for computational mechanics and CAD/CAM activities have been included.

2. A MATHEMATICAL MODEL OF A MECHANICAL SYSTEM

2.1. Equations of motion of a mechanical system - analysis

In order to formulate the equations of a mechanical model (the U system) of an engineering system we start from the Hamilton principle

$$\int_{t_o}^{t_1} (\delta(E_s - E_i) - \delta L)\, dt = 0 \quad . \tag{1}$$

To formulate the equations of the U system we shall apply the Lagrangian approach [1 , 2]. Let us derive from the Hamilton principle (Eq. (1)) two systems of differential equations.

The first one forms the differential boundary eigen problem

$$A(z) = N * z \tag{2}$$

for the differential operator

$$A(z) = \{ D^O(z^e): e=1,2,\ldots,E \} \tag{3}$$

the domain of which is defined by the relation

$$Z = \{z : C_k(z) = 0 , k=1,2,\ldots,m \} . \tag{4}$$

The symbol N denotes the block-diagonal matrix

$$N = \text{diag} [N^e] , \{e=1,2,\ldots,E \} , \tag{5}$$

where e is a finite element number.

The equations $C_k(z) = 0$ are the homogeneous boundary conditions, enumerated by the subscript k .
$D^O(z^e)$ is the matrix selfadjoint differential operator acting on the e-th finite element displacement field.

Having solved the eigen problem (2) we obtain the Schauder basis

$$B = \{b_i: A(b_i) = N_i * b_i , i \in [1,\infty) \} . \tag{6}$$

Let us now derive from the Hamilton principle the second system of the differential equations - the second order Lagrange equation system

$$d/dt (\text{grad}_{\dot{q}} (E_i)) - \text{grad}_q (E_i) + \text{grad}_{\dot{q}} (E_d) +$$
$$+ \text{grad}_q (E_s) = \text{grad}_q (L) . \tag{7}$$

Having assumed the small vibrations of the U system around the state of the static equilibrium and applying the discrete form of the solution by generalized Fourier series, finally we obtain the matrix, second order linear differential equation:

$$B(t) * \ddot{q}(t) + C(t) * \dot{q}(t) + K(t) * q(t) = F(t) . \tag{8}$$

To solve this equation we can apply the notion and the algorithms of the increasing sequence of macromodels, developped by the author in [1 , 2 , 3] , the author's concept of a block frontal approach to the hypermatrix algebra [4 , 5 , 6] as well as the computer methodology and technology developped by the author in [7 , 8] .

2.2. The notions of an U system states

Let us introduce the main notions necessary to formulate the design process of an optimum mechanical system.

To describe the properties of the mechanical system during its production, distribution, exploitation, conservation (maintenance) and liquidation one can define the following fields:

- the load vectorial field,
- the displacement vectorial field,
- the velocity vectorial field,
- the acceleration vectorial field,
- the strain tensorial field,
- the stress tensorial field.

The   s t a t e $s(r;t)$ of a mechanical system at the given moment of time is assumed to be formed of the family of (scalar, vectorial or tensorial) fields defined on the configuration $c_V(t)$ of the U system at this moment of time.

The   s p a c e   o f   s t a t e s   of a mechanical system $S(V;t)$ $s(r;t)$ at the given moment of time is assumed to be the set of values attained by the elements of the family of the state fields.

The   l i m i t   s t a t e $s_1(r;t)$ of a mechanical system at the given   moment of time is assumed to be composed of the limit values of the elements of the family of the state fields.

The limit state which is commonly used consists of the cross product of two limit states - the limit state of stresses and the limit state of displacements (deflections).

The   a d m i s s i b l e   s t a t e $s_a(r;t)$ of a mechanical system at the given moment of time is assumed to be composed of the admissible values of the elements of the family of the state fields.

The   h i s t o r y   of a mechanical system is assumed to be the family of states

$$H(U) = \{ \, s(r;t), \, t \in T \, \} \tag{9}$$

enumerated by the consecutive moments of time.

The   a d m i s s i b l e   h i s t o r y   of a mechanical system is assumed to be the family of admissible states of a system

$$H_a(U) = \{ \, s_a(r;t), \, t \in T \, \} \tag{10}$$

enumerated by the consecutive moments of time.

The   a d m i s s i b l e   s y s t e m   is assumed to be subjected to the admissible history of states.

2.3. The optimization of an U system

Let us formulate the main notions necessary to formulate the problem of an optimization of a dynamical mechanical system.

The   m e c h a n i c a l   s y s t e m s   c l a s s   is a set of mechanical systems defined in the Section 2.1. The systems are defined on the same displacement space and differ from each other only by their mechanical properties measure densities.

The   q u a l i t y   f u n c t i o n a l   of the mechanical systems class is assumed to be an arbitrary functional $J : \mathcal{M} \longrightarrow R$ defined on the mechanical properties measure density space $\mathcal{M}$ of the mechanical systems class.

The optimum mechanical system is assumed to be the admissible system, belonging to the given mechanical systems class, having the $M \in \mathcal{M}_a \subset \mathcal{M}$ properties measure density and giving the extremum of the J functional:

$$U_{opt}(M): M_{opt} \in \mathcal{M}_a \subset \mathcal{M} \ , \ J(M_{opt}) = \text{extr} \ (J(M)) \ , \ M \in \mathcal{M}_a \subset \mathcal{M} \ . \ (11)$$

2.4. The design of an optimum U system

The process of the design of an optimum mechanical system is assumed to be composed of operations of a search for an optimum mechanical system either by the optimum synthesis process or by the iterative sequence of operations: analysis - synthesis - evaluation.

2.5. The block frontal concept in the hypermatrix algebra

The main scope of the block frontal approach to the hypermatrix operations in the computation of dynamical mechanical systems W.PRZYBYŁO et al [4 , 5 , 6] consists of providing an efficient way to compute as large mechanical systems as possible (i.e. maximize - N - the number of D.O.F.) at the available central and backing store memories (CM and BSM) of a computer system. This problem still remains important, especially in the conditions of the extensive use of commonly available and inexpensive microcomputers by the large number of engineers.

W.PRZYBYŁO and A.KLEINER [6] have proposed an efficient way of realization of block frontal procedures in the hypermatrix algebra. They have reduced the amount of information to be stored at the BSM to the necessary minimum (by 60 %).

Let us ilustrate the topic by one example of a block frontal approach - the congruent transformations of inertia, damping and stiffness hypermatrices appearing in the Eq. (8), i.e. the projections of these matrices from N-dimensional vectorial space onto the n-dimensional vectorial eigen subspace (n << N) . In the following formula we denote by FRONT (.) the hypermatrix which is computed frontally in the CM only without being stored in the BSM .

$$X^r(t) = W^T * \text{FRONT} (X(t)) * W , (X = B , C , K) . (12)$$

Using these procedures only the small matrices $B^r(t)$, $C^r(t)$ and $K^r(t)$ (n * n) are stored. The large hypermatrices B(t) , C(t) and K(t) are never stored in the BSM.

2.6. The concept of an increasing sequence of macromodels

The efficient algorithms of solution of nonlinear dynamical problems still remain the challenge of computational mechanics [11].
Among the number of ideas the concept of the application of an increasing sequence of macromodels to solve:

- the analysis of nonlinear dynamical mechanical systems,
- the optimization of nonlinear systems,
- the optimum control of mechanical dynamical systems,
has been proposed in the author's own papers [1 , 2 , 3]. At present
these ideas are being carried on using the author's own algorithms of
a block-frontal approach to the hypermatrix algebra (See Section 2.5)
as well as the author's own database approach (See Section 3).

3. THE DATABASE APPROACH

In the papers [1 , 7 , 8] the computer independent software,
forming the author's own systems' generator, named FEMA, has been
described. The FEMA systems' generator has been designed and
programmed in the hierarchical form, conforming to the main postulates
of computer methodology: the independence of both the FEMA database
and the database management system from computer systems; large
productivity (minimum of the global time of processing, maximum
dimension of data sets); efficient semantic diagnostics; rational
structures of data; rational processes of formulation, processing,
storage and transfer of data; rational structure of the systems'
generator for exploitation and maintenance; interactive and real time
data processing. The main properties and advantages of the
exploitation process of the FEMA systems' generator have been
described in the papers [1, 7, 8]. Each step of the computational
process, made in the interactive way, consists of opening access to
the program-independent FEMA database, then processing the data, and
finally by closing access to the database.

The FEMA database is independent of programs. This important
property allows to apply the parallel processing for parallel
operations, described in the graph of data processing, and involves
making new data sets, modifying existing data sets, or deleting
unnecessary data sets.

The approach to the computational process is fully modular. The
data structures made and processed in the virtual paged memory are
modular. The FEMA database management system for handling the FEMA
database is fully modular. The operations on very large data
structures, like hypermatrices or regular arrays, are modular as a
consequence of applying modular algorithms and modular data
structures. These properties of the author's own approach assure the
immediate adaptation of the algorithms and programs to be exploited on
new generations of computers (e.g. matrix processor computers which
can also be equipped with the hardware property enabling parallel
processing and transfer of data between memories).

The FEMA systems' generator provides the useful and efficient tool
for the integration of any existing software for engineering
applications (See W.PRZYBYŁO et al [9] , [10]).

4. THE CHALLENGE OF COMPUTATIONAL MECHANICS

To formulate briefly the main fields of research we shall cite
Prof. O.C.ZIENKIEWICZ [11] :

"While, in principle, with sufficient computer power available, the numerical methodologies available today should be capable of answering many points, continued research is necessary to
(a) increase the efficiency of the approximation and solution procedures to deal with constantly expanding demands,
(b) to ensure that the extent of approximation is understood by the user - and if required, to improve this approximation economically to a specified standard,
(c) to derive numerical procedures for problems in which currently available methods fail or are inefficient".

5. THE CHALLENGE FROM EXPERT SYSTEMS

At present many international conferences have been focused on the emerging applications of expert systems in engineering. Expert systems can be considered to be the first applications of a new technology called the artificial intelligence [12] .

Expert systems are sophisticated computer software equipped with their own database and database management system that accumulate and manipulate knowledge to solve problems in a specific problem area. Like the real expert, these systems use the symbolic logic and heuristic rules-of-thumb to find solutions. They have the ability to learn from their own errors and from the previously solved problems.

The heart of an expert system is a powerful database containing 'corpus of knowledge', the accumulation of which requires a joint effort of people from various fields.

In general, there are three basic types of expert systems, namely diagnosis, cybernetic and design.

The diagnosis expert system starts from a set of facts through observation and is driven by the inference machine of the system. The response of the system is made by the sequence of processes of analysis and evaluation, using the knowledge base.

The cybernetic expert system is devoted to provide an intelligent decision for control parameters. The optimization of the process can be achieved by choosing appropriate strategies provided by the knowledge base, according to given constraints.

The design expert system provides an optimum solution for design projects of a given class. The design parameters are optimised by expert system using the knowledge base.

The rapid development of CAD and CAM requires the interaction of design diagnosis and control. The commonly used structural engineering programs and systems are being transformed into design expert systems by including them into powerful database management systems and filling the databases by the appropriate knowledge.

6. THE RELIABILITY OF ENGINEERING DESIGN USING COMPUTERS

6.1. The validation of Finite Element systems

All ideas have a natural rhytm to their development and application [13] . Initially they are being developped in a research environment and applied by people having deep understanding of the underlying principles and aware of the limitations.

At present the common use of commercially supported software packages by engineers have dramatically changed the scene. A number of person with no professional experience have been involved both in the production of engineering software and in the extensive use of the software in various CAD/CAM applications.

In such circumstances two danger have arisen. The first one consists of the production of unreliable software (e.g. the FEM packages working in single precision mode, or using the inappropriate finite elements). The second one consists of the use by the engineer the unreliable software or the inappropriate use of the commercially distributed software (e.g. the application of a given finite element out of its legitimate range). A CAD/CAM systems user has no time to carefully check any of programs he is going to exploit.

Any competent Finite Element analyst uses a number of "tricks" to help in modelling a given structure. The great danger is that even a competent analyst can fail to recognise the importance of features which are unfamiliar.

We can summarize these considerations by formulating two conclusions:
- the necessity of verification the commercially available software by an appropriate group of experts as well as by a large number of experienced users,
- the necessity of development of expert systems to provide to the user the "intelligent" and reliable tool for his professional training and activity.

6.2. Choosing a Finite Element Package

R.D.THOMSON has recently published an excellent paper [14] on this issue. The author fully shares his opinion so that the main ideas presented by R.D.THOMSON will be cited below.

Over the past two decades finite element (FE) analysis has evolved from a technique for large structural analyses to a tool available and familiar to most engineers. However, many engineers who might benefit from FE systems have been put off by bad initial experiences, often associated with an inappropriate choice of software. In this section we shall summarize the most important criteria which can help in making the decision on the choice of FE system grupped in subsets forming different view points [14].

1. For the financial expert:
   - How much is the package?
   - What does the licence look like?
   - How much does it cost to do a typical analysis?

2. For the systems people:
   - What machines will the package run on?
   - What minimum hardware configuration will be needed?
   - Are any peripherals needed?

- What additional software will be needed?
- What about the integration with other packages?

3. For the enthusiastic analyst:
   - What does the finite element library contain?
   - What types of problem can be solved?
   - What about large jobs?
   - What material can be specified?
   - What about pre- and post- processors?

4. For the numerical analyst:
   - The methods of solving the matrix linear equations.
   - The frontal solvers.
   - The bandwidth optimisers.
   - The database approach including restart utilities.

5. For everybody:
   - The quality assurance.

All complex software contains bugs and since the source code for a package is seldom provided, an error reporting system is essential. A package used by a number of reputable users is likely to be relatively error-free but it is worth contacting several experienced users.

7. INTERNATIONAL ASSOCIATIONS FOR COMPUTATIONAL MECHANICS

Actually there is a growing international recognition of the need for the creation of accreditation and validation procedures for commercially available CAD/CAM systems, especially for Finite Element Method systems [13] . In United Kingdom the British Government has established the National Agency for Finite Element Standards and Methods (NAFEMS). Actually several hundred home and international companies have joined the NAFEMS.

There is another important organization - the International Association for Computational Mechanics (IACM), which is devoted to stimulate the development of the computational mechanics and computer aided engineering. The Association is growing and regional Chapters are being formed. In December 1987 the United States Association for Computational Mechanics was formed. At the same time the preparatory meeting on the establishment of Asia-Pacific Chapter of the IACM took place during the CABRIDGE Conference in Bangkok.

8. FINAL REMARKS

The ideas shown in this paper are only a very brief exposition of the topic. A more detailed study will be presented at the Conference and ilustrated by a number of results of engineering applications.

REFERENCES

1. W.PRZYBYŁO: Computer methods in mechanics of large constructional systems. Technological University of Cracow, Monograph No 6, Series Basic Technical Sciences No 19, Krakow 1983, pp.1-230 (in Polish).
2. W.PRZYBYŁO: An application of the concept of an increasing series of macromodels (ISM) to the mechanics of large discrete non-linear systems. Proceedings of the International Conference on Numerical Methods in Engineering: Theory and Applications /Swansea/ 7-11 January 1985, Edited by J.MIDDLETON and G.N.PANDE, A.A.BALKEMA /Rotterdam/ Boston 1985, volume 1, pp. 253-262.
3. W.PRZYBYŁO: An Application of the Concept of an Increasing Sequence of Macromodels (ISM) to the Optimization of Large Discrete Non-Linear Systems. Proceedings of the 8th International Conference on Structural Mechanics in Reactor Technology, Brussels, Belgium, 19-23 August 1985, volume B, pp. 197-202.
4. R.WEINAR, W.PRZYBYŁO: The BANACHIEWICZ-CHOLESKI algorithm for two level hypermatrices. V Conference on Computer Methods in Structural Mechanics of Polish Academy of Sciences, Karpacz 1981, Technological University of Wrocław, Report No I-14/28/K9 Wrocław 1981, pp. 253-260.
5. W.PRZYBYŁO, R.WEINAR: Hypermatrix Block Frontal Solver for BANACHIEWICZ-CHOLESKI algorithm. Computational Mechanics'86.Theory and Applications. Proceedings of International Conference on Computational Mechanics, May 25-29, 1986, Tokyo. Editors G.YAGAWA, S.N. ATLURI. Springer-Verlag, Tokyo 1986, Volume 2, pp. 43-48.
6. W.PRZYBYŁO, A.KLEINER, A Block Frontal Approach to the Hypermatrix Algebra in an Animation of Motion of 3D Dynamical Mechanical Systems. EASEC 2. Structural Engineering and Construction. Achievements, Trends and Challenges. Proceedings of the Second East Asia-Pacific Conference on Structural Engineering and Construction, Chiang Mai, Thailand, 11-13 January 1989, Volume 2, pp. 1055-1061.
7. W.PRZYBYŁO: Relational data bases in computer aided design of technical systems. Proceedings of the International Conference on Numerical Methods in Engineering: Theory and Applications /Swansea/ 7-11 January 1985, Edited by J.MIDDLETON and G.N.PANDE, A.A.BALKEMA /Rotterdam/ Boston 1985,volume 2, pp.1011-1020.
8. W.PRZYBYŁO: Relational Databases, Parallel Processing and Minicomputer Array Processor System Algorithms in Computer Aided Engineering. Computational Mechanics'86. Theory and Applications. Proceedings of International Conference on Computational Mechanics, May 25-29, 1986, Tokyo. Editors G.YAGAWA, S.N. ATLURI. Springer-Verlag, Tokyo 1986, Volume 2, pp. 109-114.
9. W.PRZYBYŁO, T.MOKRZYCKI, A Microcomputer Integration of AUTOCAD, FEM System and 3D Graphics Using FEMA Database. EASEC-2. Structural Engineering and Construction. Achievements, Trends and Challenges. Proc. of the Second East Asia-Pacific Conference on Structural Engineering and Construction, Chiang Mai, Thailand, 11-13 January 1989, Volume 1, pp. 291-296.

10. A.JANICKI, W.PRZYBYŁO, A Concept of a Fuzzy Expert System for Structural Engineering. EASEC-2. Structural Engineering and Construction. Achievements, Trends and Challenges. Proceedings of the Second East Asia-Pacific Conference on Structural Engineering and Construction, Chiang Mai, Thailand, 11-13 January 1989, Vol.1, pp. 178-185.

11. O.C.ZIENKIEWICZ: The Challenge of Computational Mechanics. A transcript of the address given after receiving the Gauss Medal. Bulletin of the International Association for Computational Mechanics, Vol. 3 No 1, 1988, pp.5-6.

12. LIN SHAOPEI: Challenges from Expert Systems, Bulletin of the International Association for Computational Mechanics, Vol. 3 No 1, 1988, pp. 3-4.

13. A.J.MORRIS: The Process of Finite Element Validation, BENCH Mark, NAFEMS, July 1988, pp.12-14.

14. R.D.THOMSON: Choosing a Finite Element Package. BENCH Mark, NAFEMS, July 1988, pp. 20-23.

Solid Modeling With Tension

DA-PAN CHEN

Institute of Mechanical Engineering
National Chiao Tung University
Hsinchu, Taiwan, R.O.C.

Summary

Solid modeling is a process through which one creates and maintains a model of solid objects for analytic purposes. A solid model should be unambiguous and complete with respect to informations such as volumetric properties and topological properties. In a separate paper [1], the author had described how bias and tension parameters can be used to build complex surface models with geometric discontinuities. This method renders a complex surface model as one single piece. It not only simplifies the modeling process but is more complete and natural in the aspect of model topological connectivities. In this paper, further solid attributes such as the surface area integral, and the volume integral are investigated. And, the extent to which simple complex solid primitives can be used to model more complicated solids are explored.

1. Curves with Tension Control

One of the commonly used method for solid modeling is the boundary-representation (B-rep) method. In the method one builds up the boundary surface model of a solid object with a net work of spline curves. In order to see the role the tension parameter plays on the shape controlling of surface models, we start with a brief discussion of cubic interpolating curves with tension control. The conventional cubic interpolating curve has the basic form of

$$P_i(u) = (1-3u^2+2u^3)P_{i-1} + (3u^2-2u^3)P_i$$
$$+ (u-2u^2+u^3)P_{i-1}^u + (-u^2+u^3)P_i^u \qquad (1)$$

where $P_i(u)$ denotes the vector function of the ith segment of a composite curve, P_i the position vector, and P_i^u the first parametric derivative vector of the curve at the ith data point. To introduce the shape-controlling tension parameter ν, one uses the following continuity conditions [2,3] at each joint of the curve

$$P_i^u(1) = P_{i+1}^u(0)$$
$$P_i^{uu}(1) + \nu_i P_i^u(1) = P_{i+1}^{uu}(0) \tag{2}$$

In the above equations, ν_i denotes the tension value at the ith data point, $P_i^u(u)$ and $P_i^{uu}(u)$ denote the first and the second parametric derivatives of the vector function $P_i(u)$, with the parameter u normalized to range from zero to unity for each curve segment. Equations (1), (2) and free-end conditions will lead to the following set of equations for the calculation of all nodal first parametric derivative vectors

$$
\begin{bmatrix}
2 & 1 & & & & \\
1 & 4+\nu_1/2 & 1 & & & \\
& \cdots\cdots & & & & \\
& 1 & 4+\nu_i/2 & 1 & & \\
& & \cdots\cdots & & & \\
& & 1 & 4+\nu_{n-1}/2 & 1 \\
& & & 1 & 2
\end{bmatrix}
\begin{bmatrix}
P_0^u \\
P_1^u \\
\vdots \\
P_i^u \\
\vdots \\
P_{n-1}^u \\
P_n^u
\end{bmatrix}
=
\begin{bmatrix}
3(P_1-P_0) \\
3(P_2-P_0) \\
\cdots \\
3(P_{i+1}-P_{i-1}) \\
\cdots \\
3(P_n-P_{n-2}) \\
3(P_n-P_{n-1})
\end{bmatrix}
\tag{3}
$$

Other end conditions such as fixed-end and cyclic-end can be used and similar equation set will result for the calculation of the respective nodal first parametric derivatives. For instance, a four-point circle, with cyclic ends, zero tensions and equally distanced data points, is shown in Figure 1. One can see in the same figure that nodal tension values can flaten the circle into a square. The nodal tension values applied to the circle starting from the outside one are 0, 10, 100, 1000, respectively.

2. Surfaces with Tension Control

The formulation of bicubic surface patches with tension control begins with the cubic interpolating curve. With proper matrix notation, equation (1) can be rewritten as

$$
P_i(u) = [1 \ u \ u^2 \ u^3]
\begin{bmatrix}
1 & 0 & 0 & 0 \\
0 & 0 & 1 & 0 \\
-3 & 3 & -2 & -1 \\
2 & -2 & 1 & 1
\end{bmatrix}
\begin{bmatrix}
P_{i-1} \\
P_i \\
P_{i-1}^u \\
P_i^u
\end{bmatrix}
$$

$$
= UCB_i \tag{4}
$$

A bicubic surface patch is bounded by four cubic interpolating curves and its tensor-product form is

$$
P_{ij}(u,v) = UCB_{ij}^* C^T V \tag{5}
$$

where V is a single column matrix containing the four basis terms of v as those in U, and

$$
B_{ij}^* =
\begin{bmatrix}
P_{i-1,j-1} & P_{i-1,j} & P_{i-1,j-1}^v & P_{i-1,j}^v \\
P_{i,j-1} & P_{ij} & P_{i,j-1}^v & P_{ij}^v \\
P_{i-1,j-1}^u & P_{i-1,j}^u & P_{i-1,j-1}^{uv} & P_{i-1,j}^{uv} \\
P_{i,j-1}^u & P_{ij}^u & P_{i,j-1}^{uv} & P_{ij}^{uv}
\end{bmatrix} \tag{6}
$$

with similar notations to those of a cubic curve.

For a large class of mechanical parts characterized by their axial symmetry, such as those appeared in this article, the cross-derivative terms P_{ij}^{uv} in their boundary conditions matrix are zero. This reduces the surface patches to the so-called ferguson's patches and the surface model of this class of mechanical parts can be constructed by treating the patch boundary curves as a single interpolating curve with tension control discussed in section 1. Examples of surfaces with tension control are shown in Figures 2,4, and 5.

3. Modeling Accuracy and Volumetric Properties

The principal task addressed in this article is the shape fitting accuracy and the volumetric properties of the geometric models generated with the cubic interpolating curves and the bicubic surface patches with tension control.

As shown in Figure 1, a circle of a radius of 10 is fitted with four cubic curve segments with cyclic end conditions. A Gauss quadrature calculation of the arc length of the generated circle yields a results of 61.955 for the true arc length of the circle 62.832.   This discrepancy in circle fitting can be eliminated by increasing the number of the fitting curve segments.

On the other hand, for the fitting of the inscribed square to the circle, tensions are applied to the knots of the circle. As can be seen in the figure, the arcs in the circle are flatened and approach the square as the limit when tension values increase.   The Gauss quadrature yields a result of 56.569 for the true length of the boundary lines of the square with a tension value of 1000 applied at the knots.

The above fitting accuracy analysis provides a guidance for selecting appropriate tension values for complex surface fitting shown in figure 2 is an annular structure with a cross-section as shown in Figure 3.   The Gauss quadrature results of the surface area and the volume of the structure are 8167.89 and 17299.2, respectively, with error less than 0.1 percent.   One difficulty encountered in the volume integration is the sign convention for the contribution associated with each surface patch.   This difficulty can be resolved with the sign convention as depicted in Figure 3.

4. Modeling with Complex Primitives

With the techniques developed in the previous section, one can define a set of complex solid primitives for the construction of more complicated solid models. An example is given in Figure 6.

For the construction of the solid model as shown in Figure 6, we build the two complex primitives as shown in Figure 4 and 5. These primitives can be defined as a single solid using the bicubic surface patches with tension control. With the volumetric properties calculated as discussed in section 3, the solid model of the structure in Figure 6 can be constructed with appropriate combination of the primitives. Solid models constructed with the method described in this article have the advantages of being more concise and precise when compared to those with the conventional methods. Although the extent of shapes which can be rendered as a single unit by surface with tension control is not unbounded, this method certainly offers solid model practitioners a much more attractive alternative.

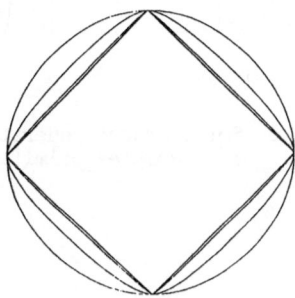

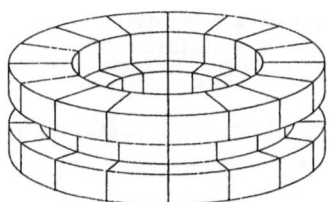

surface area = 8167.892
volume = 17299.23

Fig. 1 Shape fitting with
cubic curves

Fig. 2 A solid model with
tension control

264

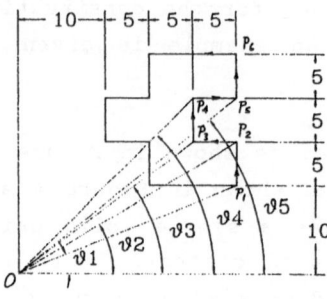

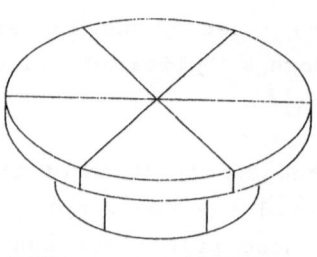

surface area = 1352.398
volume = 2104.079

Fig. 3 Sign convention
for integration

Fig. 4 Complex primitive no.1

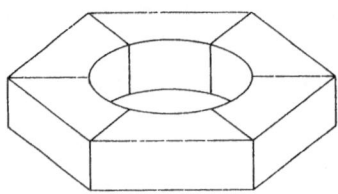

surface area = 6020.802
volume = 16440.023

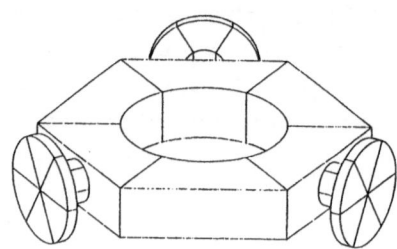

Fig. 5 Complex primitive no.2

Fig. 6 Solid object generated
with complex primitives

References:

1. Chen, D-P.; Lin, T-L.; Hsu, C-C.: Complex surface modeling with bias and tension. Computer-Aided Design, to appear (March, 1991).
2. Nielson, G.M. : Some piecewise polynomial alternatives to splines under tension. Computer Aided Geometric Design, edited by R.E. Barnhill and R.F. Riesenfeld, Academic Press, New York, 1974, pp. 210-235.
3. Barsky, B.A. : The Beta-spline : a local representation based on shape parameters and fundamental geometric measures. Ph.D. dissertation, university of Utah, December, 1981.

Integration of Design Optimization in Finite Element Analysis

FRED BAREZ

Mechanical Engineering Department
San Jose State University
San Jose, CA

ABSTRACT
Finite Element analysis is the most widely used computer based
analysis method. Design optimization algorithms recently integrated
into finite element analysis programs provide iteration capabilities
not available with manual methods. In the automated optimization, the
model is created and the loads are specified. Optimal characteristics
such as weight, shape and design are selected. The optimization
program checks the design and updates the model until it converges on
a best solution to meet the optimal characteristics and design
criteria. This paper reviews the optimization algorithms currently
used in design optimization for
weight, shape, and structural response.

INTRODUCTION
In 1960, Schmit [1] set forth a rather general approach to design
optimization, which has served as a conceptual foundation for the
development of many modern design optimization methods. It introduced
the idea and indicated the feasibility of coupling finite element
analysis and non-linear mathematical programming to create automated
optimum design capabilities for a rather broad class of design systems.

Extensive research has been performed in this field as indicated by Sheu
and Prager [2], Berke and Vankayya [3], and Haug and Arora [4]. The
non-linear programming approach to design optimization was applied to:
1) component type problems of a fundamental nature [5] and 2) system-
level structural optimization problems that involved joining finite
element analysis and non-linear programming algorithms together [6].

Design optimization can be done in different ways. Often, weight or
volume in minimize during steady-state static analysis with one load
case. Design variables in such optimizations are element properties.
Design constraints may include upper and lower limit on these
properties. Performance constraints are usually maximum allowable

stresses and deflections. Optimization also is done on problems involving multiple, simultaneous steady-state loadings. Optimizers also may work with dynamic loading on structure. A designer might perform mode analysis to set the natural frequencies within a specific range. Buckling may be a factor affecting structural members in compression. Some optimizers can perform normal-mode buckling analyses and steady-state static analysis simultaneously. Design variables in such analyses are element properties, but constraints may include buckling load factors.

In shape optimization, element properties remain constant, but design shape can change. The features of design topology become variables. When structure boundaries are changed, the finite element node point coordinates may change. Shape optimization automatically form or expand cutouts in a structure, and decrease design weight.

In this paper design optimization associated with weight, shape, and structural response will be considered.

DESIGN OPTIMIZATION PROBLEM STATEMENT

The design optimization problem can be stated as a non-linear constrained optimization in the standard mathematical form as follows: Find **X** the vector of design variables such that the objective function F(**X**) is either minimum or maximum subject to:

$$g_j(\textbf{X}) \leq 0 \qquad j\text{-}1,m \qquad inequality\ constaints$$

$$h_k(\textbf{X}) \text{-} 0 \qquad k\text{-}1,l \qquad equality\ constaints$$

$$X_i^l \leq X_i \leq X_i^u \qquad i\text{-}1,n \qquad side\ constaints$$

The objective function can be minimum weight, minimum volume, minimum or maximum displacement or frequencies, and closets match to test data.

OPTIMIZATION PROCEDURES

The basic function of an optimization procedure is to carry out an iteration process in which a search algorithm and an analysis algorithm, called the optimizer and the analyzer, respectively, are employed in obtaining an optimum function. Most optimization algorithms require that an initial set of design variable be specified. Beginning from this starting point, the design is updated iteratively. In design optimization, it is important to distinguish between analysis and design. Analysis is the process of determining the response of a specified system to its environment. Design, on the other hand, is used to mean the actual process of defining the system. Figure 1 shows the basic components and flow organization of an optimization procedure. Here the function of the analyzer is to compute values of the behavior variables which characterize the physical object's response to the input quantities. The purpose of the optimizer is to calculate a new vector of design variables on the basis of the values of the objective function and the constraints.

OPTIMIZATION APPROACHES

Although for special cases unique analysis techniques may be employed, however, the most widely optimization approaches are Approximation Concept Technique, Optimality Criteria, and Gradient Method.

Approximation Concept Technique-

Application of a general non-linear constrained optimation technique to an engineering design problem requires considerable judgement. The basic difficulties encountered are that often problems are either more or less non-linear than anticipated and that analysis times for practical design can become very long and costly. Design optimization efficiency can be improved through the use of approximation techniques [7] which include: 1) reduction of the number of independent design variables by linking and/or basis reduction; 2) reduction of the number of constraints considered at each stage by temporary deletion of inactive and redundant constraints; 3) construction of high-quality explicit approximation for retained constraint functions. The vector X, the design variables, can be expressed as a linear combination of linking vectors L_c as follows:

$$X = \sum_{c=1}^{C} L_c \gamma_c = [L]\, \gamma$$

The reduced basis concept in design space further reduces the number of independent design variables by expressing the vector as a linear combination of basis vectors R_b as follows:

$$\gamma = \sum_{b=1}^{B} R_b \delta_b = [R]\, \delta$$

Therefore **X** can be defined by

$$\boldsymbol{X} = [L] \, [R] \, \boldsymbol{\delta} = \sum_{b=1}^{B} T_b \delta_b$$

Where the vectors T_b are prelinked basis vectors and the δ_b are the reduced set of generalized design variables.

Optimality Criteria Methods-
Optimality Criteria Methods due to their relative simplicity and computational efficiency compared to mathematical methods are widely used. The most widely used optimality criteria method is the fully stressing method, which is used for minimum weight design with constraints on allowable internal stresses.

The fully stressed method of optimality criteria is exact for determinate structures subject to one load condition but is approximate otherwise. This criterion gives a simple iteration design formula, which can be efficiently applied to large structures. Other optimality criteria are used for minimum weight design with displacement, frequency, and buckling constraints. The optimality criteria is a minimum weight structural design problem with inequality constrained minimization which mathematically is defined as follows:
Find a set of independent design variables D_j, j=1,2,...,N, such that

$$g_k(D_j) \le 0 \qquad k=1,2,\ldots,M$$

and

$$W(D_j) : \text{minimum}$$

Where the objective function W is the weight and the constraint functions g_k are behavioral constraints and side constraints on the design variables. The necessary conditions for a design to be a relative or local minimum are [8]:

$$\frac{\partial W}{\partial D_j} + \sum_{k=1}^{M} \lambda_k \frac{\partial g_k}{\partial D_j} = 0 \qquad j=1,2,\ldots,N$$

$$\lambda_k g_k = 0$$

$$\lambda_k \geq 0 \qquad k = 1, 2, \ldots, M$$

Where λ_k is the Lagrange multiplier associated with constraint g_k.

Gradient Method-
The optimum design of structures in dynamic response always involves great difficulties since displacements, velocities, and accelerations are often related to the objective function and conditions of constraint, and since they are, in general, implicit non-linear functions of design variables. Among the many optimization techniques, gradient techniques using gradients of some quantities with respect to the design variables provide a powerful procedure for solving complicated optimum design problems. This technique is an effective method for dynamic response analysis which is incorporated in two procedures in obtaining optimum design, namely, step-by-step integration and modal analysis [9]. The step-by-step integration technique is suitable for both linear and non-linear systems. The equation of motion is written as:

$$M \ddot{\Delta}(t) + C \dot{\Delta}(t) + K \Delta(t) = P(t)$$

Where M, C, and K are the mass, damping, and stiffness matrices, respectively and are functions of time or constants in the linear system, while they are also functions of displacements, velocities, and acceleration in the non-linear system.

The modal analysis is one of the prevalent methods in dynamic response analysis, although it can be applied only to linear system [10]. The governing equations of motion for a linear system in the presence of a mode-shape vector ϕ_r is given by:

$$\phi_r^T M \phi \ddot{u} + \phi_r^T C \phi \dot{u} + \phi_r^T K \phi u = \phi_r^T P$$

Where $\Delta = \phi u$ and u is the general coordinate vector.

DISCUSSION
The goal of design optimization is to primarily achieve the optimum objective function such as the minimum weight, an optimized shape, or a desirable dynamic response. Three main optimization approaches are described here, however, various algorithms and procedures can be implemented in search for the optimum design. In weight minimization, optimality criteria is used commonly. The design satisfying the criteria

is guaranteed to be at least a local minimum. In this sense the optimality criteria methods fall under the category of indirect methods of optimization. The potential strength of the method is that the number of iterations need to converge to an optimum is virtually independent of the number of structural members. This property makes these methods well suited for the optimum sizing of large practical structures. The work by Khot, Berke, and Venkayya[11]; and Dobbs, Nelson[12] provide an excellent overview of the available algorithms.

In shape optimization, the availability of finite element methods has enabled the optimization of very general structural shapes. Algorithms utilizing variational approaches and perturbation techniques have successfully been developed by Tada and Seguchi[13]; and Banichuk[14], respectively. The applications of the perturbation method are not restricted to cases where the basic relations of the problems contain the small parameter in explicit form. One basic part of optimal structural design is sensitivity analysis. Perturbation method provides a tool in investigating design sensitivity analysis on optimum solutions.

Design optimization algorithms for dynamic responses have been developed based on perturbation and gradient techniques. The gradient-base formulation is divided into step-by-step integration and the modal analysis. Turner[15], and Yakasawa[16] have developed algorithms to design structures for natural frequency. The step-by-step yields itself for the study of structures under arbitrary types of loading. The application of the modal analysis method is limited to the case of linear systems. This method, however, is not computationally intensive.

CONCLUSIONS

Modern design optimization methodology, based on the innovative incorporation of finite element analysis and mathematical programming algorithm has grown rapidly. As was the case with finite element analysis, wider usage and acceptance of optimization techniques will increase with the development of quality computer programs. Optimization techniques, if used effectively, can greatly reduce engineering design time and yield improved, efficient, and economical designs. Such techniques offer an effective approach that sometimes even exhibit characteristics that one would associate with design creativity in that on occasion an unusual and unexpected solution is produced. Optimization methods give the creative designer ideas by being able to compare the best performances of various concepts quickly and convenietly. Interactive computer graphics provide effective tools for altering designs to make various choices available for final selection. The decision on which design is the best among all the trials will have to be made on the basis of a criterion or a set of criterion.

REFERENCES
1. Schmit, L. A., 'Structural Design by Systematic Synthesis, proceedings, 2nd Conference on Electronic Computation, ASCE, New York, 1960, pp. 105-122.

2. Sheu, C. Y., and Prager, W., 'Recent Developments in Optimal Structural Design,' Applied Mechanics Rev., Vol. 21, No. 10, 1968, pp. 985-992.

3. Berke, L., and Venkayya, V. B., 'Review of Optimality Criteria Approaches to Structural Optimization, Structural Optimization Symposium, AMD, Vol. 7, ASME, New York, 1974, pp. 23-34.

4. Hang, E. J., and Arora, J. S., Applied Optimal Design, John Wiley and Sons, New York, 1979.

5. Schmit, L. A., Kicher, T. P., and Morrow, W. M., 'Structural Synthesis Capability for Integrally Stiffened Waffle Plates, AIAA J., Vol. 1, No. 1, 1963, pp. 2820-2836.

6. Kicher, T. P., 'Structural Synthesis of Integrally Stiffened Cylinder,' J. Spacecraft and Rockets, Vol. 5, No. 1, 1968, pp. 62-67.

7. Vanderplaats, G. N., Numerical Optimization Techniques for Engineering Design, McGraw-Hill Book Company, 1984.

8. Berke, L., and Khot, N. S., 'Use of Optimality Criteria Methods for Large Scale Systems,' AGARD Lecture series, No. 70, on Structural Optimization, AGARD-LS-70, 1974.

9. Atrek, E., Gallager, R. H., Ragsdell, K. M., and Zienkiewicz, O. C., New Directions in Optimum Structural Design, John Wiley and Sons, 1984.

10. Fox, R. L., and Kapoor, M. P., 'Rates of Change of Eigenvalues and Eigenvectors,' AIAA J., Vol. 6, No. 1, pp. 2426-2429.

11. Khot, N. S., Berke, L., and Venkayya, V. B., 'Comparisons of Optimality Criteria Algorithms for Minimum Weigth Design of Structures,' AIAA J., Vol. 19, No. 2, 1979, pp. 182-190.

12. Dobbs, M. W., and Nelson, R. B., 'Application of Optimality Criteria to Automated Structural Design,' AIAA J., Vol. 14, No. 1, 1976, pp. 1436-1443.

272

13. Tada, Y., and Seguchi, Y.,'Shape Determination of Strucures Based on the Inverse Variational Principle/The Finite Element Approach,' <u>New Directions in Structural Design</u>, Edited by E. Atrek, R. H. Gallager, K. M. Ragsdell, and O. C. Zienkiewicz, John Wiley and Sons, 1984, pp. 197-209.

14. Banichuk, N. V .,'Application of Perturbation Method to Optimal Design of Structures,' <u>New Directions in Structural Design</u>, Edited by E. Atrek, R. H. Gallager, K. M. Ragsdell, and O. C. Zienkiewicz, John Wiley and Sons, 1984, pp. 231-248.

15. Turner, M. H.,'Design of Minimum-Mass Structures with Specified Natural Frequencies,' AIAA J., Vol. 5, No. 11, 1967, pp. 406-412.

16. Yamakawa, H.,'Optimum Structural Designs for Dynamic Response, <u>New Directions in Structural Design</u>, Edited by E. Atrek, R. H. Gallager, K. M. Ragsdell, and O. C. Zienkiewicz, John Wiley and Sons, 1984, pp. 249-266.

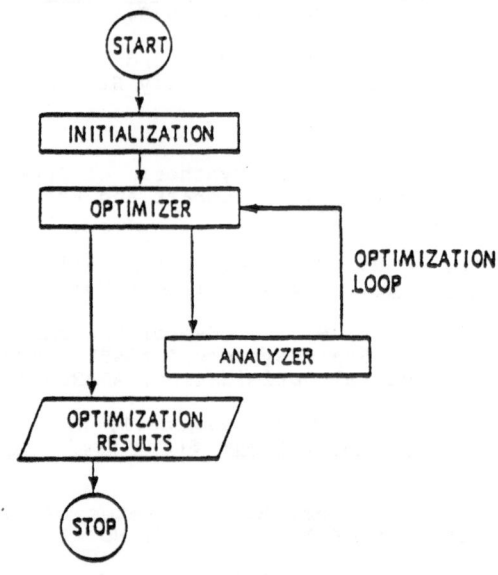

Figure 1: Basic Components and Flow
Organization of an Optimization
Procedure

Automatic Generation of Finite Element Modeling for Integrated CAD and CAE

TATSUHIKO AIZAWA

Department of Metallurgy
University of Tokyo
Tokyo, Japan

Summary

In the integrated CAD and CAE, data and program processing algorithms can be constructed on the library and database, and the whole models are automatically generated and controlled to achieve the required solutions with sufficient accuracy. Even for three dimensional problems, the finite element modeling should be translated from CAD or geometric models and must be processed so as to trace complicated mechanical behaviors appearing in designs. Direct tessellation approach is proposed and developed in the general methodology of the finite element control to deal with automatic model and related data handling.

Introduction

The finite element method or FEM has been or will be one of the most reliable, robust and powerful tools for the integrated CAD and CAE; both static and dynamic responses of final products are estimated by the finite element analysis or FEA, and reanalyses often take place in the design phase in order to attain the objective functions. In parallel with rapid increase of computer power and large reduction of required time and labors for design, the above finite element analyses should be integrated onto a system where the required FEM models are automatically generated and controlled. In past, many general-purposed preprocessors have been developed, and some of them are still widely used in the engineering circumstances of CAD and CAE; PATRAN, FEMGEN, SUPERTAB or FEMIS are wellknown systems working on the various types of computers [1]. Furthermore, new methods like boundary fitting method [2] have been proposed and developed in the analysis of fluid mechanics. Although those approaches and systems might be reliable and useful for preprocessing of FEM data, alternative methodology is necessary to realize complete automatic mesh control which satisfies the following conditions: [C1] No or very little input data are necessary to translate the object geometry to FEM models, [C2] The generated model is completely free from posterior verification or certification of whether the object region is subdivided into elements or any nodes are left to be unused, and [C3] Thus created model is easily processed for remodeling of FEM data or remapping the obtained informations by analysis.

Computational geometry [3] can afford to comprehend and describe complexity of geometry or graph in computation or construction of geometries from vertex-edge-face structure. Development of the efficient automatic triangulation algorithms in the multi-dimensional space [4,5] helps us to reconsider the meshing process from the point of computational geometrical view. Author has discussed over construction of adaptive automatic triangulation algorithm to finite element meshing [6,7], three dimensional triangulation method [8,9] and direct translation from CAD data to meshing models [10,11]. Through those studies, we have found that our developing method or direct tessellation method is superior to other heuristic approaches with respect to the following three items: 1) compatibility of node generation and triangulation algorithm with CAD data structure especially in three dimensional case, 2) transformation from triangularized subdivision to quadrilateral meshing model, and 3) data structure and handling to deal with remeshing and remapping models. In particular, conditions [C1] to [C3] can be satisfied by contiguous properties and intermediate data retrieval; through the present direct tessellation process, the objective region or geometry can be subdivided into elements which have no nodes left for disconnection, no superposed or vacant elements and no elements with inner nodes.

Data Structure and Data Handling for Integrated FEA

To reduce necessary amount of man power and FEA-time for meshing, CAD or geometric data should be processed and translated into meshing model with our processing as illustrated in Fig. 1. In the CATIA or GEOMOD systems, three dimensional geometry is represented by the surface model with vertex, edge and face list tables. While two dimensional geometry is expressed by vertex and edge lists, as shown in Fig. 2. First, node distribution is generated for boundary and inner nodes: vertices are directly stored into a set of nodes or SON-list and boundary edges are subdivided by the element length and appended into SON-list. Inner nodes inside a face and a body are created by the orthogonal grid pattern generation or the random node generation methods; conventional IN/OUT determination algorithm is utilized here to distinguish whether the current node is located inside the domain or outside. Here to be noted is the node density distribution to count the prescribed mesh size. Secondly, a family of approximately neighboring node sets or SAN(i)'s are created from SON-list; SAN(i) can be identified as an approximate set of Voronoi polygon VP(i) for each #i. The SON-list and SAN(i)'s are listed for body-1 in Table 1; in this case, the density of SAN(i) or dsan(i) is normalized to be nearly constant (dsan(i)=8 or 9). In the third stage of automatic triangulation, both Delaunay Tessellation and Voronoi Polygons are generated in the form of NDT-list and family of truly neighboring node set STN(i)'s, respectively. Table 2 lists thus obtained NDT-list with STN(i) sets. In the conventional preprocessing, only necessary data are SON-list and NDT-list; however, in the present method, 1) use of SAN(i)'s enables us to generate elements by contiguous connection of nodes with the common algorithms to every node location in two or three dimensional coordinates, and 2) commencement and termination of successive element generation is controlled by data handling and management of intermediate edge/face list, connected edge/face list and STN(i)'s. Especially, complete elimination of intermediate element boundary edge in two dimensional situation denotes that original geometry is subdivided into triangular elements without loss of nodes nor overlap of elements.

Direct Tessellation Method
Two dimensional tessellation method is described in what follows.

__(a) Standardization of nodes__   Node density distribution should be standardized into the specified number in order that 1) local density becomes uniform irrespectively of geometry and dimensions or 2) local density corresponds to the obtained error distribution.   Furthermore, a set of approximately neighboring nodes SAN(i) is created for the whole nodes 1<i<NP (NP=dense(SON-list) with the normalized density of nodes or dense(SAN(i))=NCC, where NCC is the prescribed constant number, in order that element generation should take place at the vicinity of the current node #i, irrespectively of dimensions and distances.   To be noted, connectivity of nodes with #i must be checked in order that the vicinity open set is convex or star-type, or, any line connecting two nodes of SAN(i) is also included inside thus chosen vicinity.

__(b) Generation of triangular elements__   Triangular elements are generated by connecting thus created nodes from SAN(i) with the current node #i through construction of Voronoi Polygon (VP) and Delaunay Tessellation (DT) graphs. Although some differences are present in algorithms for inner (ibc(i)=0) and boundary nodes (ibc(i)=1), generation of initial DT and construction of successive sequence of DT's are described in what follows.
__[b/1] Generation of initial DT__   #i is employed as the first node of the initial DT. Since #i has been cited several times before and dense(STN(i)) may not be zero, a set SAN should be merged in order that #j of STN(i) could be excluded from SAN if the connectivity index or idc(j) becomes unity, or, generation process of elements is already completed for #j. Both the second and the third nodes, or, #j and #k, are determined by minimization of both the distance between two nodes or dist(P(j),P(i)) and the radius of circles running through three nodes or radius(O(i,j,k)) for all nodes included in SAN(i). As beforementioned, the constructed element becomes contiguous.
__[b/2] Construction of successive DT sequence__   A sequence of DT's starting from {i,j,k} is constructed by search for Voronoi polygon VP(i); in principle, a new node will be searched until last node coincides with #j, as shown in Fig. 2-c1. For an edge starting from #i, a line line(i,m) colinear with this edge is considered, and a set of nodes S0 is defined which includes a node of SAN(i) located at the opposite side of line(i,m) against #k. A new node #k1 for the third vertex of new DT should be chosen by minimization of radius of circles running through three nodes i, m and k1 in order that new DT should become contiguous. After appending {i,m,k=k1} into NDT-list, a new edge is preceded to search for next new DT in the similar manner to the above. As illustrated in Figs. 2-c1 and c2, this iterative process terminates when the current node #i becomes a center node of polygon or when edges starting from #i are sorted.

__(c) Data management__   Through the above operations, data retrieval to both temporarily stored tables and geometric models is indispensable to obtain complete connectivity of nodes and triangular elements. Completeness in generation of elements is ascertained by the following conditions: [S1] Intermediate edge list includes only edges included in the boundary, [S2] Voronoi polygon VP(i) for each #i is generated from the temporary edge

276

list, and [S3] STN(i) is proved to be coincide with VP(i) for
each #i.

(d) Transformation into quadrilateral elements   Use of VP(i)'s
and STN(i)'s enables us to rearrange thus automatically generated
triangular elements into a sequence of class for reconnection;
at the first class (k=1) is included the elements connected with
the boundary, and in the subsequent classes (k>1) are arranged
the elements which have one/two/three common nodes with (k-1)
class elements.   Then, this reconnection process makes direct
transformation of the triangular elements or Delaunay tessellation into quadrilateral elements.

Conclusion
Finite element model control and management for reanalysis or
remeshing/remapping in the integrated CAD and CAE can be
precisely performed by the present method; since Voronoi
tessellations can be easily modified through edition or operation
of nodes, meshing models for new analyses are to be straightfor-
wardly created in the direct tessellation method.   Even in the
three dimensional situation, both the tetrahedral elements and
their modified hexahedral elements are directly created for
relatively complex geometries.

References
[1] T. Aizawa and H. Ohtubo, J. Naval Architecture (1987) 37-48.
[2] K. Miki et al., J. Comp. Phys., 53[2] (1980) 319-330.
[3] B. E. Bengtsson et al., BIT, 4 (1964) 87-105.
[4] J. L. Finney, J. Comp. Phys., 32 (1979) 137-143.
[5] M. Tanemura, et al., J. Comp. Phys., 51 (1983) 191-207.
[6] T. Aizawa, Proc. Symposium on Computational Methods in
Structural Engineering and Related Fields, 11 (1987) 221-226.
[7] T. Aizawa, Proc. Symposium on Computational Methods in
Structural Engineering and Related Fields, 12 (1988) 151-156.
[8] T. Aizawa, J. Simulation (1987) 79-84.
[9] T. Aizawa, Proc. Post Conference Seminar on CAE and Educational Aspects of Computational Mechanics (1988) 29-48.
[10] T. Aizawa, to be submitted in Computers & Structures (1990).
[11] T. Aizawa, Proc. Geometrical Science and Technologies V
(1990) 1-8.

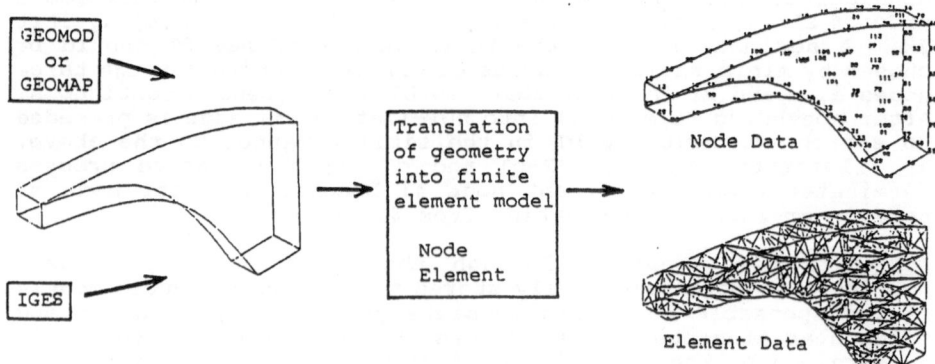

Fig. 1 Direct translation from CAD model to tetrahedral elements

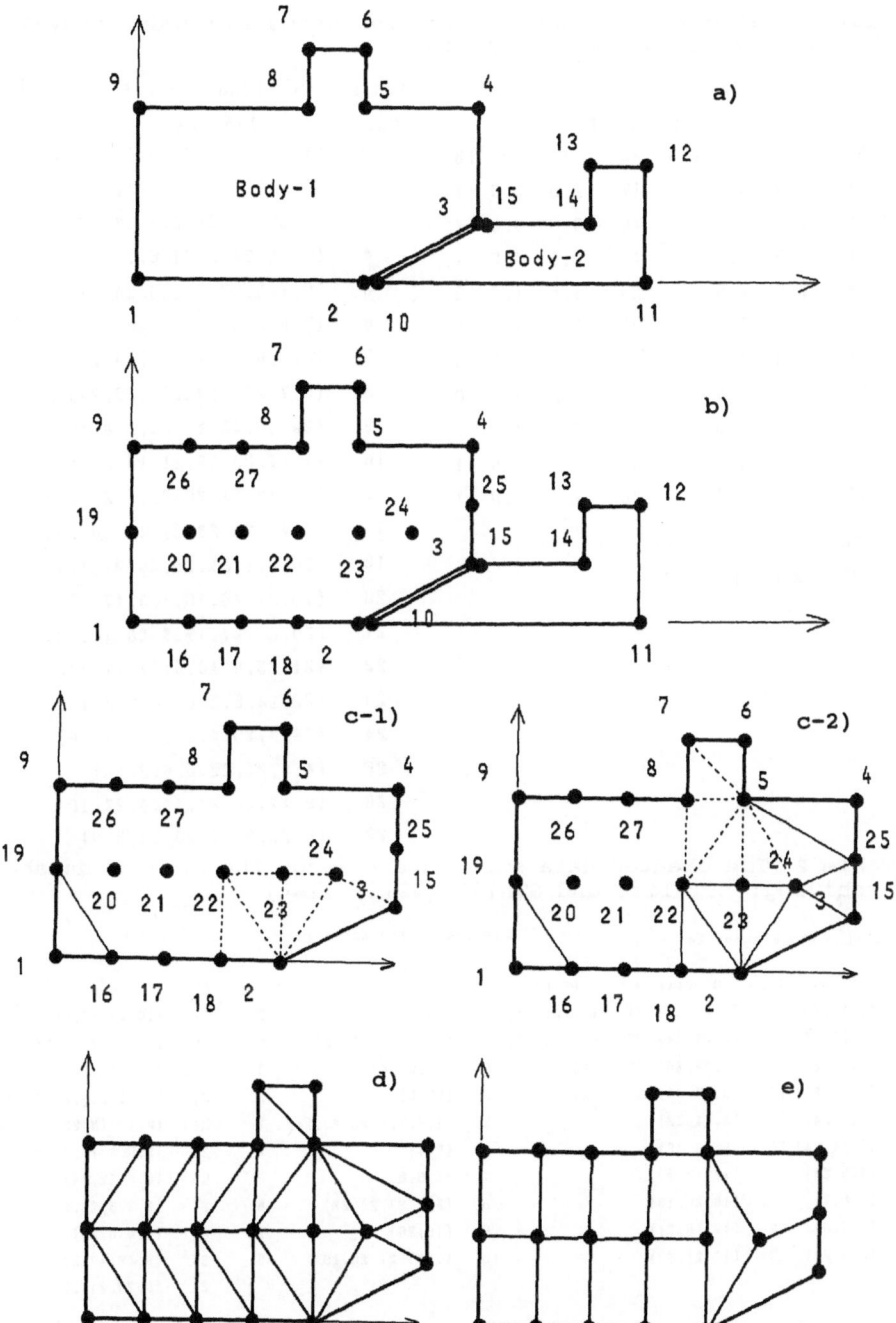

Fig. 2 Illustration of our developing procedure of automatic mesh subdivision on the basis of Voronoi Polygon and Delaunay Tessellation: a) Initial geometric model, b) Automatic node generation, c) Intermediate search for new DT's in sequence by edge control, d) Automatic triangulation and e) Transformation to quadrilateral elements.

Table 1 SON-list and SAN(i) list for geometric model body-1 in the two dimensional model in Fig. 2.

SON-list #Body=Body-1

#No_Vertex=21

No.	x	y	ibd	No.	x	y	ibd
1	0.0	0.0	1	17	2.0	0.0	1
2	4.0	0.0	1	18	3.0	0.0	1
3	6.0	1.0	1	19	0.0	1.5	1
4	6.0	3.0	1	20	1.0	1.5	0
5	4.0	3.0	1	21	2.0	1.5	0
6	4.0	4.0	1	22	3.0	1.5	0
7	3.0	4.0	1	23	4.0	1.5	0
8	3.0	3.0	1	24	5.0	1.5	0
9	0.0	3.0	1	25	6.0	2.0	1
16	1.0	0.0	1	26	1.0	3.0	1

No.	x	y	ibd
27	2.0	3.0	1

SAN set #Body=Body-1 #NCC=8

No.	SAN Data	dsan
1	{16,17,18,19,20,21,9,26}	8
2	{17,18,22,23,24,3,25,21}	8
3	{4,25,23,24,2,5,18,22}	8
4	{25,5,24,3,23,6,7,8}	8
5	{6,7,8,27,22,23,24,4}	8
6	{7,8,5,27,4,23,22,24}	8
7	{8,5,6,27,22,21,23,26}	8
8	{5,7,27,6,22,21,23,26}	8
9	{26,19,20,27,21,1,8,16}	8
16	{1,17,20,19,21,18,22,2}	8
17	{16,18,21,20,22,1,2,23,19}	9
18	{2,17,22,23,21,16,20,24}	8
19	{20,9,1,21,26,16,27,17}	8
20	{19,21,26,16,1,9,17,27}	8
21	{20,22,27,17,8,26,16,18}	8
22	{21,23,8,18,5,27,17,2}	8
23	{22,24,5,2,8,18,3,25}	8
24	{23,3,25,4,5,2,22,18,8}	9
25	{4,3,24,23,5,6,2,22}	8
26	{9,27,20,21,19,8,22,16}	8
27	{8,26,21,7,20,22,5,9}	8

Table 2 The created data structure through the present database archiving: NDT-list and STN(i) (VP(i) list).

NDT-list #No_Element=24

No.	DT Data	No.	DT Data	No.	DT Data	No.
1	{1,16,19}	11	{5,22,23}	21	{18,22,21}	
2	{2,22,18}	12	{5,23,24}	22	{19,20,26}	
3	{2,23,22}	13	{5,24,25}	23	{20,21,27}	
4	{2,24,23}	14	{8,27,20}	24	{21,27,26}	
5	{2,3,24}	15	{8,21,22}			
6	{3,25,24}	16	{9,19,26}			
7	{4,5,25}	17	{16,17,20}			
8	{5,6,7}	18	{16,20,19}			
9	{5,7,8}	19	{17,18,21}			
10	{5,8,22}	20	{17,21,20}			

STN Set #No_Nodes=21

No.	Node set	dense	No.	Node set	dstn
1	{16,19}	2	17	{16,18,22,21,20}	5
2	{18,22,23,24,3}	5	18	{2,17,22,23,21}	5
3	{2,25,24}	3	19	{1,9,16,20,26}	5
4	{25,5}	2	20	{16,17,19,21,26,27}	6
5	{4,6,7,8,22,23,24}	7	21	{8,17,18,20,22,27}	6
6	{5,7}	2	22	{2,5,8,18,21,23}	6
7	{5,8,8}	3	23	{2,5,22,24}	4
8	{5,7,27,21,22}	5	24	{2,3,5,23,25}	5
9	{19,26}	2	25	{3,4,5,24}	4
16	{1,17,21,20,19}	5	26	{9,19,20,27}	4
			27	{8,20,21,26}	4

Three Dimensional Mesh Generation: A New Approach

M.H. KADIVAR, H. SHARIFI

Mechanical Engineering Department
Engineering School
Shiraz University
Shiraz, Iran

summary
A new three dimensinal isoparametric mesh generation is proposed. This method is capable of handling supperelements with different number of nodes on their sides. Because of this capability, the number of supperelements can be reduced and it would be more convenient and time conserving. By assuminga hyphothetical plane, a pyramid supperelement can have different number of elements while it has the same number of nodes on it's sides.

1. Introduction

In the general curvilinear isoparametric mesh generations a curvilinear quadrilateral supperelement in cartesian coordinates would be transfer into a cube in ξ_η_ζ coordinates [1]. The cube would be discreticized by ξ, η and ζ =const. This discreticized cube would be transfer back to the cartesian coordinates by

$$x = \sum_{i=1}^{20} N_i(\xi, \eta, \zeta) x_i$$

$$y = \sum_{i=1}^{20} N_i(\xi, \eta, \zeta) y_i \qquad (1)$$

$$z = \sum_{i=1}^{20} N_i(\xi, \eta, \zeta) z_i$$

where N_i is the shape function associated with node i [2]. Because discretization is done by ξ, η and ζ =const., it is clear that the discretization is possible if the number of nodes on the sides along these directions are equal. Generally, there are a lot of cases which one needs different number of nodes on the sides which are parallel to ξ, η or ζ. For example, if a side plane is small with respect to the other planes of supperelement and almost

a uniform size of elements or more number of elements on one side with respect to the others are desired, then the number of nodes should be different on some sides which are parallel to either direction of theξ, η or
ζ. Two general case would be consider, one is when the number of nodes on
the sides of one plane (it is called based plane, fig.1) is different, the
other is when the number of nodes on the other four sides are different
too. For compatibility between the supperelements, the number of nodes on
the plane which is opposite to the base plane should be the same as the base
plane.

2. Number of Nodes Are Different on the Sides
2.1 Number of Nodes on the Based Plane Are Different

It is assumed that all the four sides; fig.1 ; has the same number of
nodes while the number of nodes on the base plane are

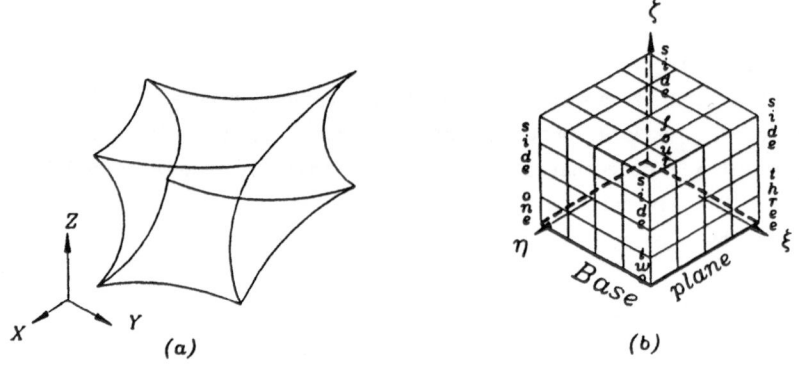

Fig.1 (a) A general superelement in X-Y-Z coordinate
(b) Discreticized superelement in $\xi-\eta-\zeta$ coordinate

all different. When the supperelement is transfer toξ , η andζ coordinates, because the number of nodes on all four sides parallel toζ
are equal, byζ=const., discretization is possible in this direction while
in ξ andη direction it is not possible. Therefore infact the problem
is changed to a two dimensional problem [3], inξ- η plane. In the
$\xi-\eta$ plane, the side with smaller number of nodes would be extended such that
the number of nodes on all opposite sides becomes equal[5] therefore the
cube is changed to a prism and the discretization is not possible but the
sides on the prism all have the same number of nodes. Therefore the prism

would be transfer to a cube in ξ' , η' and ζ' coordinates. In this new coordinate there is a cube which has equal number of nodes on all the sides which are parallel to ξ', η' and ζ', therefore by $\xi'=c_1$, $\eta'=c_2$ and $\zeta'=c_3$ the discretization can be done, while c_1, c_2 and c_3 are constants. The discreticized cube would be transferred back to ξ, η and ζ coordinates where the added part would be cancel out. In cancelling the added part in the ξ, η and ζ coordinates all the nodes which are outside the cube are moved to the nodes on the cube. The discreticized cube which has the number of nodes as it was desired, would be transfer back to the cartesian

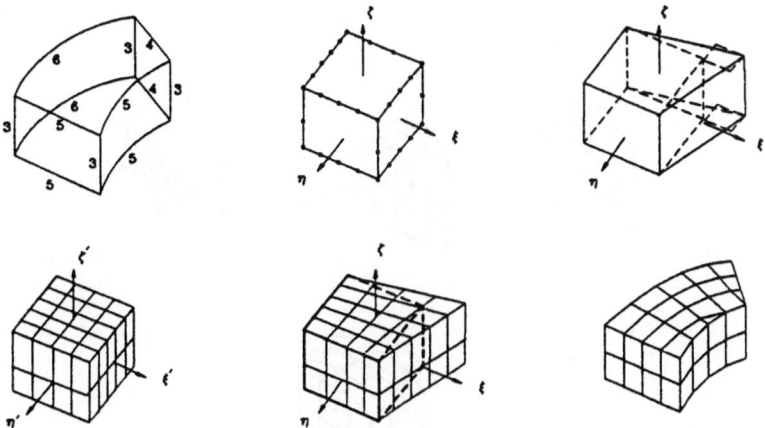

Fig. 2 *Procedure of discretization when the number of nodes on the base plane is different*

coordinates by equation 1. The whole procedure is shown in fig. 2 where as it can be seen the sides of the base plane has different number of nodes.

2.2 Base Plane and Four Sides Have Different Number of Nodes

When the four sides in addition to the base plane have different number of nodes the difference of the number of nodes on the four sides would be ignored at the first step or in another word it would be assumed that there is only two nodes on the sides. Because the sides of the base plane have different number of nodes, the supperelement would be discreticized in the same way as it was mentioned in the previous section. After discretization, in the ξ , η and ζ coordinates the sides which has smaller number of nodes would be extended such that all the four sides have equal number of nodes. The new prism would be transfer to a cube in ξ', η' and ζ' coordinate, where all of it's four sides have the same number of nodes. This cube can be easily discreticize by ζ'=const. The procedure is shown in fig. 4. After discretization the cube would be transfer to the ξ, η and ζ coordinates an

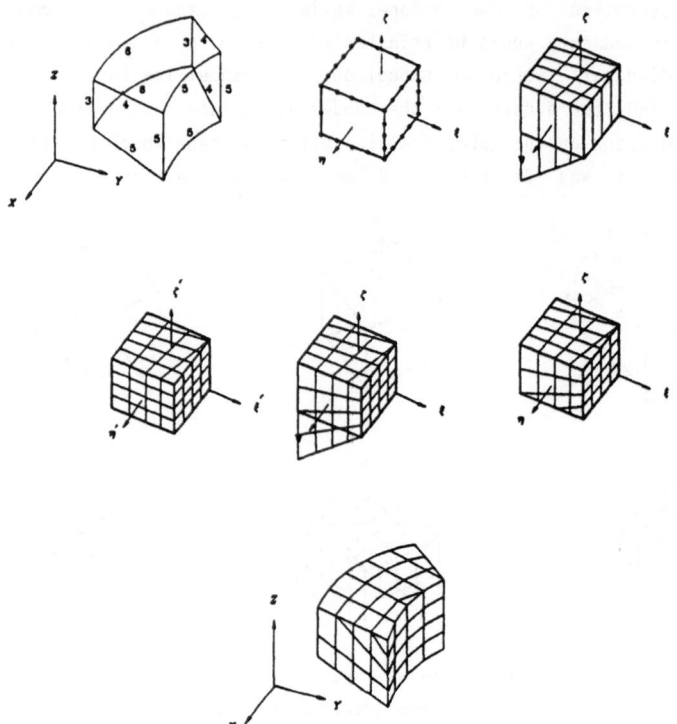

Fig. 3   *Procedure of discretization when the number of nodes on all sides are different*

the added part would be cancelled out. The remaining cube would be transfer to the cartesian coordinates and the discretization is over.

3. Refining the Discretization

In cancelling the added part, sometimes some element with a high aspect ratio would become into exit. For cancelling these elements, the aspect ratio of the elements which are adjacent to the side planes would be checked and those with big aspect ratio would be cancelled out. This procedure can be done in different ways. For example for simplicity in two dimensional case[4], four lines were drawn through 1/3 of the distance of the apex and

it's adjacent nodes in the $\xi-\eta$ plane and all the nodes which were between
these lines and $\xi=\pm1$ and $\eta=\pm1$ lines were moved on to the sides. The effect
of the inclined coefficient is shown in fig.4. In three dimensional cases
these lines would be changed to planes.

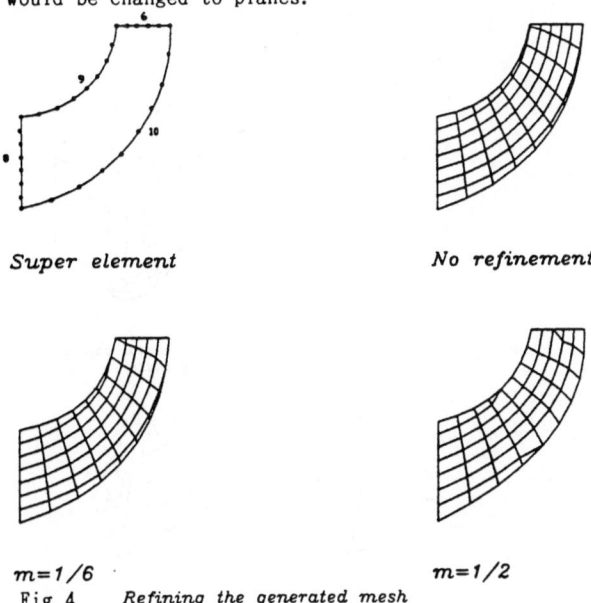

Super element

No refinement

m=1/6

m=1/2

Fig. 4   Refining the generated mesh

4. Computer Algorithm

Based on the above procedure a computer algorithm were written. This algorithm can be used easily in the same way as in a normal finite element generators, with the exception that the user is not to be worried about the
number of nodes on the sides. A typical coarse and fine mesh is shown in
fig.5. As it can be seen the supperelement number one has different number
of nodes on it's sides. Due to the proposed procedure a pyramid can be assumed as a prism with a hypothetical plane. Because this hypothetical plane
can have different number of nodes on it's sides, we can have different number of elements while the number of nodes on all sides of the pyramid are
equal. Fig. six shows a pyramid. As it can be seen the number of nodes on all
sides in all the cases are equal while the number of elements are different.

Acknowledgment

The authors would like to show their appreciation to the Shiraz University, where by their grants this research was carried out.

284

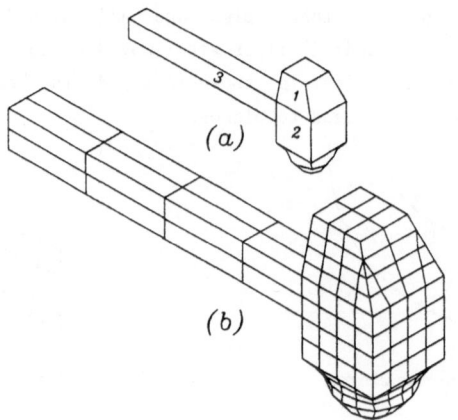

Fig. 5 *Discritization of a general 3 dimensional object*
(a) Coarse mesh - (b) Fine mesh

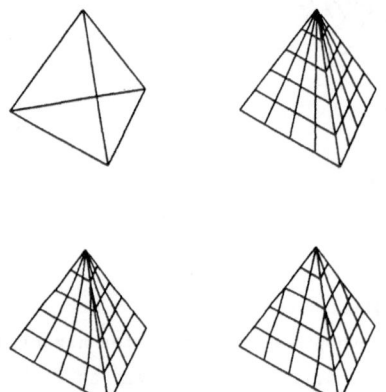

Fig. 6 *A generalize triangular tetrahedron superelement*

References

1. Zienkiewicz,O.C. and Philips, D.V., An automatic mesh generation scheme for plane and curved surfaces by isoparametric co-ordinates, Int. J .numer. methods eng.,3, 519-528(1971)

2. Zienkiewicz,O.C., The Finite Element Method, McGraw-hill, New York,1977

3.Durocher, L.L., A versatile two dimensinal mesh generator with bandwith reduction, computers & structures, vol.10 561-575(1979)

4. Kadivar,M.H., Sharifi,H., A new versatile two dimensional Mesh Generation Proceeding, Sixth international conference on applied mathematics & it's applications in engineering. 1090-1096 Swansea U.K.(1990)

Effective Modeling of Elastic Mechanical System Through Objective-Aimed Finite Element Strategies

V.H. MUCINO, W.G. WANG and J.E. SMITH

Department of Mechanical and Aerospace Engineering
West Virginia University
Morgantown, WV

Abstract

Engineering modeling plays an important role on the overall design process for a complex mechanical and/or structural system. Yet, the development and generation of a suitable finite element model for the static, dynamic or thermal analysis of the system is one of the most tedious and time consuming tasks in engineering analysis. The quality of the engineering modeling has the most significant impact on the quality of the analytical results.

Literally, enormous amounts of resources (human and computational) are dedicated and lost to the development of complex mathematical models that do not accomplish the function they are intended to fulfill, namely to produce reliable results.

In this paper a strategy for complex system modeling is presented in which the objective of the analysis and associated attributes are reflected by the model. Said attributes indicate the type of analysis required, the type of loading and boundary conditions, the type of geometry and the environment for the analysis, including the hardware, software and time frame resources. As an example, a model of a flexible internal combustion engine system based on the "Stiller-Smith Mechanism" is presented, in which a combination of solid, beam, and gap elements are used to represent the entire, 3-D system. The models presented are use to produce elastodynamic response needed for system design purposes.

Introduction

Three possible scenarios can be anticipated in a modeling exercise, one in which the complexity of the model precludes its tractability, making it virtually impossible to handle due to large number of degrees of freedom, moving boundary conditions, nonlinear material properties etc. A second possible scenario is that the model is so coarse or oversimplified, the outputs produced are no longer representative of the system's behavior and thus are meaningless. A extremely dangerous, third possibility, involves the development of a model in which the nature of the real system is somehow missmodeled, yielding results that do not correspond to the actual response of interest.

The mathematical model of a mechanical or structural system is generally conceived with a particular purpose in mind. That is, it does have a function to perform, which can be best described as follows:

> *To synthesize the features and characteristics of a mechanical*
> *or structural system to yield explicitly a particular physical*
> *response of interest through mathematical/numerical operations.*

Engineering modeling of mechanical or structural systems is a process that
establishes the relationship between system properties and parameters with the
state variables, describing the physical behavior sought. This process
requires the characteristics of the system to be expressed in terms of the
engineering features relevant in the analysis exercise.

Currently, many tools exist to assist the engineering analyst in developing
mathematical models of mechanical or structural systems. This paper focuses
on the application of finite element modeling techniques and its context in
design engineering.

Effective finite element modeling is still considered an art, heavily
dependent on the analyst's judgment to make important decisions such as the
nature and order of the elements to be used in the model, the degree of
refinement required in the analysis, the nature of the boundary conditions,
the characterization of the material properties and the appropriate modeling
of the applied loads [1]. The problem has been acknowledged by finite element
software developers by providing a wide range of mesh geometric definition
capabilities. However, most of the mesh generators require strategies that
are not in tune with current trends in the integrated computer-aided design
and manufacturing areas, particularly in the emerging concurrent engineering
environments [2].

The product of a modeling exercise is a mathematical model, which as any other
engineering product, has a function to perform. The efficiency in which the
model performs its function depends on the strategy used in the modeling
process. Ultimately, *the quality of the response sought, depends on the
quality of the model.* One may legitimately ask, what is the measure for the
quality of a model? Or, how can one measure the efficiency of a model? These
questions require a rather long answer and the example provided next
illustrates some of the aspects to be considered in developing strategies for
effective models.

The Stiller-Smith Engine System
The system used here to illustrate for effective finite element modeling
consists of an internal combustion engine system based on a rather unconven-

tional mechanism called the *"Stiller-Smith Mechanism"* [3]. Figure 1 shows the basic four cylinder Stiller-Smith Engine and the mechanism for planar rigid motion [4–6]. The basic configuration of this mechanism consist of two cross-laid reciprocating rods, with a piston mounted at each end, and two output gears mounted opposite to each other with respect to the trammel gear, which in turn is connected to the piston rods at joints A and B. An overall housing provides the support to the connecting rods and output shafts (which are rigidly connected to the output gears) through slider and journal bearings respectively. The midpoint of each connecting rod is connected to the trammel gear at joints A and B as reciprocating motion in a straight line along the longitudinal axis of each rod. The output gears and the trammel gear are mounted in such a way that the geometric centers lie on a straight line at all times. This line is always parallel to the line connecting the center O of the mechanism and the centers O_1 and O_2 of rotation of the output gears.

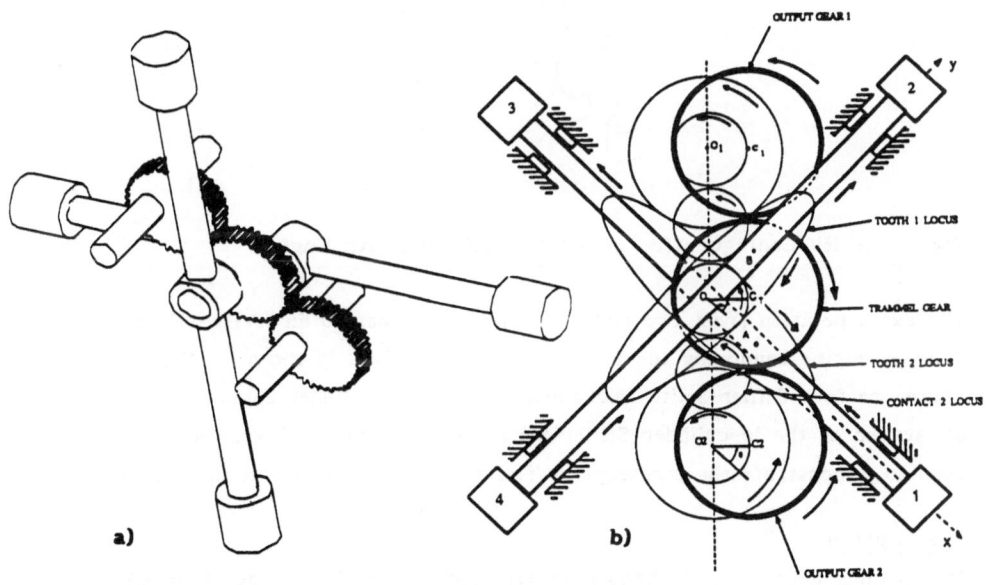

Fig. 1. a) Basic Four Cylinder Stiller-Smith Engine
b) Stiller-Smith Mechanism for Planar Rigid Motion

288

The general layout of a 4-cylinder Stiller-Smith Engine shown in Figure 1 is often referred as a "bank" of cylinders. The Stiller-Smith Engine can be built with 8 cylinders (2 banks) or 16 cylinders (4 banks) for higher horse-power demands. The banks of the engine are coupled with common output shafts. For the multibank Stiller-Smith Engine, the first connecting rod assembly which is assumed at the horizontal position is named **H** and the second rod assembly at the vertical position is named **V**. For a two-bank Stiller- Smith Engine, the connecting rod arrangement can be horizontal, vertical, horizontal and then vertical, namely, **HVHV** configuration, shown in Figure 2.

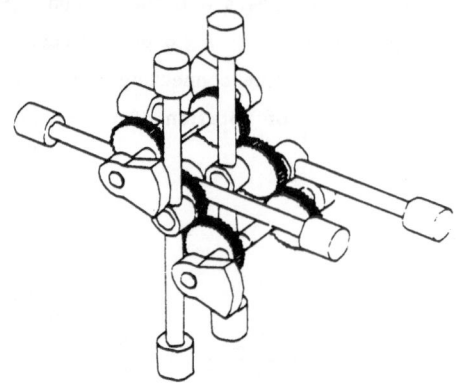

Fig. 2. HVHV Configuration of 2-Bank Stiller-Smith Engine

The extra-polation from two **HVHV** 8-cylinder engines to an **HVHV-VHVH** 16-cylinder engine employing the Stiller-Smith Mechanisms is shown in Figure 3. There are four Stiller-Smith Mechanisms for the 16-cylinder engine. The over-all system of the 16-cylinder Stiller-Smith Engine shown in Figure 4 [7], is used to demonstrate the strategy for finite element modeling in this research.

Assumptions

In designing a complex mechanical system such as the 16-cylinder internal combustion engine, the engineering dimension calculations are generally based on the assumption that the links are rigid or near rigid. However, it is well known [8] that high speed mechanisms are generally elastic rather than rigid, and the effects of link flexibility vary widely depending on the type of application and the kind of functionality required from the mechanism or system.

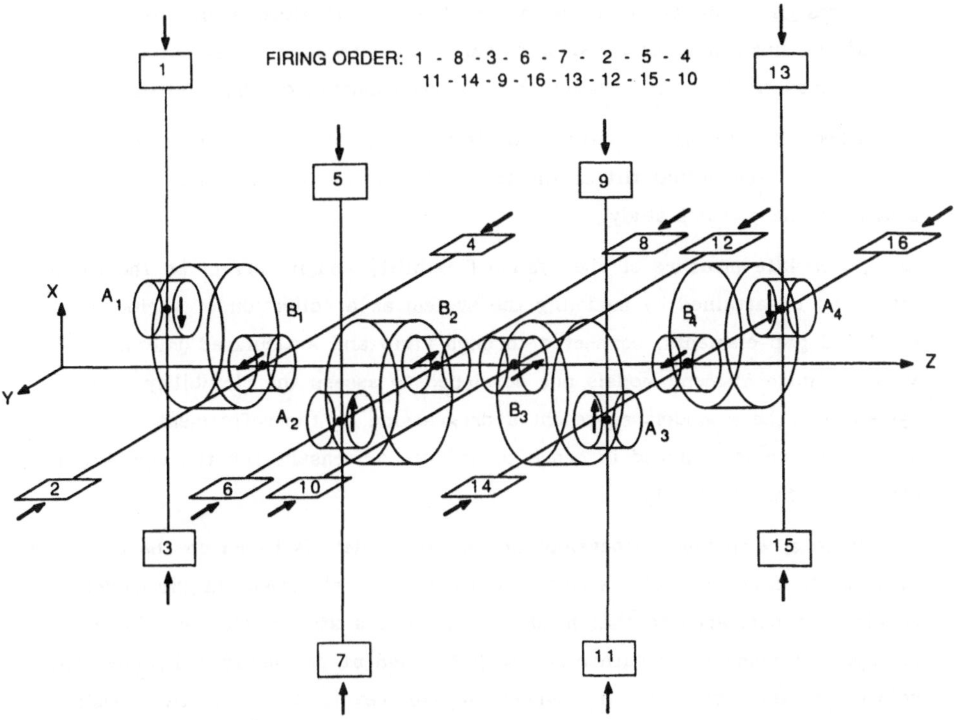

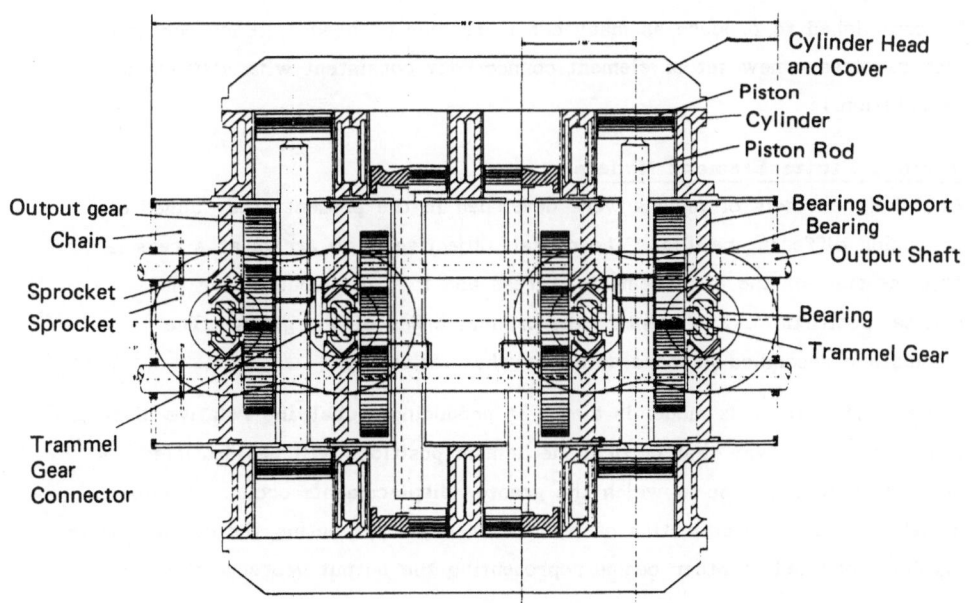

FIRING ORDER: 1 - 8 - 3 - 6 - 7 - 2 - 5 - 4
11 - 14 - 9 - 16 - 13 - 12 - 15 - 10

Fig. 3. HVHV-VHVH Configuration of 4-Bank 16-Cylinder Stiller-Smith Engine

Fig. 4. Overall System of the 16-Cylinder Stiller-Smith Engine (Side View)

In the design of an IC-engine using the Stiller-Smith Mechanism, the flexibility of the links must be considered mainly to determine its effect on the actual loads transferred between the various elements of the mechanism.

The effect of flexibility on the load distribution along the contact tooth flank and on the actual tilt of the trammel gear of the system is particularly sought in the present study.

An appropriate measure of the system flexibility and its effect on the tooth loads can be attained by modeling the system as a collection of various beam solid and gap elements, connected through rigid and articulated joints. Various finite element models are developed to assess the flexibility of the system and these models represent a progression on the refinement of the system, which is required in order to produce a measure for the significant effects of the flexibility.

The strategy employed to develop the various models is based on the fact that in order to produce a displacement solution, less refinement in the model is required as compared to that needed to produce a stress solution. Thus, the first set of models are rather coarse but aimed at producing a displacement solution of the mechanism throughout one full revolution. A critical position in the context of the study is that one would yield the largest relative displacement between the teeth of the gears. A quasi-static approach has been used to generate the results, which in turn requires that a "parametric model" be established to produce as many configurations as needed. (Each configuration requires a new set of element connectivity consistent with attitude angle or rotation.)

Various Finite Element Models

Five different sets of models were developed in the present study aimed at producing different results. The overall objective is to arrive at a reasonable measure of the displacements, stress and load distribution for the critical positions and to identify the critical areas of the gears. The models are described in more detail in what follows.

1) Beam Model: This model is aimed at producing global and relative displacements in such a way that critical mechanism positions will be identified. A critical position is one in which the greater displacements occur. A typical model consists of a collection of beam elements representing the output shafts rigidly connected to other beams representing the output gears of the

mechanism. These gears are modeled as one beam segment connecting the center of the output shaft with the trammel gear through an articulated joint representing the tooth in contact. The trammel gear in turn is modeled by means of beam segments that connect the tooth in contact with the trammel pins through rigid connections.

For different attitude angles the lengths of the segments representing the output gears would change since these gears are eccentrically mounted. Likewise, the segments connecting the trammel pins and the tooth in contact attain different lengths and orientations as the attitude angle varies in the mechanism through a full revolution. When the attitude angles are at 5, 95, 185, 275, 365, 455, 545, and 635 degrees, a maximum displacement is obtained, respectively. Figure 5 shows an example of this model for an attitude angle at 185 degrees.

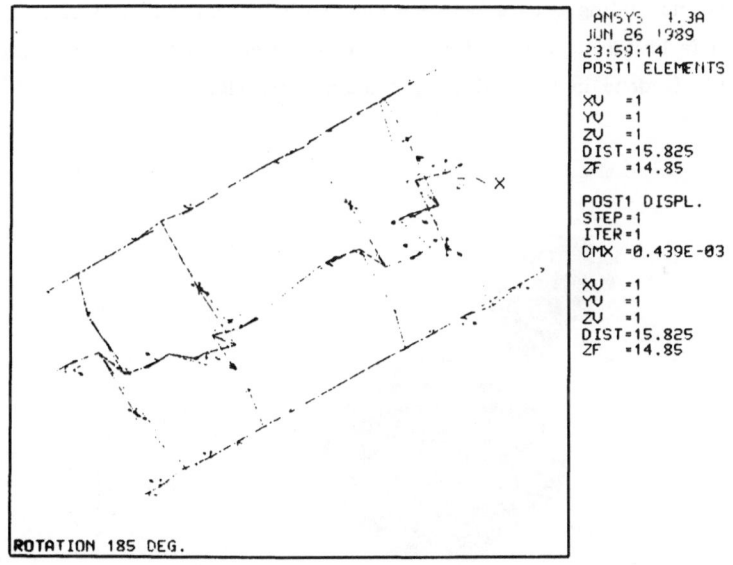

Fig. 5. Beam-element Displacement Model for Attitude Angle of 185°

2) Coarse Solid Element Model: This model is aimed at producing rough displacements within the trammel gear, as loads are applied to the trammel pins and transferred to the output gears through the meshing teeth. The model

292

consists of three objects: one trammel gear and two output gears. The trammel gear is completely elastic and is modeled in three distinctive parts: the teeth in contact, the gear body and the trammel pins. These output gears are modeled in such a way that only the portion or volume corresponding to the tooth is elastic, while the rest is considered rigid. This assumption is made to simplify the analysis, but it is reasonable to assume that the output gears do not deform in their circumference plane. Finally, these output gears are fixed at the areas where the connection with the output shaft will be. Figure 6 shows displacements and stresses around the tooth for the case considered the worst or critical. While this model alone produces both displacements and stresses, the objective of this analysis is to combine this model as a substructure with the previously described beam model. This is done in the following models. It should be mentioned that the contact between teeth is simulated through "the coupled-node condition," which allows points in contact to travel or displace together during the load application. This assumption implies that no sliding occurs between the teeth in contact. However, this assumption is better accounted for through gap elements that allow sliding to occur. This is done in the following combined models.

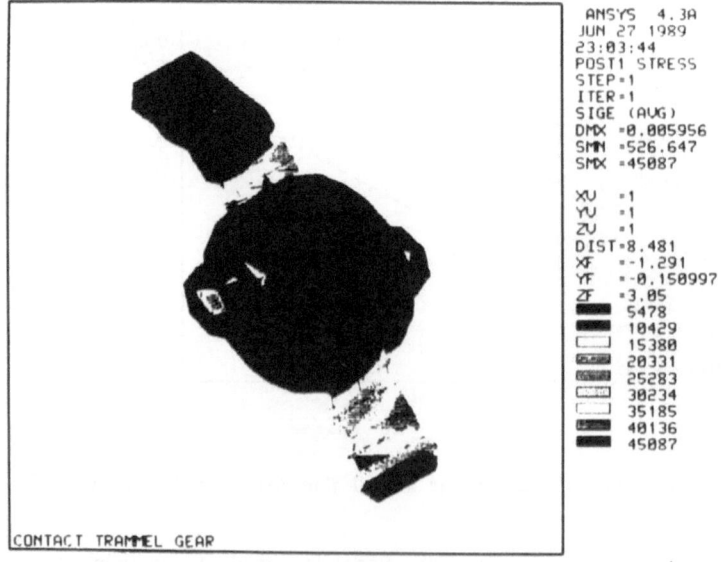

Fig. 6. Stresses for Contact Trammel Gear

3) <u>Combined Beam-Solid Model</u>: By replacing one beam element bank (out of four required for a 16-cylinder engine) with one solid element bank, a system can be obtained that further refines the displacement solution of the whole system. A combined model in which the connection between the solid elements and the beam elements is accomplished by means of the *constraint equations*, which relate the rotational displacements of the beam segments with the translation displacements of the solid elements. The contact between the gear teeth is first modeled through the coupled node, preventing the sliding condition. Results of this analysis are presented in Figure 7. It should be noted that in this model the resulting load distribution on the teeth can already be appreciated even though the model is not fully refined.

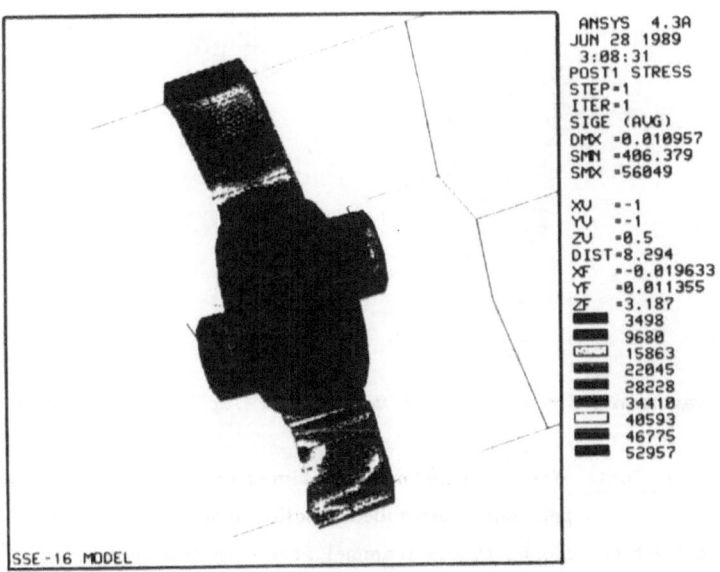

Fig. 7. Stresses for the Contact Teeth from the Combined Beam-Solid Model

4) <u>Combined Beam-Solid-Gap Model</u>: This model introduces a change in the way the contact between the gear teeth occurs. In this model, the contact is modeled through the use of "gap elements," which allow sliding and separation to occur between the two surfaces, but prevent both penetration or overlap of the teeth and tension (as if the surfaces were glued and pulled apart). While this condition enhances the representation of the contact forces, it introduces significant difficulties in the model due to the fact that the

system becomes non-linear and an iterative solution is required to produce the displacements sought. Figure 8 shows the displacements from this analysis. It is worth noting that the load distribution that results on the teeth exhibits the expected uneven shape, which in turn produces stresses that concentrate on one side of the tooth.

In order to analyze the stresses generated in the gears more closely and further identify the critical areas of load distribution, it is necessary to develop a refined finite element model on the gear teeth.

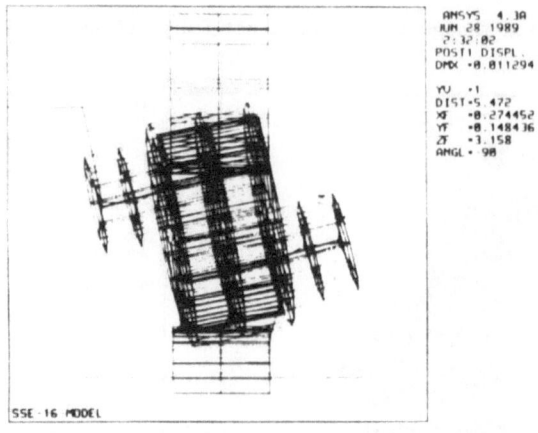

Fig. 8. Displacements from Combined Beam-Solid-Gap Model (Y-View)

5) Refined Gear Tooth Model: This model is aimed at producing more accurate results than the models previously described. Solid elements are used to model one tooth which belongs to the trammel gear. In this model the discretization is directed at critical areas, which were identified through the previous models. Similarly, the load distribution on the teeth is derived from the previous model where beam, solids and gap elements are used to simulate the whole system at a critical attitude angle of the mechanism. The refined tooth model resulting stresses are shown in Figures 9 and 10.

While the level of stresses obtained in this study are not particularly high, it is worth mentioning the design synthesis of these gears is such that the predicted critical stresses of the gears fall within the design failure criteria corresponding to the material used for the gears.

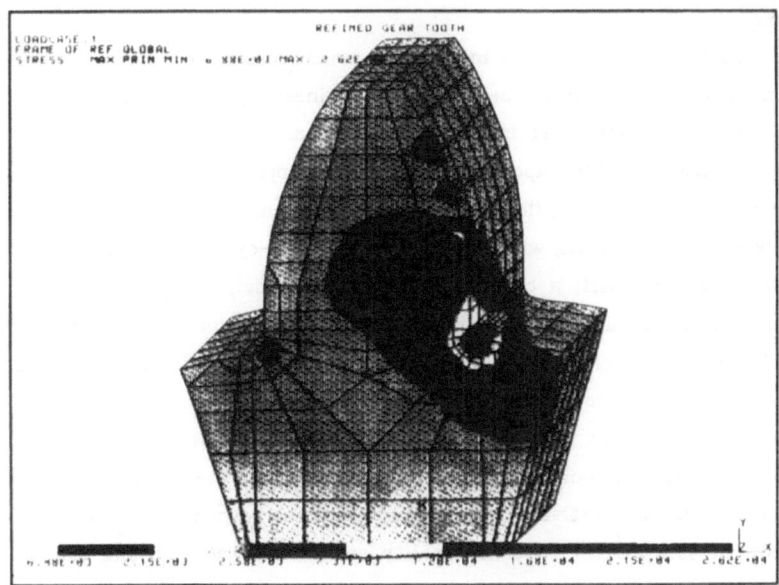

Fig. 9. Principal Stress on the Refined Tooth

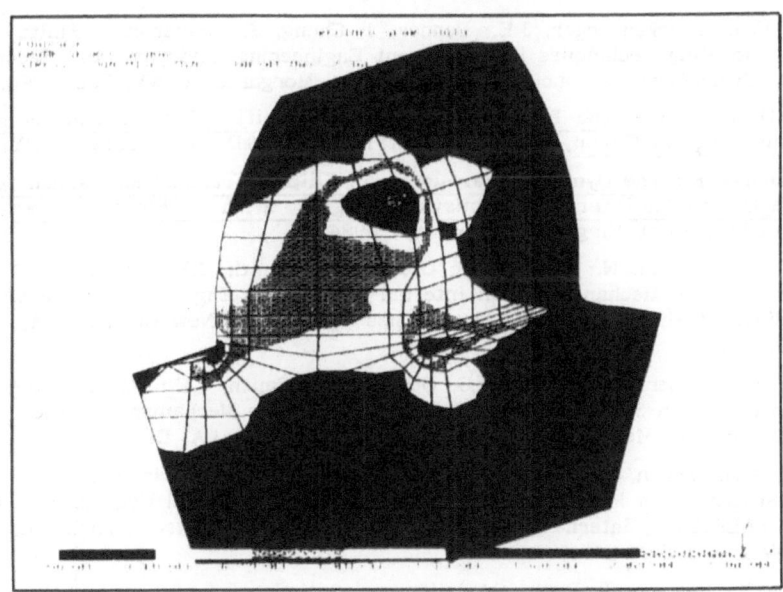

Fig. 10.   von Mises Stress on the Refined Tooth

Conclusions

Engineering modeling for a complex mechanical or structural system is an extremely important task that influences many engineering design decisions. Effective finite element modeling is still considered an art in today's engineering practice. In this paper, a sophisticated approach is developed to assess the flexibility of a complicated mechanical system from various finite element models. These models which are based on a 16-cylinder IC engine include a simple beam model, a coarse solid element model, a combined beam-solid model, a combined beam-solid-gap model and an effective critical region model. These models represent a progression on the refinement of the system, avoid either a large number of degrees of freedom, or oversimplification, arrive at a reasonable measure of the displacements, stresses and load distributions for the critical positions. The reliable results which are representative of the system's actual response of interest are produced. It is apparent that the quality analysis is achieved and the analysis efficiency is improved by using these models. The "intelligent" finite element modeling techniques are explored and illustrated.

Reference

1. Mucino, V.H.; Sneckenberger, J.E.; Benner, J; Chung, S.: "Parametric Finite Element Modeling Techniques in Concurrent Engineering Environments," Second National Symposium on Concurrent Engineering, Morgantown, WV, Feb. 1990.

2. TRUCE User's Guide, The Tridimensional Rational Unified Cubic Engine, Solid Modeling Systems Group, General Electric Co. CR&D, Schenectady, NY, 1985.

3. Smith, James E.: The Dynamic Analysis of an Elliptic Trammel Mechanism for Possible Use to An Internal Combustion Engine With a Floating Crank, Dissertation, West Virginia University, 1984.

4. Mucino, V.; Sivaneri, N.; Wang, W.G; Gokhale, H.; Smith, J.E.: Dynamics of the Stiller-Smith Mechanism in an Internal-Combustion Engine Environment. Proceedings of the 10th Applied Mechanisms Conference, New Orleans, LA, Dec.6-9, 1987.

5. Mucino, V.; Sivaneri, N.; Wang, W.G; Gokhale, H.; Smith, J.E.: Modeling the Friction Losses in Slider Bearings in the Stiller-Smith Mechanism. Proceedings of the 10th Mechanisms Conference, New Orleans, LA, Dec.6-9, 1987.

6. Mckisic, A.D.; Smith, J; Wang, W.G.; Prucz, J.: A Parametric Investigation of the Stiller-Smith Mechanism for Application in an Internal Combustion Engine. SAE880662, International Congress & Exposition, Detroit, Michigan, Feb.29-Mar.4, 1988.

7. Wang, W.G.: Parametric Design Modeling and Optimization of Unconventional Engine Systems, Ph.D. dissertation, West Virginia University, May 1990.

8. Sandor, G.N.; Erdman, A.G.: Advanced Mechanism Design: Analysis and Synthesis, Vol.2, Prentice-Hall Inc., Englewood Cliffs, N.J.,1984.

Design and Evaluation of Shock Isolation of Trailer Mounted Electronic Equipments

V. SUNDARARAMAN

Bharat Electronics
Ghaziabad, India

ABSTRACT

In the design of Trailers and flatbed trucks for transportation of mobile electronic equipments for military applications , shock levels reaching the equipment are to be limited to a specified value, over varied input shock levels encountered from cross country terrains. To achieve this, the suspension system of the vehicle needs a scientific design technique.

Starting from the basic design principles, this paper brings out a simplified design approach for selecting a suitable suspension system. The evaluation measurements done on such trailers in field trails confirm the practical utility of the simplified approach. The peak and average accelerations measured at various points in the trailer are presented in the paper. This paper presents a mobility configuration arrived at in supply of mili tary grade system by Bharat Electronics, India.

SHOCK MOTION

The acceleration-time history is considered to be the primary description of a shock motion. Various types of shock motions like acceleration impulse, half sine pulse of acceleration, decaying sinusoidal acceleration etc. are frequently encountered in practice. The time histories of velocity and displacement are derived from basic acceleration-time history by integration.

The response to a half sine acceleration pulse of a suspension system is analysed in this paper. A half-sine pulse of acceleration of duration T is shown in Fig. 1. The acceleration at any time t is given by

$$\ddot{u}_t = \ddot{u}_o \sin\left(\frac{\pi t}{T}\right) \qquad \text{for values of } 0 < t < T \qquad (1)$$

$$= 0 \qquad \text{for values of } t < 0 \text{ and } t > T$$

Integrating,

$$\dot{u}_t = \frac{T\ddot{u}_o}{\pi}\left[1 - \cos\left(\frac{\pi t}{T}\right)\right] \qquad \text{for } 0 < t < T \qquad (2)$$

$$= \frac{2T\ddot{u}_o}{\pi} \qquad \text{for } t > T \qquad (3)$$

Integrating again,

$$u_t = \frac{\ddot{u}_o T}{\pi}\left[t - \frac{T}{\pi}\sin\left(\frac{\pi t}{T}\right)\right] \text{ for } 0 < t < T \qquad (4)$$

$$= \frac{\ddot{u}_o T^2}{\pi}\left[\frac{2t}{T} - 1\right] \qquad \text{for } t > T \qquad (5)$$

These velocity and displacement time histories are also shown in Fig.1.

RESPONSE SPECTRUM

The response of a system to a shock can be expressed as the time history of a parameter that describes the motion of the system. For a simple system, the magnitudes of the response peaks can be summarised as a function of the natural frequency or natural period of the responding system at various values of the fraction of critical damping. This type of presentation is termed as a response spectrum. In the response spectrum only the maximum value of the response found in a single time history is plotted.

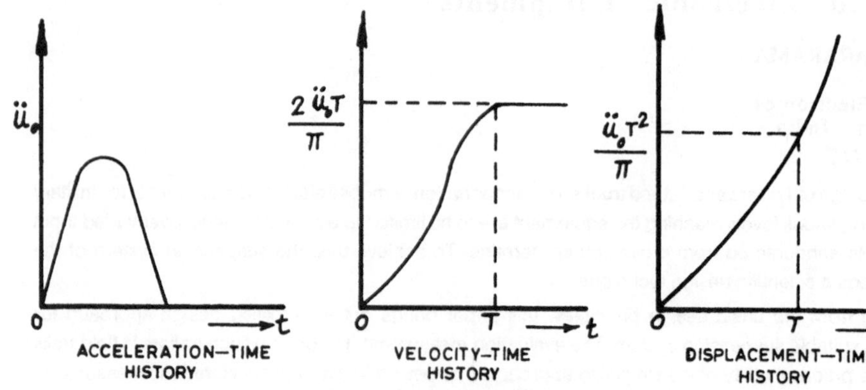

$\ddot{u}_o$

ACCELERATION—TIME
HISTORY

$\dfrac{2\,\ddot{u}_o T}{\pi}$

T

VELOCITY—TIME
HISTORY

$\dfrac{\ddot{u}_o T^2}{\pi}$

T

DISPLACEMENT—TIME
HISTORY

Fig. 1. Half – Sine Shock Motion

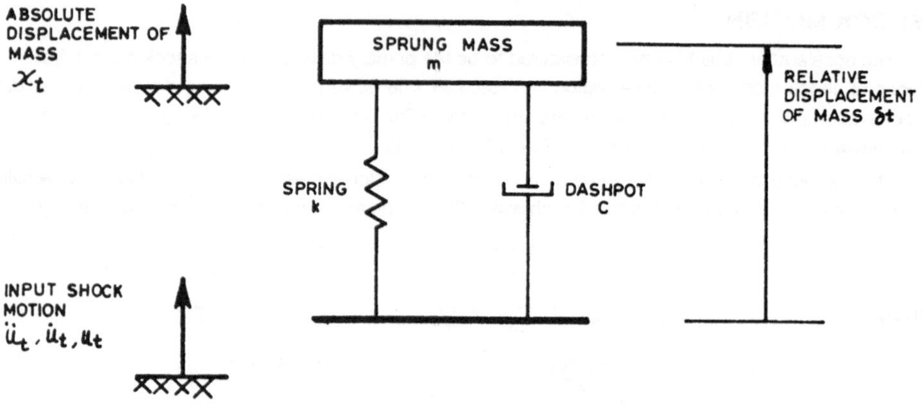

ABSOLUTE
DISPLACEMENT OF
MASS
x_t

SPRUNG MASS
m

RELATIVE
DISPLACEMENT
OF MASS δ_t

SPRING
k

DASHPOT
C

INPUT SHOCK
MOTION
$\ddot{u}_t, \dot{u}_t, u_t$

Fig. 2. Representation of a simple structure having
single degree of freedom.

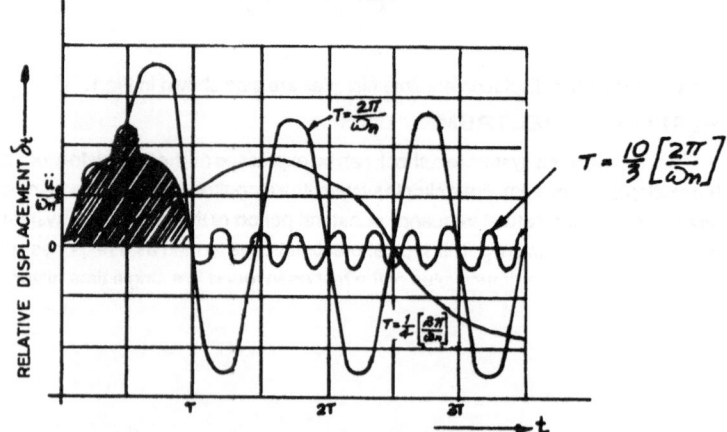

RELATIVE DISPLACEMENT δ_t

$T = \dfrac{2\pi}{\omega_m}$

$T = \dfrac{10}{3}\left[\dfrac{2\pi}{\omega_m}\right]$

$T = \dfrac{1}{4}\left[\dfrac{2\pi}{\omega_m}\right]$

T $2T$ $3T$ t

Fig. 3. Response of undamped single degree
of freedom system to the half sinusoidal
pulse acceleration.

TERMINOLOGY

Referring to the simple structure shown in Fig. 2, the following parameters are defined.

m = sprung mass

x_t = Absolute displacement of the mass m at time t

δ_t = Relative displacement of the mass m at time t

$\dot{x}_t$ = Absolute velocity of mass m at time t

$\dot{\delta}_t$ = Relative velocity of mass m at time t

$\ddot{x}_t$ = Absolute acceleration of mass m at time t

K = Spring stiffness

ω_n = undamped natural frequency of the sprung mass system. $= \sqrt{\frac{K}{m}}$

c = Damping coefficient

ζ = Fraction of critical damping $= c/2m\,\omega_n$

ω_d = Damped natural frequency $= \omega_n\sqrt{1-\zeta^2}$

$\ddot{u}_t$ = Input shock at time t

T = Duration of shock pulse

ω_p = Forced frequency $= \pi/T$

δ_{max} = Maximum relative displacement.

$\ddot{x}_{max}$ = Maximum acceleration of sprung mass.

EQUIVALENT STATIC ACCELERATION

The equivalent static acceleration is that steadily applied acceleration, expressed as a multiple of the acceleration of gravity, which distorts the structure to the maximum distortion resulting from the action of shock.

For the simple structure of Fig. 2, the relative displacement response δ indicates the distortion under shock condition. The corresponding distortion under static conditions in a 1g gravitational field , is

$$\delta_{st} = \frac{mg}{K} = \frac{g}{\omega_n^2} \tag{6}$$

By analogy, the maximum distortion under the shock condition is

$$\delta_{max} = \frac{Aeq\,g}{\omega_n^2} \tag{7}$$

Where Aeq. is the equivalent static acceleration in units of gravitational acceleration

$$Aeq = \frac{\delta_{max}\,\omega_n^2}{g} \tag{8}$$

RELATION BETWEEN VELOCITY ü AND δ MAX

Referring to fig.2, the kinetic energy of the sprung mass relative to the support is $\frac{1}{2}m\ddot{u}^2$

The work done on the isolator system is $\int_0^{\delta_{max}} F_s\left[\dot{\delta},\delta\right]d\delta$

Where $F_s[\dot{\delta},\delta]$ is the force of the spring system at any point δ

Equating $\int_0^{\delta_{max}} F_s\left[\dot{\delta},\delta\right]d\delta = \frac{1}{2}m\ddot{u}^2$

For a linear spring with zero damping

$$\int_0^{\delta_{max}} K\delta\,d\delta = \frac{1}{2}m\ddot{u}^2$$

$$or \quad m\dot{u}^2 = K\,\delta_{max}$$
$$\dot{u} = \omega_n\,\delta_{max}. \tag{9}$$

CALCULATION OF RESPONSE SPECTRUM

The relative displacement response of a structure (Fig.2) resulting from a shock defined by the acceleration $\ddot{u}_t$ of the support is given by the Duhamel integral

$$\delta_t = \frac{1}{\omega_d} \int_0^t \ddot{u}_t\, e^{-\zeta \omega_n (t-x)} \sin \omega_d (t-x)\, dx \tag{10}$$

Substituting the value of $\ddot{u}_t$ for a half sine shock pulse of acceleration defined by eqn.(1) in eqn. (10), the relative displacement response for an undamped system becomes,

$$\delta_t = \frac{\ddot{u}_0}{\omega_n^2} \left[\frac{\omega_n \frac{T}{\pi}}{1-(\frac{\omega_n T}{\pi})^2} \right]\left[\sin \omega_n T - \frac{\omega_n T}{\pi} \sin \frac{\pi t}{T} \right]$$
$$\text{for } o < t < T \tag{11}$$

$$= \frac{\ddot{u}_0}{\omega_n^2} \left[\frac{\omega_n \frac{T}{\pi}}{1-(\frac{\omega_n T}{\pi})^2} \right] \cdot 2 \cdot \cos \frac{\omega_n T}{2} \cdot \sin \left[\omega_n (t - \frac{T}{2}) \right]$$
$$\text{for } t > T \tag{12}$$

ANALYSIS

The displacement δ_t Vs time t is shown in Fig.3 for various values of ω_n. The value of 't' for which δ_t is maximum is found differentiating eqn.(11) and equating to 0 i.e. $\delta_t = 0$ i.e. $\cos \omega_n t - \cos \omega_p t = 0$
When $\omega_n \neq \omega_p$ the maximum value of δ_t does not occur during the period $o < t < T$; the maximum value occurs after $t = T$; in other words the first maximum in the response of a simple structure with natural frequency less than π/T occurs during the residual response.

The maximum value being $\quad \delta_{max} = \frac{\ddot{u}_0}{\omega_n^2} \left[\frac{\omega_n/\omega_p}{1-(\omega_n/\omega_p)^2} \right] 2 \cos \frac{\omega_n T}{2} \tag{13}$

Also the residual response is sinusoidal with constant amplitude given by the eqn.(13).
The time at which δ max, occurs is given by $\quad \omega_n \left[t - \frac{T}{2} \right] = \frac{\pi}{2}$
$$or\, t = \frac{T}{2} + \frac{\pi}{2\omega_n} \tag{14}$$
Also from eqn (8) and (13), we get
$$Aeq. = \frac{\ddot{u}_0}{g} \left[\frac{2(\omega_n/\omega_p)}{1-(\omega_n/\omega_p)^2} \right] \cos \frac{\omega_n T}{2} \tag{15}$$
Eqn.(15) gives the value of equivalent static acceleration for an undamped system with $\omega_n < \omega_p$
Plotting $\frac{Aeq}{\ddot{u}_0/g}$ for various values of ω_n we get response spectrum as shown in fig.4, with $\zeta=0$
The trace for ζ= 0.05,0.5 & 1.0 are also incorporated in the fig.4
The value of the slope of the curve given by fig.4 for ζ = 0 is calculated by differentiating eqn (12) w.r.t. ω_n

The slope $= \frac{\ddot{u}_0 T}{2g} \sin\left(\frac{\omega_n T}{2} \right) \left[\frac{2(\omega_n/\omega_p)}{1-(\omega_n/\omega_p)^2} \right] + \frac{2\ddot{u}_0}{g} \cos\left(\frac{\omega_n T}{2} \right) \frac{d}{d\omega_n} \left[\frac{\omega_n \cdot \omega_p}{\omega_p^2 - \omega_n^2} \right]$

Slope at origin is $= \frac{2\ddot{u}_0 \omega_p}{g} \left[\frac{1}{\omega_p^2 - \omega_n^2} \right]_{0,0} = \frac{2\ddot{u}_0 T}{\pi g}$, ignoring the term $(\omega_n/\omega_p)^2$

Therefore, for values of ω_n/ω_p <0.5, the response spectrum is very nearly linear.

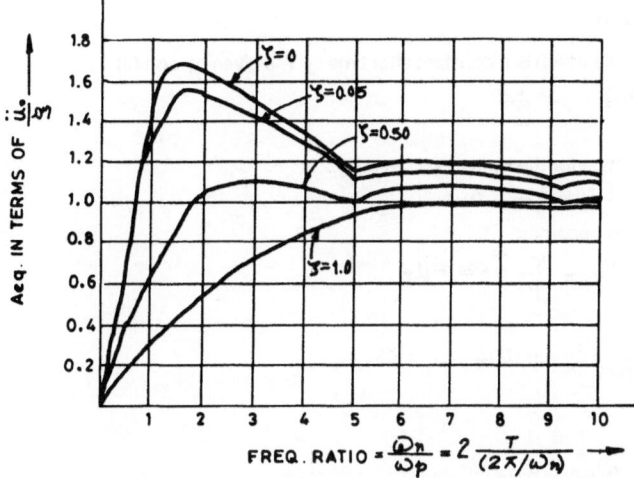

Fig. 4. Response spectrum–maximum value of response for half sinusoidal pulse acceleration.

Hence Aeq. $= \dfrac{2\ddot{u}_o T}{\pi \zeta} \cdot \omega n$ (16)

Also , velocity of the shock pulse which is a constant after time t = T is given by eqn (3)

Substituting in eqn (16)   Aeq. $= \dfrac{\dot{u}\,\omega n}{\zeta}$ (17)

The differential equation of motion for the system shown in fig.2 is

$$m\ddot{x}_t + F\left[\dot{\delta},\delta\right] = -m\ddot{u}_t$$

For a linear spring the equation becomes

$$\ddot{x}_t + \dfrac{c}{m}\dot{\delta}_t + \dfrac{K}{m}\delta_t = -\ddot{u}_t$$

For t > T, $\ddot{u}_t = 0$

Hence, $\ddot{x}_t + 2\zeta\omega_n\dot{\delta}_t + \omega_n^2\,\delta_t = 0$

When $\zeta = 0$

$$\ddot{x}_{max} = -\omega_n^2\,\delta_{max}.$$ (18)

EVALUATION

Based on the above simplified approach design was undertaken for transporting a mobile electronic equipment housed in a Fiber Reinforced Plastic Container of net weight 3000 kg to meet type III mobility requirement as per MIL-M-8090. The suspension system has been designed for a shock isolation of input shock level of 40g of 10msecs.

The trailer thus manufactured with a rated payload was put under field trials and shock measurements were undertaken at the axle and on the bed using B&K accelerometers and readings recorded on a magnetic tape. An extract of the readings are as follows

		Vertical Acceleration of Rear Bed (g)	
Terrain	Speed	Maximum RMS value	Frequent peaks
On track	10 KMPH	0.70	2.0
On track	15KMPH	0.88	2.2
On track	20KMPH	1.20	3.2
On highway	40 KMPH	1.10	3.0
On Highway	50 KMPH	1.1	2.2

CONCLUSION

A simplified approach for the design of suspension system mentioned above was found to be useful in arriving at a trailer for transportaion of electronic equipments needing cross country mobility requirements. Such systems are under production at Bharat Electronics, India.

REFRENCES

1.   Shock and Vibration handbook by Cyril M Harris and Charles E. Crede.
2.   MIL-M-8090
3.   JSS 55555

Chapter V

Workcell Operations

Introduction

The problems encountered in design, simulation, operation, and monitoring of manufacturing cells are analyzed by the papers included in this chapter. The authors of the first paper propose an algorithm to minimize the inter-cell traffic, whereas the second paper is concerned with cost considerations for cell design. Both address the important issues of the Group Technology approach to multi-product manufacturing settings.

The third paper shows several applications of CAD/CAM in the textile industry, specifically improving the design of ring frame cam, optimizing picking cam, and computer-aided textile design. Similarly, the next paper is concerned with the textile industry, for it presents an integrated software package aiding design and manufacture of cams for single spindle automata.

The next three papers focus on the tool replacement problems. The simulation tool, aimed at monitoring of the working of a cutting tool at a machining center within a FMS environment, was developed by the authors of the fifth paper. Finding the realistic number of tool replacements in practical conditions is made possible here. A similar problem is addressed in the next paper, namely the method of determining a suitable time instant of grinding wheel redressing. The method is based on the so-called Group Method of Data Handling technique. The seventh paper proposes a method to predict tool wear on the basis of spectral analysis of tool vibrations.

The eighth paper presents a design of an industrial-grade multi-channel temperature controller. The energy savings resulting from application of the controller in the sugar industry are given.

The last paper discusses use of various types of sensors which will detect the approach of the robot arm close to a human - a problem very important for safety of personnel in roboticized factories.

Group Technology: Cell Formation Using Simulated Annealing

J.M. PROTH

INRIA-Lorraine-SAGEP Project
Cescom
Technopole Metz
Metz, FRANCE

ABSTRACT:

System layout is a key problem for the design of manufacturing systems. The first stage of the layout process is the formation of manufacturing cells. In this paper, we propose a powerful algorithm with an objective to minimize the inter-cell traffic. We propose a real-life example to illustrate our approach.

KEY WORDS: Simulated Annealing, Group Technology, Manufacturing System Design, Optimization, Clustering.

1. INTRODUCTION

The manufacturing system layout problem is drawing more and more attention of the manufacturers. The main reasons are that a well designed layout usually leads to reduction of manufacturing design, implementation and running costs. In our opinion, the first stage of any layout process should be the partitioning of the manufacturing system on hand into cells. Various approaches have been proposed to solve this group technology (GT) problems, in particular in [1], [2], [4] and [5]. As we can see in the literature, some of the algorithms aim at concentrating the operations performed on a part in a unique cell [1], while others aim at minimizing inter-cell traffic [2]. Some authors also simplify the problem in order to meet well-known mathematical models like linear programming or assignment problem models.

In the following, we propose a powerful algorithm based on simulated annealing. The remainder of the paper is organized as follows. In section 2, we present a brief overview of the basis of simulated annealing. Section 3 is devoted to the manufacturing cell design problem. In the last section, we present the real-life application of the algorithm.

2. BASIS OF SIMULATED ANNEALING

We owe simulated annealing to Kirkpatrick et al. [3], whose work is based on the analogy with the behavior of a large set of atoms with regard to the temperature. At a high temperature, the energy of the set of atoms is high. When the temperature decreases, the energy of the system evolves depending on the cooling speed. When this speed is very high, the system reaches a frozen state and the energy remains high. This final state is called "cahos". If the cooling speed is low, the energy decreases with the temperature and, at the frozen state, the mean value of energy is minimal. This final state is called "crystal".

It has been shown that the probability of the energy of the system being e at temperature T in the crystal state is given by:

$$Pr\{e/T\} = exp\{-e/K_B T)\} / U(T)$$

where:

U(T) is the standardization constant

K_B is the Boltzmann's constant

T is the temperature

Metropolis et al. [6] introduced algorithm 1 to find a "crystal" state when the temperature T is known.

Algorithm 1

1. Choose an initial state s_0 and compute the related energy e_0
2. Perturb s_0 in order to obtain $s_1 \in H(s_0)$, in the neighborhood of s_0.

Compute e_1, energy related to s_1.

3. Test:

    3.1. If $e_1 \leq e_0$ set $s_0 = s_1$

    3.2. If $e_1 > e_0$ set $s_0 = s_1$ with the probability $P = \exp[-\Delta_e/(K_BT)]$

4. Go to 2

We stop the computation after some computational step.

Combinatorial optimization uses a similar algorithm with conjunction to a process of lowering the temperature.

Note that solving a combinatorial optimization problem P consists of finding $s^* \in S$ such that:

$$f(s^*) = \underset{s \in S}{\mathrm{opt}} \; f(s) \tag{1}$$

where:

    S is the set of feasible solutions

    f is an objective function $f: S \to \mathbb{R}$

    opt represents either the **maximum** or the **minimum**, depending on the problem.

In the previous explanation concerning simulated annealing, let us replace the terms **state** and **energy** with **feasible solution** and **objective function** value respectively. Furthermore, let us introduce:

    L, which is the number of consecutive steps of the computation using the same temperature.

    $r \in (0, 1)$, which is used to lower the temperature every L computational steps.

    ε, small positive number used to stop the computation. These parameters are provided by the user of the algorithm. Algorithm 2 is used for solving combinatorial problems when the goal is to minimize an objective function.

Algorithm 2

1. Generate an initial feasible solution s_0 and compute $f(s_0)$
2. Introduce L, r, and ε
3. Introduce an initial temperature T
4. Set M = L
5. Generate $s_1 \in H(s_0)$, neighborhood of s_0, and compute $f(s_1)$
6. Compute $\Delta = f(s_1) - f(s_0)$
7. Test:

    7.1. If $\Delta \leq 0$, set $s_0 = s_1$

    7.1. If $\Delta \leq 0$, set $s_0 = s_1$ with the probability $\exp(-\Delta/T)$

8. Set M = M - 1
9. Test:

    9.1. If M > 0, go to 5

9.2. If $M \leq 0$:

     9.2.1. Set $T = rT$

     9.2.2. If $T > \varepsilon$, go to 4

10. Print the best solution among all those that have been visited

Note: In the case when the goal is to maximize the objective function, we just have to change $\Delta \leq 0$ and $\Delta > 0$ into $\Delta \geq 0$ and $\Delta < 0$ in the previous algorithm.

3. MANUFACTURING CELL DESIGN

The problem at hand is to group machines into manufacturing cells in order to minimize inter-cell traffic, assuming that the number of machine in a manufacturing cell is limited (otherwise, the solution consisting of grouping all the machines in the same cell would be optimal).

In the following, $\mathcal{M} = \{M_1, M_2, ..., M_m\}$ is the set of machines, $\mathcal{P} = \{P_1, P_1, ..., P_n\}$ is the set of part types. The unique routing associated to a part type P_k is also called P_k. A weight u_k is assigned to each part type P_k. It represents, for instance, the average number of parts of type P_k manufactured during a one year period. The traffic between M_i and M_j, denoted by t_{ij}, is computed as follows:

$$t_{ij} = \sum_{k=1}^{n} u_k \, q_{kij} \tag{2}$$

where q_{kij} is the number of times M_j follows M_i or M_i follows M_j in the routing P_k.

Let $s = \{C_1, C_2, ..., C_q\}$ be a partition of $\mathcal{M}$ and $f(s)$ the total inter-cell traffic related to s, i.e.:

$$f(s) = \sum_{(i,j)\in E(i,j)} t_{ij} \tag{3}$$

where:

$E(i,j) = \{(i, j)$ such that $i \in C_k, j \in C_r, k, r \in \{1, 2, ..., q\}, k \neq r\}$

We also assume that the number of machines in a cell C_i is upper bounded by an integer N.

$$\text{card}(C_i) \leq N \; ; \; i = 1, 2, ..., q \tag{4}$$

Thus, the problem to solve is as follows:

     find $s^* \in U$ such that:

$$f(s^*) = \underset{s\in U}{\mathcal{M}in} \; f(s) \tag{5}$$

where U is the set of partitions of $\mathcal{M}$ verifying (4).

In order to apply algorithm 2, we have to specify how to define $s_1 \in H(s_0)$ for $s_0 \in U$.

Partition s_1 is obtained by:

(i) either removing a machine from a cell and placing it in another cell, assuming that constraint (4) is verified

(ii) or exchanging two machines located in two different cells.

At each step of the computation:

1. We choose at random one of the previous strategies

2. If strategy (i) is selected:

    2.1. Choose at random (uniform probability) the cell where the machine will be removed and the cell where the machine will be placed

    2.2. Test:

        2.2.1. If the number of elements in the cell where the machine will be placed is equal to N, go to 3

        2.2.2. Otherwise, choose at random (uniform probability) a machine in the first cell and place it in the second cell

3. If strategy (ii) is selected:

    3.1. Choose at random (uniform probability) a pair of cells

    3.2. Choose at random (uniform probability) one machine in each of the selected cells

    3.3. Exchange the selected machines

4. A REAL-LIFE APPLICATION

We applied the previous approach to an industrial example with 292 machines and 460 part types. The weight w_i ($i = 1, 2, ..., 460$) was equal to unity. The number of machines in a cell is limited to 10.

The parameter values were $T = 500$, $\varepsilon = 0.01$, $r = 0.95$ and $L = 30$. The algorithm was executed five times. The results are presented in table 1: A SUN 3/60 has been used for the computation.

Table 1: A numerical example

Example number	Results		
	Number of cells	Inter-cell traffic	Computation time (sec.)
1	30	40830	441
2	31	43537	402
3	32	40609	402
4	31	41707	428
5	31	41842	422

5. CONCLUSION

An important property of simulated annealing is to reach various good results (i.e. various results with a low traffic) starting from the initial feasible solution. Thus, this approach provides to designers a set of good solutions among which he can make his final choice (on qualitative basis, for instance).

The difficulty associated with the simulated annealing approach is the choice of the parameter values: unfortunately, there is no procedure to obtain the best values for T, ε, r and L.

REFERENCES

[1] GARCIA H. and PROTH J.M.,
"A New Cross-Decomposition Algorithm: the GPM. Comparison with the Band Energy Method", Control and Cybernetics, vol. 15, n° 2, pp. 115-165, 1986.

[2] HARHALAKIS G. NAGI R. and PROTH J.M.,
"An Efficient Heuristic in Manufacturing Cell Formation for Group Technology Applications", International Journal of Production Research, vol. 28, n° 1, pp. 185-198, 1990.

[3] KIRKPATRCK S., GELATT C.D. and VECCHI A.,
"Optimization by Simulated Annealing", Science, vol. 220, n° 4598, May 13, 1983.

[4] KUMAR R.K., KUSIAK A. and VANNELLI A.,
"Grouping of Parts and Components in Flexible Manufacturing Systems", European Journal of Operations Research, 24, pp. 387-397, 1986.

[5] McAULEY J.,
"Machine Grouping for Efficient Production", The Production Engineer, pp. 53-57, Feb. 1972.

[6] METROPOLIS N., ROSENBLUTH A., ROSENBLUTH M., TELLER A. and TELLER E.,
"Equation of State Calculations by Fast Computing machine", J. Chem. Phys., 21, pp. 1087-1092, 1953.

Cost Considerations for Cell Design in Group Technology

KENNETH R. CURRIE

Department of Industrial Engineering
Tennessee Technological University
Cookeville, TN

Abstract

Group Technology exploits the similarities of part types to form part families and machine cells so that a particular family of parts is completely manufactured by a particular cell of machines. Group Technology has found many applications in multi-product, small lot-size manufacturing settings, where the benefits range from reduced tooling and set-up costs to reduced inventories and shorter throughput time. Previous research has focused on cell design as a mathematical exercise of partitioning a binary machine-component matrix to minimize the inter-cellular travel of parts between cells. This paper will highlight recent research papers that have attempted to incorporate cost considerations into the design of manufacturing cells. Specifically, a technique referred to as MAPFLO (Multi-Attribute Part Family LOading) will be presented which is one method that attempts to establish the link between cell design and an economic benefit/cost tradeoff.

Introduction

Group Technology (GT) has received a resurgence in interest and applications due to the introduction of Just-In-Time (JIT) and Flexible Manufacturing Systems (FMS). Both JIT and FMS rely on the saving of effort and planning achieved from grouping similar parts to be produced in a cell of interrelated processing and handling equipment. The underlying principle of GT is a very old one that can be stated very simply; "Applying a common solution methodology to a group of similar problems". The difficulty in applying GT is defining which solution methodology is appropriate and how similar (or dissimilar) should the problems be. Or in terms of manufacturing discrete parts, by what criteria should one define *similar* parts, and to what extent are the parts and machines included in a cellular layout. These two problems are interrelated since the benefits of GT are dependent upon how part families and machine cells are formed.

Research Issues in Cell Design for Group Technology

The literature dealing with cell design issues has focused primarily on a binary matrix comprised of components and their corresponding machine routings (referred to as a machine-component matrix), and restructuring the matrix so that it is partitioned into part families and machine cells (see King [1], Chan and Miller [2], and Seifoddini and Wolfe [3] for examples). This particular approach has several faults as detailed below:

1.   No consideration is given for differing capacities and demands of equipment selected for a particular cell.

2.   Implicitly assumed is design similarities coincide with manufacturing similarities. In many cases this is not true and can result in significant losses usually associated with GT with respect to tooling and setup savings.

3.   In many cases the cell design is "all or nothing", in that all parts are grouped into a cell, and there is no allowances for prioritizing changes from a functional layout to a cellular layout.

4.   Economics has played a very minor role and has only been used in minimizing the inter-cellular movement of parts.

Recent research has taken a more holistic approach to cell design which incorporates several of the components that are missing from the machine-component approach. Choobineh [4] incorporated both costs and machine loading into an integer programming model to design manufacturing cells. Ballakur and Steudel [5] also incorporated machine capacities in loading machine cells. Very often the economic costs and benefits of GT have been ignored, primarily because they are so difficult to quantify. The benefits of GT are particularly difficult to estimate in advance due to the lack of empirical data of the potential savings due to GT. Whatever benefits are achieved through GT are dependent upon the potential for savings, the type of cellular layout used (i.e. GT flowline, GT cell, hybrid cell, etc.), and the composition of the parts and machines that represent the family of parts produced in a particular machine cell.   A new approach to cellular design was developed by Currie [6] called Multi-Attribute Part Family LOading (MAPFLO). MAPFLO incorporates both machine capacities as well as benefit/cost tradeoffs into a decision support tool to aid in the design of part families and machine cells.

MAPFLO Approach to Cell Design

MAPFLO is comprised of three segments; 1) Part Family Formation, 2) Machine Cell Loading and, 3) Cell Justification. The focus of this paper will be the third segment dealing with cell justification, however the process by which part families/machine cells are formed plays an integral part in determining whether a particular cell is economically justified. Therefore, before addressing the issues of cell justification, background information concerning part family formation and machine cell loading as performed in MAPFLO will be discussed.

Part families are formed by considering similarities among part types based on design as well as manufacturing characteristics. Using a unitized distance metric proposed by Anderberg [7], called a *disagreement index*, the relative similarity of one part to another is used in an hierarchical complete linkage cluster analysis. The development of the disagreement index permits only appropriate mathematical operations on mixed variable types such as nominal, interval, ordinal, and binary variables. The major benefit of the part family formation segment is the ability to weight design and/or manufacturing characteristics to determine the sensitivity of part family formation to design similarities, or manufacturing similarities, or both simultaneously. A more detailed description of the application of the disagreement index and part family formation is presented by Currie and Creese [8].

The process of cell loading uses the part groups formed from the part family formation stage and assigns machines to cells based on two assumptions; 1) only one part family is allowed to operate in a given cell and , 2) none of the operations are performed outside the cell. The first assumption limiting cell membership to only one family is not very restrictive, and flexibility occurs by relaxing the threshold value to form larger families which may consist of one or more smaller family groupings. The second assumption, limiting the operations to only one family members, is established so the benefits derived by GT are not diluted by introducing dissimilar parts that may increase setup times or reduce throughput. The net result at the end of the cell loading phase is the formation of a number of machine groups, each capable of producing a particular family of parts entirely within the machine group. The final step is to determine through economic justification which if any of the machine groups should be formed, and correspondingly which parts should be produced by cellular manufacturing.

Cell Justification Using MAPFLO

Machine cell justification is a trade-off between the benefits of manufacturing in a cellular layout versus the costs associated with under-utilization and movement of equipment. Under-utilization of equipment is created by the assumption that none of the operations required for a part family can be performed outside the cell (no inter-cell movement). This results in the purchase of additional equipment to satisfy the machine requirements of those parts not produced within a cell. The benefits used in the MAPFLO analysis are inventory reductions, scrap and rework reductions, and reductions in material handling. Reduced setup times due to design similarities in a part family are offset by an increase in the number of setups to take advantage of the benefits of the JIT philosophy which translates into a "make to order" schedule. Non-quantifiable savings such as decreased throughput times, improved exit quality, and higher worker morale should be analyzed from a strategic standpoint.

The expected benefits of a particular part family, if produced within a cell, is calculated using the following components; 1) the demand of each part in the family, 2) the expected percent reduction (as estimated by the user) for scrap, inventory, and material handling costs, and 3) the current scrap, inventory, and material handling costs on a per part basis. The costs of establishing a particular machine group are based on the annualized equipment costs of purchasing new equipment and moving existing equipment. The number of machines of a particular machine type required for a part family, plus the number of machines of the same type required for the remaining parts, is subtracted from the total number of machines available to determine the costs for additional equipment. Tables 1 and 2 illustrate the calculation of benefits and costs, respectively.

Although the benefits as calculated in Tables 1 and 2 are static, the user is able to change any of the expected percent savings and key cost parameters to test for solution sensitivity. Examples of cost parameters that the user has the ability to alter in MAPFLO are expected utilization loss of equipment capacity due to machine breakdowns and interruptions, lost opportunity cost or holding cost of the inventory, and changes in annual volume of one or more parts. The costs of introducing a

Part No.	Annual Volume	Unit Cost	Current Costs			Projected After-Tax[4] Savings Due to GT			TOTAL BENEFIT
			Ind.Lab.[1]	Inventory[2]	Scrap[3]	Ind.Lab. (50%)	Inventory (40%)	Scrap (50%)	
1	37,000	$ 9.00	$8,325	$12,488	$49,950	$2,498	$2,997	$14,985	$20,480
4	22,000	$10.00	$5,500	$ 8,250	$33,000	$1,650	$1,980	$ 9,900	$13,530
[1]Indirect Labor Costs = 2.5% of the annual product cost [2]Inventory Costs = 15% of the average inventory cost [3]Scrap Costs = 15% of the annual product cost [4]After-Tax Savings = (Savings)(1 - Marginal Tax Rate=40%)									

Tab. 1. Annualized After-Tax Calculation of GT Benefits for a Part Family

particular part family are calculated assuming that all remaining parts and machines will be grouped into a "remainder cell" consisting of the remaining machines plus the purchase of additional machines needed to satisfy demand due to under-utilization of equipment. A minimum acceptable net benefit is used to determine justification for cellular manufacturing, and the family with the highest acceptable net benefit is justified for cellular manufacturing. This incremental approach to cellular justification is repeated until all part families are justified for cellular manufacturing or there are no more cells that satisfy the minimum acceptable net benefit

Equipment Cost Component	Estimate of After-Tax Cost	Annualized Discount Factor	Annual After-Tax Cost
Initial Cost	$96,000	$(A/P,i,7)=0.2054$	$19,718
Annual Overhead Contribution	10% of first cost (after taxes) $(0.10)($96,000)(1 - 0.4)=$5,760$	1.0	$5,760
Salvage Value	5% of first cost= $(0.05)($96,000)=$4,800$	$(A/F,i,7)=0.1054$	($ 506)
Annual Depreciation 1st yr 2nd yr	^{3} $(0.1429)($96,000)(0.4)=$5,487$ $(0.2449)($96,000)(0.4)=$9,404$	$(P/F,i,1)(A/P,i,7)=.1867$ $(P/F,i,2)(A/P,i,7)=.1698$	($1,024) ($1,597)
. . 7th yr	. . $(0.0893)($96,000)(0.4)=$3,429$	. . $(P/F,i,7)(A/P,i,7)=.1054$	. . ($ 361)
TOTAL			$19.445

[1] Interest rate is assumed to be 10% [2] Marginal tax rate is 40%
[3] Current MACRS depreciation schedule with a recovery period of 7 years

Tab. 2. Annualized After-Tax Calculations for a $96,000 Piece of Equipment

Valve Manufacturer Results

The MAPFLO program has been used to evaluate the justification of GT in a valve manufacturing facility producing very useful results. Two part families, Gate Valves and Sillcocks, were justified for cellular manufacturing which comprised approximately 42% of the 118 different part types analyzed. The total expected annual net benefit from introducing these two cells was $42,800. The part family of

316

Sillcocks was found to be extremely sensitive to the cost parameter estimate of equipment utilization. The use of the disagreement index helped in highlighting part families with similar manufacturing characteristics yet dissimilar design characteristics, and vice versa. One specific part family was observed to have very similar design similarities to another group of parts, but because the manufacturing characteristics were dissimilar the two groups of parts were not justified for cellular manufacturing. If a machine-component matrix approach had been applied to this particular problem it could have been easily partitioned, but not all part families are economically justified for cellular manufacturing due to small production volumes (resulting in low expected benefits) and/or equipment inefficiencies (increased costs).

Future Directions

The research future for GT and cell design is understanding or estimating the relationship of savings and costs to the interdependence of part family formation, machine cell design, and economic justification. Increasing emphasis must be spent on assessing the strategic importance GT has on throughput, quality, and labor relations, and not so much on algorithmic reorganization of a machine-component matrix.

References

1. King, J. R. Machine Component Grouping in Production Flow Analysis: An Approach Using a Rank Order Clustering Algorithm. Int J of Prod Res. 18 (2) (1980) 213-232.
2. Chan, H. M. and Milner, D. A. Direct Clustering Algorithm for Group Formation in Cellular Manufacture. J Mfg Sys. 1 (1) (1986) 66-75.
3. Seifoddini, H. and Wolfe, P. M. Application of the Similarity Coefficient Method in Group Technology. IIE Trans 18 (3) Sept 1986 271-277.
4. Choobineh, F. A Framework for the Design of Cellular Manufacturing Systems Working Paper, Department of Industrial and Management Systems Engineering, University of Nebraska-Lincoln (1985).
5. Ballakur, A. and Steudel, H. J. A Within-Cell Utilization Based Heuristic for Designing Cellular Manufacturing Systems. Int J Prod Res. 25 (5) (1987) 639-665.
6. Currie, K. R. Multi-Criteria Part Classification and Machine Cell Formation in a Cellular Manufacturing Environment. Unpublished Ph.D. Dissertation: West Virginia University (1988).
7. Anderberg, M. R. Cluster Analysis For Applications. New York: Academic Press (1973).
8. Currie, K. R. and Creese, R. C. Justification of Cellular Manufacturing Using Multi-Attribute Part Family Loading - MAPFLO. Justification Methods for Computer Integrated Manufacturing Systems. Elsevier Press (1990) 203-219.

Application of CAD/CAM in the Textile Industry

P. B. JHALA

Machinery Design Division
Ahmedabad Textile Industry's Research Association
Ahmedabad, India

Abstract

In the textile industry for the manufacture of textiles as well as machinery, the CAD/CAM technology is receiving wide spread acceptance. It has been found an immensely useful tool in design and development of textile machinery. In the paper a few cases of such applications namely improved ring frame cam, optimal design of picking cam and computer aided textile design which were undertaken in the R&D work of ATIRA are briefly described. All these designs are applied in industrial plants.

Introduction

CAD/CAM is becoming more and more powerful tool for the manufacturing industry offering several benefits such as improved product design, increased productivity, higher utilization and better quality control. In the textile industry too for the manufacture of textiles as well as machinery the CAD/CAM technology is receiving wide spread acceptance. Already some beginnings have been made with the Computer Integrated Manufacturing (CIM) system in textile mills from planning through product development right upto the actual production process. CAD/CAM is also being found an immensely useful tool in design and development of textile machinery.

In the R&D work at ATIRA[1] the computer is often used to design and develop textile mechanisms as well as to simulate textile weaves. To cite a few examples (i) A cam of special computer-aided design for the ring frame enables more yarn to be wound in a compact form on the package. (ii) On the loom, the three dimensional picking cam to propel the shuttle was computer-designed and computer-plotted so that different profiles could be experimented with to achieve best results. (iii) Computer-aided textile designing enables quick translation of design

---

1. ATIRA is the author's institution

ideas into woven design on personal computer. All these designs are applied in industrial machines and plants and they are in use in large numbers in textile mills.

DPM Cam Design for Ring Frame

In the spinning departments of textile mills, conversion of the roving into yarn is done on ring frame. In this machine as shown in Fig. 1, yarn spun is wound in layers on a package over a profile which is conical in shape. Such a movement called chase is obtained by imparting an oscillatory movement to the ring rail (A) with the help of a cam (C). It is desirable that the overall yarn content in the package (B) must be large since it reduces doffing cost at ring frames, increases efficiency in winding machines, raises loomshed efficiency and reduces

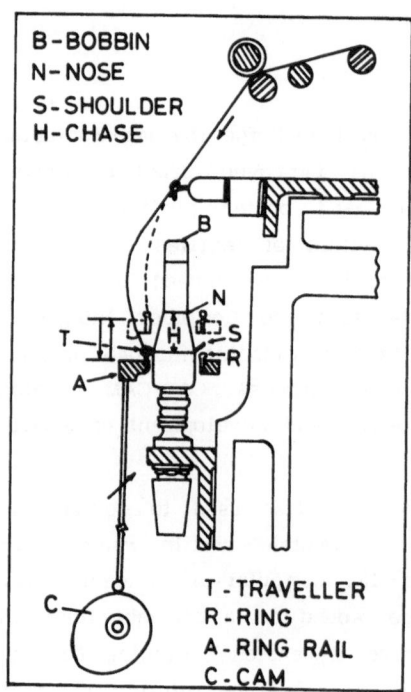

Fig.1. Build up mechanism

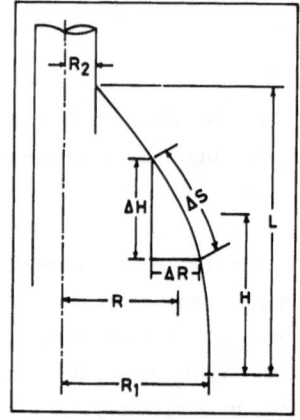

Fig.2. Package geometrical details

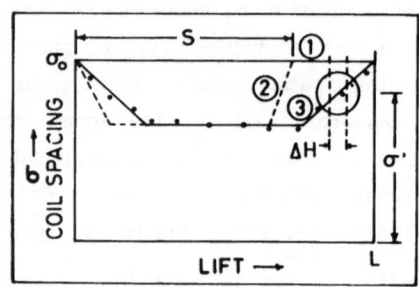

Fig.3. Transformation for DPM design

hard waste. At the same time, the problem of sloughoff should not arise. These have conflicting requirements a compromise between which is struck in DPM Cam design by densely packing the middle region of the chase to get high yarn content and providing wide spacing in the nose and shoulder region whose geometrical details are shown in Fig. 2.

When this complex coil spacing pattern on a package of varying diameter has to be translated into cam design, the analytical solutions become extremely complicated and in actual design a semi-empirical approach had to be resorted to, which however, was found to provide sufficient rigour for practical design purposes [1]. In this approach design starts with known uniform coil spacing (UCS) profile which is successively adjusted in such a way as to be converted in stages towards the DPM profile. The actual procedure is as follows : Refering to Fig. 3 for the UCS design the successive ΔH are calculated for small and constant angular displacement of $\Delta\theta$. This being a UCS design will give a constant coil spacing σ_o throughout. Considering any small region as circled, the required spacing for DPM cam is σ' ($<\sigma_o$), which means that the cam movement, instead of being ΔH, must be $\Delta H'$ ($<\Delta H$) given by $\Delta H' = \Delta H \, \sigma'/\sigma_o$. After making this adjustment the total lift ($\Sigma\Delta H'=S$) will become less than the specified chase length L. To adjust for this, an overall transformation is made in the ratio of L/S that is, $\Delta H^*= \Delta H' \, L/S$. This gives $\Sigma\Delta H^*=L$.

The calculations involved were complicated and voluminous and hence the aid of coputer was resorted to and not only were data on the cam design parameters and coil spacings, but also information directly useful in the fabrication of the cam were also computed. The DPM cam design gives upto 25% higher yarn content and is now in use in 180 mills accounting for more than a million spindles.

Optimum Picking Cam Design for Loom

In conventional looms, shuttle propulsion which is known as picking, is one of the most crucial operations which accelerates the shuttle to a speed of 12 m/s in about 25 ms involving an increasing acceleration reaching to a peak of 1000 m/s^2, that is about 100g. The picking mechanism as shown in Fig. 4 is operated by a three dimensional cam which should give the desired velocity and smooth acceleration to the shuttle with low torque on the cam shaft. But the picking cams of most of the conventional looms are deficient in this regard.

320

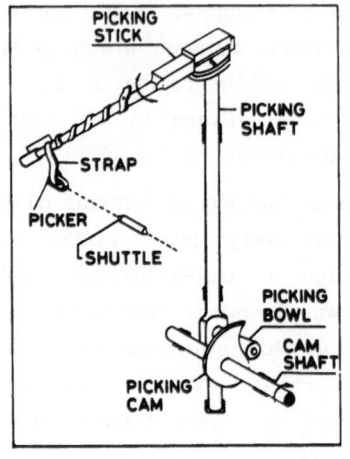

Fig.4. Overpick mechanism

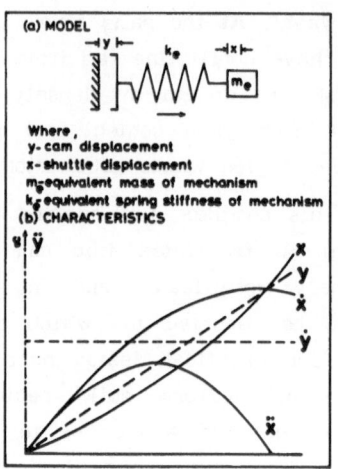

Fig.5. Dynamics of mechanism

Table.1. Characteristics of cam and shuttle movements.

Types of Cams/ Characteristics		Linear	Cosine	Polynomial
CAM	y	$p\omega t$	$a(1-\cos\omega t)$	$q\omega t+r\omega^2 t^2+s\omega^3 t^3$
	$\dot{y}$	$p\omega$	$a\omega\sin\omega t$	$q\omega+2r\omega^2 t+3s\omega^3 t^2$
	$\ddot{y}$	$\longrightarrow\infty$	$a\omega^2\cos\omega t$	$2r\omega^2+6s\omega^3 t$
SHUTTLE	x	$p\omega\left(t-\dfrac{\sin nt}{n}\right)$	$\dfrac{a}{n^2-\omega^2}\left[n^2(1-\cos\omega t)-\omega^2(1-\cos nt)\right]$	$q\omega\left(t-\dfrac{\sin nt}{n}\right)+r\omega^2\left[t^2-\dfrac{2}{n^2}(1-\cos nt)\right]$ $+s\omega^3\left[t^3-\dfrac{6}{n^2}\left(t-\dfrac{\sin nt}{n}\right)\right]$
	$\dot{x}$	$p\omega(1-\cos nt)$	$\dfrac{a\omega n}{n^2-\omega^2}\left[n\sin\omega t-\omega\sin nt\right]$	$q\omega(1-\cos nt)+2r\omega^2\left(t-\dfrac{\sin nt}{n}\right)$ $+3s\omega^3\left[t^2-\dfrac{2}{n^2}(1-\cos nt)\right]$
	$\ddot{x}$	$p\omega n\sin nt$	$\dfrac{a\omega^2 n^2}{n^2-\omega^2}\left[\cos\omega t-\cos nt\right]$	$q\omega n\sin nt+2r\omega^2(1-\cos nt)$ $+6s\omega^3\left(t-\dfrac{\sin nt}{n}\right)$

For optimising the cam design, a detailed dynamic model of the picking mechanism as shown in Fig. 5 was constructed and cams of linear, cosine and polynomial profiles were designed [2]. The characteristics of cam and shuttle movements from the mathematical equations in Table 1 were calculated for these three types of cams using computer. To obtain functionally correct and accurate profile, the cams were plotted on a computer using the principle of envelopes [3]. The actual procedure is as follows : Referring to Fig. 6, for the overpick cam mechanism, the

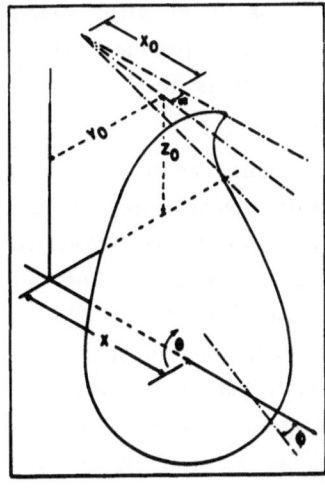

Fig. 6. Geometry of overpick
mechanism

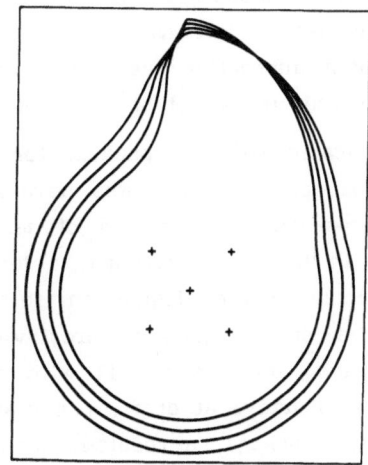

Fig.7. Computer-plotted
polynomial picking cam

follower is normally cone-shaped and swings parallel to the XY plane
with its fulcrum on the YZ plane. X axis represents the axis of rotation
of the picking cam. The cam profile is obtained as the envelope of
the cone for a given relationship between θ , the angle of cam rotation
and ϕ , the angle made by the projection of the axis of the follower
on the XY plane with the positive direction of the X-axis. For plotting
in two dimensions, the cam profile is obtained as sections corresponding
to various X-values. At any given X value, the cross-sections of the
cone in the YZ-plne are obtained for various θ values and related
ϕ values. A typical profile of the picking cam plotted on computer
is given in Fig. 7. The computer data were also used for the manufacture
of cams as per the plot. The polynomial cams which give the same
shuttle velocity from both the sides and provide smooth pattern of
strain and torque was chosen as the optimum. The polynomial cams are
in use in more than 25 mills.

Computer Aided Textile Design

Today's textile market is dynamic needing frequent changes in designs
and color of the fabric. The fabric designing based on conventional

point-paper technique and sample weaving on handloom with manual heald lifting has limitations in terms of creativity and productivity. Use of computer interactive graphics and machine control is the judicious choice to enhance the output of a textile designer.

A dedicated software package for textile designing on an IBM compatible personal computer [4] was developed to suit woven design requirements of three different sectors of textiles namely Handloom, Powerloom and Mill. In the first software, a textile designer can choose the grid size and input woven design on this grid using soft keys. The special features of the package are, viewing of warp and weft face design, rearranging warp and weft with many permutations and analysing fabric design to arrive at draft, peg plan, weft cross-section etc. In the second software package, a resultant woven design can be arrived at by feeding in data on draft and peg plan which is a case of reverse engineering. The special features of this package are combining draft and peg plan in any desired manner keeping in view the loom constraints. There is good demand for these softwares from the textile industry.

Acknowledgements

The author is grateful to the Director, ATIRA, for his kind permission to present this paper at this conference. The work reported here was done by several researchers at ATIRA over the years. I am thankful to all of them and in particular to Mr. M. Ratna Prabhu and Mr. V.S. Jadeja.

References

1. Ratna Prabhu, M.; Lunkad, M.N. : Design of builder motion cams in textile ring frame. Proc. of the Sixth World Congress on Theory of Machines and Mechanisms, Delhi, Wiley Eastern Ltd., New Delhi, 1983.

2. Jhala, P.B.; Venkataramanan, C.G. : Optimisation of Picking cam design through dynamic studies. Proc. of the Fifth World Congress on Theory of Machines and Mechanisms, ASME, New York, 1979.

3. Narasimham, T.; Trivedi, G.H.; Venkataramanan, C.G. : Computerised plotting of shedding and picking cams. Proc. of the 17th Tech. Conf., SITRA, Coimbatore, India, 1976.

4. Jadeja, V.S.; Jhala, P.B. : Computer simulation of fabric weave and colour. Resume of Papers, 28th Tech. Conf., SITRA, Coimbatore, India, 1987.

CAD/CAM of Cams for Use in Automatic Lathes

P.C. PANDEY, N.K. MEHTA and AATUL WADEGAONKAR

Department of Mechanical and Industrial Engineering
University of Roorkee
Roorkeee, India

Abstract

The present work deals with the development of integrated computer software for the design and manufacture of cams for single spindle automats. The software developed for this purpose comprises of 3 modules; responsible for cam layout, selection of best base curve and CNC machining. The performance of the software has been illustrated by means of a typical example.

Introduction

Traditional method for design and manufacture of cams, for automatic lathes involves:

(a) Calculation of cam rise and rotation for each operation to be completed (cam layout).

(b) Design analysis of cam profile, and

(c) Manufacture of cams.

This procedure is lengthy. Furthermore, new sets of cam are needed whenever the component design changes. Application of integrated CAD/CAM approach for cams appears to be ideal.

Present work deals with the development of computer software for :

(1) design and layout of cams required for machining of components on single spindle automatic lathe of Traub make (Model A 25) and

(ii) Development of software for CNC milling of cam profile based on the design data.

In order that CAD software is valid for a group of parts, a master component having 9 different machining features was considered (Fig. 1). From this component it is possible to obtain more than 500 different part designs.

Past Work

As such, no published literature on integrated CAD/CAM systems for cams to be used on automats is available. However,a number of references to published work concerning computer applications to cam design/manufacture can be found [1-5].

In addition to the usual criteria employed for the design of cams for conventional applications the cams for automats should have following features:

The various cams for production of a component be arranged such that, all the operations on the blank are carried out in minimum time . Due consideration be given to balancing of forces, overlapping of operations, machining accuracy, surface finish etc.

CAD of Cams

A computer program has been developed in COBOL, for execution on DEC2050 computer system, for the computation of cam layout data and analysis(Fig.2). The program uses 3 database files and 2 input data files. A brief description of these files are:

DATA II DAT database file contains data on 9 selected operations regarding depth of cut, permissible feeds for five different work materials, namely brass, aluminium alloy B-1113 alloy, structural steel and stainless steel.

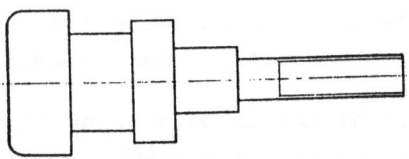

Fig.1: Composite Component

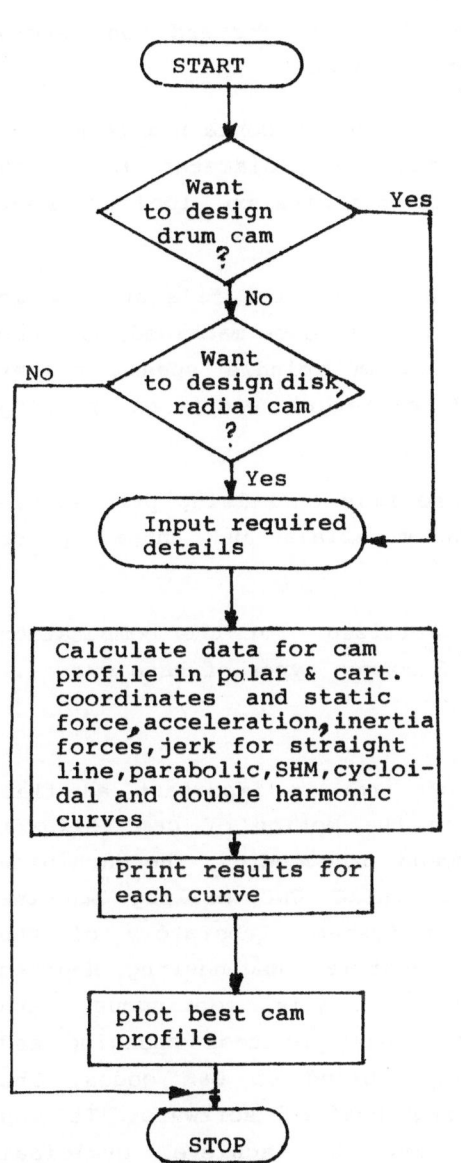

Fig.2 Flow chart for Cam Design and plotting.

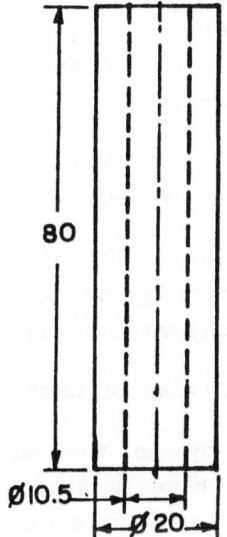

Material:Steel 4140
Stock bar   dia.:25.0mm

Fig.3: Automotive Engine
       Valve Guide.

DATA 33 DAT: Database file consists of forward and return spindle speeds of the threading attachment.

DATA 44 DAT is data input file and should contain information concerning the operation to be performed, diameter and length to be machined and the ratio of cam rise to tool movement etc.

DATA 55 DAT: This is the second data input file and should contain information regarding material to be machined, maximum bar size, production rate, maximum spindle speed, number of turning operations, drill diameter, width of parting tool etc.

DAT 66 DAT: This is a data base file containing permissible cutting speeds for different materials and thread pitch values.

The cam layout data is next utilized for the computation of acceleration, jerk etc. for 5 common types of base curves.

CNC Milling of Cams

Accuracy in the manufacture of cams is important as this would have a direct bearing on the motion of cam follower system and the quality of component produced[6]. The machining of cams has been planned on TRIAC CNC milling machine available in the Manufacturing Systems Laboratory of the Departmen of Mechanical and Industrial Engineering, Roorkee University. The machine has a 3-axis continuous path controller with facilities for automatic tool changing and uses standard programming format based on G&M codes. The NC program was developed using Denford software. It was not possible on this system, to achieve cycloidal interpolation therefore milling of cycloidal profile was programmed using linear interpolation. The step size for interpolation was decided on the basis of permissible profile error and keeping in view the fact that the interpolation data has to be fed manually to the processor.

Results

The cam design data sheet for the component in Fig.3 is presented in table 1 whereas the table 2 gives the follower velocity, acceleration and Jerk for different base curves. A typical cam profile is shown in Fig.4.

In view of the limitations of the CNC system a fully integrated CAD/CAM system could not be achieved. Furthermore, the accuracy of the manufactured cam profile was poor because the step size of 2° used for interpolation did not to yield accurate results.

References

1. Pusztai, J., and Sava, M., Computer Numerical Control Reston, Virginia, 1983.

2. Frisch J. and Renyuan, Fei, Computer Aided Design and Production of Plate Cam Contours, Proc. 13th Conf. on Prod. Res. & Tech., May 1985, Washington (DC).

3. Stockmann, P., and Wallier, G., Combination of CAD and CAM in Machine Tool Manufacturing, Prof. 15th Int. MTDR Conf., Sept. 1974, Birmingham (UK).

4. Grewal, P.S., Computer Simulation of Cam System Dynamics and Sensitivity Studies, Proc. 4th Int. Conf. CAD, CAM, Robotics & Factory of Future, Dec. 1989, New Delhi.

5. Mishra, O.P., Computer Aided Profile Design of Cams for Single Spindle Automatic Lathe, M.E. thesis, Univ. of Roorkee, 1989.

6. Norton, R.L., Effects of Cam Manufacturing Methods on Dynamic Performance of Follower Acceleration Waveforms, Proc. 13th NAMRC, May 85, Berkeley, California.

Table 1: Cam Design Data

Component: Automotive Valve guide (Fig. 3.0)
Production rate   110 pcs/hr

	Operation				
	Turning 1	Parting	Drilling	Hold cam rise 1	Hold cam dwell 1
Dia.(mm)	20.0	0.00	10.5	0.00	0.00
Length of machining (mm)	80.0	0.00	82.0	0.00	0.00
Tool travel (mm)	81.0	14.50	21.6	10.00	81.00
Cam throw(mm)	54.0	21.60	83.0	10.0	10.0
Cam rise(throw) over degrees	40.0	101	132	5.0	40.0

Note: Main spindle rpm = 1800
      Effective degrees available in Prod. cycle = 310

Table 2: Dynamic Characteristics of base curves

Follower motion	Max. accel.(mm/sec^2)	Maximum jerk(mm/sec^3)
Straight line	infinite	infinite
Parabolic	1.9444	2.888
Simple harmonic	2.3959	2.3959
Cycloidal	3.0521	0.2395
Double harmonic	-4.7549	4.7549

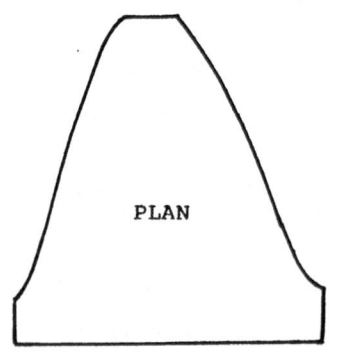

RISE : 5.4000 CM for 40 DEGREES
DWELL FOR 10 DEGREES
FALL: 5.4000 CM FOR 50 DEGREES
THICKNESS OF CAM PLATE : 1.0000CM
MIN WIDTH OF CAM PLATE : 1.0000CM
MAX WIDTH OF CAM PLATE : 6.4000CM
RADIUS OF CAM MOUNTING DRUM: 3.0000CM

RECOMMENDED MATERIAL FOR CAM:
        HARDENED STEEL/HARDENED TOOL STEEL
(CAM ROTATES COUNTER-CLOCKWISE)

PLAN

ELEVATION

SIDE VIEW

Figure 4 : Drum Cam

An Objective SIMTOOL in FMS

C. ESWARA REDDY and O. V. KRISHNAIAH CHETTY
Department of Mechanical Engineering

DIPAK CHAUDHURI
Engineering and Management
Indian Institute of Technology
Madras, India

Summary
A simulation model `Objective SIMTOOL' is developed to monitor
the working of a cutting tool at a machining centre in an FMS
environment to find number of tool replacements due to the effect
of important parameters of tool viz., life, wear, temperature and
breakage. Unlike previous models the SIMTOOL considered both part
and volume flexibilities with combination of turning, drilling,
taping etc., normally performed on a job. The simulation output
shows that most of the time, the tool fails before its predetermined life due to either wear or temperature or breakage or combination. This necessiates the tool monitoring to find realistic
number of tool replacements for a job and whenever a part variety
or volume changes for an effective Tool Requirement Planning.

Introduction

FMS is generally considered capable of processing a variety of
parts on a random basis and can accommodate the production
facility with low volume/high variety of part mix to one with
medium volume/low variety of part mix [1]. Advances in Flexible
Manufacturing require updated machinery and methodologies for the
design and integration of the functional subsystems. The full
exploitation of **Flexibility** is obtained by gathering quantitative
process information with simulators to give detailed duplication
of the dynamic behaviour of physical systems. Production is
dictated by cutting tools, cutting conditions, process monitoring
and tool control strategies. In automated manufacturing system it
is important to keep the machine running than to maximise cutting
rates[2], which leads to shorter tool life and thus requiring more
cutting tools. Hence, paramount to have modules which will
automatically determine the various cutting tools and assess the
need of resetting/replacement. In Flexible Manufacturing System
(FMS), tool replacement can be triggered by one of the following:

* completion of predetermined tool life,
* primitive tool wear or breakage detected by touch
  sensors/vision systems
* excessive tool temperatures developed
* poor surface quality produced on the job due to metallurgical
  defects on tool or job, monitored by image processing systems.

The tool monitoring is an essential activity of Tool Management. Gray et al[3] discussed the tool management issues at (a) individual tool level(b) Machine level and (c) System level The machine group level was not mentioned. Walter has observed that the effective tool management could help unit production time down by 30% and a 70% reduction in tool preparation time and introduced `Walter TDM' (Walter Tool Data base Management)[4].

Literature Survey

Age (tool life) and wear were considered in diffusion threshold simulation model[5], where as tool life and load variation on motor due to wear were monitored in simulation for tool change in drilling operation[6]. In the FMS environment, the replacement of tools was estimated through a post-processor simulator which showed the tool changes due to product variety were less compared to the tool wear in rough turning and finish turning[7]. This may hold good in a typical case but cannot be generalised[3].`KALMAN' filtering was used to estimate breakages in milling operation [8] but this was limited to milling. Tool flow simulation model estimated the tool changes due to wear and breakage in turning, considering the change in part variety[9]. But the changes in volume flexibility was not mentioned.

Though the individual operations were considered to find the replacements of tools due to either wear or breakage the combination of different operations normally performed on a job in an FMS environment was not considered. Moreover the rise in interface temperature which has an impact on subsequent working of tool was not considered. Rise in temperature beyond the design limit increases crater wear, causes plastic deformation of tool nose and increases local stresses and also results in sudden failure of high speed steels[10]. The carbide tools subjected to intermittent cutting like in milling may lead to thermal cracks on the face[10]. The tool may not require immediate replacement on attaining allowable temperature limit but if allowed to continue, the high interface temperature will not only cause sudden failure of the tool but also deform the work surface.

The present paper investigates the effects of tool parameters, viz., life, wear, temperature and breakage, in tool replacement considering the combination of turning, drilling and tapping on a variety of parts with different volumes. An Objective type of simulation model called `SIMTOOL' is developed to find the number of replacements on real time basis.

System Description

On arrival of a job, the tools required for its production are made available. The loading of tools can be in two different ways based on (1) Part Dominant System, (2) Tool dominant system. The sequence of operation of a job and the tool numbers are shown in Table 1. The tool is identified by considering the operation and the material. Three digit and five digit codes are incorporated to identify the job and the tool respectively, in the data-base.

228   ---->   Job Number.

14030   ---->   First two digits indicate the tool material and the next three digits indicate the operation. For eg. 14 for Carbide & 030 drilling operation.

The machining database consists of data relating to cutting tools for different materials like, HSS, coated HSS, carbide, carbide coated and ceramic; tool parameters like predetermined life, wear and temperature limits and cutting angles, [11], [12] & [13].

Data Generation

The tool life, tool wear, rise in temperature and breakage of a cutting are considered and estimated using standard equations.

Tool life: Cook [14]

$$T = AV^{-B} t^{-C} b^{-D} \qquad ----------(1)$$

Where T = tool life (min) V = cutting speed (m/min)
         t = feed rate (mm/rev) b = depth of cut (mm)
A,B,C,D are empirical constants and $B > C > D$

The wear: Juneja and Sekhon [15]

$$W = \left[\frac{k_2 P}{\sigma_0} \right] 1 \qquad ----------(2)$$

where W = flank wear volume, k_2 = proportionality constant
       P = normal load σ_0 = compressive yield and
       1 = the sliding distance

Tool temperature: Ghosh and Malik [16]

$$= \theta_v = KU_C \sqrt{\frac{Vt_1}{k\rho c}} \qquad ----------(3)$$

θ_v = overall temperature rise U_C = specific energy
V = cutting speed in m/sec K = Proportionality constant
ρ = density of the material C = sp. heat of the material
k = thermal conductivity of the mtrl., t_1 = uncut thickness

332

Breakage: Single injury model, Ramalingam and Watson [17]
$$F(t) = 1 - \exp(-t/\lambda) \qquad \text{-----------}(4)$$
F(t) = function of time `t' λ = mean tool life

OBJECTIVE SIMTOOL

Table-1 also shows the tools required for a job with their number
After loading the tool 14010, it is monitored on real time basis
by updating process times, wear and temperature using respective
equations. The tool breakage is monitored to find the failures
before reaching its life using equation(4). The updated values of
these parameters are continuously compared with the standards in
machining data-base. If any one of these exceeds the limits or in
case of breakage whichever occurs early a signal is given to host
computer indicating the necessity of tool replacement. It is
replaced with another tool and machining is continued. The
replacements are added-up due to individual parameters and
Combined Total Tool Replacements(CTTR) are found. The new tool is
loaded on to the spindle for operation and all the parameters are
being initialized and monitoring is continued. The cycle is
repeated for all operations on a piece and the volume of the job
completed. The procedure is repeated to produce different variety
of jobs with different volumes, after loading the magazine with
respective tools. The reusage of tools is possible (i) after
regrinding the HSS tools and (ii) depending on the active edges
of carbide tools. This is dealt in detail elsewhere[18].

Assumptions
1.   All tools to process a job are available in the magazine.
2.   If a tool is to be replaced, the next tool is
      readily available in the magazine [18]
3.   The time of replacement of a tool is negligible.
4.   The tool is replaced with an ATC
5.   Machine never fails, and the stoppage of machining is due
      to either replacement of a tool or completion of a job.
6.   Each tool can perform only a specified operation.

Input: Job No. Tool No., speed, feed, depth of cut, length of cut
volume and job topology. These are given interactively, but in
practice through CAPP. The identification and selection of an
available specified tool is through the `EXPERT TOOL'[19].

Output: The results ofjob whose sequence is given in Table-1, are
shown in Table-2. The simulation is carried-out for a volume of
1000 pieces of each job. The SIMTOOL package is written in
`TurboC' to run on a IBM compatible PC/AT.

Table(1)   Sequence of operations on a job

OPRN	Rough	Finish	Square	Round	Taper	Drill	Tap	Partoff	Chamfer
Tool No	14010	14011	14013	14014	14016	14030	14035	14017	14018

Table(2)   Output Results

Replace due to	14010	14011	14013	14014	14016	14030	14035	14017	14018
Life	0	1	0	0	0	0	0	1	1
Wear	35	28	50	48	34	47	57	14	28
Temp	5	1	9	6	2	5	8	0	1
Break	21	18	18	20	12	16	20	5	8
Total	81	48	77	74	48	68	85	20	38
Scrap	21	18	18	20	12	16	20	5	8
Total tools reqd.	14	08	13	13	8	68	85	4	7

CTTR = 220 (Assuming only single point tools have 6 active edges)

Results and Conclusions

The results indicate that tool replacements are depending on the nature of operation. This is due to decrease in hardness of the tool with increased stress and temperature. Except in part-off, chamfer and finish, the other tools could not reach their theoretical estimated life.

The SIMTOOL is developed and estimated number of tool replacements of different tools, used for various operations of a job during the specified volume of production. Unlike earlier models on tool replacements, this paper investigated the effects of important parameters viz., life, wear, temperature and breakage on tool replacements. The management can take preventive measures to over come the failures and decide total tools to be made ready at the machining centre/tool crib. This helps to study in advance the possible behaviour of a tool on a machining centre for an effective Tool Management. Thus facilitates to plan an uninterrupted production scheduling of FMS. When `SIMTOOL'is combined with `Design of a tool delivery system by Expert Simulation'[18], then it gives realistic number of tools required, leading to an Tool Requirement Planning (TRP). The analysis and performance of the SIMTOOL can be augmented with application of Petri Nets[20].

References

1. Pavel Tomek, Tooling Strategies Related FMS Management, The FMS Magazine, April (1986), 102-107.

2. Jack Hollingum, Tooling Highway Proves It-out for Yamazaki, The FMS Magazine, July, (1989), 122-126.

3. A.S.Gray, A. Seidman and K.E. Stecke, Tool Management in Automated Manufacturing: A Tutorial, Proc.of 3rd ORSA/TIMS Int. Conf. on Flex. Mfg. Systems. Aug(1989), 93-98.

4. Dr. Ing. Bernd Brodbeck, Cutter Manufacturer Finds Right Tool, Integrated Manufacturing Systems, Jan (1990), 31-34.

5. Charles J. Conrad and N. Harris McClamroch, The drilling problems: A stochastic modeling and control example in Mfg IEEE Trans. on Auto. Control, Vol.AC-32,11(1987), 947-958.

6. Vestnik Mashindstroeniya, Soviet Engineering Research, Vol 65, 6, (1985), 38-40.

7. A.S.Carrie and D.T.S.Perera, Work Scheduling in FMS Under Tool Availability Constraint,IJPR, V24, 6(1986) 1299-1308.

8. Shozo Takata, M. Ogawa, P. Bertok, J. Ootsuka, K.Matushima and T. Sata, Real-time Monitoring System of Tool Breakage using `Kalman' filtering, Robotics and Computer Integrated Manufacturing, Vol. 2, No 1, (1985), 33-40.

9. D.T.S. Perera and A.S.Carrie, `Simulation of Tool Flow within Flex. Mfg. Systems'. Proc. of 6th Int. Conf. on Flex. Mfg. Systems, Nov.(1987), 211-222.

10. E.M.Trent, Metal Cutting, II Ed, Butterworths, London 1984.

11. Machining Data Hand Book, Vol.I & Vol.II, Machinability Data Centre, Metcut Res. Asso. Inc., Cincinnati, Ohio,1980.

12. J.A. Swartley-Loush (Ed), Tool Materials for High-speed Machining, Proc. of Conf. on Adv in Tool Materials for Use in High Speed Machining, Soc. of Carbide and Tool Engineers, Scottsdale, Arizona, Feb(1987) 25-27.

13. Tool Engineers Hand Book, 2nd Edition, ASTME, SME National Technical Publications Committee, Soc. of Mfg. Engineers.

14. N.H. Cook, Tool wear and Tool life, Jl. of Engg. for Indus. ASME Trans., (1973), 931-937.

15. B.L. Juneja and G.S. Sekhon, Fundamentals of Metal Cutting and Machine Tools, Publ: Wiley Eastern Ltd, New Delhi,1987.

16. Amitabh Ghosh and Ashok Kumar Malik, Manufacturing Science Publ: Affiliated East-West Press Pvt. Ltd, New Delhi, 1988.

17. S.Ramalingam and J.D.Watson, Part 1, Single-Inj. Tool life Model, Jl. of Eng. for Indus, ASME Tran, Aug(1977) 519-522.

18. C. Eswara Reddy, O.V.Krishnaiah Chetty and Dipak Chaudhuri Design of Tool Delivery System by Expert Simulation, Accepted for publication and presentation at VI National Conven. of Comp. Engrs., Sep(1990), Tiruchirapalli,India.

19. C. Eswara Reddy, O.V.Krishnaiah Chetty and Dipak Chaudhuri EXPERT TOOl in Flex Mfg. Systems, Accepted for publication and presentation at Int. Conf. on Automation, Robotics & Computer Vision, Sep 18-21,(1990), Singapore.

20. C. Eswara Reddy, O.V.Krishnaiah Chetty and Dipak Chaudhuri Appls. of Petri nets in Tool Management, Working Paper, Dept of Mech Engg, Indian Inst of Tech. Madras,India, 1990.

SAMPLE PROGRAM LISTING:

```
   clrscr();
printf("\n Be careful about input file");
printf("\n Give Name of the Machining data File \n");
   gets(&infile);
printf("\n Give Total No of Jobs \n");
   scanf(" %d",&n);
printf("\n Give No of Processes in each job \n");
   scanf(" %d",&p);
   fp = fopen(infile,"r");
      if(fp == NULL){
printf("File is not found");
            exit(0);
   }
   fp2 = fopen("simtool.out\o", "w");
   for (i=0; i<n; i++){
       for(j=0; j<p+1;j++)
         fscanf (fp," %d,&data[i][j]);
         fscanf(fp,"\n");
   }
        ....
        ....
   clrscr();
   pi=4.0*atan(1.0);
   w=k/(cut_spd*pow(feed,b)*pow(dp_cut,c));
   ad= 1.0/a;
   tool_life= pow(w,ad);
printf("\n Tool life = %f, minutes", tool_life);
        ....
        ....
   for (n=1, trp=0; n<1000; n++{
     total_cut_time+= pce_cut_time;
     if (if total_cut_time>= tool_life){
       trp += 1;
       total_cut_time=0;
       total_temp =0;
       total_wear =0;
       }
        }
        ....
        ....
   else if (total_wear>= wear_lim)
        {
       trpwear =trpwear+1;
       total_cut_time=0;
       total_temp=0;
       total_wear=0;
       }
        ....
        ....

      CTTR = (trp + trpwear+trptlf+trptemp+trpbrk);
printf("\n Total No. of Replacements:\n %d", CTTR);
fprintf(fp2,"\n Total No. of Replacements:\n %d",CTTR);
      return(x);
       }
}
```

A Methodology for Automating the Redressing of the Grinding Wheel

A. C. S. KUMAR
JNTU College of Engineering
Hyderabad, India

U.R.K. RAO
Indian Institute of Technology
New Delhi, India

Abstract

Wheel wear is an important factor which influences the productivity of the grinding process. Monitoring the wheel wear
and determining the wheel life assume considerable significance in the automation of the grinding process. Hence suitable
methodologies to determine the instant of redressing the grinding wheel, based on the process information, are called for.
This paper presents one such method wherein the redress life
of the grinding wheel is modelled in terms of the process
parameters. The model is based on the Group Method of Data
Handling (GMDH) technique. The data required to build up the
model are generated by suitable experimentation. The tangential
force criterion is employed to determine the wheel life. The
configuration that may be employed for automatic control of
the wheel dressing, based on this wheel life model, is suggested.

Introduction

Trends in manufacturing indicate that the grinding process
will be increasingly employed in the factories of the future.
An important factor which influences the productivity of
the grinding process is the wheel wear. During grinding, it
becomes necessary to redress the grinding wheel periodically
in order to ensure a satisfactory level of grinding performance.
The volume of the workpiece material per unit active wheel
width which can be ground between two successive dressing
operations is termed as the 'redress life' of the grinding
wheel [1]. There is need to develop suitable and comprehensive
models for wheel life in terms of the process variables, so
that these models may be employed in systems for automating
the redressing of the grinding wheel.

Some wheel life models were attempted earlier by several in-
vestigators. Aerens and Peters [1] provided the specific wheel
life curves in their grinding charts, but these curves were
given for a specified set of grinding conditions only. Trmal

and Kaliszer [2], and Verkerk and Pekelharing [3] proposed
certain relationships for the redress life of the grinding
wheel as a function of the specific metal removal rate. Other
important parameters such as the wheel speed, grain size,
grinding fluid etc. also merit consideration for a comprehensive
model of the wheel life. Nagasaka et al [4] identified a novel
method of modelling the wheel life by means of a statistical
technique called the 'Group Method of Data Handling (GMDH)'.
This is a good technique for modelling a stochastic system
from a small amount of data. This GMDH technique is employed
in this paper to model the wheel life in terms of some of the
pertinent variables of the grinding process.

Criterion for Wheel Life

The wheel redress life that is determined, evidently depends
upon the criterion employed to determine the instant at which
the wheel is to be redressed [2,3,5]. Features of the grinding
process such as the specific grinding energy, residual stress
induced in the workpiece, thermal changes in the work surface
etc. are considerably affected by the tangential force. Therefore, for applications involving materials susceptible to
thermally induced damage in grinding, and for cases where
close tolerances are to be met, a suitable limit on the tangential force may be prescribed as the criterion for determining
the wheel life. This criterion is employed in the present
investigation to model the wheel life.

Experimentation

Plunge grinding tests are performed on En 31 Steel on a hori-
zontal spindle surface grinder. Standard dressing conditions
for finish grinding practice [6] are adopted for the tests.
Three proprietory fluids A, B, and C, as given below, are
used in the experimentation, with 2.5% concentration and a
flow rate of 2 litres per minute.
A - Glycol based synthetic grinding fluid.
B - Synthetic grinding fluid with E.P. additive.
C - Soluble oil (transparent type).
The following parameters are chosen as the independent variables

in the model.

X_1=Hardness of the workpiece material (HRc)

X_2=Wheel grain size

X_3=Wheel peripheral speed(m/s)

X_4= Work speed (m/min)

X_5=Wheel depth of cut (μm)

X_6=Grinding Fluid

Various combinations of these input variables are formed for each test as per the experimental design. A listing of the data set used in the experimentation is shown in Table 1. The tangential force values are recorded throughout each test at regular intervals of 75mm³/mm of the volume of work material removed.

Establishing the Wheel Life Data

Graphical plots are made of the tangential force intensity(F_t') vs. specific volume of the workpiece material ground, for each test. A sample plot drawn for the test No.8 is shown in Fig.1. The general trend of these curves is similar to that reported in Ref.[5]. A limiting value for F_t' may be fixed, based on the practical requirements of the process and the quality characteristics demanded of the workpiece, to signal the end of useful life of the grinding wheel. However, in the present analysis which aims at illustrating the method to develop a wheel life model based on the tangential force criterion, a typical value of 8 N/mm is set for F_t'. From the graphical plots mentioned above, values of the specific volume of work material corresponding to the force intensity of 8 N/mm are obtained for each test. These are the wheel redress life values required, and are shown in the first column of Table 2.

Modelling the Wheel Redress Life

Using these values, the wheel life is modelled in terms of the chosen controllable variables, by means of the GMDH technique. The modelling procedure is similar to that reported in Ref.[7]. The wheel life model equation, corresponding to the overall minimum mean square error of 0.0095 obtained in the third layer, is given by the following.

$$\Delta v' = u_1 = 0.030 + 1.291\ z_3 - 0.242\ z_6 - 0.173\ z_3^2$$
$$- 1.743\ z_6^2 + 1.827\ z_3 \cdot z_6$$
$$z_3 = 0.045 + 0.953\ y_2 + 0.192\ y_3 - 0.033\ y_2^2$$

Test No.	Workpiece Hardness (HRc)	Wheel grain size	Wheel speed (m/s)	Work speed (m/min)	Wheel depth of cut (μm)	Grinding Fluid
1.	25	46	26.63	8.73	2	A
2.	50	60	19.89	8.97	4	A
3.	60	80	23.54	8.00	6	A
4.	35	46	19.47	11.29	6	B
5.	45	60	22.34	11.29	2	B
6.	35	80	26.80	5.91	4	B
7.	25	46	23.56	4.36	6	C
8.	45	46	22.49	6.19	4	C
9.	60	60	26.20	6.86	4	C
10.	60	80	19.83	10.11	2	C

Table 1. List of the experimental data.

Test No.	Wheel redress life (mm³/mm)	
	Observed value	Estimated value
1.	1397.63	1406.81
2.	449.64	461.13
3.	663.61	615.53
4.	315.00	303.27
5.	412.50	390.56
6.	450.00	520.61
7.	444.91	446.86
8.	487.50	436.74
9.	898.80	820.52
10.	1241.84	1230.37

Table 2. Observed and estimated values of wheel life.

$$- 1.144\ y_3^2 + 1.021\ y_2 \cdot y_3$$
$$z_6 = -0.072 + 0.817\ y_2 + 0.036\ y_5 - 0.159\ y_2^2$$
$$- 0.247\ y_5^2 + 0.706\ y_2 \cdot y_5$$
$$y_2 = -0.438 + 0.084\ x_2 + 1.235\ x_3 + 3.201\ x_2^2$$
$$+18.930\ x_3^2 - 16.659\ x_2 \cdot x_3$$
$$y_3 = -0.066 + 0.110\ x_3 + 0.319\ x_4 + 31.707\ x_3^2$$
$$-0.314\ x_4^2 + 18.877\ x_3 \cdot x_4$$
$$y_5 = -0.119 - 0.803\ x_5 + 0.329\ x_6 + 1.463\ x_5^2$$
$$-0.216\ x_6^2 - 0.471\ x_5 \cdot x_6 \tag{1}$$

The values of wheel redress life for each test, estimated from the equation (1) are shown in the second column of Table 2. A very good agreement is noticed between the observed and estimated values of wheel life, and that reveals a high level of accuracy of the model.

Application of the Wheel Life Model to Automate Wheel Dressing

The wheel life model developed can be employed to automate the redressing of the grinding wheel. A microprocessor-based controller into which the wheel life model is built, may be employed for this purpose. The real time data of the process variables may be fed from the grinding system to the controller. From the knowledge of the actual volume of work material removed at any given time, and the wheel life estimated from the model, the controller signals redressing of the grinding wheel. This configuration can be incorporated in the adaptive control systems whereby the dressing operation is performed at the appropriate time, depending upon the values of the process variables.

Conclusions

1. A mathematical model for the redress life of the grinding wheel is developed by means of the GMDH technique. The tangential force criterion is employed to determine the wheel life.

2. A method of applying this wheel life model to automate the wheel dressing is suggested.

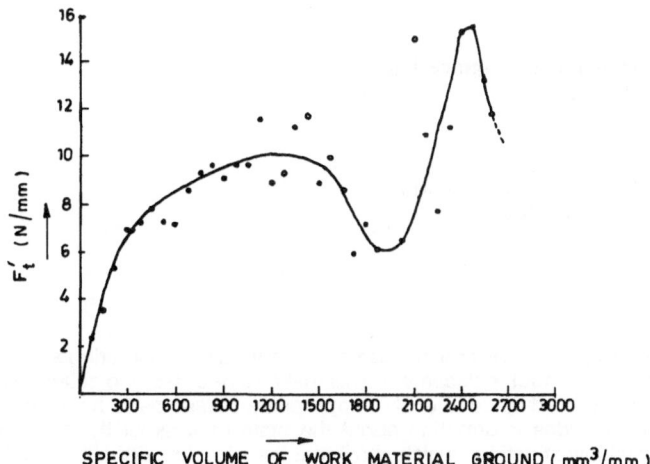

Fig. 1. Tangential force intensity curve for test No. 8

References

1. Aerens, R., Peters, J.: Using grinding charts for optimaliz-
   ing grinding conditions. C.R.I.F., Heverlee. (Private communication).

2. Tramal, G.; Kaliszer,H.: Optimization of a grinding process
   and criteria for wheel life. Proc. 15th Int. Mach. Tool Des.
   and Res. Conf., Birmingham (1974) p.311.

3. Verkerk, J.; Pekelharing,A.J.: The influence of the dressing
   operation on productivity in precision grinding. Ann. CIRP
   28 (2) (1979) p.487.

4. Nagasaka, K.; Kita, Y.; Hashimoto, F.: Identification of
   a model of grinding wheel life by the Group Method of Data
   Handling.Wear 58 (1980) p.147.

5. Pande, S.J.; Lal, G.K.: Grinding wheel life and economics.
   J. Engg. Prod. 4 (3) (1981) p.18.

6. Machining Data Handbook 2, Machinability Data Centre, Metcut
   Res. Assoc. Inc., Ohio (1980).

7. Kumar, A.C.S.; Dubey, S.P.; Mehta, A.K.; Rao, U.R.K.: Modelling of grinding wheel life by the Group Method of Data Handling technique together with sensitivity analysis. Wear 127
   (1988) p.179.

Experimental Investigations on Tool Vibrations in Turning for On-Line Tool Wear Monitoring

D.N. RAO
Department of Mechanical Engineering
A.U. Engineering College
Visakhapatnam, India

P.N. RAO and U.R.K. RAO
Department of Mechanical Engineering
Indian Institute of Technology
New Delhi, India

Summary

There is an increasing trend towards the use of unmanned manufacturing systems. These systems demand automatic tool replacement. This makes it necessary to estimate the instant of tool failure sufficiently in advance to enable appropriate measures to be effected. Cutting tool vibration signal provides information about the state of wear of the tool. This paper presents experimental investigations which show that the vibration signal is sensitive to tool wear in certain narrow bands of frequencies. A method is proposed to use this information in predicting tool wear on-line.

Introduction

Replacement of cutting tool is necessary when the tool has lost its useful life. The undesirable effects of using a cutting tool beyond its life are, non-achievement of workpiece tolerances and/or surface finish and even machine tool damage. The consequences of these effects are lower production rate and higher production cost. The increasing use of FMS and the general move towards minimally manned manufacture necessitate the use of automatic tool replacement systems. It has been reported that a manufacturing systems with an on-line tool wear sensor can have cost savings upto 9-15% and cycle time savings upto 3-5% [1].

Currently, research is in progress world over, to develop reliable and practicable methods of on-line assessment of tool wear. The work reported so far can be classified into 'direct' and 'indirect' methods. Direct methods measure flank or crater wear directly, interrupting the cutting process, thus requiring more production time. Indirect methods assess the tool wear from the measured values of process variables without interrupting the cut. Because of their non-interference with the cutting process, indirect methods are preferred to direct methods. Of the many indirect methods reported, vibration signal monitoring has been of much interest during recent years. Several investigations [2,3,4] viewed that the vibrations of a cutting tool in stable conditions are due to the friction between the tool flank and the workpiece. It has been reported [5] that at low cutting speeds the unsteady nature of the built-up edge creates vibrations whose amplitudes vary slightly with the cutting conditions.

In their work, Pandit and Kashou [6] studied the vertical vibrations of the tool holder in the frequency bands of 4-5 kHz, 8-10 kHz and 14-16 kHz. They found that the vibrational spectral density initially decreased with the increase of wear, reached a minimum corresponding to critical tool wear and then continued to increase. This confirmed the observations made earlier by Weller et al.[2]. Del Taglia et al.[4] and Martin et al.[8] expressed that the narrow frequency bands in the range of 0-2.5 kHz which contained a small percentage of the total power were sensitive to tool wear. Recently, Jiang et al.[7] proposed frequency band- energy method for in-process monitoring of tool wear. The work was concentrated in the low frequency range of 0-510 Hz. They claimed that the micro-breakage stage can be predicted by this method with a success rate of 80-90%.

Work has been in progress to find a suitable method by which the cutting tool vibration signals may be used for detecting the tool condition. Minyoung Lee et al.[9] after reviewing the work relating to vibration signals, reported so far, concluded that the relationship between tool wear and tool vibrations might help in developing a reliable tool monitoring system based on low-frequency system vibration signals. However, such a relationship between tool wear and vibration signals is yet to be established.

In the present investigation, the effect of tool wear on tool vibration signals in low frequency bands is studied experimentally. A quantitative relationship between wear and amplitudes of vibration signals as cutting progresses is established for the experimental results obtained. A method is proposed where by this relationship may be used for on-line prediction of tool wear.

The Experiment

The experimental set-up shown In Fig. 1 consists of a piezoelectric dynamometer (KISTLER 9257A) seated in place of the tool post. A tool holder - WIDAX 270 SCP 2525R is mounted on the dynamometer. An accelerometer (B&K 4371) is fixed on the rear end of the tool holder to pick up the tool vibrations in the direction of radial cutting force. The vibration signals are amplified by a charge amplifier (B&K 2635). The amplified signals are recorded on an instrument tape recorder (RACAL, 7 channel) and analyzed off-line using a spectrum analyzer (HP 3582A).Tool flank wear is measured at different stages of machining using a tool maker's microscope.

A HMT lathe (NH22) is used for turning EN8 steel of hardness number 190 on the Vicker scale (with a load of 20 kgs. using a pyramidal indentor). The uncoated carbide inserts used are - SPUN 120308, grade:P20-P30, normal rake:6 deg., front clearance angle 11 deg., make : WIDIA.

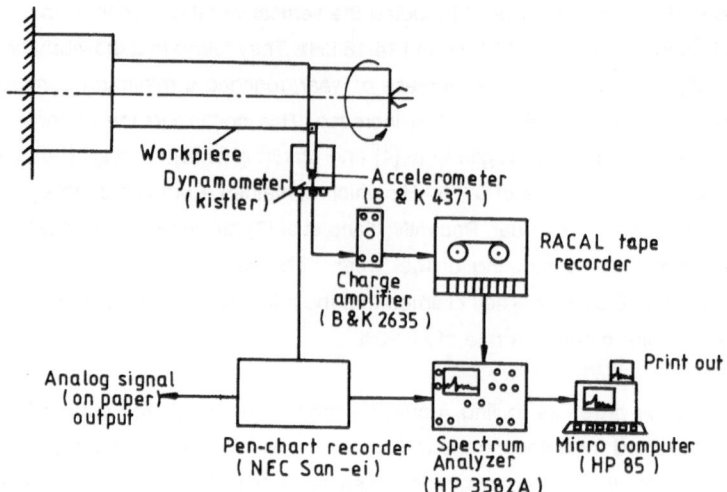

Fig.1. Experimental set-up used.

Selected cutting parameters are - average cutting speed = 110 m/min (in a range of 100-120 m/min), feed = 0.2 mm/rev, depth of cut = 1 mm. Tool wear criterion is - average flank wear (VB) of 0.3 mm.

Results and Discussions

The recorded tool vibration signals are analyzed in the frequency band of 0-5 kHz. Using the built-in filters of the spectrum analyzer, the narrow frequency bands (in the lower frequencies) which are sensitive to tool wear are identified and isolated. At different instants of time and tool wear, the vibration amplitudes in the frequency bands of 102-107.5 Hz. are extracted from the recorded signals. Fig. 2 shows the pattern of the amplitudes at 106.8 Hz. as the tool wear and cutting time progress. The variation of vibration amplitudes with the progress of tool wear is represented in Fig. 3. The relationship between the tool wear and the vibration amplitudes as cutting progresses may be depicted as shown in Fig. 4 obtained using the software SYGRAPH. The relationship obtained employing the statistical software SYSTAT is:

$$\text{amplitude} = -9139.154\, t_w + 139050.446\, t^2_w - 440092.339\, t^3_w$$

(for time between 0 to 7 min.)

and

$$\text{amplitude} = -12877.919\, t_w + 114384.286\, t^2_w - 231166.249\, t^3_w$$

(for the time 7 min. onwards)

where t_w = tool wear in mm

amplitude = vibration amplitude in μV.

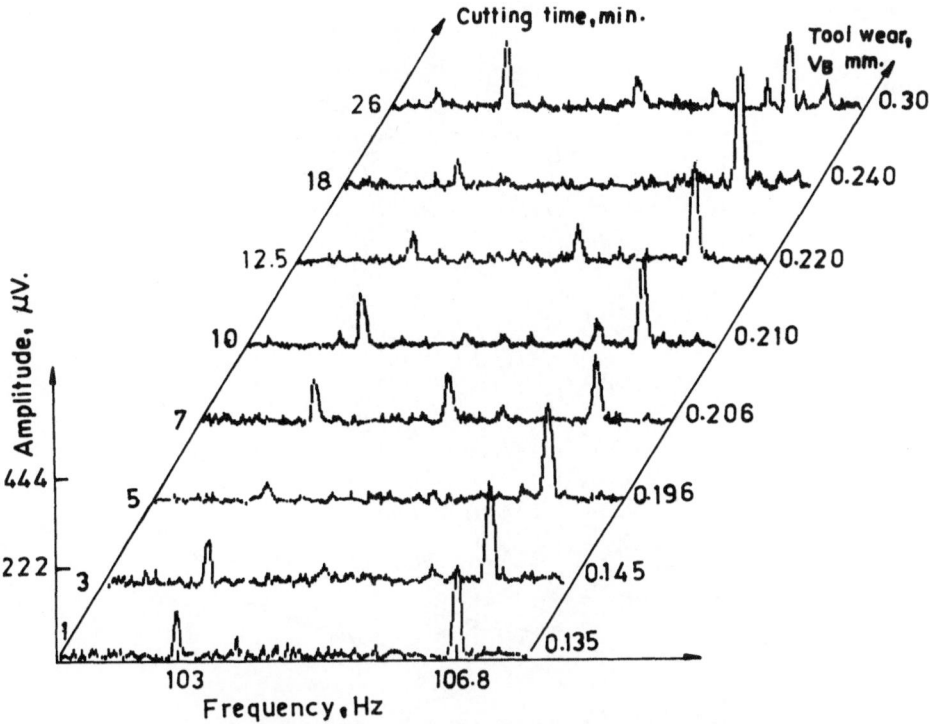

Fig.2. Vibration Amplitudes at different instants of tool
wear and cutting time in the lower frequency band.

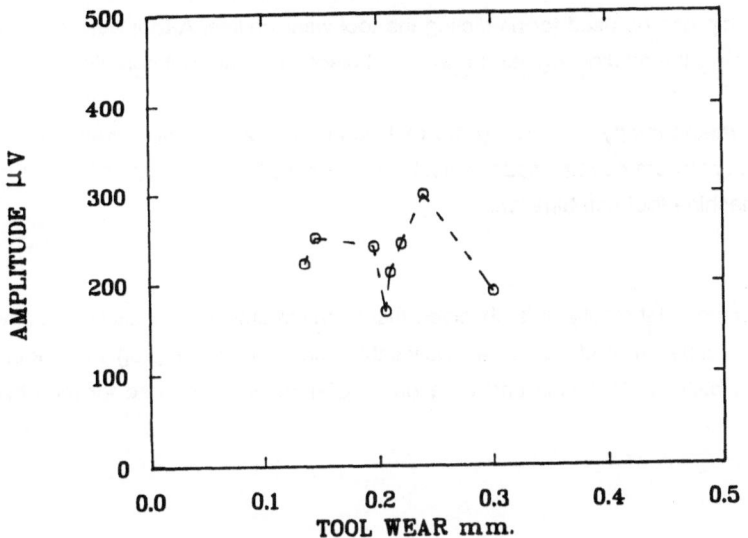

Fig.3. Variation of vibration amplitude with the progress of tool wear.

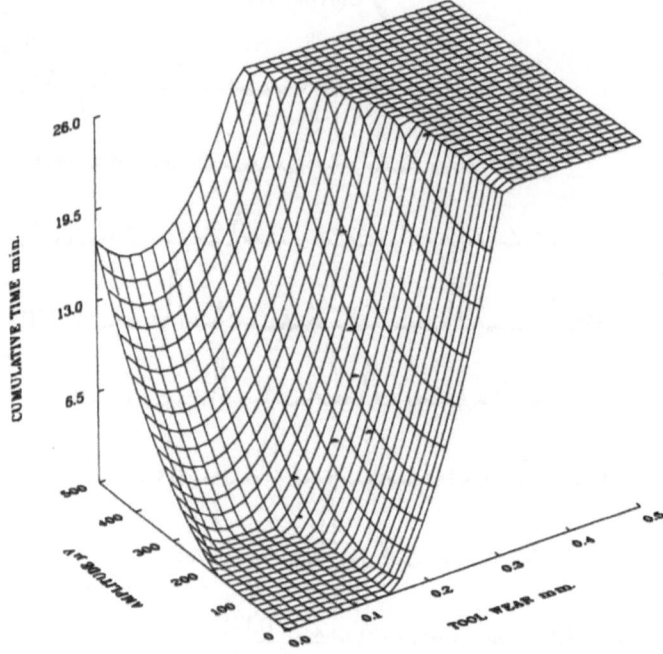

Fig.4. Tool wear - Amplitude - Cumulative Time Relationship.

This relationship can be used for predicting the tool wear on-line. A tool wear data file can be created using the relationship for the tool-workpiece-machine tool combination. Fig. 5

explains the methodology for sensing the tool condition from on-line vibration amplitude data. A built-in software can be used to select the frequency band corresponding to the tool-workpiece-machine tool combination.

Conclusions

From the experimental results, it is observed that vibration signal is sensitive to tool wear. The amplitude values at 106.8 Hz. show a noticeable variation with the progress of tool wear. A relationship between tool wear and vibration amplitudes is established for the observed results.

A methodology is proposed for on-line monitoring of tool wear using the above relationship based on a PC and built-in software.

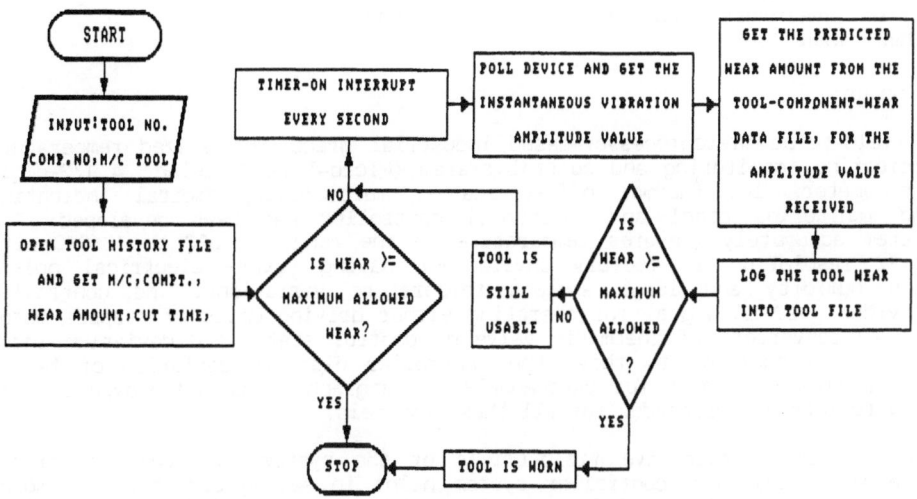

Fig.5. Proposed methodology for predicting tool wear from on-line vibration amplitude data.

References
1. Sampath, A. and Vajpayee, S.: On-line tool wear prediction using AE. 12th AIMTDR conf., IIT Delhi (1986) 269.

2. Weller, E.J., Schrier, H.M., Welchbrodt, B.: What sound can be expected from a worn tool? Trans.ASME. J.Eng.Ind. 91 (1969) 525.

3. Kronenberg, M.: Machining Science and Applications. Pergamon Press (1966) 306.

4. Del Taglia, A., Portunato, S. and Toni, P.: An approach to on-line measurement of tool wear by spectrum analysis Proc.17th Mach.Tool Des.&Res.Conf.(1976) 141.

5. Micheletti, G.F, Koenig, W. and Victor, H.R.: In-process tool wear sensors for cutting operations. Ann.CIRP 25/2 (1976) 483.

6. Pandit, S.N. and Kashou, S.: A data dependent system strategy of on-line tool wear sensing. Trans.ASME. J.Eng.Ind. 104(1982) 217.

7. Jiang, C.Y., Zhang, Y.Z. and Hu, H.J.: In-process monitoring of tool wear stage by the frequency band energy method. Ann. CIRP 36 (1987) 45.

8. Martin, P., Mital, B. and Drapier, J.P.: Influence of lathe tool wear on the vibrations sustained in cutting. Proc. 15th Int.Mach.Tool Des.&Res.Conf.(1974) 251

9. Minyoung Lee, Charles E.Thomas and Douglas G.Wildes: Prospects for in-process diagnosis of metal cutting by monitoring vibration signals. J.Mat.Sci. 22(1987) 3821.

μp-Based Industrial Grade Multi-Channel Temperature Controller For Sugar and Allied Industries

H. SINGH, S.M. SHARMA and C.R.K. PRASAD

Electronic Systems Area
Central Electronics Engineerng Research Institute
Pilani, India

Abstract

A field-tried microprocessor-based industrial grade centralized temperature indicating, monitoring and control system (Micro-TEIMAC) using 4 1/2-digit thermometer, low thermal offset scanner multiplexer, digital indicating and monitoring panel and 24-channel controller has been developed. The system accurately measures temperature in the range of 0°C to 1260°C with 0.1°C resolution in factory environment having severe electrical noise, high humidity and large ambient temperature variations. The controller provides 4-20 mA signal for operating either driving units of stepper motor or I/P converters of pneumatic valve or control status and deviation indicators. In addition to this, the controller displays deviation of temperature from set point for 24-channels in bargraph forms and provides facility to introduce recorder for all these channels.

The results of extensive field-trials of the system, at sugar factories, have shown that the controller system helps in saving considerable amount of energy apart from improving operation of the process. It is worth mentioning here that heating of juice of a typical 100 TCH sugar factory by 1°C requires energy costing Rs.50,000/- approximately per season and saving of 1% of energy amounts to a profit of Rs.1.65 lakhs per season. The economic figures for production of Micro-TEIMAC system are given. It has been shown that the payback period for the system is one season for sugar industry.

Introduction

In India, there are more than 400 sugar factories which produce more than 9 millions tonnes of white sugar per year. Most of these sugar industries are still using traditional methods of production which results in inefficient utilization of energy and lower productivity. In recent years the importance of instrumentation for controlling and monitoring process parameters has been realised in various industries. These can be most profitably used in modernising the sugar factory too and reaping the full benefits it can offer in terms of by products utilization such as for paper, fertilizers, chemicals and electric generation and providing better opportunities of employment in villages leading generally to betterment of life in rural areas.

In sugar factories, large volume of juice is heated every day at different stages, underheating the process results in incomplete process and over-heating results in loss of sugar by inversion. In addition to this overheating of juice consumes extra fuel. It has been calculated [1] that heating of juice by 1°C in a 100 TCH factory requires fuel worth Rs.50,000 per season per stage. Since the juice is heated at several stages in a sugar industry, the financial benefits, which may be drawn by using reliable and accurate temperature control system, may be found manifolds.

Central Electronics Engineering Research Institute, Pilani has developed an accurate and reliable microprocessor-based industrial grade 24-channel temperature controller (Micro-TEIMAC) which has been successfully field-tried in a few sugar factories of U.P. and Maharashtra. A unique feature of the Micro-TEIMAC system is the twenty-four channel controller. It pro-vides 4-20 mA PID controller current output and helps in controlling the temperature of all the twenty-four channels automatically either with the help of stepper motor driven low torque valves or pneumatically controlled high torque valves and manually with the help of control status and devia-tion indicators by providing visual indication to the valve operator. In addition to this controller displays actual deviation of process temperature from set point in the form of bar graph for all 24-process points and has the facility to connect recorder to all points.

Micro-TEIMAC System

The block diagram of Micro-TEIMAC system is shown in Fig. 1. It comprises of i) 8086 microprocessor-based microcomputer, ii) Precision 4 1/2-digit industrial grade thermometer, iii) Low offset 24-point scanner multiplexer, (iv) Digital indicating end monitoring panel, v) 24-channel controller, vi) Control status and deviation indicator, vii) Stepper motor operated low torque valve, and viii) Electro-pneumatically controlled high torque valve

Details about the sub-systems (i) and (iv) mentioned above are given in reference [1]. The rest of the sub-systems are briefly discussed below.

Twentyfour Channel Controller

A multi-purpose modular 24-channel microprocessor-based controller is developed. It consists of twentyfour independent interchangeable printed circuit boards and a power supply. Each board is having two modes, manual and auto. When the processor is cut off from the controller or when the

350

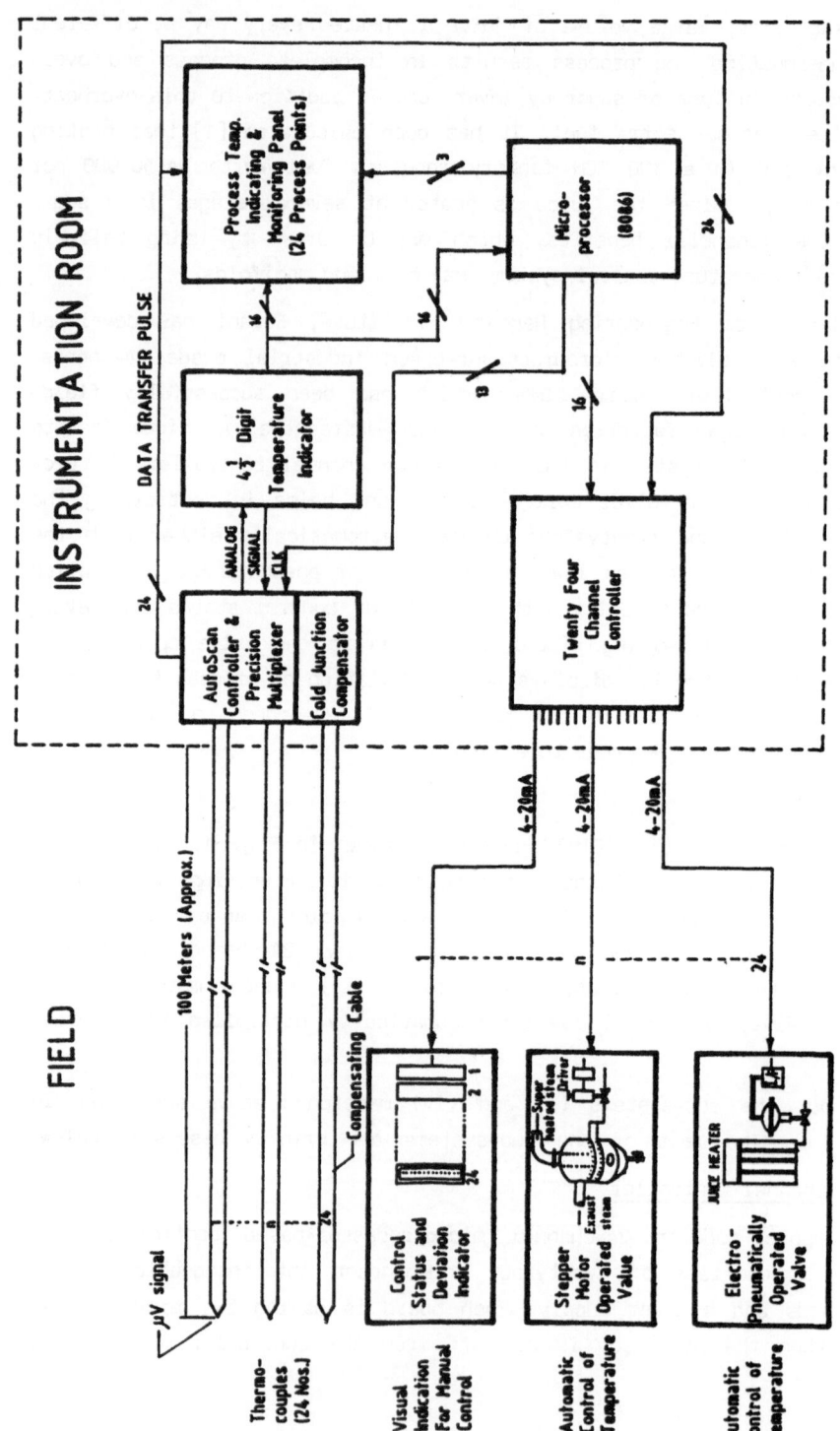

Fig.1. SCHEMATIC DIAGRAM OF MICRO-TEIMAC SYSTEM

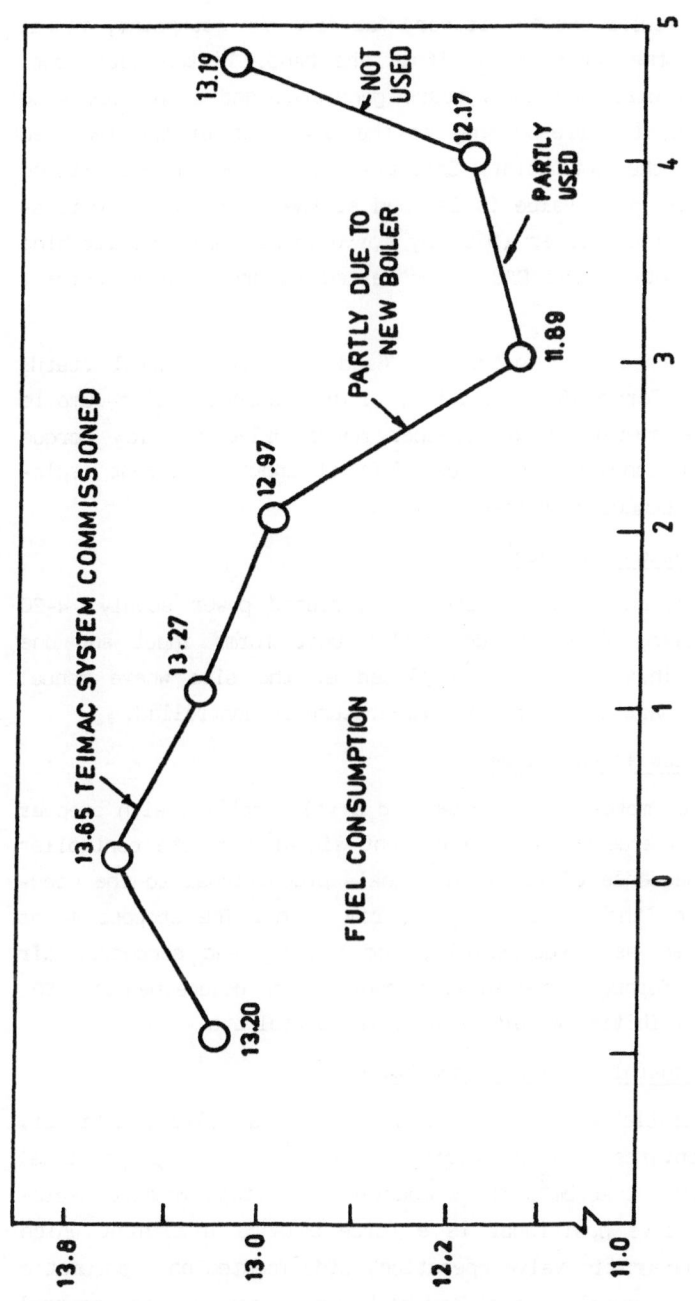

FIG.2 ENERGY SAVING OVER A COUPLE OF YEARS

temperature of any particular channel is very far from the set point, manual mode is used to bring the temperature within the band. In the auto mode, microprocessor controls directly. Each board generates and sends out 4-20 mA standard current output corresponding to the deviation of the measured temperature value from the set point. Software takes care of calculating the manipulated variable. This value is latched at the input of the corresponding digital-to-analog converter (DAC) by corresponding channel latching signal. The analog output of the DAC is converted to the 4-20 mA current form.

This manipulated 4-20 mA current output is used to drive control status and deviation indicator for manual operation, or IP transducer of pneumatic valve for high torque operations or stepper motor drive for low torque operations. The current converter provides 1500V DC input to output isolation in interfacing the standard process signals.

Control Status and Deviation Indicator

It is a twenty LED indicator having its own isolated power supply. 4-20 mA current signal, coming from the controller unit forms input and one LED glows at a time. This indicator is placed at the site where manual valve exists with which particular process temperature is controlled.

Stepper Motor Operated Low Torque Valve

It consists of a stepper motor drive system and a valve coupled with stepper motor. The drive system accepts 4-20 mA current signal from the controller and generates proper sequence of position signals proportional to the modulus of the deviation of the control signal from 12 mA. The stepper motor operated valve has been used successfully for primary and secondary air control in a sulphur furnace, condensates control in desuperheater, SO_2 control in sulphiter, imbibition water temperature control etc.

Electro-pneumatically Controlled High Torque Valve

It consists of a pneumatic valve, I/P transducer and a valve positioner. The I/P transducer converts 4-20 mA current signal in to a proportional pneumatic signal of 0.2-1.0 kg/cm^2. It is mounted in a field mounted explosion proof case. The valve positioner is a force balance instrument which avoids slugishness in pneumatic valve operation. Side mounted on a pneumatic valve, it utilises an air supply and a feed back cam to position the control valve in accordance with the air signal from the I/P transducer. To increase operational stability, the pilot is a bleed type which balances between supply and exhaust air volume in normal condition. This type of valve have been used for control of temperature of raw juice, sulphite juice and clear juice.

Results of Field-Trials with Micro-TEIMAC System

The Micro-TEIMAC system was installed and commissioned covering a number of most important temperature points including juice clarification, evaporators, pan crystallizers, boilers and sulphur furnaces. The system has worked reliably in the industrial environment. All the temperature data points value has been simultaneously displayed on the display panel and updated after every 24 seconds. The control status and deviation indicator for each process point is mounted near the manually operated valve of the process point. The operator finds the presence of these indicators near the control valves very useful in the process of maintaining the temperature within the limits.

Type 'K' thermocouples of industrial grade has been employed for accurate measurement of not only the higher side of temperature range but also the lower one such as that of injection inlet water and crystallizers. The system has been rechecked at the end of each crushing season and its accuracy is found to be within ± 1°C over the entire temperature range with a resolution of 0.1°C. It is worthwhile to note that a 40 uV signal corresponding to 0.1°C is being transmitted over a maximum length of 100 meters of compensating cable to the Micro-TEIMAC system which is located in the environment of sugar mill where ambient temperature rises upto about 50°C in summer and where severe electrical disturbances and mechanical vibrations are present. In spite of such adverse environmental conditions there is no deterioration in accuracy of the system and no jumping of fractional decimal digit of the indicators took place.

The measurements of energy consumption in a particular sugar factory are shown in Fig. 2. As the number of process temperature monitoring points were increased and automatic control of temperature of important points was done, the steam consumption in the factory decreased. The rate of decrease in steam consumption was further enhanced by efficient running of a newly installed boiler. During third year of field-trials the Micro-TEIMAC system was used for a short period and so there was a slight rise in fuel consumption.

Economic Data

    I.   Minimum Economic Unit   10 Units/year (single shift)

II.	Fixed Capital:		
	Bldg. Furniture & Fixtures	Rs. 50,000 (Bldg. on rent)	
	Plant/Equipment/Machinery	Rs. 7,50,000	
III.	Working Capital (Mfg. cost of 90 days)	Rs.15,50,000	
IV.	Cost of production per unit	Rs. 4,28,000	
V.	Summary		
	1. Selling price per unit	Rs. 5,10,000	
	2. Total revenue	Rs.51,00,000	
	3. Total cost of production	Rs.42,80,000	
	4. Annual return	Rs. 8,20,000	
	5. Return on investment	35%	

Figures are illustrative only for estimates. 2% energy saving amount to Rs.3.00 lakhs per season in a sugar factory having crushing capacity of 100 tonnes of cane per hour.

Conclusions

Micro-TEIMAC system which has been field-tried in different sugar factories, works reliably. It precisely measures the temperature of 24 points with an accuracy of ± 1°C and resolution 0.1°C. There are twentyfour control signals (4-20 mA) available for automatic control of twentyfour process points either using pneumatically controlled valves or stepper motor operated valves. For manual control of valves, the 4-20 mA signal is used to drive control status and and deviation indicators which provide visual indication of temperature status to the operator. Measurements of energy consumption in a particular factory shows that the Micro-TEIMAC system has helped in reducing steam consumption by 2%. In addition to this the accurate control of temperature has also helped in reducing inversion losses leading to increase in sugar production.

Acknowledgements

Thanks are due to Dr. W.S. Khokle, Director, CEERI Pilani for his kind encouragement during the development and field-trials of the system.

References

1. H. Singh, S.M. Sharma and S.S. Ahluwalia, "Auto-TEIMAC system for sugar and allied industries", Research and Industry (India), Vol. 30, September 1985, pp 265-277.

Use of Sensors for Safety of Personnel in Robotic Installations

K. GHOSH

Ecole Polytechnique
Montréal, Québec, Canada

J.-J. PAQUES

I.R.S.S.T. du Québec
Montréal, Québec, Canada

Y. BEAUCHAMP

Université du Québec à Trois-Rivières
Trois-Rivières, Québec, Canada

ABSTRACT

With the widespread use of robots in complex, highly-integrated installations such as Flexible Manufacturing Systems, assuring adequate safety of personnel is becoming a major problem. The most vulnerable people are the programmers and the maintenance technicians, because they have to work close to active robots. Safety of personnel can be assured if a reliable sensor can be developed which will detect the approach of the robot arm close to a human. However, the system should not be affected when the effector approaches the workpiece during the normal operation of the robot. Various types of sensors (such as ultrasonic, capacitive, infrared, microwave, etc.) could be used, though each of them has certain limitations. Our investigation indicates that the capacitive sensors are the most promising.

INTRODUCTION

Industrial robots are now being increasingly used in a variety of industries. The automotive industry uses a very large number of robots in operations such as spot welding and spray painting. Flexible manufacturing systems will become very important in the future and robotic manipulators constitute a key element of such systems. In case of many automated systems, the workers do not have to come near the dangerous areas of the machines when they are in operation and hence there is reduced risk of accidents. But in case of robots, there are certain tasks that require human intervention with the machine in active or semi-active state. Examples of such operations are programming, set-up, adjustments, fault identification and certain maintainance activities. During the last few years, there have been many serious accidents involving robots and in just one country, Japan, ten fatalities have already been recorded in accidents involving robots[1]. Fixed barrier guards around robot workplaces has been recommended by robot safety standards in the United States[2] and many other countries. Use of such perimeter fencing with interlocked doors ensure that no unauthorized persons can come into the robot work envelope and this has resulted in some reduction of the accident rate. But for activities such as programming, the technician has to work inside the enclosed space with the robot in an active state. It will be a significant help in improving the safety at the robotic workplace, if a reliable instrument system could be developed to detect human intrusion into a specified volume around the effector and around the forearm of the robot. Such means of detection do not exist at this time and a number of researchers are now working on this problem. Various principles of detection are now being investigated and we will discuss some of the research work that is interesting from a practical standpoint.

ULTRASONIC SENSORS

In ultrasonic detection systems, a sound wave with a frequency of over 20 KHz is emitted. The time taken for the emitted wave to be reflected back to the sensor is measured. This measurement indicates the distance between the robot arm and the human.
This type of sensors has been abandoned because of several problems. False triggerings occur because of echos. The electrostatic type of ultrasonic sensor is less expensive, but its repeatability is poor. The angle of the emission cone is limited to 30° and hence only a limited space is covered[3].

INFRARED SENSORS

Any object whose temperature is greater than absolute zero emits radiation. Since radiation emitted by the human body falls within a well-defined spectral range, it is theoretically possible to detect the intrusion of humans into a robotic workplace by using infrared sensors. Unfortunately, many other objects such as electric light bulbs emit radiation in the same spectral range as the human body. Therefore, infrared devices are subject to frequent false triggerings. These devices have been examined by a few researchers[3], who do not recommend their use because of the above-mentioned reasons.

MICROWAVE SENSORS

Microwave sensors are based on radiation and reception of electromagnetic radiation in the GHz range. Because of the Doppler effect, the frequency of the reflected wave will differ from that of the emitted wave in relation to the speed of the object reflecting the radiation. The reflected wave is compared with the initial emitted wave.

The U.S. Bureau of Mines Research Center in Minneapolis, Minnesota, recently tested three types of warning sensors to be used when heavy trucks are backed up in mines[4]. They tested infrared, ultrasonic and microwave sensors. The microwave sensor system was most effective, especially because it was little affected by environmental conditions. A 10.525 GHz wave was used and the movement of an object in relation to the sensor was detected by the Doppler effect. When the truck was backing up, the system was able to detect the presence of a person at a distance of four to six meters.

The same detection principle was examined at R.P.I.[3] for use in a robotic work cell. However, the results were not that promising in this case, because the system does not detect low speeds and displacements that are lateral to the sensor.

SENSITIVE SKIN SYSTEMS

Some researchers have tried to develop a flexible film for the robot arm that will emit a signal in case of deformation. In one such project at National Bureau of Standards[5], the aim was to develop a magnetoresistive film composed of layers of rubber and mylar that would incorporate tiny sensors. This system is sensitive to contact pressure and a signal is emitted that could be used to stop the movement of the arm. A row of flat conductors in the sensitive skin creates a magnetic field that is translated into an electric current by magnetoresistive sensors. When pressure is exerted on the sensitive skin, the conductors and sensors are forced nearer to each other and hence the electrical signal provided by the sensors increases. The sensitivity of the sensitive skin can be varied by altering the rigidity of the rubber used. Thus, a pressure detection range of 30 N/m^2 to 2 X 10 N/m^2 can be obtained.

CAPACITIVE SENSOR SYSTEMS

Principle

Metallic elements placed on the robot arm constitute, with the earth, a
condenser whose dielectric is the ambient air of the robot. When a
person enters the condenser's electrical field, the capacitance of this
field varies. This variation is measured by an electronic circuit.
This type of device has attracted considerable interest among
researchers and further technical details are provided in the following
paragraphs.

Applications in Robotic Work Cells

General Motors Research Laboratory conducted a research project from
1982 to 1984 in order to develop a safety system for robotic work cells
using a capacitive sensor. A digital method was used to study the
electrical field about an antenna. It was decided to place the antenna
(four metallic wires) parallel to the forearm of a PUMA 560 robot. By
using an impedance bridge with a sensitivity of 5.5V/pF, it was possible
to detect human intrusion from a distance of 60 cm around the robot's
arm. A few prototypes of this system were tested in some General Motors
plants. Although the results were satisfactory with regard to
sensitivity, false triggering (and consequent interruption of production
activities in the cell) soon caused the production personnel to lose
interest.

In a research project carried out by the Rensselaer Polytechnic's Center
for Manufacturing Productivity and Technology Transfer on the detection
of human intrusion into robotic work areas, several types of sensors
were examined. Recently, they carried out more advanced work on the
capacitive sensor[6]. Their observations on the design of the
antenna are as follows:

- the greatest sensitivity is obtained with aluminium and not a
  copper or steel antenna;
- the surface area of the antenna should be maximized;
- the antenna should have a maximum number of plane surfaces; and
- the antenna should be small so as not to hamper the movements
  of the robot.

At R.P.I., A Cincinnati Milacron T3 robot (an anthropomorphic robot) was
used. The antenna was mounted on the forearm behind the wrist. An
impedance bridge was designed in order to be able to detect a very weak
deviation (of the order of 10 pF) of the capacitance. Human intrusion

was detected by comparison of signatures. An IPM PC-AT computer was used and it had in memory the robot's variation in capacitance for its cycle without human intrusion (ideal signature). The signature made by the robot was entered in the computer by an A/D card which compared this signature with the ideal one and human intrusion was detected from 45 cm. However, this comparison is only possible in case of repetitive movements (such as during production operations) and not during programming or maintenance. Another method that the researchers are currently studying is the spectrum filter. It was discovered that human movements are found in a 4 to 9 Hz frequency range of the bridge output signal, while robot movements are found in frequencies lower than 2 Hz. Therefore, it could be possible to detect human intrusion by filtering frequencies.

Research work on the use of capacitive sensors for detecting the intrusion of humans into robotic workcells is now being carried out at the Robotics Laboratory of Ecole Polytechnique. We are using the Proxagard capacitive system[7] on a PUMA 560 robot. The immediate aim is to optimize the design of the antenna to protect the technicians during programming and similar operations. In addition, we are looking at ways to better distinguish between the presence of humans and inanimate objects in a dielectric field.

Observations Concerning Capacitive Sensor Detection Systems

The shape of the field developed by the antenna is well-adapted for use in robotic work cells, but the problem of false triggering remains. It is absolutely necessary to eliminate false triggering before the system can be put into routine use in the industry. This has not yet been achieved and further research and development work must be carried out to improve the performance.

In sum, a few problems that detract from the correct operation of capacitive sensors are as follows:

- variations in the temperature and humidity of the ambient air cause variations in the characteristics of the dielectric (air) of the condenser formed by the antenna and the earth, and hence the measured value of the capacitance;
- walkie talkies and other radio communication devices may interfere with the electronic system of the sensor;
- the industrial environment can cause a deposit of dust or grease on the antenna, thereby creating a variation in capacitance;

– the variation in resistance between the body of the person and the ground could affect the operation of the system.

CONCLUSION

Robots will become an essential component of integrated automated manufacturing systems in the future. The development of a reliable sensor system for detecting the intrusion of humans into robotic work areas will greatly improve the safety conditions. Many types of systems are possible candidates, but the capacitive system is the most promising one at the present time.

REFERENCES

1. Nagamachi, M.: Ten Fatal Accidents Due to Robots in Japan. Proc. 1st International Conference on Ergonomics of Advanced Manufacturing and Hybrid Automated Systems, Louisville, Kentucky. Amsterdam: Elsevier Science Publishers. 1988. Pp. 391–402.

2. American National Standard for Industrial Robots and Robot Systems, ANSI/RIA R15.06–1986. New York: American National Standards Institute, Inc. 1986. 12 p.

3. Derby, S. et al: A Robot Safety and Collision Avoidance Controller. Proc. Robots 8 Conference. Dearborn, Michigan: Society of Manufacturing Engineers. 1984. Pp. 21–33 to 21–43.

4. Johnson, G.A. et al: Improved Backup Alarm Technology for Mobile Mining Equipment. Information Circular 9079. Washington, D.C.: Bureau of Mines, U.S. Dept. of the Interior, 1986. 19 p.

5. Vranish, J.M.: Magnetoresistive Skin for Robots. Paper No. MS84–506. Dearborn, Michigan: Society of Manufacturing Engineers. 1984. 17 p.

6. Millard, D.L.: An In-Situ Evaluation of a Capacitive Sensor-Based Safety System for Automated Manufacturing Environments. Paper No. MS89–301. Dearborn, Michigan: Society of Manufacturing Engineers. 1989. 12 p.

7. Operation and Maintenance Manual of Proxagard PC110. Brookfield, Connecticut: Gordon Eng. Corp. 1989.

SECTION B: DEVELOPMENTS IN APPLIED ROBOTICS AND AUTOMATION

Chapter VI

Industrial Applications

Introduction

A CAD-based methodology using 'implementation space' has been developed and tested in the first paper to determine the physical implementation of a robot in an existing industrial environment. An optimization criterion enables the user to minimize the number of robots necessary to accomplish all considered tasks. In the second paper, an expert system for the selection of an industrial robot has been developed from a survey of industrial practices. A knowledge base and rule set allow functional parameters to be matched against the available robot characteristics. In the third paper, an industrial robot can perform product assembly without special part feeders and fixtures. Feedback from a machine vision system is used with CAD-based part descriptions to achieve automated assembly. In the fourth paper, a robotics system that can locomote a vertical surface is shown to be feasible. Four vacuum cup legs and a PC-based controller enable the robot to climb curved as well as straight surfaces. In the fifth paper, the design of a semi-automatic manipulator to clear ore jams that risk human life and shut down mines has been conceived. This remotely controlled telemanipulator would recover its cost within one year.

Techno-economic feasibility that pick-and-place robots with vision systems can substantially improve milk plant operations has been established in the sixth paper. Manual handling associated with washing and filling bottles would be eliminated. In the seventh paper, the controller for a robot assembly cell that effectively integrated the robot controller with the workcell controller was developed. The cell controller performed real-time monitoring of the robot controller to track the robot operations. The eighth paper states that product manufacturing for various reasons needs a CAD model of existing generic data by direct image scanning. The ninth paper shows the use of the Kalman filter for extraction of a rigid body's motion from a sequence of object images, and the last one presents some optimization techniques of standard mathematical routines for high-level language software.

Chapter VI

Industrial Applications

Introduction

Determining the Workspace Design of Robotized Cells in Pre-Determined Environments

LOUISE CLEROUX

Department of Industrial Engineering
Ecole Polytechnique
Montreal, Quebec Canada

Summary

We propose the use of implementation spaces[1] to establish a CAD based methodology for the implementation of a robot in rigidly pre-determined environments (e.g.: assembly lines). We use the three module of the McDonnell Douglas graphic system, and the proposed methodology was applied to the ASEA IRB-6 robot. The determined possible implementation area enables the robot to accomplish all assigned tasks, considering all the constraints on its location. The procedure involves three main steps: (1) constructing all necessary implementation spaces (IS); (2) obtaining the reduced implementation space(s) (RIS), which only takes into account all the tasks to be accomplished; (3) obtaining the possible implementation space(s) (PIS), which take(s) into account all additional constraints on the location of the robot. The methodology is transferable to other animation softwares such as Catia or Robcad, and is used to validate a numerical recurrent approach currently being developed.

Introduction

For robotized cells, the workspace design process depends on the rigidity of the environment: (1) some environments can be modified, and the cell is designed around the robot; (2) for rigidly pre-determined environments, the robot must be retrofitted to the line or cell. Most softwares then require a tedious trial and error approach to implement the robot. Few books give concrete information on how to determine possible locations of a robot in a particular environment, and many authors overlook the physical aspect of implementation [1-4].

Data Gathering for Creation of Implementation Spaces

The different steps of the data gathering, necessary to the creation of the IS of every task to consider, are shown in fig.1. All steps are accomplished on the Robotics module.

---

1 We refer to the domain of implementation (location and orientation) possibilities as the robot's implementation space.

Inverse kinematics often offers multiple solutions. The multiplicity depends on the mechanical construction of the arm and wrist. Revolute joints and nonzero link length parameters can cause as many as thirty two solutions [5]. With the ASEA IRB-6, $d_j=0$ $(j=1,\ldots, 5)$ and the allowable ranges on the joints preclude the access to most multiplicities: wrist multiplicities remain accessible. They produce a layered IS constructed from two basic ISs (fig.2). Being determined by the elbow, the vertical cut of the basic IS is unique regardless of the task pose. This explains the results obtained for different task orientations.

Data gathering may thus be simplified. The gathering of three points per circle arc suffices. Nonetheless, determining the inflection points requires special care. Fig.2 exemplifies the data to gather. Given the length of the arcs in the top view, and as a precaution, the number of points recorded is greater than three.

Influence of the Task Orientation
Having determined the IS, following a data transfer to the Unigraphics module for the given task (fig.3), a study of the influence of the task orientation on the IS was carried out. Six supplementary tasks were studied.

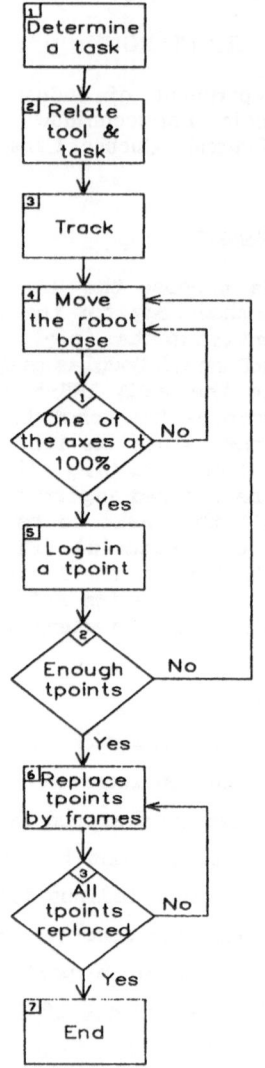

Figure 1 Data Gathering Flow Chart

The basic shape of the IS is similar across tasks 1, 3, 4, 5 and 7. Only the radial dimensions differ. Tasks 3 and 7 produce a single solution (fig.4). Conversely, tasks 4 and 5 (fig.3) have a space constructed from two basic ISs. Contrary to tasks 3, 4, 5 and 7, tasks 2 and 6 yield a skewed and distorted IS. Fig.5 shows that axis 5 points upward or downward depending on the position of the base around the task. The difference in the position of axis 4 of the robot explains the vertical shear of the IS. It is no longer generated by a pure rotation of the basic vertical cut. This cut maintains its vertical orientation, and the projection of the implementation space is elliptical (fig.5).

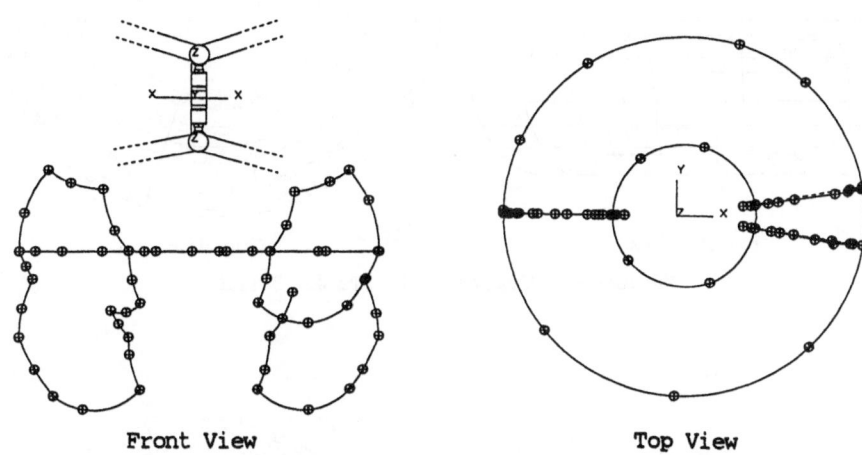

Front View   Top View

Figure 2   Data gathering

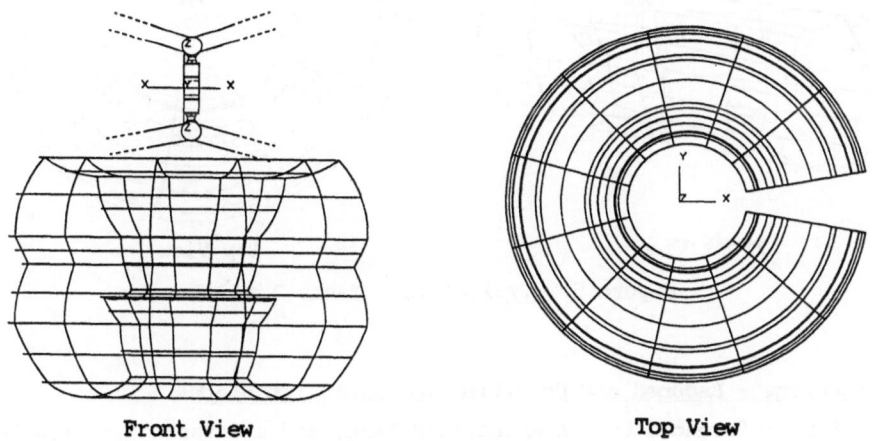

Front View   Top View

Figure 3   Typical IS - Tasks 1, 4 and 5

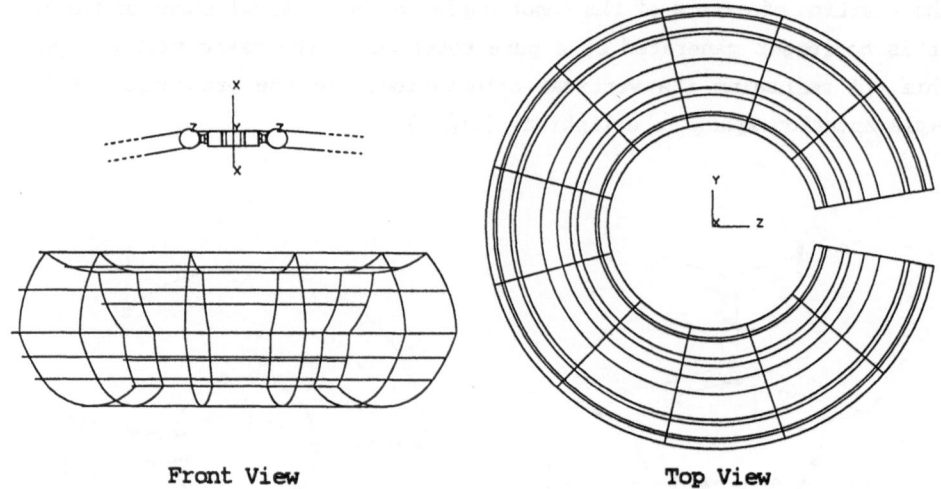

Front View   Top View

Figure 4   Typical IS - Tasks 3 and 7

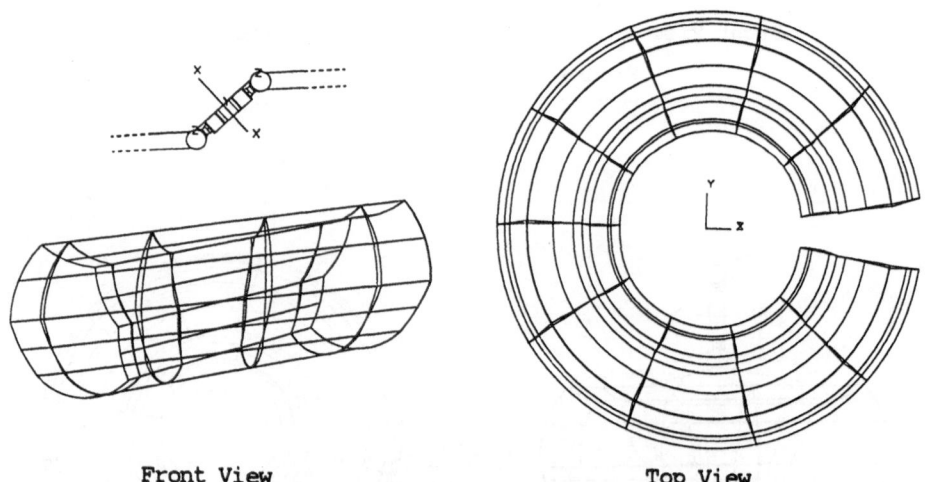

Front View   Top View

Figure 5   Typical IS - Tasks 2 and 6

Applications - Reduced and Possible Implementation Spaces

Each IS is referenced to its associated task, and all tasks are situated
in a common frame (fig.6a & c). The intersection(s) of all the associated
spaces is(are) obtained to determine if the tasks may be executed by the
same robot. The RIS(s) is obtained in free space.

After obtaining the RIS(s) with Unisolids, the tasks and the RIS(s) are added to the workcell. The floor, walls, and other machines are located, to visualize the space truly available. It is also necessary to consider every other aspect that may constrain the location of the robot (safety measures, ...) and thus create a set of virtual limitations. Subtracting all of these from the RIS, a PIS is obtained (fig.6b & d). The robot has to be located within this PIS. This ensures that all tasks are accessible while respecting the environmental constraints.

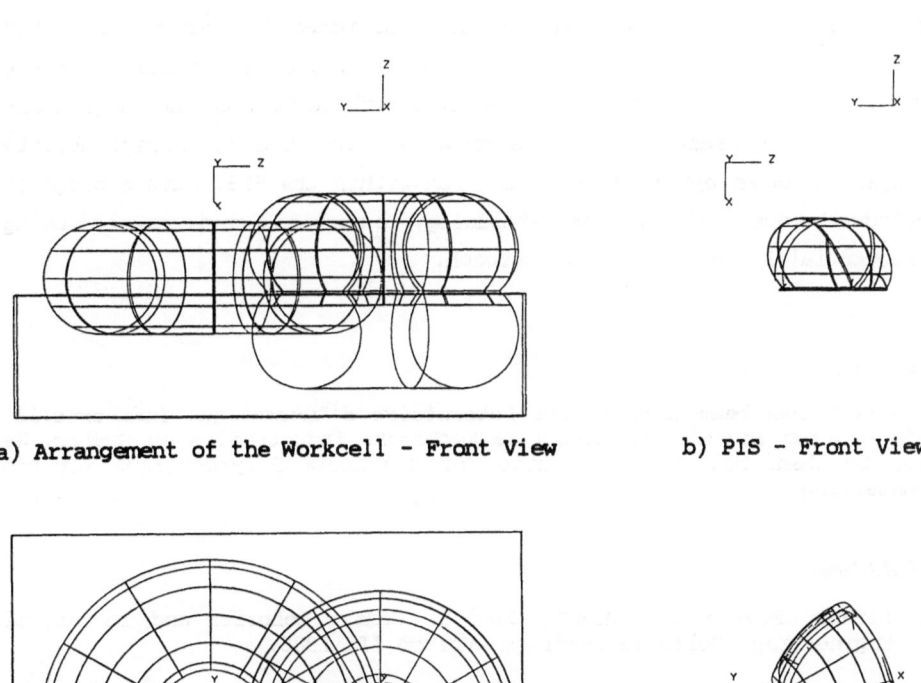

a) Arrangement of the Workcell - Front View   b) PIS - Front View

c) Arrangement of the Workcell - Top View   d) PIS - Top View

Figure 6   Workcell with 2 tasks

Conclusion

A new methodology was developed and tested. (1) It provides manufacturing industries with a systematic method to determine the physical implementation of robots in existing facilities. (2) It is applicable to evaluate which, or what combination minimize the number of robots necessary to accomplish all considered tasks, in the material selection process. (3) It contributes to the comparison of layouts, by offering tradeoffs as to the number of robots required.

The proposed methodology maintains most of its advantages in semi-rigid environments where (1) tasks are rigidly positioned, but can be reoriented by flexible fixturing, or (2) the relative situations of the tasks are rigid, but the group of tasks can be moved with respect to the environmental limits. Current research encompasses the use of manipulability ellipsoids as an optimization criterion within the PIS. Future projects include the adaptation of the methodology to nonrigid environments, using a sequential PIS maximization approach.

Acknowledgements

This work has been done at the *Laboratoire d'Automatique Industrielle*, INSA of Lyon and at the *Centre de Recherche Industrielle du Québec*. We wish to thank Mr. Guy M. Cloutier of the *École Polytechnique* for his counselling.

References

1. FISHER, Edward L.; MAIMON, Oded Z. (eds): Robotics and Industrial Engineering: Selected readings. Volume II, 1986.

2. JUTARD, A.; LIEGEOIS, G.: Application et mise en oeuvre des robots industriels. École Polytechnique de Montréal, Centre de cours intensifs, Mars 1987.

3. Proceedings of: Robots in the automotive industry, An International Conference. Bedford, England: IFS Publications 1982.

4. WARNECKE, H.J.; SCHRAFT, R. D.: Industrial Robots: Application Experience. Bedford, England: IFS Publications Ltd 1982.

5. CRAIG, John J.: Introduction to Robotics: Mechanics and control. Addisson-Wesley Publishing Company 1986.

Judicious Selection of a Robot for an Industrial Task - An Expert System Approach

SURENDER KUMAR and ALOK VARMA
Production Engineering Department
Birla Institute of Technology
Mesra, Ranchi, India

ABSTRACT

An expert system for judicious selection of a robot for an industrial task has been developed. The expert system has been realised to meet the user's requirement. However further development is needed to satisfy the requirements of multiple constraints.

1. INTRODUCTION

An expert system is a computing system that embodies the knowledge and reasoning process elicited from people renowned for their high level of expertise in their specific knowledge domain. India being a developing country, financial resources are very limited, therefore scarce money must be utilised in a better way, specially for those works/ projects :

a)   Which have a high probability of success

b)   Which gives good return on investment

c)   Which have a short payback period.

Moreover it is also unreasonable to expect manufacturers to carry out field trials with untried robots since improper selection could be prohibitively expensive.

Therefore the objective of this paper is to develop an expert system for the selection of an industrial robot. A knowledge base and rule set have been developed from a survey of industrial practices which allow functional parameters to be matched against the available robot characteristics.

370

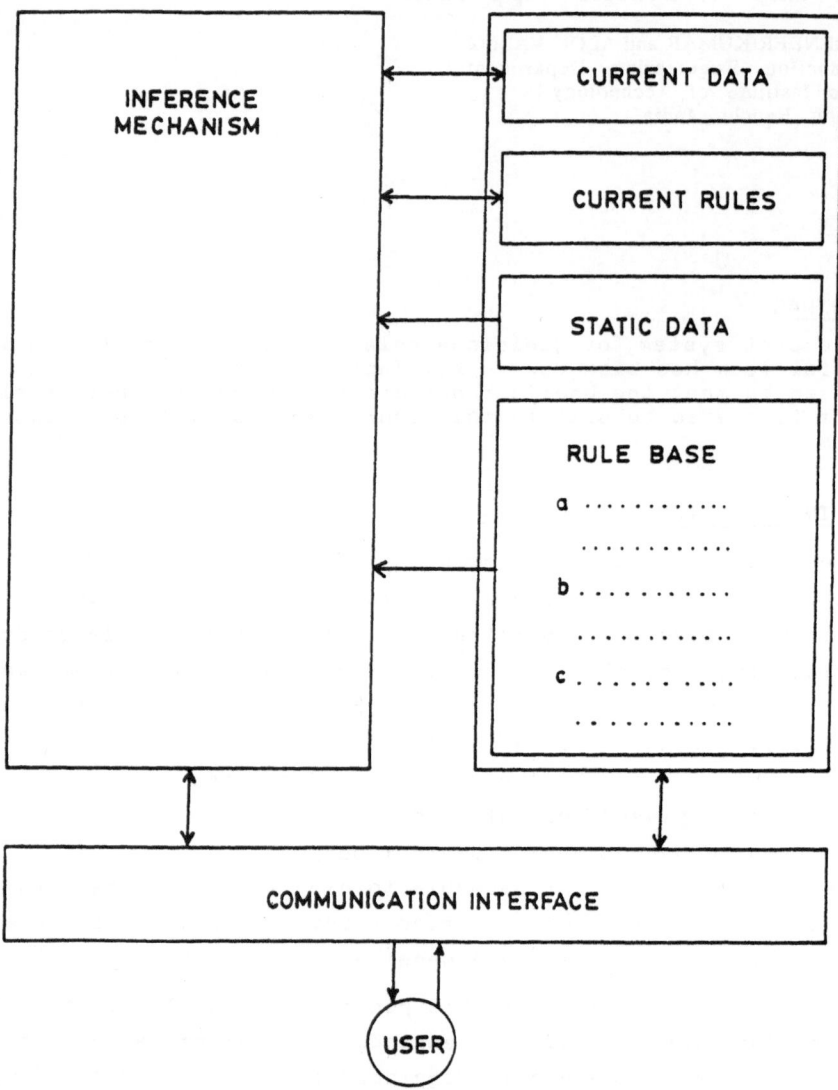

FIG 1 STRUCTURE OF DEVELOPED SYSTEM

2. EXPERT SYSTEM

2.1 Conceptual Idea :

Artificial intelligence has been responsible for the knowledge-based systems. The area of expert systems investigates methods for constructing man-machine systems with specialized domain-specific problem solving expertise. An expert system is a programme, which has a wide base of knowledge in a restricated domain, and uses complex inferential reasoning to perform tasks, which a human expert could do. Expert systems instead of attempting to design new and better algorithms, find out how human experts handle tasks in their respective domains of expertise and build systems that can manipulate this specialised knowledge.

Although expert systems vary in structure they have in general two distinct parts, a domain specific base consists of factual knowledge, heuristic knowledge and procedural knowledge, the latter two being the production rules of the system. The inference engine contains the operating rules or the rules that control the production rules. Inference engines are characterised by the way rules are applied. If it looks at facts and from them tries to reach a goal then it is forward chaining. If it starts with the intended goal and then looks for the facts that fit, it is backward chaining. Forward chaining tends to give better control of the rule ordering process as all are accessed, while backward chaining, in only using the rules required to satisfy the goal, tends to be faster in operation.

All expert systems depend on expert elicitation and two main approaches are used. Firstly, based on the assumption that experts can articulate their knowledge, a programme of interviewing is carried out. This approach is prone to error as experts are often better at doing than analysing why. The second approach, knowledge induction, relies on second party analysis of case studies that the expert has worked on and is often preferred.

2.2. Development Steps:

The major steps for expert system development are :

a) Problem selection

  * Requirements for development,

  * Characteristics required for deciding the appropriateness of the approach,

```
data  type                  = record
name of robot               : string   < 25 >
Velocity                    = integer (mm/sec.)
repeatability               = real (± mm)
payload                     = integer (Kg.)
cost                        = longint (Rupees)
```

The above programme has been implemented using TURBO PASCAL
version 5.0 on the IBM-PC.

A consultation is initiated by a querry that can be inputted
automatically. The inference process is fired by a querry that
is automatically called once the required data has been entered.
As the inference process proceeds a data file is generated that
hold the current processing status. The querry is in the form
of procedure rule, "IF querry anything, THEN check anything"
and when the statement querry best robot is read, the floating
variable 'anything' assumes the value best robot. A combination
of forward and backwardchaining is used to reach a conclusion.
This process of forward and backward chaining is continued until all of the hierarchy of buyer objectives have been evaluated. Although for each objective, every robot that meets the
criteria is recorded, only the ones that meet all of the objectives are considered for the choice of best robot. At the
end of the querry the list of best robot is given in report
form. The following is an example of the type of logic rule
construction used.

"IF robot include some robot and velocity of some robot is
something and repeatability is something and payload is something THEN best robot includes some robot".

4. CONCLUSION

The developed system serves as an aid to the user in decision
making. It prepares and selects the most promising solutions
while the final choice is made by the user. The system is based
on expert system technique within which the engineering knowledge base, inference mechanism and communication interface
are integrated. It has been implemented using TURBO PASCAL
ver 5.0 on the IBM -PC.

The developed system has been successfully tested on several
trial problems. Although the knowledge base is not complete
yet considerable progress has been made in implementing the
domain knowledge required for selection of industrial robots.
Further improvements need to the made in the areas of multiple
constraint satisfaction.

It has however demonstrated the relative ease by which expert
knowledge and inference can be captured and if given the right
application such systems should be effective in industries.
Some of the major drawbacks in this particular field are the
limited logical functions and the required data base. However
these facts can be improved if such packages are to be considered for industrial use.Therefore, by utilising the facili-
lities and expertise available both at B.I.T., Mesra and Industrial complexes useful research programme may be mounted.

5. REFRENCES

!.   Mileham A.R., Buchnell K., Hunt L.,"Expert System Appr-
      oach to Machine Tool Selection", Advances in Manufactu-
      ring Technology II, pp. 112-116.

2.   Varma A., Kumar Dr. S., "Knowledgeable Expert System for
      Flexible Manufacturing", Proceedings International Con-
      ference on CAD,CAM, Robotics and FOF, 1989.

3.   Rolston David W., "Artificial Intelligence and Expert
      System Development", McGraw Hill Proole Co., 1988,
      New York.

4.   D. Janaki Ram, L.V. Prasad and O.R.K. Rai, "Material
      Selection for Components using an Expert System", Pro-
      ceeding Int. Conference on CAD, CAM, Roboticsand FOF,
      New Delhi, 1989 .

5.   Kumar Dr. S., Anand A., Varma A., "An expert system
      Approach For Selection and Evaluation of Rolling
      Lubricants", National Seminar On Steel Rolling Lubri-
      cants, Ranchi, 1989

Fixtureless Robotic Assembly Workcell

LARRY BANTA
Mechanical and Aerospace Engineering
West Virginia University
Morgantown, WV

THOMAS BUBNICK
Cincinnati Milacron Inc.
Cincinnati, OH

ABSTRACT

The assembly of manufactured products represents an area of increasing interest for the application of robotics and automation. Most applications currently require the construction of special parts feeders and precise fixtures to hold the assembly. These fixtures add cost and lengthen setup times, making the implementation of robotic assembly less attractive, especially for short production runs. Furthermore, robot programming must usually be adjusted to the workspace by trial and error, adding to the development costs.

In this research, a technique has been developed for coplanar assembly operations without special fixtures and parts feeders by incorporating feedback from a machine vision system to the robot. Vision feedback from one camera is used to guide the robot in picking up randomly oriented parts from a moving conveyor. The robot moves the part to an inverted light table for grasping error correction and assembly. The positioning of new pieces is done relative to the existing pieces, irrespective of their location in the assembly area. The relative assembly calculations make use of part description data derived from CAD drawings of the pieces. These techniques have been demonstrated at West Virginia University.

INTRODUCTION

The objective of this work is to reduc% workcell setup time, reduce or eliminate fixturing costs and improve workcell flexibility. We describe a system which successfully performs the assembly of a puzzle by grasping randomly ordered and randomly placed pieces from a moving conveyor, detects and corrects for grasping and prior part placement errors, and completes the assembly process reliably and accurately. The research reported here employed visual feedback only, but the ideas apply to other sensing mechanisms as well.

Recent advances in robot controller flexibility and machine vision technology have made the use of vision feedback commonplace in robotic workcells. The most common configuration is to place the camera on or near the gripper--a configuration which allows high accuracy and simple coordinate transformations between the camera and end effector reference frames [1,2,3,4]. For small robots however, the mass of the camera may be many times greater than that of the parts (e.g. integrated circuit chips) or perhaps even greater than the payload of the robot. The system described here places the cameras off the robot to allow use with any size robot and workcell configuration. This approach requires careful calibration between the camera and robot. Two techniques have been adopted to minimize calibration errors. The first is a simple but effective method for automatically calibrating portions of the workspace, which has been reported in reference [5]. The second technique is the primary subject of this paper, and involves the use of relative, rather than absolute assembly commands for the robot. This technique is an attempt to emulate humans in the assembly process, by moving pieces into position relative to those already in place, rather than attempting to place each piece of an assembly in a precisely fixed Cartesian position each time. This method automatically compensates for robot grasping errors, prior part placement errors, certain kinematic equation errors and part-to-part variations. Motion planning is done using simple relational data derived from CAD descriptions of the parts.

TECHNICAL APPROACH

The assembly of a puzzle was chosen as the target task. The puzzle is composed of four unique pieces which approach the robot one at a time in random orientation on a conveyor, and enter the field of view of Camera 1, located above the conveyor. Figure 1 shows the workcell layout. The vision processing system then uses the image data received from Camera 1 to identify the part and to calculate the part's centroid location and orientation angle. Several images are taken in rapid succession, and the motion of the part centroid is used to calculate the part's velocity, thus eliminating the need for installing encoders or tachometers on the conveyor. The position, orientation and velocity information are used by the PC to generate the robot commands required to grasp the piece.

A second video camera is employed to detect grasping errors and to provide position feedback during the assembly process itself. An assembly table was designed for

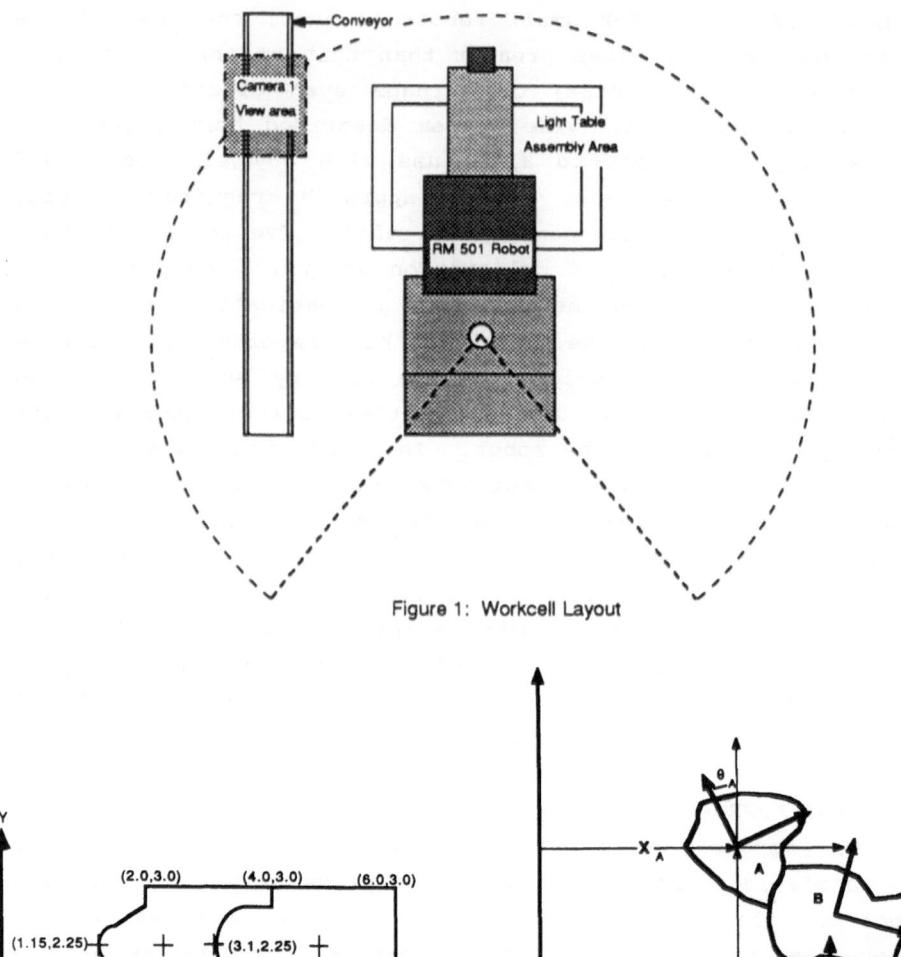

Figure 1: Workcell Layout

Figure 2: Puzzle pieces showing centroid and principal points

Figure 3: Part coordinate frames

this project which is basically an inverted light table. The table has a translucent top which, when illuminated from the outside provides a sharp silhouette of any flat, opaque objects placed on the work surface. Camera 2 was mounted inside the assembly table in such a manner that the translucent surface was in its view. The lighting was arranged so that any shadows cast on the translucent surface by the robot were eliminated from the image processing by setting the threshold and noise reduction levels of the vision processing system.

Grasped parts are moved to the assembly area in the field of view of Camera 2 where final assembly takes place. The first part is placed without attempting to correct for grasping errors. Its position and orientation are then determined by Camera 2 as the basis for the assembly of the rest of the puzzle. As subsequent parts are brought to the assembly area, they are placed on the table a small distance from their final assembly position. Camera 2 determines the position and orientation of the part, and passes the information to the PC. The PC consults a database containing the puzzle assembly information, and calculates the moves needed to bring the piece to its final assembled position relative to the piece(s) already on the table. The move commands are relative to the part's current position, thus grasping errors are automatically compensated. When the part is in place, the robot releases it and returns to the conveyor to await the next piece.

RELATIVE ASSEMBLY PROCEDURE

Figure 2 shows the four pieces of the puzzle in their assembled positions along with the two primary features used for positioning the parts. Parts were identified by using a combination of part image area and perimeter length. Part position was determined based on the location of the centroid and what GE calls the "principal point". The principal point is a rather noisy indicator of part orientation, but was the best among the options available using the GE system. Each part was assigned a coordinate frame with the origin at the centroid and the x-axis running through the principal point.

Regardless what features or algorithms are used, the parts and their relative positions in the finished assembly can be described by a simple vector $x = [x,y,\theta]^T$, with x,y the location of the part centroid and θ the orientation, as shown in Figure 3. The assembly data for the four parts of the demonstration puzzle are listed in Table 1.

Table 1: Puzzle Assembly Location Data

	Part 1	Part 2	Part 3	Part 4
x_c	2.56	4.71	2.60	4.88
y_c	2.25	2.25	0.75	0.75
θ_p	0	0	0	0

Two criteria are important in choosing location features: the features must give unambiguous part position information, and the feature values should be calculable from CAD information. Note that this representation must be supplemented by additional information regarding constraints on the assembly process itself, such as approach directions, threading operations, etc. Such constraints can always be expressed as intermediate positions in the assembly process, and can be easily incorporated into the relative assembly method.

Given the position/orientation vectors of the assembled parts in a reference coordinate frame, the position of any part relative to the others can be expressed by a homogeneous coordinate transformation:

$$x_A = T_A^B \, x_B \tag{1}$$

where the T matrix is a homogeneous transformation matrix given by

$$T_A^B = \begin{bmatrix} \cos\theta_{AB} & -\sin\theta_{AB} & 0 & x_{AB} \\ \sin\theta_{AB} & \cos\theta_{AB} & 0 & y_{AB} \\ 0 & 0 & 1 & 0 \\ 0 & 0 & 0 & 1 \end{bmatrix} \tag{2}$$

where $\theta_{AB} = \theta_A - \theta_B$, $x_{AB} = x_A - x_B$ and $y_{AB} = y_A - y_B$. Thus the relationship between part 1 and part 2 is given by T_2^1, that between part 1 and part 4 is given by T_4^1, etc. The T matrices are simple to construct from the data in Table 1.

Likewise, any piece placed on the assembly table can be located in the coordinate frame of the camera viewing the part. Suppose part number 3 is placed on the assembly table first. The vision system is commanded to take a picture and return x_3. If part number 2 is retrieved next, its final assembly position in camera coordinates can be calculated by the transformation

$$x_2 = T_2^3 \, x_3 \qquad\qquad (3)$$

with the transformation matrix calculated from the data in Table 1. The workspace coordinates and the robot coordinates are related through a known coordinate transformation found in the autocalibration process described in [5].

Part number 2 cannot be moved directly into its final assembly position without first correcting for gripping errors, part-to-part tolerances or other errors that might cause interference between the parts. Instead, the piece is moved to a position close to its final assembly position, and is placed on the assembly table. The vision system takes a picture and returns the actual position of Part number 2, x_{2a}. The relative moves required to assemble the piece are calculated. The translations are then executed to complete the assembly. The final assembly must be done in two steps: a rotation to align the part properly and then one or more translations to mate it. The rotation must be done first, since rotation of the robot wrist will not necessarily rotate the part about its centroid. Rotating the part about a point other than the centroid will change the translational movements required for mating.

EXPERIMENTAL VERIFICATION

This technique has been demonstrated at West Virginia University using a Move Master II RM-501 micro industrial robot, a conveyor system, a Zenith AT personal computer, and a General Electric Optomation II vision processing system with two digital/video cameras.

The system was able to reliably assemble the puzzle using

the techniques described. Tests were run to assess the workcell's ability to detect and correct for grasping error in a quantitative fashion. In these tests, a piece of the puzzle was placed in the robot gripper manually with grasping errors intentionally induced. The robot was then commanded to move the piece to an assembly position, where the vision system calculated the actual centroid and orientation. The robot was then commanded to execute the relative moves required to bring the piece to the final assembly position, and a second picture was taken to assess the accuracy of the corrected position. This procedure was repeated 35 times. The results of the tests are presented in Table 2. They demonstrate a dramatic reduction in the part placement errors resulting from the vision feedback.

Table 2: Experimental Data

	x-error mean/std (inches)	y-error mean/std (inches)	θ-error mean/std (degrees)
initial	-0.02/0.03	0.12/0.42	-0.42/0.67
corrected	0.01/0.02	0.08/0.08	-0.10/0.60

REFERENCES

[1] Fu, K. S., Gonzalez, R. C., Lee, C. S., Robotics: Control, Sensing, Vision, and Intelligence, McGraw-Hill Inc., 1987.

[2] Hong, T. H., Shneier, M., "Describing a Robot's Workspace Using a Sequence of Views From a Moving Camera", National Bureau of Standards, Washington, D. C. 20234.

[3] Burgess, D., Hill, J., and Pugh, A., "Vision Processing for Robot Inspection and Assembly", Proceeedings of the SPIE 1982 Robotics and Industrial Inspection Conference, San Diego, CA, (SPIE : 1982), Vol 360 pp 272-279.

[4] Agin, G. J. and Duda, R. O., "SRI Vision Research", Proceedings of the 2nd USA-Japan Computer Conference, Tokyo 1985, pp 113-117.

[5] Banta, Larry, Bubnick, Timothy, and Bubnick, Thomas, "Automatic Eye-Hand Calibration for Visually-Servoed Robots", Proceedings of the 1990 ASME International Computers in Engineering Conference, Boston, MA, (ASME: New York, 1990), in press.

Design of a Wall-Scaling Robot for Inspection and Maintenance

BEHNAM BAHR and SAMI MAARI

Department of Mechanical Engineering
National Institute for Aviation Research
Wichita State University
Wichita, KS

Abstract

This paper describes the design of a new wall-scaling robot for the inspection and maintenance of various structures, such as aircraft, or nuclear power plants is. This robot is capable of climbing straight or curved surfaces in two directions, up/down or left/right in strokes of 55 millimeters. The load carrying capability of this robot is about 100 kilograms. Thus, it can carry a variety of non-destructive testing devices for inspection, paint removal, and de-riveting devices for maintenance. The robot uses vacuum suction cups for sticking to the surface of an object and can be integrated with a vision sensor system for guidance and visual inspection. The positioning of the robot is achieved by the use of a joy-stick or a PC computer. With this portable robot, it is possible to a) inspect areas that are within easy reach, b) program the robot to follow a specified path while inspecting or maintaining the structures, and c) remotely guide it to desired locations. This robotic system will minimize human error, increase the productivity of an inspector, and reduce the cost of inspection and maintenance.

Introduction

Mobile robots are being developed for many purposes, such as inspection tasks in nuclear power plants. These mobile robots are required to have various locomotive functions similar to human beings. That is to, rotate, avoid obstacles, realize the location, and so on. One of the most recent studies is on the development of a wall-climbing robot to achieve special tasks such as reaching highly radioactive environments for the inspection of several sections of a nuclear power plant [1-3]. Another development is a gantry robotic system for inspection of a small military aircraft (F-111) which costs 25.9 million dollars [4]. However, there is no other automated technology in the field of aircraft inspection, most of the inspection is done manually. Therefore, development of wall-climbing robotic systems aircraft structures [5]. In this paper the recipe for construction of a typical wall-scaling robot is given.

Methods of sticking to the surface

To compel an object to stick to a wall, two methods can be used. The first method is to use a magnet mounted on the object, as a provider of a sticking force on a ferrous wall. The second method is to create a vacuum pressure between the object and the wall using a fan; however, a high vacuum pressure cannot be reached by this method. All wall-climbing robots use one of these two methods. On the other hand if compressed air is sent to the inlet of the ejector, high vacuum will result, making it possible to reach values of -508 torr or higher vacuum. Therefore, it is possible to use the ejector/suction cup assembly in the design of a wall-scaling robot. Since very high vacuum is achieved, the payload of the prospective robot is high relative to its weight.

Description of the wall-scaling robot

The robot has four legs mounted at 90° from each other as shown in Fig 1. The gear shaft brackets and the motor mounts are mounted on a central disk, under which a central suction cup of 203mm in diameter is mounted. This central disk is at the base of the robot. Even though the central suction cup is not vital to the design, it gives a larger safety factor under working conditions.

The four legs have exactly the same design and each leg has two degrees of freedom. The first degree is a rotational one, which enables the leg to rotate around the gear shaft. This rotational motion is transmitted from a gear head DC motor to a worm gear, then to the rotational linkage which is a part of the leg. The second is a longitudinal one, which enables the leg to extend along the cylinder axis. The longitudinal motion of the leg is created by sending compressed air to the ports of a pneumatic cylinder which causes a rod to extend or retract. The rotation of the cylinder rod around the cylinder axis is blocked. The leg has two suction cups of 11.53mm in diameter, and each is equipped with an ejector.

Figure 1 The wall-scaling robot.

Pneumatics of the robot

The pneumatic circuit for control of the robot is shown in Fig. 2. The robot has four pneumatic cylinders and nine ejectors and suction cups. The compressed air is sent to the pneumatic control circuit at a pressure of 100 psi to a manifold where it will be distributed to the cylinders and ejectors. Each cylinder has its own individual pressure line. One pressure line is sent to the leg and distributed to two lines to provide air to the two ejectors mounted on the suction cups of that leg. A 3-way 24-volt DC solenoid valve and a pressure gage is utilized to monitor the value of the pressure of the flow going to the ejectors. The pressure line of each cylinder is equipped with a 4-way valve which enables the cylinder rod to extend or retract.

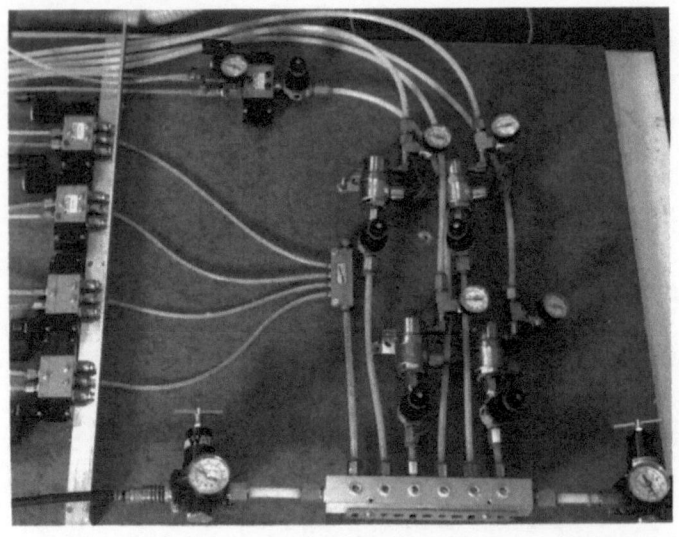

Figure 2   pneumatic circuit of wall-scaling robot.

Robot Interface

Each of the motors is connected to the power source through two
normally open relays. One of these relays, if closed, will rotate
the motor clockwise, while the other, if closed, will rotate the
motor counterclockwise. Thirteen air pressure lines, a 24-volt
DC power line, an electrical ground and a 25 pin computer cable
are channeled through a rubber hose, 38mm in diameter which is
fitted to the robot.   The computer cable is used to transmit
signals to the relays which control the motors and also receives
the feedback position signals from the robot.

The robot is maneuvered by a PC computer which is linked to a
logical network for receiving the feedback and the status of the
robot at any time. A program written in C language is used to
send signals of "fprint" and receive signals "fscan" through the
data bus of a 286 PC computer.   After comparing the inputs, an
appropriate signal would be sent to the 17 relays controlling the
motors and the valves of the pneumatic circuit. After achieving
the permanent prompt, a new feedback will be read by the program
and a new prompt decision will be issued to complete one step.

The motion of the wall-scaling robot

The motion of the wall-climbing robot is depicted in Fig. 3. The left column shows legs #1 and #3, while the right column shows legs #2 and #4. Each row represents the commands issued for each leg throughout the motion. The motion starts assuming that all the suction cups are sticking to the wall (the central suction cup as well as the radial suction cups). The following sequence presents a simplistic approach for one step up a wall.

A) All of the suction cups are sticking to the wall and the cylinders are in their retracted positions.

B) Stop the vacuum in the suction cups of the leg #1.

C) Rotate leg #1 counterclockwise to reach the limit.

D) Extend leg #1 to reach the limit.

E) Rotate leg #1 clockwise.

F) Stop the rotation when the suction cups reach the wall.

G) Start the vacuum in the suction cups of leg #1 and then stop the vacuum from the suction cups of leg #2 and #4.

H) Rotate leg #2 and leg #4 away from the wall, when they reach the limit release the central suction cup.

I) Exchange the position of the compressed air on the cylinders of legs #1 and #3, from the lower port to the upper port on leg #1, and from the upper port to the lower port on leg #3. After achieving this state, the body of the robot will move one step up.

J) Start the vacuum in the central suction cup again.

K) Rotate legs #2 and #4 to the wall.

L) Stop legs #2 and #4 individually when their suction cups reach the wall.

M) Start the vacuum in the suction cups in legs #2 and #4.

N) Stop the vacuum from the suction cup of leg #3.

O) Rotate leg #3 clockwise to reach its limit.

P) Retract the rod of leg #3.

Q) Rotate leg #3 counterclockwise.

R) Stop the rotation when the suction cups of leg #3 reach the wall.

S) Start the vacuum in the suction cup of leg #3.

386

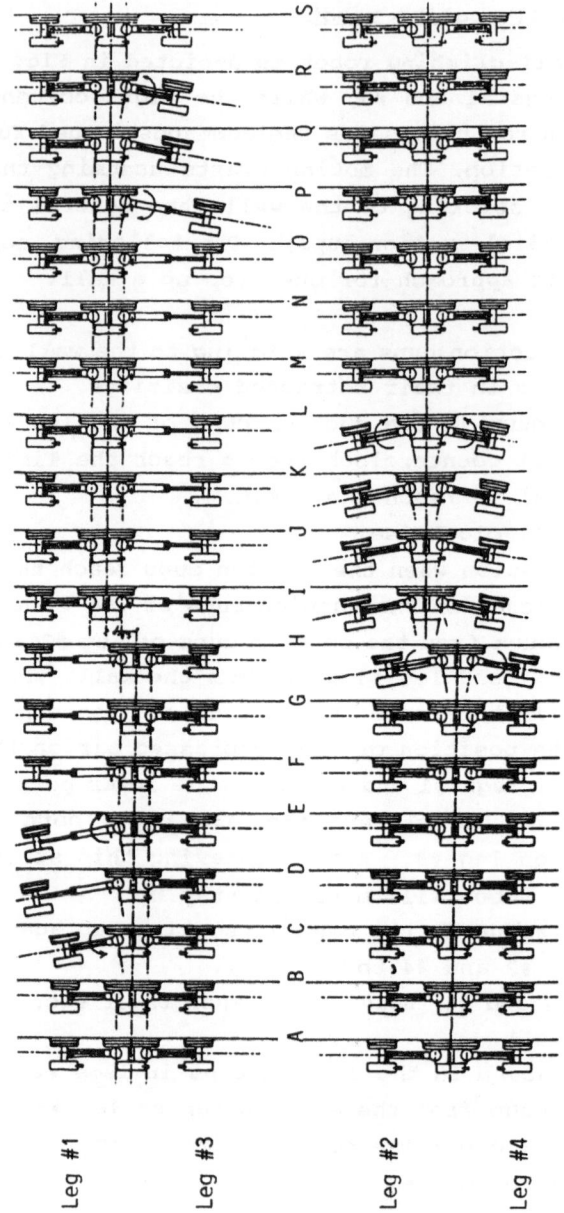

Figure 3 The motion of the wall-scaling robot.

CONCLUSION

Therefore, with this robot it is possible to carry a variety of non-destructive testing equipment , de-riveting devices, and paint removal systems. In addition, this robot or a variation of it can be used in many other applications such as the inspection of hazardous areas in nuclear power plants, ship maintenance, cleaning windows of sky scrapers, painting storage tanks, etc. It is clear that a robotic system can be used for aircraft inspection and maintenance. The work which has been done to this point is an introduction to prove the feasibility of such a robotic system. Further work needs to be done to optimize the mechanical design or to develop a faster system with more intelligence.

REFERENCES

1. Sato, K., Fukagawa, Y., and Tominaga, I. "Inspection Robot for Tank Walls in Nuclear Power Plants," Conference of Remote System and Robotics in Hostile Environments Pasco, Washington, March 29- April 2, 1987.

2. Hirose, S., "Wall Climbing Vehicle Using Internally Balanced Magentic Unit," Tokyo Institute of Technology, Japan, March 13, 1987.

3. Sato, K., Watanabe, M., Fukagawa, Y., and Morita, H. "On-Wall Traveling Robots for Nuclear Power Plants," Proceeding of the ANS Third Topical Meeting on Robotics and Remote Systems, Charleston, South Carolina, March 13-16, 1989.

4. Henderson, B. W. "USAF Expects Robotic Inspection Facility to Cut Maintenance costs," Aviation Week & Space Technology, March 13, 1989.

5. Bahr, B., Maari, S. "Robotics: Another Choice for Inspection of Aging Aircraft,"Proceeding: AIAA/FAA Joint Symposium on General Aviation Systems, April, 1990.

A Telemanipulator for Hazardous Mining Operations

M. R. UDAYAGIRI, T. R. RANGNATH, K. C. S MURTY, S. RAGHUNATH
Central Electronics Engineering Research Institute
Pilani, India

PRAVEEN DHYANI
BM Birla Science and Technology Centre
Statue Circle, Jaipur, India

SUMMARY

The paper presents design, control and functional aspects of an application specific heavy duty telemanipulator for operations in hazardous mining areas which are either inaccessible or involve high risk to human life. One of the problems faced in a typical copper mine is the jamming of ore pass tunnel due to uneven ore boulders. The manipulator performs the implantation of dynamite cartridges in the boulder(s). The designed manipulator is semiautomatic and has seven degrees of freedom and is controlled from a remotely located control room.

The jam takes place within 15 meters from the bottom of the pass and the total weight of the jammed ore is around 60-80 tons. The diameter of the ore pass ranges from 4 meters to 6 meters and has loose and uneven terrain. The jam is cleared by blasting the jammed boulder(s). The manipulator is required to enter from bottom of the ore pass, access the jammed location, scan the area, locate the weak points in the jammed boulder(s) with the operator's assitance, drill holes at the identified points, implant dynamite cartridges, and return back to the safe area.

Apart from eliminating risk to human life, utilisation of the presented manipulator is going to decrease the shut-down period of the mine (due to jams), thereby increasing the productivity. It would also reduce the present explosive consumption by 70% thereby recovering the cost of the manipulator within one year's time. The manipulator can be used in other heavy duty mining operations also by changing the end effectors only.

INTRODUCTION

Manipulators with several degrees of freedom are being used commonly for handling routine and complex operations. Second and third generation robots have added programmability and varied amount of

---

*   Central Electronics Engineering Research Institute,
    Pilani-333031, India

+   BM Birla Science & Technology Centre,
    Statue Circle, Jaipur-302005, India

intelligence for more versatile operations. Though robots with high degree of intelligence are available, some applications like operations in inaccessible and remote areas require remote control of the manipulator, less autonomy to the robot and more guided operations by the operator. One such application was encountered in mining industry where the ore having boulders of irregular dimensions get jammed in the tapered portion of the shaft. The major operations to clear the jam are that the rough location of the jam has to be estimated; the manipulator has to approach the location as nearer as possible, observe the cause of jam, decide the location for drilling and fixing of explosives and return back. Each phase of the above operation is complex and has to be performed in a dusty and non conducive environment. A possible solution is framed by the use of a Telemanipulator with unique design criteria. The details of the problem and a conceptual design of the manipulator are explained below.

2. PROBLEM

Copper ore is drawn from different levels and is moved through a vertical opening (ore-pass) to reach the crusher chamber as shown in Fig.(1). The ore falls down through the pass under gravity. The pass has irregular dimensions with an average diameter of 6 meters (tapered from 8 meters to 4 meters) the walls are highly uneven due to caving at different places.

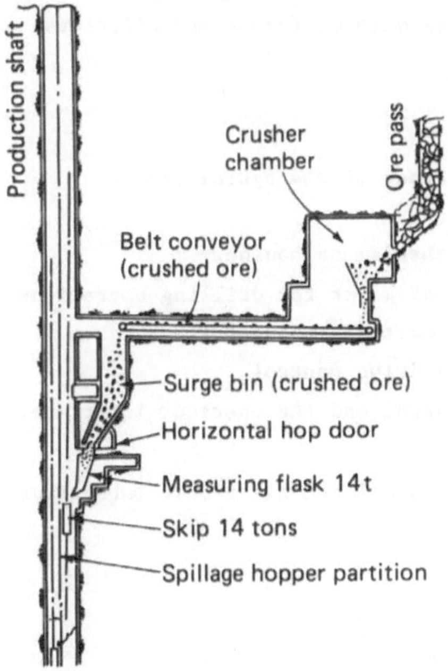

FIG. 1   PRODUCTION SHAFT HOISTING SYSTEM

Production shaft

Crusher chamber

Ore pass

Belt conveyor (crushed ore)

Surge bin (crushed ore)

Horizontal hop door

Measuring flask 14t

Skip 14 tons

Spillage hopper partition

The ore pass tapers at its lower end where it is slanted at about 60 degrees; as a result some boulders get stuck and block the passage. No ore can then pass to the crusher and movement of the ore to the production shaft is stopped. Presently whenever a jam occurs, the level at which the jam occured has to be estimated. The jam can occur any where in the slanted portion of this ore pass, but generally it is at about 15 meters from base. The exact location of jam cannot be precisely estimated.(The jam may be due to a single boulder or due to multiple number of boulders.) Present practice of jam clearrance is manual and is purely a hit and trial process. Hence more explosive energy gets wasted and there is no possibility of clearing the jam in one explosion. A series of such trial explosions have to be made with sufficient gap to allow the explosive gas to escape causing heavy loss of production.

The first objective of the design is to avoid any human risks by totally automating all clearing operations from a remote location. The second objective is to drill shot holes on the jammed rock surface for planting explosives at multiple locations so that the explosive energy is not wasted and the jammed boulder give way in one trial. The third objective is to reduce the time lag in clearance of jams and improve productivity by reducing the break downs in the continuous process of ore production. Moreover the design has to take care such that the same system can handle similar heavy duty operations not only in ore pass jam clearance but other areas like stopping with different end effectors.

3.DESIGN DETAILS

The major functions of the system are

a) access mechanism to boulder
b) alignment of wrist for drilling operations on boulder
c) explosive cartridge plugging
d) drives and drive control
e) remote Control and the operator interface.

The mechanical and electronic subsystems are explained as below:

3.1 MECHANICAL SYSTEM

A solution to the problem of accessing the boulder can be by providing mobility to the manipulator. One possible solution is by designing a mobile vehicle which can go up 60 degrees slant and get locked near the jam, do the operations and then come down. But such a vehicle cannot be built as there is no plain terrain and no guided path can be installed since it will get immediately damaged by the passing of ore in normal conditions and also due to loose and uneven terrains. Another possible solution is to install a guide temporarily but the dynamics for such solution are not practical. The solution with six legged robot going up the terrain is possible, but it is highly complex and not cost effective. Hence, it is proposed to have the base of the vehicle stationed at the face of the tunnel firmly clamped. A telescopic arm (fig.2) controlled by the operator will be pneumatically operated to reach the boulder. Another degree of freedom is provided by a supporting telescopic plunger by which angle of elevation can be changed. Hence, two degrees of freedom are provided at the vehicle base. The two degrees of freedom are sufficient enough for the tip of the arm to reach upto the boulder.

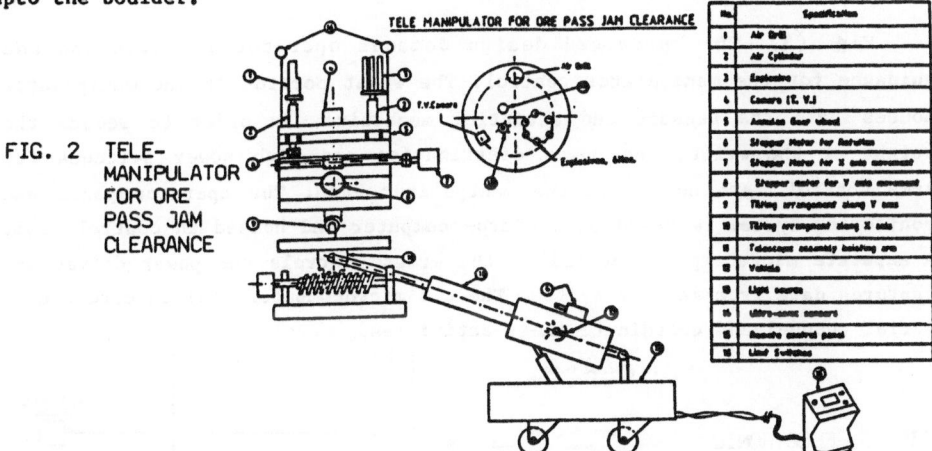

FIG. 2 TELE-
MANIPULATOR
FOR ORE
PASS JAM
CLEARANCE

The wrist mechanism provides four degrees of freedom to align the drilling mechanism parallel to the boulder. Tilting mechanism along X axis is provided by worm gear system controlled by a stepper motor. Similiar rotation in Y direction is provided. In addition to the above two degrees of

392

freedom, X,Y movement of the plate are necessary for aligning the drill to specified location. X,Y movements are through two stepper motors. By the above four degrees of freedom, the TV camera fixed at the periphery of circular wrist plate can be brought to the desired location.

The main operation of the manipulator is to drill and plug the cartridges one by one in all the identified locations on the boulder(s). The operations are activated by pneumetic system. An air drill, a TV Camera and dynamite cartridge holding and inserting system are mounted on the circumference of the wrist plate. They are at 120 degrees apart. Once the TV Camera is aligned, the plate is rotated by 120 degrees to get the drill to the same position. Similarly six cartridges are placed on the circumference of a smaller revolving chamber. Hence TV Camera, drill and dynamite cartridge plugging assembly are placed on annulus wheel at 120 degrees apart. The wheel is rotated precisely by stepper motor. Cartridge holding chamber is similarly rotated. The mines are already equipped with compressed air lines which will be effectively utilised for pneumatic cylinders.

3.2 ELECTRONIC SYSTEM

Fig. (3). The proposed design demands operator intervention and guidance for the manipulator control. The wrist portion of the manipulator houses all the sensors and actuating mechanism. In order to reduce the weight of the wrist, the power distribution and high power switches for Arms Control are housed at the manipulater base. The operator panel and monitoring system is based on a micro- computer and housed in control room. A seperate microcomputer housed on the wrist controls the power drives and captures data from sensory system. The two microcomputers communicate over a serial channel and coordinate their activities.

FIG. 3 ELECTRONIC
        SUB SYSTEM

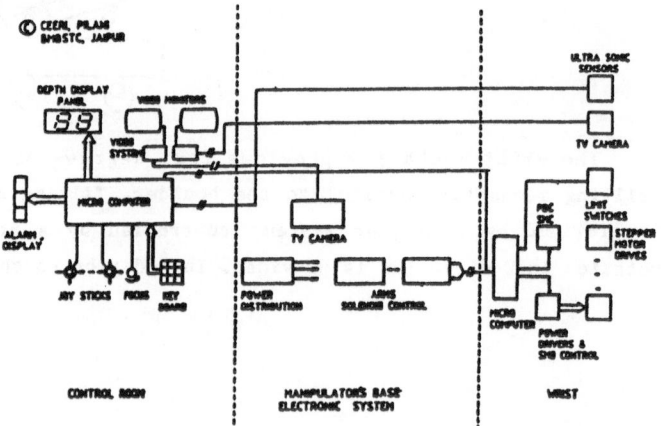

The basic sensing elements mounted on the wrist are to measure proximity to the boulder, the shape and protect collision. Ultrasonic sensors are mounted to monitor the proximity of Jam area. The proximity sensor signals are directly fed to the microcomputer at the control room as the proximity is monitored at the control panel and necessary manual controls are operated. A highly sensitive TV Camera is mounted on the wrist with remote control and necessary degrees of freedom for visual monitoring of jammed area. The video signals are also fed directly to control room. The microcomputer in the wrist will generate necessary control signals for power drive control of stepper motors and DC servos , solenoid operation for drillings, alignment operations etc. The limit switches are housed on the wrist to protect the wrist from accidental collision with a boulder. Light source is also mounted on the wrist.

The base of the manipulator houses another TV Camera to monitor the movement of the wrist and the telescopic arm. The control panel is microprocessor based and monitors the Jam and wrist operations over two video monitors and the distance of the wrist to jam. Operator has human friendly controls through Joysticks for wrist operations. Necessary commands are generated by the microcomputer based on operator control and are communicated to the computer on wrist. Operator can take up remedial action from the alarm on the panel due to limit switch operations.

4 CONCLUSIONS

A study on modernisation of mining has revealed that the complexities involved in clearance of jams can effectively be done by telemanipulators. In this particular application Economic analysis has revealed that productivity can be improved by reducing jam period to one sixth. Large savings of the explosives is expected with telemanipulators than with conventional methods. A payback period of one year is estimated in this modernisation process.

5. ACKNOWLEDGEMENTS

The authors wish to thank Director , CEERI and Director, BMBSTC. The authors also acknowledge Khetri Copper Complex personnel for discussions and problem definition.

Adoption of Robotic System for Inter-Station Handling Operations for Nagpur Milk Scheme, India

J.P. MODAK and R.D. ASKHEDKAR
VRCE
Nagpur, India

A.V. PESHWE
Datta Meghe Polytechnic
Nagpur, India

Summary

This paper discusses and establishes the techno-economic feasibility of robotizing some of the manual handling operations in processing of milk at Nagpur Milk Scheme, Nagpur.(India). Technical criterian for adoption of robot proposed by Groover was adopted. Economic analysis revealed that Two SCROBOT-ERIII robots with vision system and computer can be installed at two work stations for (i) feeding empty bottles to bottle washing machine and (ii) lifting filled bottles from chain conveyor and placing them in crates. These robots replace 56 workers and requires initial capital investment of Rs.11.8 lakhs which will be recovered in 1.46 years.

Processing of Milk at Nagpur Milk Scheme

Fig. 1 shows the plant layout for milk processing at Nagpur Milk Scheme. The processing of milk is carried out in three phase,i.e.(I)Raw material receiption(II)Milk storage and processing and(III)Bottle washing and filling. In phase I raw milk from villages is brought in cans to work station S_1 and is conveyed to tipping point 1. Cans are tipped manually. In phase II the milk is pumped through the chilling unit 3 to storage tank at a higher level. The milk flows from storage tank 4 to balance tank 5 by gravity. The chilled milk is pumped from balance tank to pasteurizer 7. The output of pasteurizer is connected to storage tank 4 from where it flows to milk filling units 8 and 9. In phase III used bottles are received and conveyed to bottle washing machine 10. The automatic filling units 8 and 9 receive cleaned bottles and fill them with milk and seal them by alluminium foils. These filled bottles are, then, conveyed to storage room 11. These bottles are distributed to customers on next day.

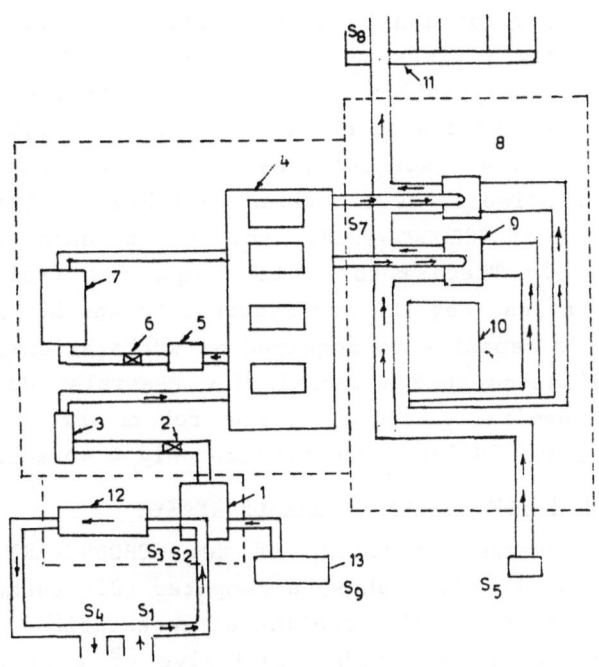

1. Tipping point
2. Pump
3. Chilling unit
4. Storage tanks
5. Balance tank
6. Pump
7. Pasteurizer
8. Filling m/c 1
9. Filling m/c 2
10. Bottle washer
11. Store
12. Can Washer
13. Tank for Returned milk

Dotted lines indicates different phases of processing

Fig.1 Plant layout for milk processing

Table I indicates activities carried out at various workstations and the number of workers employed at each workstation.

Cost Analysis of Present System :

The cost of processing of milk with present system of manual handling is Rs.1.45 per litre and the average quantity of milk processed per day is 90,000 litres. Manual handling utilizes service of 75 workers. The average wage payment of workers inclusive of the cost of facility is Rs. 1200.00 per month and hence the annual cost of manual handling is Rs. 10.80 Lakhs.

Selection of Robot System

Technical criteria for adoption of robot proposed by

Groover[1]suggests that robot can be adopted for workstations S5,S6,S7 and S8 as they satisfy the above criteria to the fullest extent [2]. RTM analysis[3]revealed that operations at workstations S5,S6,S7 and S8 can be robotized by placing 1,5,5 and 1 average robots at these workstations of different makes of robots presently available, as tabulated in Table 2. It is proposed to place two SCROBOT-ERIII robots at each of workstations S6 and S7. These robots will require a vision system and a computer at each of the workstations S6 and S7. As robots of high pay load capacity are required at workstations S5 and S8, their installation though technically feasible, does not seem to be economically sound. These robots ask for initial investment of Rs. 20 lakhs and replace only 4 workers.

Cost Analysis of Manual-Cum-Robotic Handling System

The cost of robotic system consisting of two SCROBOT-ERIII (3.40 lakhs) vision system (1.7 lakhs) a computer (0.4 Lakhs) and two motors (0.20 lakhs) to be installed at workstations S6 and S7 comes out to be Rs.11.18 lakhs (inclusive of cost of installation). Expenses for electricity and maintenance are neglible. The robotization of work at workstations S6 and S7 will lead to saving of salaries of 56 workers amounting to Rs. 8,06,000 per annum. Thus installation S6 and S7 requires an initial investment of Rs.11.80 lakhs and leads to an annual saving of Rs. 8.06 Lakhs.

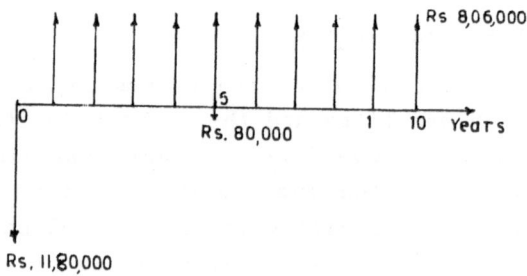

Fig. 2   Cash flow diagram

Fig.2 shows the cash flow diagram for the proposal of installing robotic system at workstations S6 and S7. For the economic analysis, the life of robot and motor is taken as 10 years and 5 years respectively. It is also presumed that they will have no salvage value at the end of their life.

The economic evaluation of proposal is as under :

1. Payback period (without interest rate) = 1.46 years

2. Pay back period (with interest rate of 18%) = 1.86 years

3. Internal rate of return is 68.3%

Conclusion

The robotization of material handling at work stations S6 and S7 for processing of milk in Nagpur milk scheme is technically feasible and economically viable.

TABLE-1 **WORK STATIONS AND FUNCTIONS PERFORMED**

.NO.	Work locations	ACTIVITY	No.of workers
1.	Work station No.1	Cans are received & placed on chain conveyor	4
2.	Work station No.2	Cans are emptied in the tank	2
3.	Work station No.3	Empty cans are lifted & placed on the chain conveyor for can washing machine	1
4.	Work station No.4	Washed cans are loaded on the trucks for next delivery	2
5.	Work station No.5	Empty bottles with crates are placed on chain conveyor for bottle washing machine.	4
6.	Work station No.6	Feeding of empty bottles to the bottle washing machine.	28
7.	Work station No.7.	Packed milk bottles are lifted from the chain conveyor and placed in the crates.	28
8.	Work station No.8	This work station is in cold storage room. Crates are lifted & placed on platform	4
9.	Work station No.9	Returned milk bottles are emptied and sent again for processing.	2

TABLE - 2 **TECHNICAL SPECIFICATIONS OF ROBOTS**

S.N.	SPECIFICATIONS	SCROBOT -ER III	KAWASAKI ROBOT LINE UP	ADEPTONE ROBOT	C R S ROBOT SYSTEM
1.	Model	Scrobot ER - III	PH 560	Adeptone Robot	C R S Plus
2.	No of Axes	5 plus gripper	6	5	[5 + 3] DC Servos
3.	Construction	Articula-ted arm	Revolute Type	-	-
4.	Load capacity	1 Kg	2.5 Kg	20 Lbs	2.2 Lb
	Weight	16 Kg	95 Kg	400 Lbs	-
	Repeatability	.5 mm	10.1 mm	.001"	.13 mm
	Max speed	330 mm/ sec.	500 mm /sec.	-	-
	Actuators	6 DC servo motors with closed loop control	Electric DC servo motor	-	DC servo motor
5.	Working Envelop	Body joint 340 dg. Shoulder joint 85 deg. Elbow joint 150dg Pitch joint 150dg Rolled joint unlimit-ed Max.rad. of operation 610 mm.	Waist rotation (JT1) 320 dg Shoulder rot. (JT2)250 dg Elbow rot.(JT3) 270dg Wrist rot.(JT4) 280dg Flange rot. (JT6) 520dg	Joint motion Joint 1 300dg. Joint 2 294dg Joint 3 Standard Joint 4 554dg	Joint speed Base 60dg/s Shoulder 60dg/s Elbow 60dg/sec Wrist 180 dg/s Tool 180dg/s

References :

1. M.P. Groover, M. Weiss, R.N. Nagel & N.G. Odrey, Industrial Robotics, Mc.Graw Hill Book Company, 1987 (IInd Printing)

2. A.V. Peshwe and J.P. Modak, MTM and RTM analysis of interstation handling operations in a milk processing plant. Proceedings of Industrial Problems in Machines and Mechanisms - 88 (I PROMM - 88), Jan.11-12 (1989) 171-183.

3. S.Y. Nof and A. Lechtman, The RTM Method of Analyzing Robot work, Industrial Engineering, (April 1982), 38-48.

Integration and Realtime Monitoring of Robotic Controllers

SUDHAKAR R. PAIDY
Department of Industrial and Manufacturing Engineering
Rochester Institute of Technology
Rochester, NY

MICHAEL SHEA
Research and Development
Optical Gaging Products, Inc.
Rochester, NY

Abstract

Many robotic controllers, especially the older generation, are limited in their capabilities to aid integration of material handling activities in an automated manufacturing environment. This paper will present a realworld case-study of integrating an IBM 7545 robot into a manufacturing cell. An AML program is designed to control the robot for loading and unloading operations at an automatic assembly station. The programs also maintain data necessary to track the robot operations in realtime. A C program running on a microcomputer (PC-AT) is used to poll the robot controller for that realtime data and act as a realtime monitoring and control station. Programs running on a Concurrent Computer 3280 minicomputer that control the overall assembly operations also interact with the robot controller concurrently to provide the detail of the assembly steps. A software protocol is developed to define high level interaction among the minicomputer, microcomputer and the robot. This paper will also present the issues, problems and some resolutions in integration of such complex manufacturing cell environments.

Introduction

There are five general functionality levels that can be acquired in a robot controller: sequence control mechanisms, playback machines, controlled path(computed trajectory) robots, adaptive robots and intelligent robots. The capabilities at these levels vary from allowing only predefined motions(alterable only at a very high cost) in sequence controlled machines, to only needing to supply start/end point information to the controller(the controller then determines its' optimum path between the points) in the controlled path robots, to the robot having enough intelligence to avoid unexpected obstacles or have learning capabilities in the adaptive and intelligent robots[1].

The IBM 7545 robot falls into the middle category of a controlled path robot. Given the limited capabilities of this robot, we were faced with two options: replace the robot with a newer, more intelligent robot or provide a method by which it would be possible to achieve the level of integration needed using existing equipment. This paper will outline the hardware interfaces and software protocols developed to allow a high degree of intelligent interaction in an assembly operation among a Concurrent Computer Corporation's (CCC) model 3280 mini-computer, IBM 7545 industrial robot and an IBM PC-AT. The issues, problems and some resolutions in the integration of the complex manufacturing cell environment will also be presented.

Primary objective of this study was to integrate an existing robot into an already operational manufacturing cell. We will describe the manufacturing cell and its operations first. The requirements for the capabilities of an assembly robot will be established. The hardware topology selected to integrate the robot will be described. A two phase, three way interaction among the cell's host computer, robot controller and a microcomputer will be discussed. Then software protocols developed to support such interaction will be presented.

Manufacturing Cell

The manufacturing cell occupies about 600 sq. ft. area and is comprised of two CNC machining centers (a mill and a lathe), a vision/inspection station and a Load/Unload station(Dock). These four stations and four assembly stations are connected by a conveyor system. The two closed loop conveyors are placed end to end in an 'L' shape configuration. Figure 1 shows a schematic of the entire facility. The main conveyor loop connects the four

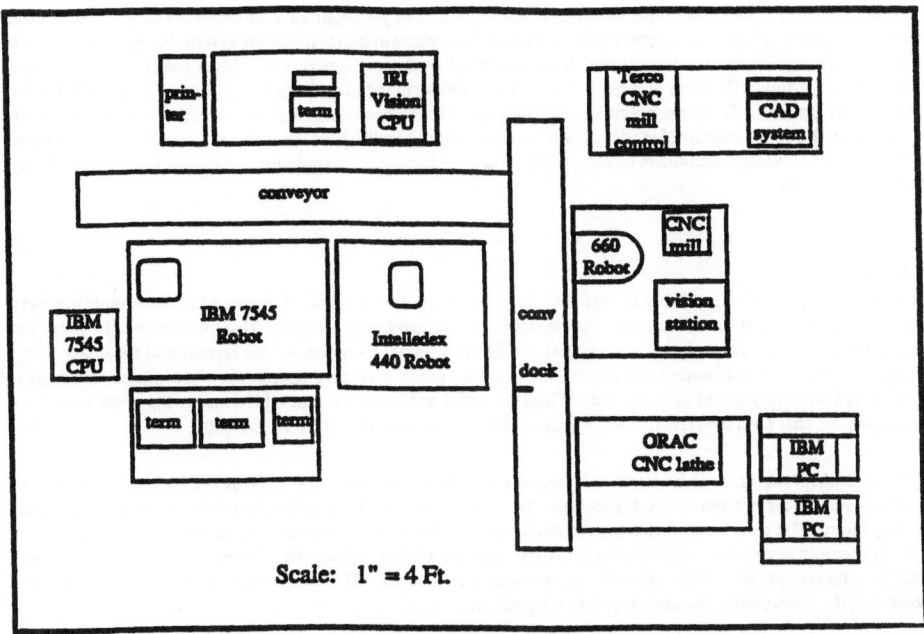

Figure 1: Schematic of the manufacturing cell.

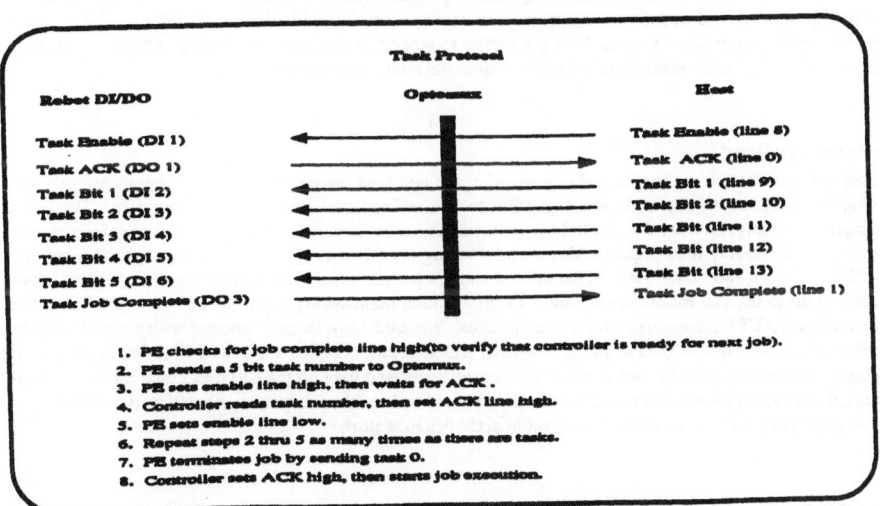

Figure 2. Host - Robot Task Protocol

stations identified above while the secondary conveyor that is perpendicular to the main loop connects the assembly stations. These conveyors together act as a buffer and material transport system for up to ten 8 in x 8 in. size pallets which hold materials and fixtures needed in the cell operations. Each pallet carries a Radio Frequency tag for identification purposes. All pallets circulate in the main loop except for the ones requested by the assembly stations. These are automatically transferred to the assembly loop when needed and returned back into the main loop. An industrial grade robot provides material handling for the mill center and the vision station. A robot and an elevator mechanism developed by Industrial Engineering students provide load/unload capability for the lathe center.

Assembly Station

When a part required for assembly is available within the manufacturing cell it is routed to an assembly station. Once the part arrives at the station, the robot is notified of its arrival by the host. The robot moves the part from the pallet on the conveyor and places it in a fixture. Multiple parts/components may be required for an assembly. Assembly operations are carried out one step at a time. When assembly operations are complete, the host is notified and an empty pallet is requested. When the pallet arrives at the assembly station, the robot will move the assembled part onto the pallet. This sequence is repeated for each part to be assembled.

If the assembly station will be used to process many different parts, the control program will contain all assembly sequences needed for each part type. Even with only one part in the station, the control program can be very large. The size is not dictated so much by the number of parts, but rather that one program has to deal with all possible assembly operations for every part type the station will handle. Given this, and adding multiple fixtures , the complexity of the control program increases as well. The danger of having the control program exceeding the controllers' memory capacity then arises.

If the control program is broken up into several programs by part type, it is possible to achieve the level of sophistication needed without exceeding memory limitations. Each program would have common moves/operations as well as the unique sequences of moves for the individual part . In order for this approach to work effectively, the controller must have the ability to automatically load or unload a particular part program from memory. The robot controller, as received, does not have this capability.

Hardware Topology

A three way interaction among the robot controller, the cell host computer and a microcomputer was used to provide the robot controller with the capability to load/unload part programs automatically. The cell host computer, obviously, dictates the part to be processed. The robot controller does not have either local storage or capability to selectively load required files. A microcomputer (IBM-AT) is used to interact with the robot via a RS-232 serial port(4800 baud rate). This microcomputer provides the necessary storage for the files related to all parts in the cell. The robot does not have additional communication ports to be able to link to the cell host. However it has 32 TTL compatible Input/Output lines. We used 16 of these to interact with digital Input/Output interface of an OPTOMUX data acquisition system which in turn is linked via a RS-232 (9600 baud) port to the cell host. Though the cell host can provide file storage, increased load on the cell host could not be justified. A dedicated microcomputer could provide a faster download and realtime response. As will be described later, this microcomputer is also used to monitor the robot activities in realtime.

Overall Software Architecture

In order to develop the software for the station, the actions of each piece of equipment are defined. The implementation fell into two phases; definition of part type and specific assembly sequences. After completion of an assembly, the robot waits for notification of the next part type to be processed by the station. This notification comes from the cell controller. After receiving the next part type the robot controller notifies the microcomputer monitor of the part type requested. The microcomputer compares the requested part type to the current part and downloads the new program to the controller if necessary. A separate program for each part type exists on the microcomputer's hard disk. An AML program contains all possible assembly sequences for the part. Then when a part is required for a particular assembly, the assembly sequence can be constructed from the commands stored in the file. These phase I operations are illustrated in Figure 3.

Phase 2, illustrated in Figure 4, involves the specification and execution of assembly sequences. Once the control program has been loaded into the robot controller, the cell controller downloads the specific command sequence to the control program. When the entire assembly operation has been sent, execution begins. The robot executes only the command sequences specified by the cell controller. While the robot is running, it provides active realtime status messages to the cell controller through the Optomux interface.

The microcomputer is used to passively monitor the state of the assembly operations. This is done by querying the robot controller instead of the AML control program. The controller provides information to the microcomputer regarding program variables which are used to determine where the robot is in the execution of the command sequence.

Protocol Description

In developing the software solution to the problem at hand the system was broken down into logical groups. Programs had to be written for the cell controller, robot controller and microcomputer. The programs for the robot controller were written in IBM proprietary robot control language AML(A Manufacturing Language). These programs were divided into subroutines for receiving the incoming task numbers , message passing, executing the job and the actual tasks.

In order to be able to download assembly sequences to the control program, a communication protocol was developed. This protocol allowed the AML program to communicate with the host over the robot's DI/DO lines to Optomux. In defining the protocol, eight I/O lines (6 input to and 2 output from the robot) were used for passing command/task numbers(see Figure 2). Five inputs to the robot were used to specify the command number with the remaining input for an enable line. The host uses the enable line to tell the robot that there is a command number on the DI lines to be read. The two outputs are used for the acknowledge and job complete lines. The acknowledge line is used by the robot to tell the host that it has successfully received the command number, while the job complete line is used to indicate the completion of execution of an assembly sequence. With a five bit command number the host is able to specify up to thirty-two distinct tasks. From our experience, this is a sufficient number of tasks to perform any job.

In downloading and executing a job, the host first checks the job complete line of the robot. If the line is low the robot is still running the previous job. Once a high is detected the host sends the new part type to the Optomux and then sets the enable line high. The part type is a pre-defined number that notifies the IBM PC-AT which control program file needs to be loaded on the controller. The control program waits for the enable line to go high before it reads the input lines and stores the part type. The monitor program on the AT, waits for the control program to receive the part type. Once the part type has been received by the control program, it passes it on to the monitor. The communication between the robot and the monitor is done across the serial port using the built in robot controller communication's functions. If a new program needs to be loaded, it then deactivates the current robot control program, and loads the new program requested by the host and activates it. The new control program then sets the job complete line high to notify the host it has received the part type and is ready to proceed. The host, on detecting the job complete line high sets the enable line low. The first task number is then sent to the Optomux. The sequence of setting the enable high, waiting for acknowledge, setting enable low, sending the next task is repeated until the entire job has been specified.

Version 4.0 of the AML language used on the IBM 7545 robot controller has only three data types; counters,constants, and points. The constant data type is set at initialization and cannot be modified; the point type is made up of the x, y, z and rotational coordinate of the gripper. The counter type is the only data type that can be modified during runtime. The methods to modify counters, which are available in this version of AML, are increment or decrement by one, or setting it to some specific value. Since the language has no array type, a

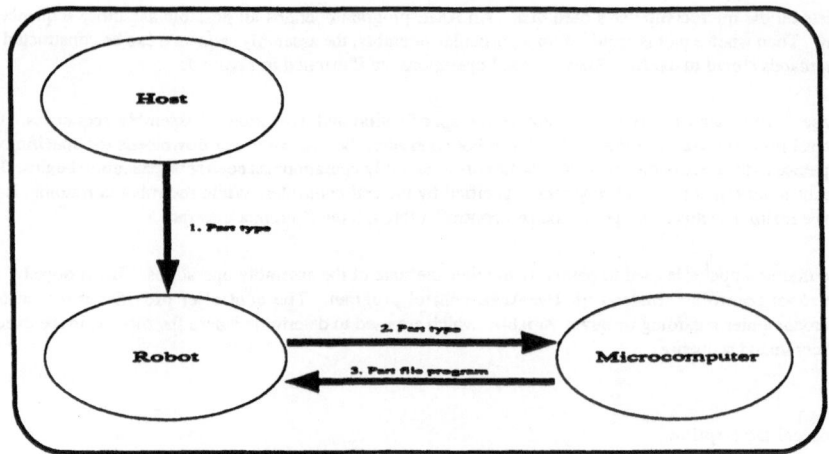

Figure 3: Phase 1, Definition of Part Type

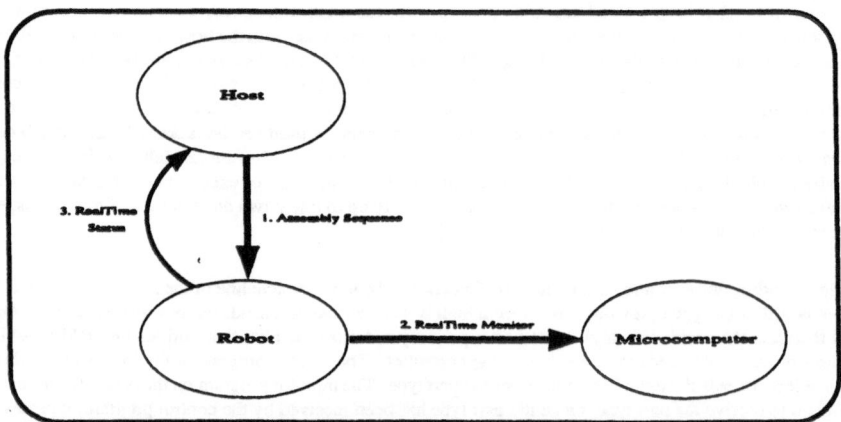

Figure 4: Phase 2, Specification and Execution of Assembly Sequences

counter has to be declared to store each command in the job. This increases the size of the program. With no array capability there is no real looping capacity available as well. The impact of this was that much code had to be duplicated. To fit into the finite amount of memory available on the controller, it was decided to limit the number of tasks per job to fifteen to prevent running into the memory barrier.

Once execution begins, the AML program starts sending status message numbers to the host through the Optomux. Another eight I/O lines (6 outputs from and 2 inputs to the robot) are used, with five outputs being allocated to the message number. Similar to the task number protocol, the five bit message number allows up to thirty two different numbers. The meaning of a number would depend on the part type being handled. For example, if part type #1 was being used in an assembly, the message number 1 may be defined as waiting for the part to arrive at the assembly station. With a different part type, message number one may indicate the robot has reached a certain stage in the assembly.

The third piece of the system is the software developed to run on the IBM PC-AT. The AT is used as a realtime monitoring station of the robot, providing such information as total number of tasks in a job, current task being executed, number of tasks left, at what point in the current task the program is in. To fulfil this function, software was developed to maintain communication with the controller through the serial port. This software took advantage of the built in communication capabilities of the robot controller. There are five different command types native to the robot controller; read, execute, data, new program and end. The first two types provide to an external program a method to obtain the details or actions of the particular control program being run. The read command supplies to the external information on such things as internal program variable status', DI/DO line status' and machine status. The first two would allow an external program to determine where the internal program is in its execution. The third command would be used more to query the robot controller about any possible error conditions (i.e. destination point out of work envelope, transmission error).

Summary

The purpose of this project was to increase the level of intelligence of the robot controller so that it would have the capability of accepting assembly jobs from the cell controller automatically. This goal has been achieved by connecting the robot controller to the cell controller and the PC-AT. By using the developed communication protocol we have provided the controller with enough intelligence so that it is capable of functioning efficiently within the manufacturing cell.

References

[1]. Critchlow, A.J., Introduction to Robotics, MacMillan, 1985.

Acknowledgment

The Authors would like to thank Raymond Corporation for donating the robot, Dr. Richard Reeve - Head of Industrial Engineering- for providing the resources needed, Mr. William Gallacher - IE technician - for setting up the hardware interfaces, Mr. Juan Hun for his work related to assembly on the cell host, and Mr. Carl Tesavis for his development of a C-library for the Optomux device.

On the Applications of Part Image Reconstruction Systems in Automated Manufacturing

SAEID MOTAVALLI
Department of Industrial Engineering
The Wichita State University
Wichita, KN

BOPAYA BIDANDA
Industrial Engineering Department
The University of Pittsburgh
Pittsburgh, PA

SUMMARY

This paper outlines the reverse engineering methodology to automate the process of CAD model reconstruction for existing parts and prototypes. Three-Dimensional (3-D) geometric information is extracted from a part surface by a non-contact Part Image Reconstruction System (PAIRS) developed by the authors. For each scanned part a CAD model consisting of a 3-D polygon mesh, along with the orthographic projections is created. This is in the form of a wire frame CAD model.

The real power of PAIRs will become apparent when such systems move away from the islands of automation concept and interface with other manufacturing automation software. This paper explores some possible interfaces.

1.0 THE MANUFACTURING SEQUENCE

Manufacturing can be defined as the set of interrelated activities needed to produce a given part. These activities (consisting of product design, production planning, production, etc.) are traditionally performed in a sequential manner. This sequential model is used by a large majority of U.S. industries. Any changes in product or process are reflected by changes in the product design. This is obviously time consuming and can lead to delays in manufacturing, because the sequential chain of activities must be repeated. More and more industries are striving to move towards the concept of *simultaneous engineering* to reduce the lead time between product design and the time the product is actually produced. Part Image Reconstruction Systems (PAIRS), based on the concept of *reverse engineering* can play an important role in reducing this lead time.

2.0 NEED FOR PART IMAGE RECONSTRUCTION SYSTEMS & REVERSE ENGINEERING

There are many instances where one-of-a-kind parts such as prototypes or custom-built parts need to be reproduced. Moreover, the design of the existing manufactured parts may require periodic modifications to update and improve. CAD models of these parts are often not available.

However, the creation of a CAD model is desirable and often necessary under the following circumstances [1]:

1. New Design: The design process of a new product does not always start from a CAD model. A prototype is often built first. Once the design is approved, measurements are made (either manually or with the use of a contact probe). The extracted data is then manually entered into a Computer Aided Engineering (CAE) based system for further analysis. This process has two disadvantages: it is time consuming and is a potential source of measurement errors.

2. Modify Existing Design: In some instances the design of an existing product has to be modified. The modification process and design improvements are best performed on a CAD model. However, CAD models for many existing products are not available. Here, Part Image Reconstruction Systems can play an important role in reducing design time.

3. Design of Large Items: Precise measurements of large parts are often not possible with traditional metrological equipment. Here, Part Image Reconstruction Systems can help by mapping the part surface in the form of CAD model. This CAD model can now be scaled and modified as needed.

4. Worn or Broken Parts: When a one-off part breaks or is worn, and the engineering drawing is no longer available, Part Image Reconstruction Systems can be used to create the CAD model. The CAD model can now be used to manufacture the clone of the worn /broken part.

The first step in this aspect of reverse engineering involves the extraction of geometric data from the surface of existing parts or prototypes to input into a Computer Aided Design system. The design of the part is then analyzed and improved using the CAD system. The improved design is stored in a CAD data base. The geometric information contained in the CAD data base along with other manufacturing information (materials, roughness, etc.) is then used to create a process plan. An NC part program is also created using the part geometry. The NC code and Process Plan are then used to manufacture the part.

The aim of integration of PAIRS with other manufacturing software is to automate the above process by directly scanning a prototype and automatically creating a CAD model, Process Plan and NC part program to manufacture the part.

3.0 THE PAIRS SYSTEM

The PAIRS system developed consists of the following parts[1,2,3]:

a) A non-contact optical range sensor which uses a single stripe of laser light and a video camera connected to a computer to acquire three-dimensional (3D) data from the surface of the part, and

b) A software interface that links the acquired data to an existing 3D CAD system.

The PAIRs system captures and stores the digitized image of the part and laser line, on the part surface. This image is sent to the computer. The image is then searched pixel by pixel with regard to intensity, to locate the (x,y) coordinates of the lighted points on the surface. After the image is scanned, the table is rotated in discrete increments (chosen according to the complexity of the part). The process is repeated until the table is rotated 360°. Thus, for each point on the laser lighted line a set of (x,y,z,r) coordinates are computed. The (x,y,z) coordinates are used to create the 3-D polygon mesh, and (r,y) are used in edge detection and boundary recognition to create orthographic projections of the scanned part.

The (X,Y,Z,R) coordinates of each scanned point along the lighted line are stored in a data file. This data base is used as input to an interface software which modifies the data and creates an AUTOCAD data exchange file (DXF) to represent the 3-D polygon mesh of the object in AUTOCAD. The 3-D polygon mesh can be used to shade the image and represent it as a 3-D solid model of the object.[1,2,3]

Figure 1 details the different steps in creating a CAD model using the PAIRs and associated software. The data acquired by a PAIRs must first be smoothed. Next, orthographic projections are created. This ensures that the data file is small enough to be realistically displayed in a CAD system. The CAD interface transforms this data into a format compatible with the CAD package, before the CAD model can be displayed on a computer screen.

4.0 INTEGRATION OF PAIRS WITH MANUFACTURING SOFTWARE: THREE CASE STUDIES

The language of CAD is geometry based, in that the drawing is typically stored in a data base as a collection of entities such as line, arc, point, etc. A CAM data base, however, is based on part features such as face, taper, chamfer, groove, ring, etc. These features are used in process planning for interpretation of the manufacturing process. Machining requirements and tolerances are also necessary information in a manufacturing data base. Thus, a major effort in the integration of CAD/CAM is in the creation of a common data base for CAD and CAM. The first step in this process is the extraction of part surface features from the CAD data base. These features along with the machining requirements of different surfaces and also the required materials and tolerances are used in process planning and production of the part.

An example of a generic part feature extraction system for rotational parts is contained in Reference 4. This part feature recognition system is able to distinguish features of a part based on topological information stored in a CAD data base. In order to store part features, a data structure was created containing information on turning features and surfaces, part geometry, non-turning features and surfaces, coordinate dimensions and tolerances, and also raw material specifications. The input to this system is a 2D CAD drawing of the part surface. The CAD file used in this system is in IGES format. Since the IGES format can be directly translated into the DXF format, the output of PAIRS can be directly used as an input to this system.

Turbo CAPP[5] is an example of a Computer Aided Process Planning system. This system contains its own part feature extraction algorithm. However, a feature recognition system (similar to that described earlier) can also be used for extraction of surface features. Turbo CAPP checks design consistency of geometric tolerancing and dimensioning. It also interacts with an experienced user to update the acquired knowledge. The input to this system is a 2D CAD drawing (AUTOCAD DXF file) such as PAIRS output. The geometric characteristics of the part extracted form the CAD data file along with the machining parameters stored in the knowledge base of the system are used to create a generic numerical control part program (BCL). The output of this system can be directly used to produce the part.

Yet another manufacturing software system that can be readily integrated with PAIRS is a Computer Aided Fixture Selection (CAFS) system.[6] The input to this system is also a 2-D wire frame CAD data base (output of PAIRS) along with the output of a process planning system. This system is capable of selecting fixtures for concentric rotational parts. The system uses a set of rules to choose the best type of fixtures for the manufacture of a part. It consists of the following modules:

1. A set of standards for the engineering drawing on the CAD system to include manufacturing planning information,

2. A feature extraction module, to extract part surface features from the CAD data base, and

3. A rule based module for the selection of fixtures.

The standards added to the CAD data base consists of a title block and a manufacturing information block. The manufacturing block contains a list of surfaces to be

410

machined as well as tolerance specifications. Dimensional and geometric tolerances are also added using ANSI 14.5 standards. These standards can be easily added to the output of PAIRS since it used the DXF format as input for part feature extraction.

Part feature extraction is implemented in a two-stage algorithm. First, entities are extracted from the CAD data base, and then the part features are extracted from the list of entities. Some of the features needed for each surface are: length of the part, diameter of the part, presence of holes in part, presence of threads in part, cross sectional geometry of part. Feature extraction is done by matching the list of entities to the information contained in the manufacturing information block of the drawing.

The output of this system is a list of all feasible fixtures with all feasible holding surfaces. Then an algorithm based on dynamic programming optimizes the fixture selection by minimizing the load/ unload and setup time for the process. Thus, a set of the best set of fixtures for the overall process is selected.

5.0 CONCLUSIONS

In this paper the elements of the developed Part Image Reconstruction System are briefly described. Potential integration with other research based manufacturing software is also detailed. The integration process requires minimal effort. These integrated systems make it possible to shorten the product development cycle by going from a preliminary prototype part to a final manufactured product. This is especially important because of the trend towards shorter life cycles for the new products. We believe that further development of these integrated systems will have a substantial impact in the process of integration of CAD and CAM.

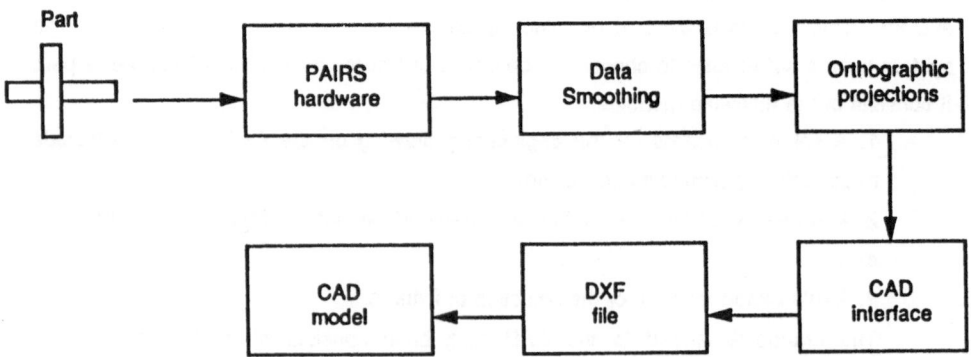

Figure 1: Steps Involved in the Creation of a CAD model using PAIRS

6.0 REFERENCES

1. Motavalli S., Bidanda B., "A Part Image Reconstruction System for Reverse Engineering of Design Modifications", Technical Report #90-4, Department of Industrial Engineering, University of Pittsburgh, Pittsburgh, PA 15261 (submitted to *Journal of Manufacturing Systems*)

2. Proceedings of the IIE Society for Integrated Manufacturing Systems Conference, Atlanta Georgia, November, 1989, "Building CAD Models with Non-Contact Techniques", Motavalli S., Bidanda B., pp. 680-684.

3. Proceedings of Fourth International Conference on CAD/CAM Robotics and Factories of the Future, New Delhi, India, December 1989, "Reverse Engineering Using Structured Lighting", Motavalli S., Bidanda, B., pp 47-56.

4. Rong-Kweili, "A Part-Feature Recognition System for Rotational Parts", International Journal of Production Research, Vol. 26, No. 9, (Sept. 1988).

5. Wang H., Wysk R. A., "Intelligent Reasoning for Process Planning," Computers in Industry, Vol. 8 (1987), pp. 293-309.

6. Bidanda B. & Cohen P.H., "Development of a Computer Aided Fixture Selection System for Concentric Rotational Parts," Proceeding of Symposium on Advances in Integrated Product Design and Manufacturing, ASME 1990 Winter Annual Meeting, Dallas, Nov. 26-27, 1990.

Kalman Filter Application to Tridimensional Rigid Body Motion Parameter Estimation from a Sequence of Images

R. VASQUEZ and J. MAYORA

Department of Electrical and Computer Engineering
University of Puerto Rico
Mayaguez, PR

Summary

The Kalman Filter is applied to the problem of extracting position and motion parameters from a sequence of images of a moving object. The kinematic model of the moving object and the camera model are developed to construct the filter structure. Experiments are performed to calibrate the camera and to test the filter with actual data.

Introduction

The purpose of this paper is to show the application of Kalman Filter theory to the problem of extracting the motion parameters from a sequence of images of a rigid body which is moving in a three dimensional space. The objective is to apply the Kalman Filter equations to the system of a moving object using the state variable approach. The system is described by a set of equations that specify the position and orientation of the body in space and how these coordinates are changing in time.

The images provide the necessary data to drive the algorithm of the Kalman Filter. Some feature points are extracted from the images, and their positions in the image coordinate system are used to update the motion parameters estimation.

The paper is divided into four sections. Section I shows a review of related papers. Section II deals with the set of equations that describes the system of the body in three dimensional motion. Section III develops the application of the Kalman Filter theory to this problem and Section IV describes the experiment and some results. Conclusions are shown at the end of the paper.

I. Literature Review

The problem of extracting motion information from a sequence of images has received much attention. Several techniques have been used to obtain this information. Most of the research has been oriented to finding methods to solve the set of simultaneous nonlinear equations that arise from the description of the body motion. Roach and Aggarwal [1] used the modified Levenberg-Marquardt finite difference algorithm to solve those equations. Tsai and Huang [2] used a variable transformation to change the estimation of the parameters to a solution of a sixth-order polynomial equation. Tsai et al. [3] used the same system in [2] but solve it using the singular value decomposition method.

The Kalman Filter method has recently been applied to this problem. Broida and Chellapa [4] used it to estimate the object motion parameters from noisy images. Their goal was to develop a technique that could cope with noise in the data; for this reason, they used a planar object moving in a plane perpendicular to the image plane. This setting produces a line on the image that stretch in accordance to the movement of the planar object. Also they used computer generated images to test their filter. The work in this paper is different from [4] in the sense that the object is tridimensional and its movement is also tridimensional. We used real images taken by a video camera.

The above approaches to the problem of extracting motion information from a sequence of images use points displacements from one image to the following and construct nonlinear equations that relates these displacements on the images to the real world. The Kalman Filter approach uses a model of the movement and the measuring instrument to generate an iterative algorithm specific to that system. This approach could be implemented in real time for using robots. In order to apply the Kalman Filter to this inherently nonlinear system, it is necessary to use the Extended Kalman Filter. In this case, the filter may diverge due to linearization imposed over the nonlinear system.

II. Kinematic Equations

The kinematic equations that describe the body motion are shown in this section. The motion is restricted to as to be uniform, i.e., the body is moving with constant speed. In order to write these equations, a reference fixed (camera) frame of coordinates must be used; also there is a frame fixed to the body. The body position and orientation are established by the position and orientation of the body frame origin and the orientation of this frame with respect to the fixed reference frame. Since the motion is tridimensional, there are three position coordinates and three angles of orientation. Also, these variables are changing, producing three components of speed and rotation. A total of twelve state variables are needed to describe the body motion. To describe the orientation of the body frame, orientation angles are used to form a matrix known as the orientation matrix. The resulting orientation matrix is

$$R(\gamma,\beta,\alpha) = \begin{matrix} C\beta C\gamma & -C\beta C\gamma & S\beta \\ C\alpha S\beta S\gamma+C\alpha S\gamma & -S\alpha S\beta S\gamma+C\alpha C\gamma & -S\alpha C\beta \\ -C\alpha S\beta C\gamma+S\alpha S\gamma & C\alpha S\beta S\gamma+S\alpha C\gamma & C\alpha C\beta \end{matrix} \tag{1}$$

where S stands for sine and C stands for cosine.

The coordinates of any point on the object can be referred to the fixed (camera) frame or to the body frame. The equation that relates both coordinate systems is:

$$r = R(\gamma,\beta,\alpha) * r' + O \tag{2}$$

where r is the position vector of the point with respect to the fixed frame, r' is the position vector of the same point with respect to the body frame, and O is the position vector of the origin of the body frame with respect to the fixed frame.

The system is described by a set of equations in discrete time form. The equations relate the state (O,R) of the system at one time with the state at a later time. Displacements are constants because an uniform motion is supposed. The equations are:

$$Ox(k+1) = Ox(k) + DOx(k)$$

$$Oy(k+1) = Oy(k) + DOy(k)$$

$$Oz(k+1) = Oz(k) + DOz(k)$$

$$\alpha(k+1) = \tan^{-1}[-r_{23}(k+1), r_{33}(k+1)]$$

$$\beta(k+1) = \tan^{-1}[r_{13}(k+1), \text{sqrt}(r^2{}_{11}(k+1) + r^2{}_{12}(k+1))] \tag{3}$$

$$\gamma(k+1) = \tan^{-1}[-r_{12}(k+1), -r_{11}(k+1)]$$

$$DOx(k+1) = DOx(k)$$

$$DOy(k+1) = DOy(k)$$

$$DOz(k+1) = DOz(k)$$

$$D\alpha(k+1) = D\alpha(k)$$

$$D\beta(k+1) = D\beta(k)$$

$$D\gamma(k+1) = D\gamma(k)$$

where the D's stand for "delta". These are the displacements from one time frame to the next. The index (k) indicates the time frame number. The orientation angles are given in terms of the components of the orientation matrix at the time frame (k+1). To find the orientation matrix at time frame (k+1), a rotation matrix R(k+1,k) is right multiplied by the orientation matrix at time frame (k).

The rotation matrix is formed using the orientation displacements, i.e., orientation angles deltas. The rotation matrix has the same form of the orientation matrix with the difference that the displacements angles are used.

In addition to the system description, the equations that relate the measures taken from the system and the state variables are needed. In this case the measuring instrument is the video camera. The image is formed by a transformation from three dimensions to two dimensions known as Central Projection. The coordinates of any point in the image plane (X,Y) are related to the object points (camera frame) coordinates (x,y,z) by the following equations:

$$X = -f^*(x/z) \qquad Y = -f^*(y/z) \tag{4}$$

where f is the focal length of the camera lenses.

In order to avoid complexity in the measuring equations, the equations of the central projection are rewritten so the measures always equal zero. The measuring equations are:

$$m_1 = z^*X + f^*x = 0$$
$$m_2 = z^*Y + f^*y = 0 \tag{5}$$

These equations are in terms of the fixed (camera) frame coordinates (x,y,z); it is necessary to change them in terms of the state variables. This is done using equation (2).

$$\begin{matrix} x \\ y \\ z \end{matrix} = \begin{matrix} r_{11} & r_{12} & r_{13} \\ r_{21} & r_{22} & r_{23} \\ r_{31} & r_{32} & r_{33} \end{matrix} \ * \ \begin{matrix} x' \\ y' \\ z' \end{matrix} \ + \ \begin{matrix} Ox \\ Oy \\ Oz \end{matrix} \tag{6}$$

Because the object is rigid, the coordinates of any point of this body with respect to the body frame are constants. By choosing the reference points on the body, these coordinates are fixed and the only variables in equation (6) are the state variables. The components of the rotation matrix are in terms of the orientation angles. By replacing the (x,y,z) in equations (5) for the ones in equation (6), the measurement equations are changed in terms of the state variables.

III. Kalman Filter

The Kalman Filter equations for this system are shown in this section. The discrete Kalman Filter was developed for linear systems that have the following model:

$$X(k+1) = H(k+1,k)^*X(k) + u$$
$$M(k+1) = G(k+1)^*X(k+1) + v \tag{7}$$

where X is the state variable vector, H is the transition matrix, M is the measure vector, and G is the matrix that relates measures to state variables; u and v are uncorrelated noise vectors.

In our case the vector X of state variables consists of the three coordinates for object position (Ox,Oy,Oz), three orientation angles (α,β,γ), three linear displacements (DOx, DOy, DOz), and three rotational displacements $(D\alpha,D\beta,D\gamma)$. This set of state variables change in time following equations (3). This set of equations corresponds to the transition function h[X(k)], which is nonlinear. This function is linearized using Taylor's expansion about the most recent estimate of the state vector. This approach is known as the Extended Kalman Filter. The same procedure is applied to the measurement function g[X(k+1)] which relates the measures of the system and the state variables.

IV. Experiment

An experiment was designed to test the filter. The experimental setup consisted of a rectilinear track, a frame where the object could be mounted, and a video camera. The frame has three degrees of freedom in order to orient and rotate the object at any angle. This frame is mounted in the rectilinear track and could be moved over it in controlled increments to simulate motion. The video camera was connected to a frame grabber board inside a computer. The frame grabber comes with software to acquire images and manipulate them.

The filter requires some parameters of the camera. These are focal length, distortion factors and scale factor. In this research we used the calibration procedure designed by Tsai [5]. This procedure uses at least seven non-coplanar reference points of the object. The results of the calibration experiment are shown in Table 1.

Table 1. Calibration results.

Focal length	1100.02
Scale Factor	.837877
Distortion Factor 1	1.37e-9
Distortion Factor 2	-1.24e-14

Several images sets of different motion types were taken to experiment with the filter. The first set consisted of translation only, a second set contains translation and rotation, a third and fourth set consisted of rotation only, near and far from the camera. A total of seven images were taken for each set. A Sobel filter was used to enhance the images in order to obtain the coordinates of the feature points. Ten feature points were used from each image. The results obtained for the second set are shown in Figures 1, 2, 3, and 4. These figures show the difference between the correct value of the state variable at each image frame and the value of the state variable calculated by the filter after processing the corresponding image. The difference is referred as error.

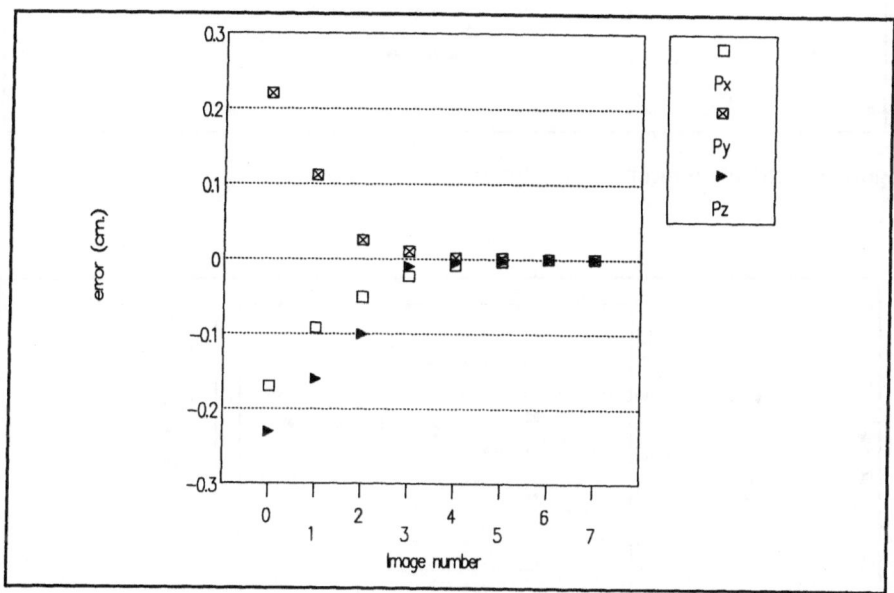

Figure 1. Estimation error for position.

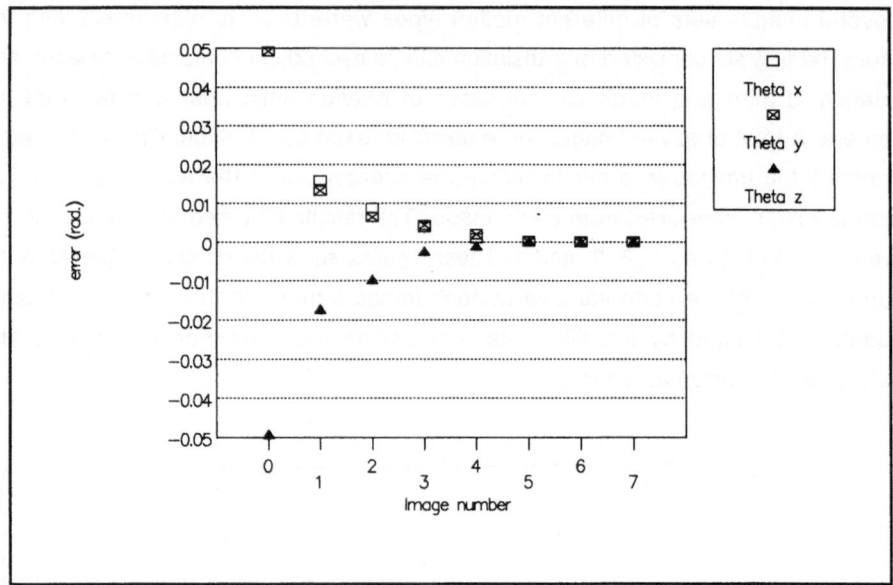

Figure 2. Estimation error for orientation.

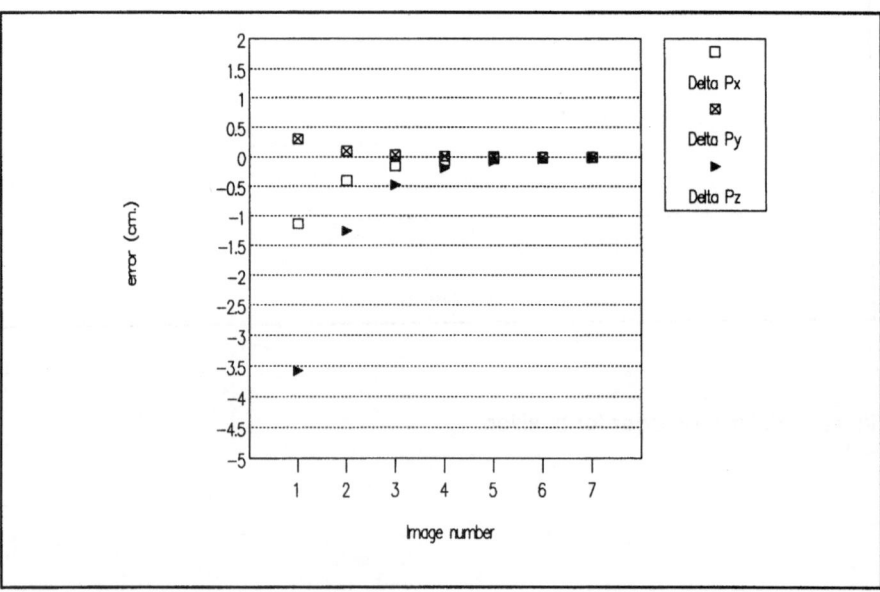

Figure 3. Estimation error for traslation.

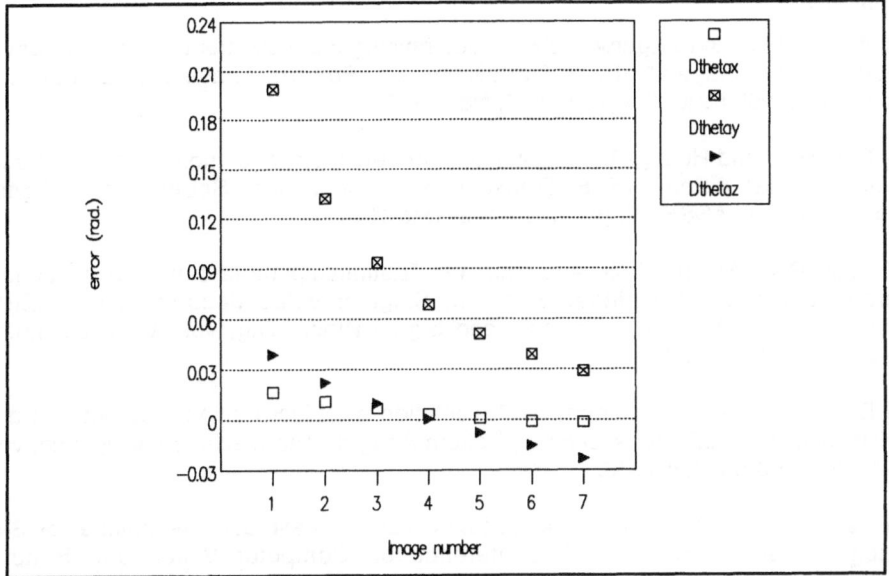

Figure 4. Estimation error for rotation.

V. Conclusion

From the results of the experiment we can say that the filter is capable of extracting motion parameters from a sequence of images. In order to get good estimates from the filter, it is necessary to adjust the values of the covariance matrix P, Q, and S. This adjustment is known as training. Several sets of values for these matrices were tried until we got one that shows good results.

The filter had no problem estimating the position and orientation of the object, but displacements were more difficult to estimate. Altougth it takes more images to find motion parameters using the Kalman Filter, it is better than the other methods cited because it isn't necessary to find the displacements of the feature points in the image before the actual estimation by the filter is performed. This facilitates the integration of the measuring system to the filter structure. Also the filter makes adjustments to its estimates every time a new image is available.

A method to ensure the convergence of the filter is the subject of future research, as well as, a complete integrated system that contains the image acquisition and image processing subsystems.

References

[1] Roach, J.W. and Aggarwal, J.K. , "Determining the movement of objects from a sequence of images," IEEE Transaction on Pattern Analysis and Machine Intelligence, vol. PAMI-2, pp.554-562, Nov 1980.

[2] Tsai, R.Y. and Huang, T.S., "Estimating three-dimensional motion parameters of a rigid planar patch," IEEE Transaction on Acoustics, Speech, and Signal Processing, vol. ASSP-29, pp.1147-1152, Dec 1981.

[3] Tsai, R.Y., Huang, T.S. and Zhu, W., "Estimating three-dimensional motion parameters of a rigid planar patch, II: Singular value decomposition," IEEE Transaction on Acoustics, Speech, and Signal Processing, vol. ASSP-30, no.4, pp.525-534, Aug 1982.

[4] Broida, T.J. and Chellappa, R., "Estimation of object motion parameters from noisy images," IEEE Transaction on Pattern Analysis and Machine Intelligence, vol. PAMI-8, pp.90-99, Jan 1986.

[5] Tsai, R.Y. , "An efficient and accurate camera calibration technique for 3-D machine vision," Proc. IEEE Conference on Computer Vision and Pattern Recognition, Miami, Florida, pp.364-374, 1986.

Optimization Techniques for Mathematical Routines Available through High-Level Source Code

S. ROY and A. CHAUDHURI

Department of Computer Science
M.S. University of Baroda
Gujarat, India

1   INTRODUCTION

1.1   THE NEED FOR PERSONALIZED ROUTINES

High-level language packages, today provide us with standard mathematical routines, whose competence has become an important criterion for determining the efficiency of complementary library software. When it comes to writing software for computer-graphics' or statistical applications, it is highly desirable that available math-functions should exhibit a certain level of efficiency in terms of both accuracy and execution time. Available math software packages indeed demonstrate such capabilities, but, under certain situations, we may expect our software to behave a bit differently to suit our applications; But, clearly, it is not directly possible to modify any object-routine. It becomes necessary, therefore, to develop personalized algorithms, which not only provide us with the desired services but are fast and accurate as well. Thus, it becomes imperative to employ certain optimization techniques, which would introduce in our software, a satisfying compromise between speed and precision.

1.2   THE BASIS FOR DEVELOPING MATH ROUTINES

The process of custom-tailoring software involves two stages. Initially, one must realize that the usage of any high-level language in place of direct assembly-code indeed slows down our programs. Therefore, we should stress on the specific nuances of languages, which tend to generate efficient code; for e.g., the usage of \=, *= etc. operators (in C), help the compiler to generate better results [a]. Moreover, the inherrent features of the digital computers by themselves may be exploited to suit our purposes; e.g. since integer-math routines execute faster, they may be used in place of slower floating point-routines wherever feasible. Furthermore, if a math-co-processor is available (alongwith the existing microprocessor), execution time is drastically reduced. But it needs to be stressed that the mere physical presence of an '87 chip is by no means sufficient : our software should be capable enough to take suitable advantage of the same.

The second stage in the development process involves the usage of mathematical formulae and algorithms, customized to suit digital computer applications.

Formulae borrowed from mathematical texts cannot be coded directly : they are modified (sometimes drastically), to enable their usage in programs. The difference between actual formulae and the resultant computer algorithms can be readily felt when we study the operational nature of the same. for e.g., much of the math-routines depend on formulae whose functional basis is characterized by the usage of approximations. Such routines often rely on the quality of initial approximations to produce optimum results. Moreover, the range of values that may be fed to an approxiamting algorithm is often limited. Hence, for developing math-routines (through high-level language tools), other than hardware and software issues, the theoretical aspects of algorithms should also be considered.

2 THE NATURE OF MATH-ROUTINES
2.1 STANDARD MATH-LIBRARY FUNCTIONS

Any standard math-library provides the following services which help us to perform mathematical calcualtions and conversions.

(a) Power and logarithmic functions are used for finding the roots and natural-logarithms of arguments.
(b) Conversion routines are used for transforming ASCII values into equivalent integer or real numbers and the reverse.
(c) Modification routines round off real numbes to the nearest largest or least integers, or return the remainder on division etc.
(d) Trigonometric functions determine the sine, cosine, tangent etc. values of the input arguments.
(e) Miscellaneous routines other than the above are available for services like random-number generation, or for controlling mathco-processor operations or to determine math-domain errors.

As a part of our study we have developed the software for most of these routines. Due to the limited space available, a detailed discussion of the source-code is not feasible. Therefore, the following subsections describe the general optimization techniques we have employed, to generate the math-routines.

2.2 OPTIMIZATION TECHNIQUES
2.2.1 CHOICE OF APPROPRIATE FORMULAE

As mentioned earlier, mathematical routines depend (much) on approximations to calculate the required values. The choice of the appropriate approximating algorithm plays an important role in the process of generating accurate values at sufficient speed.

For e.g., the square-root of a number may be worked out by using a progressive series of integers until the square of one number equals the argument whose root is to be taken. In terms of computer speed, this process is quite slow. But, the usage of Newton's approximation

root = 1/index * (radicannd/approx.+ (index-1)*approx.) i.e.
square-root = 1/2*(radicand/approx.+approx.) (1)[b]

certainly provides us with a faster operation.

2.2.2 CHOICE OF AN INITIAL APPROXIMATION

The efficiency of any formula using approximations depends highly on the quality of the initial approximation taken. The closer this approximation is to the actual value, the fewer times it is necessary to work through the formula.

For example, the square-root function works well if the initial approximation is chosen as half the radicand. But, since we know that computers work well with integer math, Newton's approximation can be used to return the integral part of the root to the actual floating point-routine (supplying the final answer). As seen from the speed-table (appended at the end), the final version of the square-root function Sqrt2(), works approximately four times faster than its predecessors. This is quite an achievement in terms of enhanced execution speed.

Furthermore, the correctness of approximations is essential, since, beyond a certain range of values, approximating-formulae behave erratically. As an example, we know that,

$$sum = arg1 + arg2 + arg3 + ... + argn \qquad (2)$$

shall always work correctly if all the arguments lie within the operating range of the computer. But, as mentioned earlier, approximating routines do not offer such a facility of responding to a wide range of input values.

2.2.3 MODIFICATION OF INITIAL (AND FINAL) VALUES

This technique is used in conjunction with the one employed for chosing the initial approximation. For example, to determine the natural logarithms of an argument, we have used a polynomial approximationn routine, which estimates logarithms to the base 10 and uses a quick final conversion rouine to arrive at the desired result. Initially, modification procedures have been used to convert the input arguments into values between 1 and 2, (since the function works best with such approximations).

This provided us with an extremely fast log function; But when we needed more accurate results, we used other modification routines (though we compromized on execution time). Similar techniques have been employed, (especially) for circular functions.

2.2.4 DIRECT PROGRAMMING

One may question the inclusion of routines like mult(x,y) for multiplying two numbers, in place of x*y, in certain compilers. Such cosmetic applications should be avoided. Good precision and execution speed depend much on the efficiency of the sub-steps used. The slightest error in any such steps may multiply into a larger inaccuracy and result into the loss of siginificant digits. Therefore, for both speed and accuracy, direct programming may be used, since this would be definitely faster than calling other functions.

As an example, it would be better to use

```
y    +=    0.001329882 * x * x * x * x * x * x;
```

rather than the more elegant ;

```
y    +=    0.001329882 * pow(x,6);
```

while coding the algorithm for generating cosine values. (Moreover, elegance is a quality identified only at the source level but it is definitely lost, when it comes to bits and bytes)!

2.2.5 DEVELOPING FUNCTIONS USING OTHER LIBRARY ROUTINES

As opposed to the previous techniques, certain functions can be derived from existing ones. Such a choice should be made only if it does not contribute significantly to the loss of execution speed. A trivial example is the coding of cotangent, cosecant etc. values through tangent, cosine etc. functions.

2.2.6 CUSTOM TAILORING

Certain applications do not depend heavily on the accuracy of computed values. Under such circumstances, the programmer should resort ot custom tailoring techniques which in turn depend on his wish for either speed or accuracy (or both). As an illustratuon, consider the graphics application of plotting arcs and curves on the screen. We have found that instead of using the commercially available circular functions which are fast and accurate upto (needless) 16 decimal places, the incorporation of faster and less accurate routines, reduces program-execution time (with the same graphical effect).
Normally, we use the following categories of functions to approach any application-problem : slow and very accurate, fast and enough accurate, very fast and just accurate ! We should use custom tailoring techniques only when the currently available software fails to satisfy any of the above needs.

2.2.7 THE LOOK-UP TABLE

When we didn't have calculators to aid us, we referred to standard sine, cosine etc. tables. A similar operation may be performed, using the computer. A look-up table can be generated using any standard math-routine. Thereafter, only a bit of code is required for returning the values in the table to the calling functions. Moreover, a general program-generator routine may be used to directly produce the required table and complementary code.
The concept behind such procedures is to develop look-up tables for functions which arrive at accurate values simply through an inspection procedure but at greatly enhanced speeds. Furthermore, if the values in the table are not accurate enough, then one may use a more precise routine to generate the table again or use a screen-editor to modify the required digits. It should be noted, however, that lookup tables operate within a limited range of values. Moreover, since standard functions are efficient enough, only special applications would require the simplistic sophistication of inspection-tables.

3 CONCLUSION

The speed-table on the next page, describes how some of the math routines, which we have developed, corelate with each other and with those available with a standard math-library. It can be seen that it is difficult to approach the efficiency exhibited by tight-professionally written software. Nevertheless, it has been amply demonstrated that while coding mathematical algorithms through high-level languages, certain techniques may be adopted to derive the optimum from our routines.

4 BIBLIOGRAPHY

(a) **Brian W. Kernighan**, Dennis M. Ritchie, The C Programming language, Prentice Hall, 1989, pp 40-52,250.
(b) **Robert J. Traister**, BASIC to C Conversion Manual pp. 29-99.
(c) **William S. Dorn**, Daniel D. McCracken, Numerical Methods with FORTRAN IV Case Studies, John Wiley & Sons, 1972., pp 313,416-420.
(d) **Erwin Kreyszig**, Advanced Engineering Mathematics, Wiley Easte university Edition, pp 651.
(e) **Ralston Anthony**, Herbert S. Wilf, Mathematical Methods for Digital Computers, Volume : 2, Wiley'67.
(f) **Turbo C (Ver. 2.0)**, User's guide Reference Manual, **Borland International**, 1800 Green Hills Road, P. O. Bos 660001, Scotts Valley, CA, 95066-001 USA. Turbo C is the registered TM of Borland.

SPEED - TABLE : MATH ROUTINES

Sr. No.	Function used.	Time rqd.(sec) for 1000 calls	% age efficiency	Remarks about the source-code
1.	Sqrt1	2.91208712	2.0	Sq.root:series
2	Sqrt2	0.65934065	8.33	" " :Newton
3.	Sqrt3	0.71428571	7.69	" " :int."
4.	sqrt	0.05494504	100	STANDARD (*)
5.	Log1	0.38461538	28.57	Nat.log:approx.
6.	Log2	0.54940549	20.0	" " "
7.	Log3	0.43956044	24.99	" " "
8.	log	0.10989011	100	STANDARD (*)
9.	Exp1	0.32967033	33.33	Exponent:poly.
10.	Exp2	0.16483516	66.65	" "
11.	exp	0.10989011	100	STANDARD (*)
12.	Power1	0.38461538	57.14	inaccurate,fast
13.	Power2	0.43956044	49.99	accurate,slow
14.	pow	0.21978022	100	STANDARD (*)
15.	Sine1	0.60439560	9.09	polynomials
16.	Sine2	0.32967033	24.99	faster
17.	Sine3	0.10989011	49.99	dedicated ver.
18.	sin	0.05494506	100	STANDARD (*)
19.	Cosine1	0.49450549	9.19	polynomials
20.	Cosine2	0.21978022	20.6	faster.
21.	Cosine3	0.10989011	41.36	dedicated ver.
22.	cos	0.05495406	100	STANDARD (*)
23.	Tangent	0.16483516	27.57	sine _ cosine
24.	tan	0.05494504	100	STANDARD (*)
25.	Cotangent	0.16483516	N.A.	cosine _ sine
26.	Secant	0.05494504	N.A.	1 _ cosine
27.	Cosecant	0.16483516	N.A.	1 _ sine
28.	Arccosine	0.16483516	35.33	using std. for.
29.	acos	0.10989011	100	STANDARD (*)
30.	Arcsine	0.16483516	33.33	using std. for.
31.	asin	0.10989011	100	STANDARD (*)
32.	Arctangent	0.10989011	100	polynomial
33.	atan	0.10989011	100	STANDARD (*)
34.	Arccotan.	0.10989011	N.A.	std. formula
35.	Arcsecant	0.16483516	N.A.	std. formula
36.	Arccosec.	0.16483516	N.A.	std. formula
37.	Sinh	0.274725275	42.37	std. formula
38.	sinh	0.16483516	100	STANDARD (*)
39.	Cosh	0.21978022	49.99	std. formula
40.	cosh	0.16483516	100	STANDARD (*)
41.	Tanh	0.54945054	30	std. formula
42.	tanh	0.16483516	100	STANDARD (*)
43.	Coth	0.54945054	N.A.	std. formula

* These routines have compiled using Turbo C compiler (ver.2.0), and have been run on an IBM AT'386 compatible equipped with a '387 math-coprocessor.

Chapter VII

Task Performance

Introduction

A mathematical model of an end effector grip on a cylindrical peg has been applied in a control algorithm in the first paper for fine tuning the position of the end effector. The paper states that successful implementation of this approach increases facilitation of high precision assembly tasks. In the second paper, the effect of errors in link length, joint misalignment, and joint encoder offset on the accuracy of a closed-loop robot is numerically computed. The effect due to link/joint parameter errors on the dynamic performance of a robot is presented in the third paper in terms of the deviations in robot manipulator torque for several link/joint parameter errors. An analytical assessment of the positioning error of a robot end effector caused by kinematic errors in the link/joint parameters has been performed in the fourth paper for several point-to-point end effector tasks. In the fifth paper a lump sum calibration approach is described that corrects for the tolerance buildup within a manipulator arm and the location of the tooling fixtures by utilizing a patented infra-red sensory device mounted on the end of the manipulator to locate tooling balls within the work area. In the sixth paper, an essential component for quality control is stated to be a low cost recalibration system that can be applied on-line during the production cycle. An expert system has been developed in the seventh paper to aid the designer in evaluating the most appropriate robot-hand kinematic structural concept from a large number of possible alternatives. The applicability of existing expert system techniques such as rule-based systems to the conceptual design automation of mechanisms is shown. In the eighth paper, the pose estimation problem, i.e., to determine the position and orientation of an object based on sensor data, is solved using simplified sensor and object models and an algorithm developed on the basis of a triangulation scheme. Such a grasping strategy will ensure grasping, increase the stability of the grasp, and provide a grasp that will either not require further manipulation of the object or will make manipulation of the object easier.

Sensing and Analysis of End-Effector Forces for Precision Assembly

ANTHONY DE SAM LAZARO, ECHEMPATI RAGHU, and BERAT GUROCAK

Department of Mechanical and Materials Engineering
Washington State University
Pullman, WA

Introduction:

Although programmable automation has been used both in manufacturing and in assembly work for nearly three decades, its effectiveness in precision assembly has only recently become evident due to improved accuracy and repeatability. The principal weakness of earlier robotic systems was the absence of sensors for fine-tuning and hence there was often an error in the position of the end-effector/gripper. Earlier efforts were directed at developing better hardware and control systems, which greatly increased the unit price of the robots and reduced, therefore, their cost-effectiveness. Development of machine vision and tactile sensors was the next step, but it was by no means the best because of the enormous cost. Also, programming was complex and time consuming.

An alternate approach to the use of vision systems is described in this paper. The mathematical model of force-closure grip on a cylindrical peg will be first discussed. This will be followed by an analysis of the forces/deflections during the assembly process. The results of computer simulations carried out will be reported in a separate paper.

Background:

As mentioned earlier, using force-feedback as a method of control in industrial robotics is still relatively uncommon, when compared to other means of sensing and control. In prosthetics, and in remote controlled manipulators used in space technology force sensing or hybrid controls using force and position control techniques [1-4] were developed. One of the applications which was studied in considerable detail was peg-in-hole assembly [5-6], which typifies about 35% of the assembly process in manufacturing [5]. Insertion of a peg in a hole using only positional control may not always be feasible since the clearances are generally closer than the accuracy and repeatability of the manipulator. Studies of the position-force control and the part geometry have been undertaken successfully [7]. However, the forces sensed at the points of gripping of the peg, in the event of an unsuccessful insertion, have not been directly related to the direction vector required to re-position the manipulator. Analysis of a model of the active and reactive forces on a peg during unsuccessful insertion could generate the required vector, as will be discussed in the next section. This vector may then be translated into positional control commands based on the manipulator geometry.

Modeling of a Force-Closure Grip on a Cylindrical Peg:

Assumptions: A uniform cylindrical peg has been selected. It is assumed that the peg itself is rigid and undergoes negligible elastic deformation during the insertion process. This assumption will be established to be valid during the finite element study to be reported separately. It is further stipulated that coefficient of friction (μ) between the gripper jaws and the peg is the same at each point of contact. The insertion is attempted when the axes of the peg and hole are parallel/co-axial. A double v-block jaw configuration has been selected after considerable deflection analysis of other forms.

Analysis of Forces: Three super-imposed systems of coordinates have been defined to study the force action between the gripper and the peg. These are (see Figure 1):

- World coordinates (X,Y,Z) { with unit vectors $\hat{i}, \hat{j}, \hat{k}$)} - to define the environment.
- Polar coordinates (x,y,z) { with unit vectors $\hat{p}, \hat{q}, \hat{r}$)} - to define the peg.
- Jaw coordinates (x_j, y_j, z_j) { with unit vectors $\hat{u}, \hat{v}, \hat{w}$)} - to define the jaw forces.

In the polar coordinate frame, the in-plane moment of the external force ($\vec{M} = \vec{a} \times \vec{F}$) is given by:

$$\vec{M} = -Fa\sin\theta\hat{q} + Fa\cos\theta\hat{r} \qquad (0 < a < a_p) \qquad (1)$$

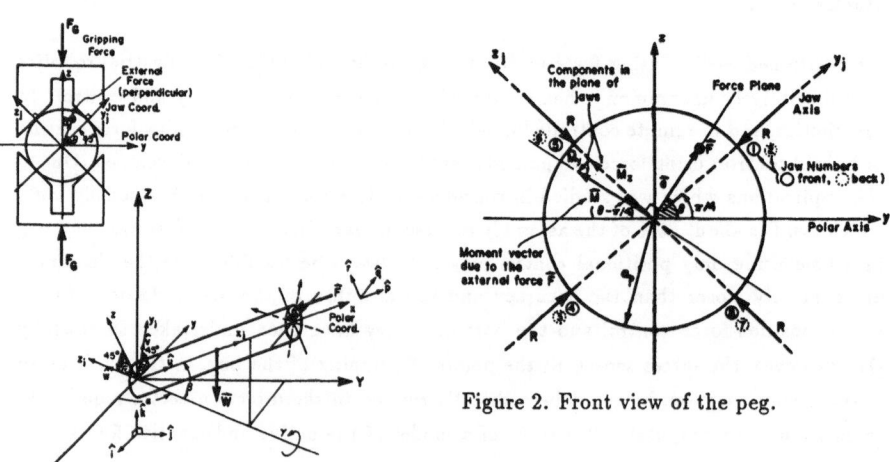

Figure 2. Front view of the peg.

Figure 1. Model of a force-closure grip on a cylindrical peg.

Since the jaws and peg are symmetrical, $\phi = \pi/4$, and equation (1) may be transformed as (see Figure 2):

$$\vec{M} = Fa\cos(\theta - \phi)\hat{w} - Fa\sin(\theta - \phi)\hat{v} \tag{2}$$

Using equation (2) the static balance equations in the x_jz_j and x_jy_j planes can be written. Depending on the attitude of the peg, its weight will effect the static balance and must be accounted for in the calculations. The weight vector $\vec{W}$ is always in -z direction in the world coordinate frame. Its effect expressed in the polar coordinate frame (peg) is:

$$\vec{w}_p = [R_{wp}]\vec{W}$$

where $\vec{w}_p$ is the weight vector expressed in the polar coordinate frame, and $[R_{wp}]$ is the orthogonal rotation matrix (from world coordinates to the polar).

Now $\vec{w}_p$ can also be transformed into jaw coordinates by $[R_{pj}]$, another orthogonal rotational transformation matrix, from polar to jaw coordinates.

$$\vec{w}_j = [R_{pj}][R_{wp}]\vec{W}$$

where $\vec{w}_j$ is the weight vector expressed in the jaw coordinate frame.

The point of application of the force $\vec{F}$ could be either in the upper $(0 < \theta < \pi)$ or in the lower half plane $(\pi < \theta < 2\pi)$. In the former case, (Figure 3) conditions of static equilibrium dictate that in the x_jy_j plane:

$$R_3\hat{v} - (\sum_{i=5}^{8} f_i)\hat{v} + \vec{w}_j\hat{v} - 2F_{GJ}\hat{v} + R_4\hat{v} = 0 \tag{3}$$

$$R_B\hat{u} + (f_2 + f_1)\hat{u} - (f_3 + f_4)\hat{u} + \vec{w}_j\hat{u} - \vec{F}\hat{u} = 0 \tag{4}$$

and

$$M_z\hat{w} + R_3\ell_1\hat{w} - (f_3a_p)\hat{w} + (\vec{r}_w \ \times \ \vec{w}_j) - (f_1a_p)\hat{w} - (f_6 + f_7)\ell_1\hat{w}$$
$$-(f_5 + f_8)\ell_2\hat{w} - F_{GJ}\ell_2\hat{w} - (f_4a_p)\hat{w} - (f_2a_p)\hat{w} + R_4\ell_2\hat{w} - F_{GJ}\ell_1\hat{w} = 0 \tag{5}$$

where: $\vec{r}_w = (\ell_p/2)\hat{u}$, F_{GJ} is the gripper force in the jaw coordinate frame, and f_i and R are the frictional and the reaction forces respectively, on the peg at the points of contact (numbered 1-8 as shown in Figure 3). Similarly, in the x_jz_j plane (see Figure 4):

$$R_7\hat{w} - (\sum_{i=1}^{4} f_i)\hat{w} + \vec{w}_j\hat{w} - 2F_{GJ}\hat{w} + R_8\hat{w} = 0 \tag{6}$$

$$R_B\hat{u} + (f_6 + f_5)\hat{u} - (f_7 + f_8)\hat{u} + \vec{w}_j\hat{u} - \vec{F}\hat{u} = 0 \tag{7}$$

and:

$$-M_y\hat{v} - R_7\ell_2\hat{v} + (f_7a_p)\hat{v} + (f_2 + f_3)\ell_1v + (f_1 + f_4)\ell_2\hat{v} + (\vec{r}_w \ \times \ \vec{w}_j)$$
$$+ F_{GJ}\ell_2\hat{v} + (f_5a_p)\hat{v} - R_8\ell_2\hat{v} + F_{GJ}\ell_1\hat{v} + (f_8a_p)\hat{v} + (f_6a_p)\hat{v} = 0 \tag{8}$$

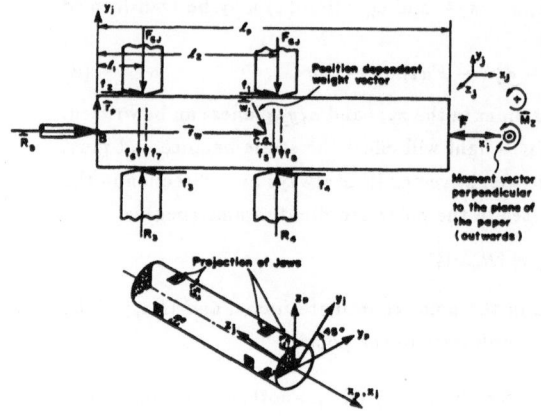

If the force $\vec{F}$ is in the lower half of the peg ($\pi < \theta < 2\pi$) the same analysis holds true, with the exception that the frictional forces are reversed. Knowing the gripping force, weight of the peg and the coefficient of friction μ the model can be solved for a configuration of eight contact points on the rotational surface and one on the end face. This model is used in the next section to design a gripper, where in measurable deflection is required in order to sense the position of the force $\vec{F}$.

Figure 3. Force balance on the peg in the $x_j y_j$ plane.

Deflection Analysis for Design of Gripper Jaws: The forces on each jaw are the reaction in the normal direction and friction in the tangential direction to the surface, as shown in Figure 5. The stress distribution due to the reactive force R in yz plane is the sum of the stress due to bending moment of R_y and the compressive stress due to R_z. The compressive stress distribution over the cross sectional area is given by:

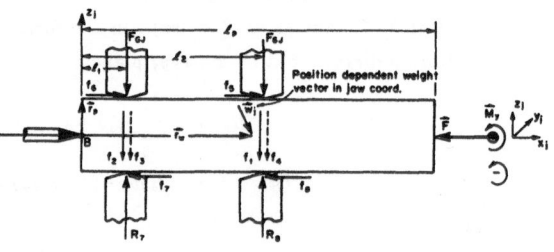

Figure 4. Force balance on the peg in the $x_j z_j$ plane.

Figure 5. Deflection analysis of jaw.

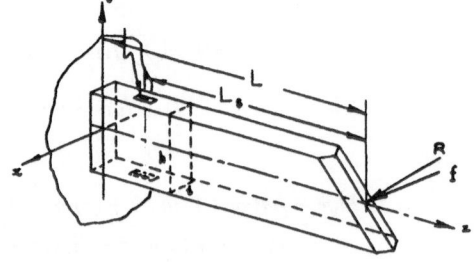

$$\sigma_c = R_z/(h \cdot t) \tag{9}$$

The bending moment may be obtained from the equations:

$$\sigma_b = Mc/I = [(R_y L)h/2]/[th^3/12]$$
$$= \pm(6R_y L)/(th^2) \quad \text{'+' for the upper half; '-' for the lower half} \tag{10}$$

Since the reactive force is neither axial nor longitudinal, there is a resultant shift in the neutral axis, and it does not coincide with the geometric axis of the jaw. The total stress at the location L_s, where strain-gages will be fixed is:

$$\sigma_{Tz} = [(6R_y L_s/th^2) \pm (R_z/th)] \quad \text{'-': for the upper half; '+': for the lower half} \tag{11}$$

The strain in the x-direction:

$$e_z = \sigma_{Tz}/E_{Al} \tag{12}$$

where E_{Al} is the modulus of elasticity of Aluminum. The total deflection in the yz plane may be determined similarly.

Computer Simulations and Results:

The static balance equations (3) through (8) could be expressed in terms of the gripper force F_{GJ}, the coefficient of static friction μ_s, and the reactive forces at the points of contact, $R_i(i = 1-8)$. Four contact points (shown diagrammatically as F_{GJ}) are assumed to apply a gripping force, while the reactions are at the other four points, namely R_3, R_4, R_7, and R_8. The five unknown quantities here are the reactions at the contact points 3, 4, 7 and 8 and at the base. Four equations are used to solve these four reactions, and the reaction at the base is determined by rewriting the equations and solving for it:

$$R_B = F + \mu_s[R_3 + R_4 - 2F_{GJ}] - w_x \tag{13}$$

In order to determine the magnitude of the reactive forces at the points 3, 4, 7, and 8 at various angles θ, and also the resultant strains due to both the reactions and friction at all points, a computer simulation program was written in FORTRAN. The input variables include dimensions, mass and orientation of the peg and its position relative to the jaws, the position and magnitude of the force $\vec{F}$ and F_{GJ} and the dimensions and material of the jaws. The results generated with an external force of 500 N, on an aluminum jaw of 20 x 10 mm cross section, will be reported separately.

Conclusion:

In this paper a mathematical model of the force-closure grip on a cylindrical peg has been presented. The results of this study are applied in a control algorithm for the fine tuning of the position of the end effector. The direction of the resultant force vector is being utilized to determine the direction that the manipulator has to be moved and the amount is ascertained with fuzzy control. Successful implementation of this approach would facilitate high precision assembly tasks without the accompanying escalation in equipment costs.

Aknowledgements:

The authors wish to acknowledge the facilities provided by the Department of Mechanical and Materials Engineering at Washington State University. They also thank Ms. Jo Ann Hicks for the preparation of the final manuscript of this paper.

References:

1. Whitney, D.E., "Historical Perspective and State of the Art in Robot Force Control," *International Journal of Robotics Research*, Vol. 6, No. 1, Spring 1987, pp. 3-13.
2. Rothchild, R.A. and Mann, R.W., "An EMG-Controlled Force Sensing Proportional Rate Elbow Prosthesis," *Proceedings of the Symposium on Biomedical Engineering*, Milwaukee, WI, 1966.
3. West, H., and Asada, H., "A Method for the Design of Hybrid Position/Force Controllers for Manipulators Constrained by Contact with the Environment," *Proceedings of the IEEE Conference on Robotics and Automation*, 1985, pp. 251-259.
4. Raibert, M.H. and Craig, J.J., "Hybrid Position/Force Control of Manipulators," *Transactions of the American Society of Mechanical Engineers Journal of Dynamic Systems, Measurement and Control*, Vol. 102, 1981, pp. 126-133.
5. Potkonjak, V. and Vukobratovic, M., "Dynamics of Manipulation Mechanisms with Constrained Gripper Motion Part I and II," *Journal of Robotic Systems*, Vol. 3, No. 3, 1986, pp. 321-334.
6. Inoue, H., "Force Feedback in Precise Assembly Tasks in Artificial Intelligence: An MIT Perspective," Vol. 2., Winston and Brown, eds. MIT Press 1979, pp. 219-241.
7. Strip, D.R., "Insertions Using Geometric Analysis and Hybrid Force-Position Control on a Puma 560 with VAL-II," *Proceedings of AAAI-87*, July 1987, pp. 695-698.

Accuracy Test From Kinematic Parameter Errors in a Closed-Loop Robot

CHENG Y. LIN and ALOK K. VERMA
Department of Mechanical Engineeirng Technology
Old Dominion University
Norfolk, VA

LOUIS J. EVERETT
Department of Mechanical Engineering
Texas A&M University
College Station, TX

[Abstract]

As it is impossible to construct a manipulator without any parameter errors during the manufacturing processes, the major objective of this paper is to see the effects of these errors on the accuracy of a closed-loop robot. A simulated robot with a 3D constraint is used in this numerical test. Three types of parameter errors are studied: (1) link length, (2) joint misalignment, and (3) joint encoder offset. In this test, a small variation is added to one of the link parameters, the accuracy is thus obtained by using forward kinematic transformations. A numerical method is proposed to calculate the dependent variables for each set of joint encoders.
The results are also compared to an open-loop robot having the same parameter errors.

[Introduction]

Although an industrial robot can servo each joint to a desired joint position, it has no direct knowledge of the tool location. Typically, a kinematic model which maps the tool location to joint positions is used. The kinematic model of a robot is generally formulated through homogeneous transformation. One of the well-known methods is the Hartenberg-Denavit convention [1], in which one variable and three parameters for each joint are used. The parameters including the link lengths, the twist angles between joint axes, and joint offsets are called kinematic parameters. From Mooring's tests [2], even slight misalignments of the major joints can result in significant positioning error on the end effector. Therefore, although an industrial robot has good repeatability, it is unable to predict its tool pose accurately. A number of kinematic calibration algorithms have been proposed and a few have been tested and verified in the laboratory [3-6].

In recent years, a closed-loop robot using a direct-drive device has been developed. It has some advantages when compared to an open-loo [7]. In this research, the accuracy tests resulting from kinematic parameter errors in a closed-loop constraint will be studied. A simulated robot shown in Figure 1 with a 3D constraint is used in this numerical test. The results are also compared to an open-loop robot having the same parameter errors. Three types of parameter errors are studied: (1) link length, (2) joint misalignment, and (3) joint encoder offset.

[Kinematic Transformation for a Closed-Loop Robot and Results]

As shown in Figure 1, the driving motors are located at joints 1 and 5, and the dependent variables are located at joints 2, 3, and 4. If T_o and T_c represent open-loop and closed-loop transformations, then

$$T_o = A_{w1}A_{12}A_{2t} \tag{1}$$

$$T_c = A_{12}A_{23}A_{34}A_{45}A_{51} = I \tag{2}$$

Here A_{w1} is the homogeneous transformation from the world frame to joint 1, etc. The matrix I is the identity matrix. If the measured pose is expressed as T_m the pose error dT relative to the world frame can have the form:

$$dT = T_m - T_o \tag{3}$$

Equation 3 can be further expressed relative to the tool frame, it has the form [8]:

$$\delta T = T_o^{-1}(T_m - T_o) = \begin{bmatrix} 0 & -\delta z & \delta y & dx \\ \delta z & 0 & -\delta x & dy \\ -\delta y & \delta x & 0 & dz \\ 0 & 0 & 0 & 0 \end{bmatrix} \tag{4}$$

Here $(\delta x, \delta y, \delta z)$ and (dx, dy, dz) represent the differential orientation and position errors respectively. To find the dependent variables in a closed-loop constraint, one can get the algorithm by setting: (1) $T_o = T_c$ and (2) $T_m = I$. Using the least square method, it can have the form:

$$-S^t[T_c^{-1}(1-T_c)]S[\frac{\partial T_c^{-1}}{\partial V_j}] =$$

$$\sum_{k=1}^{nd} S^t[\frac{\partial T_c^{-1}}{\partial V_k}]S[\frac{\partial T_c^{-1}}{\partial V_j}]dV_k \qquad (5)$$

Here S[] is a selecting operator used to pick up these three orientation and three position errors; S'[] is the transpose of s[]; V_j is the dependent variable on joint j, and k is from 1 to the number of dependent variables nd in a closed-loop constraint. One can have a set of dependent variables from equation 4 by making an initial guess for each V_k in equation 4.

The position errors can be obtained through the following steps:

(1) Create a robot.

(2) Select a set of encoder readings.

(3) Calculate the dependent variables using equation 5.

(4) Calculate T matrix and set it to be T_m shown in equation 1.

(5) Make a small change on the closed-loop parameters and calculate dependent variables.

(6) Calculate T matrix and set it to be T_o.

(7) Calculate the position error between Tm and T_o.

(8) Repeat from step (2) through step (7) for another set of encoder readings.

The link lengths in a closed-loop constraint are properly selected to have a wide range of encoder values. According to Bajpai and Paul [9], the link lengths are set to be 1.5, 5, 4.5, and 3 respectively. The results for this test are shown in Figures 2 through 6. In those figures, error is defined as following:

$$Error_i = \sqrt{(\bar{x}-x)_i^2 + (\bar{y}-y)_i^2 + (\bar{z}-z)_i^2} \qquad (6)$$

where $(\bar{x},\bar{y},\bar{z})_i$ and $(x,y,z)_i$ are the end-effector positions calculated from the corrected and uncorrected parameters respectively. The performance index of a robot accuracy RA is defined as following:

$$RA = \sqrt{\frac{\sum_{i=1}^{m} Error_i^2}{m}}$$ (7)

where m is the number of positions used in the test. The simulation results are shown as following:

[Link Length]

Figure 2 shows the result of the error surface when the length L_1 is changed from 1.5 to 1.505. The other parameters remain unchanged. The RA and maximum position error in this case are $2.403*10^{-2}$ and 0.168 respectively. One can see that the error surface is very smooth except in a small area where the error suddenly jumps to a large value. As shown in Figure 3, the error surface for an open-loop robot is flat. Both the RA and maximum error are 0.005 which is equal to the length increment.

[Encoder Offset]

Figure 4 shows the result of the error surface when the encoder values are added by a small offset of 0.005°. The RA in this case is only $1.656*10^{-3}$ and the maximum error is $9.459*10^{-3}$. For an open-loop robot, the error surface is similar to Figure 3. The RA is $1.757*10^{-3}$ and the maximum error is $1.875*10^{-3}$.

[Joint Misalignment]

Figure 5 shows the result of the error surface when a joint misalignment of 1° is made on joint 1 of the closed-loop robot. The RA in this case is $1.88*10^{-1}$ and the maximum error is 0.1745. Figure 6 shows the result of the open-loop robot. The RA in this case is $1.234*10^{-3}$ and the maximum error is 0.1745.

[Conclusion]

From the simulation tests shown above, the error caused by the link lengths in a closed-loop constraint may affect a robot's accuracy more significantly than that of an open-loop robot. This is because at some particular configuration, the dependent variables may have a significant change due to a small change on a link length. A small encoder offset may not affect a robot accuracy significantly. However, the maximum error for the closed-loop robot is larger than the open-loop robot. Similar to that of an open-loop robot, the error caused by the joint misalignment can significantly affect the accuracy performance for a closed-loop robot. Although one can eliminate this error by constructing a two dimensional constraint, it may have internal stress on each joint if there is a slight joint misalignment in the closed-loop constraint. For the identification algorithm and the number of independent variables to be identified in a closed-loop constraint, one can refer to the papers published by Everett and Lin [10,11].

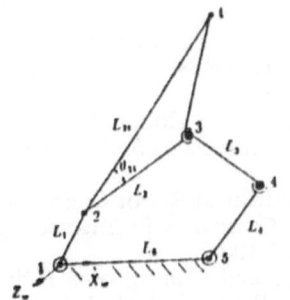

Fig. 1 A Closed-Loop Robot with
an RRSSR Constraint

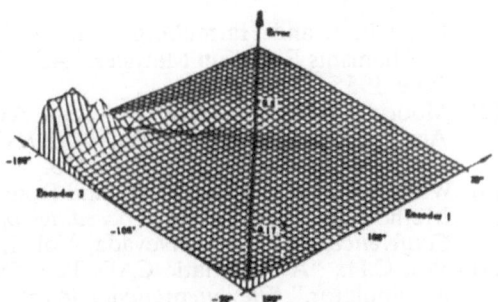

Fig. 2 Error Plot with an error of 0.005
on L_1 of the Closed-Loop Robot.

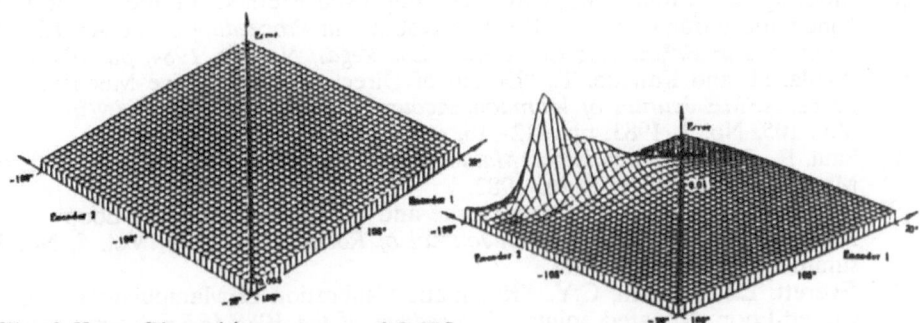

Fig. 3 Error Plot with an error of 0.005
on L_1 of the Open-Loop Robot

Fig. 4 Error Plot with an Error of
0.005° on Encoders of the
Closed-Loop Robot

440

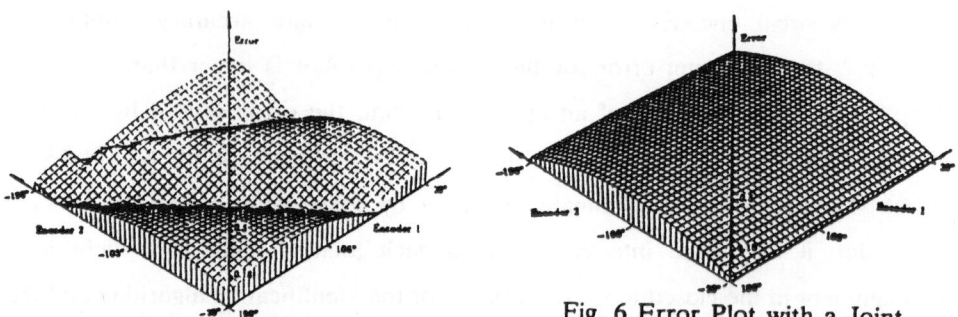

Fig. 5 Error Plot with a Joint Misalignment of 1° on Joint 1 of the Closed-Loop Robot.

Fig. 6 Error Plot with a Joint Misalignment of 1° on Joint 1 of the Open-Loop Robot

[References]
[1] Denavit, J. and Hartenberg, R.S.,"A Kinematic Notation for Lower-Pair Mechanisms Based on Matrices," *ASME Journal of Applied Mechanics*, Vol. 77, June 1955, PP. 215-221.
[2] Mooring, B.W.,"The Effect of Joint Axis Misalignment on Robot Positioning Accuracy," in *Proceedings of the ASME Conference on Computers in Engineering*, Chicago, Illinois, 1983, pp. 151-155.
[3] Whitney, D.E., Lozinski, C.A., and Rourke, J.M., "Industrial Robot Calibration Method and Results," in *Proceedings of the ASME Computers in Engineering Conference*, Las Vegas, Nevada, Vol. 1, 1984, pp. 92-100.
[4] Wu, C.H., "A Kinematic CAD Tool for the Design and Control of a Robot Manipulator," *The International Journal of Robotics Research*, Vol. 3, Spring 1984, pp. 58-67.
[5] Hsu, T., W., and Everett, L., J., "Identification of the Kinematic Parameters of a Robot Manipulator for Positional Accuracy Improvement", *Proceeding of the 1985 Computers in Engineering Conference and Exhibition*, Boston, Mass., August 1985, pp. 263-267.
[6] Mooring, B.W. and Tang, G.R., "An Improved Method for Identifying the Kinematic Parameters in a Six-Axis Robot," in *Proceeding of the ASME Computers in Engineering Conference, Las Vegas, Nevada, 1984, pp. 79-84.*
[7] Asada, H. and Kanade, T., "Design of Direct of Direct-Drive Mechanical Arms,"*ASME Journal of Vibration, Acoustics, Stress, and Reliability in Design*, Vol. 105, No. 3, 1983, pp. 312-316.
[8] Paul, R., *Robot Manipulators: Mathematics, Programming and Control.* Boston, Massachusetts: M.I.T. Press, 1982.
[9] Bajpai, A. and Roth, B., "Workspace and Mobility of a Closed-Loop Manipulator." *The International Journal of Robotics Research*, Vol. 5, No. 2, summer, 1986, pp. 131-142.
[10] Everett, L.J. and Lin, C.Y., "Kinematic Calibration of Manipulators with Closed-Loop Actuated Joints," *Proceedings of the 1988 IEEE International Conference on Robotics and Automation*, Vol. 2, April 1988, pp. 792-797.
[11] Lin, C.Y. and Everett, L.J., "A Complete Kinematic Model for a Closed-Loop Constraint in Robot Calibration," *The Winter Annual Meeting of the American Society of Mechanical Engineers*, DSC-Vol. 14, December, 1989, pp. 57-62.

The Effect of Robot Kinematic Parameter Errors on Joint Torques

JING TIAN and JOHN E. SNECKENBERGER
Mechanical and Aerospace Engineering
West Virginia University
Morgantown, WV

Abstract

Research on improving robot positioning accuracy due to link/joint parameter errors has concentrated mostly on the effects of these errors on the kinematic performance of the robot. This paper extends this error-related analysis to the dynamic performance of the robot, by presenting the deviations in robot manipulator torque for several cases in which link/joint kinematic parameter errors exist. The deviation in manipulator joint torque due to robot link/joint kinematic parameter errors was studied for a two-link revolute robot manipulator. Algorithms have been developed that successfully compute such joint torque deviations. A simple robot task cases was studied. The results shows the contribution of joint torque deviation due to the small kinematic parameter errors.

I. Introduction

One of the main research efforts devoted to robotics is to investigate the effect of the robot link/joint kinematic errors on the robot end-effector positioning accuracy [1]. In the recent years, robot manipulators are being required to perform more complicated tasks instead of simple pick and place motions. The rapid development of CAD/CAM systems, off-line robot programming, and use of sensor guided robots have also required that robot end-effectors move more accurately in Cartesian task space. This robot position accuracy study is concentrated on robot kinematic model correction and improvement, mainly because the majority of robot manipulators used in industry are controlled based on robot kinematics.

Robotic research has indicated that a robotic system is basically a dynamic system, requiring fast motions and mechanical configurations with strongly coupled subsystems. The control task is thus essentially dynamic. It is necessary to compute joint torques, based on the robot dynamics model, to be applied at the joints to achieve the desired motion [2]. The primary focus of robot dynamics has been on the development of the computationally efficient robot system dynamic equations. There has also some research efforts devoted to robot dynamic model enhancement and compensation when dynamic parameter (link mass, center of mass, etc) errors exist [3]. In the case where manipulator link/joint kinematic errors exist due to manufacturing process, it would be of general robot design interest to appreciate the effect of such errors on robot joint torque dynamics. This paper will present a preliminary investigation focused on this issue. A sample case of a two link planer robot manipulator with consecutive near parallel joints is studied. A kinematic analysis that considers link/joint parameter errors is formulated, and the resultant joint torque deviations due to kinematic errors presented.

II. Formulation of Kinematic Equations

Manipulators consist of nearly rigid links connected by joints which allow relative motion of neighboring links. For the so called "open-loop" manipulator, links are serially connected. By using the homogeneous transformation matrix A_i to represent relationships between various coordinate frames that are assigned to the mechanical linkages to represent a manipulator, the mapping of the manipulator joint space and end-effector Cartesian space can be kinematically represented as:

$$T_N = A_1 A_2 \ldots A_N \qquad (1)$$

The homogeneous transformation matrix A_i uses four parameters to describe the kinematic characteristics of a link/joint as defined by Hartenberg and Denevit (H-D) [4] and later expanded upon by Paul [5] and Craig [6]. Mathematically, A_i is a function of a_i, d_i, α_i and θ_i, defined as link length, link offset, link twist and joint angle, separately. The first three are fixed parameters after the manipulator is designed and manufactured. The last one is a variable to control joint rotations. The work presented here uses the formulation given by Craig. This type of link coordinate system is easily defined and conveniently located if the joint axes are perfectly parallel or orthogonal.

When kinematic errors exist, A_i will be a function of original four parameters (nominal values) plus another four parameters which describe the link/joint errors, namely Δa_i, Δd_i, $\Delta \alpha_i$ and $\Delta \theta_i$. This small error setup is not true in case where the manipulator uses two consecutive near parallel axes configurations [7]. In such a case, it is possible that the small joint error occurs which happen to make the consecutive axes intersect. There will then be no "common normal" between the axes, thus link length a_i will vanish at the intersect point and link offset d_i becomes very large. If such an error is very small, the coordinate system location approaches infinity, such that H-D procedure results in very poorly conditioned transformation matrices. One way to resolve the problem is by post multiplying the A_1 matrices by an additional rotation $Rot(k, \beta_1)$ [8], where the axis of rotation k depends on the real manipulator configuration (normally k is about the y axis). Therefore the small variations of the joint/link parameters of the two consecutive near parallel links can be modeled by small variations in the five link/joint parameters a_1, d_1, α_1, θ_1 and β_1. This model is only necessary in the case of two consecutive near parallel joints. The result transformation matrix in such case is: (where β_1 is the rotation around the y axis)

$$A_1 = \begin{bmatrix} C\theta_1 C\theta_1 & -S\theta_1 & 0 & a_{1-1} \\ S\theta_1 C\alpha_{1-1}C\beta_1 + S\alpha_{1-1}S\beta_1 & C\theta_1 C\alpha_{1-1} & S\theta_1 C\alpha_{1-1}S\beta_1 - S\alpha_{1-1}C\beta_1 & -S\alpha_{1-1}d_1 \\ S\theta_1 S\alpha_{1-1}C\beta_1 - C\alpha_{1-1}S\beta_1 & C\theta_1 S\alpha_{1-1} & S\theta_1 S\alpha_{1-1}S\beta_1 + C\alpha_{-1}CS\beta_1 & C\alpha_{1-1}d_1 \\ 0 & 0 & 0 & 1 \end{bmatrix} \quad (2)$$

A two link manipulator with consecutive near parallel revolute joints is used as an example to demonstrate the current research. The manipulator configuration, link frame assignments and associated kinematic properties are shown in Figure 1. This type of link arrangement is a common part of industry robot configuration. A large variety of robot manipulators possess this type of link/joint arrangement, such as links 2 and 3 of the PUMA robot, links 2 and 3 of the Adept assembly robot, etc. The five parameter transformation matrix as shown in equation (2) is applied to study the small kinematic parameter variations due to manufacturing. Small kinematic joint/link kinematic errors are assumed to exist in the manipulator system and the effects of those errors on the manipulator joint torques are studied.

III. Joint Torque Deviation

A robot manipulator essentially is a dynamic system. For high speed and high precision motions, manipulator system dynamics plays a very important role. The dynamic equation derived for the manipulator system are commonly based on the Lagrangian formulation and the Newton-Euler formulation. It has been proven that both formulation are equivalent. The Newton-Euler iterative equations are used in this research to calculate the required joint torques, which use outward iterations to compute link velocities and accelerations and inward iterations to compute joint forces and torques. Given the specified robot end effector trajectory, attention is focused on calculating the required vector of joint torques when manipulator link/joint kinematic errors exist.

This research follow the common practice of calculating manipulator required joint torques. First the desired manipulator end-effector path is specified, then based on this Cartesian path the required joint angles are calculated by using robot nominal inverse kinematics (use nominal kinematic values). The trajectory planner is used based on linear function with parabolic blends, then a trajectory generator is applied to generate a smooth path. Finally, the iterative Newton-Euler dynamic equations are employed to compute the require torques at each joints. Because the nominal kinematic parameter values used in the calculation, these set of required manipulator torques are considered as nominal torques.

In order to determine the effects of link/joint parameter errors on the joint torques, the actual manipulator dynamics is also computed with those kinematic errors using the transformation matrices shown in equation (2). By using the same procedure to generate robot joint path, the actual robot kinematic parameter values are substitute in the Newton-Euler dynamic equations to compute manipulator actual dynamics. This is a straightforward calculation.

Dedicated algorithms have been implemented for the above purposes. These include a general utility function program to plan and generate a joint space trajectory based on linear functions with parabolic blends. Manipulator nominal kinematic transformation matrix using link/joint parameters, actual (error included) kinematic transformation matrices to predict end effector positioning errors (in direct fashion), nominal inverse kinematics, and manipulator dynamics based on Newton-Euler iterative equations have also been programmed. Computation of the manipulator joint torque deviation is then obtained in direct fashion; ie, the differences between the computed robot nominal joint torques and the computed robot actual joint torques that include joint/link kinematic parameter errors is the joint torque deviation for the manipulator.

A robot end-effector moving in a straight line was used as an example for this study. Joint one and two were required to move simultaneously at moderate speeds of 50 cm/sec. It was assumed that there were 2mm errors in link length and offset (a_1 and d_1), one degree errors in joint twist and

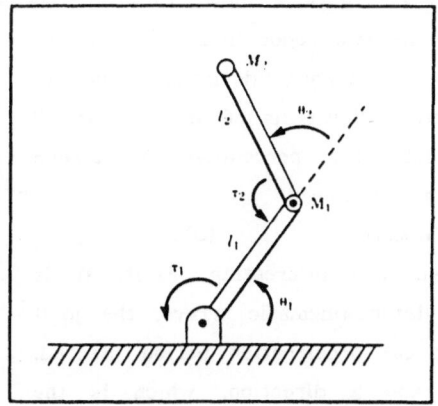

Fig. 1 Two Link Robot Configuration

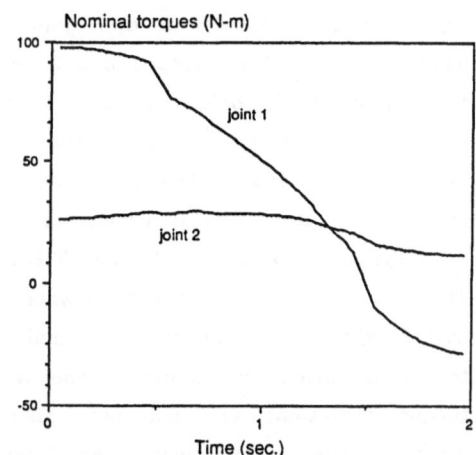

Fig. 2 Nominal Torques Applied to Joints 1 and 2.

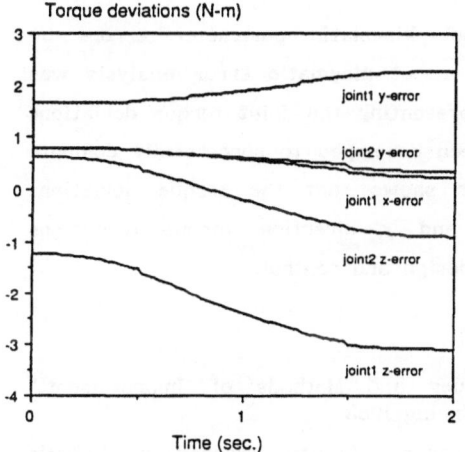

Fig. 3 Torque Deviations at Joints 1 and 2

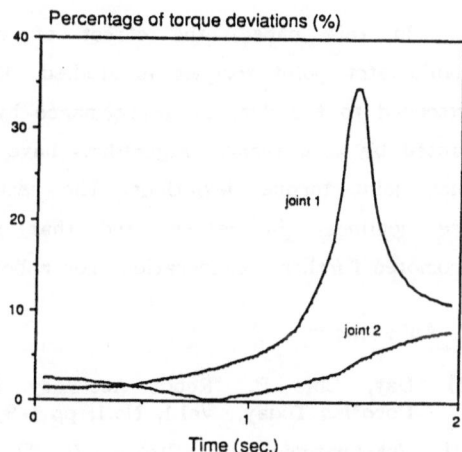

Fig. 4 Percentage of Torque Deviations

angle variable $(\alpha_1, \beta_1, \theta_1)$ at both joints 1 and 2. Figure 2 shows the required nominal torques for joints 1 and 2 to move the two robot links according to the given trajectories. Figure 3 shows the torque deviation (due to link/joint parameter errors) in the x, y, and z directions of joints 1 and 2 in their coordinate frames. Figure 4 plots the percentage of torque deviations in the z direction based on following equation:

$$T_{percent\ deviations} = \left| (T_{error} - T_{nominal}) / T_{nominal} \right| \qquad (3)$$

The results of this straightforward analysis are interesting. First, it is noticed that under the assumed small manipulator kinematic errors, the joint torque deviation are nonlinear and large at the deceleration period. Second, torque deviations are not only shown in the z direction, which is the direction the torque drive the links, but also shown in the x and y directions. Those extra torque in x and y direction certainly propose some extra wear on robot joint couplings, and require extra consideration give robot manipulator design and manufacturing.

IV. Conclusion

In this paper, the effect of robot kinematic parameter errors on manipulator joint torques is studied. The robot kinematic error analysis was extended to the dynamic performance by presenting the joint torque deviations caused by such errors. Algorithms have been developed to successfully compute such joint torque deviations. The results shows that the torque deviations are nonlinear in nature, and that x and y direction torque deviations promoted further considerations for robot design and control.

V. References

[1] Day, Chia P. "Robot Accuracy Issues and Methods of Improvement," Robotics Today,, Vol.1, No.1, pp 1-9, Spring 1988

[2] Vukobratovic, M., Stokic, D. "Is Dynamic Control Needed in Robotic Systems, and, if So, to What Extent?" The International Journal of Robotics Research, Vol.2, No.2, pp 18-34, Summer 1983

[3] Leahy, M.B.Jr, and Saridis, T.N. "Compensation of Industrial Manipulator Dynamics," The International Journal of Robotics Research, Vol.8, No.4, pp 73-84, August 1989

[4] Denavit, J. and Hartenberg, R.S., " A Kinematic Notation for Lower-Pair Mechanisms Based on Matrices," ASME Journal of Applied Mechanics,pp 215-221, June 1959

[5] Paul, R.P., Robot Manipulators: Mathematics, Programming, and Control,

MIT Press, 1981

[6]   Craig, J.J., <u>Introduction to Robotics: Mechanics and Control,</u> Addison-
      Weslsy Publishing Company, Inc., 1986

[7]   Mooring, B.M., "The Effect of Joint Axis Misalignment on Robot
      Positioning Accuracy," ASME Proc. of International Computers in
      Engineering Conference, Vol.2, pp 151-155, 1983

[8]   Veitschegger, W.K. and Wu, Chi-haur, "Robot Accuracy Analysis Based on
      Kinematics," IEEE Journal of Robotics and Automation, Vol.RA-2, No.3,
      pp 171-179, September 1986,

Kinematic Error Budgeting to Obtain the Best Feasible Task Performance for a Specified SCARA Manipulator

TONY M. LAMB and JOHN E. SNECKENBERGER
Mechanical and Aerospace Engineering
West Virginia University
Morgantown, WV

ABSTRACT
The positioning accuracy of the end effector of a robot manipulator with respect to its workplace has become significantly more important as the number of industrial and commercial processes that require robotic assembly have increased. Various technical approaches have been proposed and/or investigated for increasing the accuracy of the robot's end effector. It is well appreciated that geometric errors in the kinematic links and joints of the robot manipulator are obvious contributors to the positioning error of the end effector. This paper develops the analytical assessment of the positioning error of the robot end effector caused by kinematic errors in the link/joint parameters that describe the manipulator geometry. This assessment is performed for several point-to-point end effector tasks within the robot's preferred workspace. A kinematic error budgeting program is used to determine the best feasible end effector performance for a given SCARA manipulator based on an error evaluation of the desired versus actual paths for specific link/joint errors.

Explanation of ABL Robot
A Selective Compliant Assembly Robot Arm (SCARA) robot manipulator is used at Allegany Ballistics Laboratory (ABL). The robotic arm is an IBM 7540 Model A02 Robotic System [1]. The IBM 7540 System is a continuous path robot and consists of a manipulator, controller, control panel, and a computer system.

The IBM SCARA robot has a four degree of freedom manipulator with an R-R-P-R configuration where R stands for a rotational joint, and P stands for a prismatic joint. The two revolute arm joints consist of the θ_1 axis and the θ_2 axis. They give the manipulator two degrees of freedom through a double swivel motion. The vertical prismatic joint along the Z-axis provides the third degree of freedom for the arm. The fourth degree of freedom is the wrist rotation about the roll axis. The fourth axis comprises the single axis for the wrist motion while the first three axes comprise the arm motions.

Actual Specifications For Construction
The IBM 7540 Model A02 robot has a tight manufacturing tolerance range. The various specifications and dimensions [2] necessary are listed in Table 1.

Table 1

A_1 = 630 ± 0.1 mm,	A_2 = 400 ± 0.1 mm,	D_2 = 100 ± 0.1 mm
D_{3max} = 250 ± 0.1 mm		θ_1 Range: 0° to 200° ± 1°
θ_2 Range: 0° to 160° ± 1°		θ_4 Range: -180° to 180° ± 1.5°
Repeatability: +0.05 mm (For constant temp., load, velocity)		

These specifications and dimensions yield a reachable workspace in a four degree of freedom subspace. The center of the θ_1 axis was chosen as the (0, 0) location. The home position is the location from which the manipulator calibrates its servo and stepper motors for proper positioning and orientation.

Coordinate Frames and Link/Joint Parameters

The four axes of the manipulator arm and wrist are all parallel and vertical. The parameter alpha is ideally zero for each link since the joint axes are all parallel. The positive direction of the z-axes was chosen as up. The θ_3 home position has been chosen to be 90°. The manipulator variables, therefore, are θ_1, θ_2, D_3, and θ_4, with the nonzero parameters being A_1, A_2, D_2, and θ_3. The link/joint parameters and variables are summarized in Table 2.

Table 2

Joint$_i$	θ_i	D_i	A_i	α_i
1	θ_1	0	A_1	0
2	θ_2	$-D_2$	A_2	0
3	90°	$-D_3$	0	0
4	θ_4	0	0	0

The minus signs on the D_2 and D_3 parameters indicate negative numbers since they are in the negative z-direction.

Kinematics of the SCARA

The direct kinematics solution is determined using the homogenous transformation matrix, proposed by Denavit and Hartenberg [3]. The matrix describes the relative translation and rotation between link/joint coordinate systems, using the D-H parameter representation. When α_i = 0, then for the two revolute joints, θ_i = θ_1 and θ_i = θ_2, as well as prismatic joints, D_i = D_3, $^{i-1}A_i$ reduces respectively to:

$$^{i-1}A_i = \begin{bmatrix} C_i & -S_i & 0 & A_iC_i \\ S_i & C_i & 0 & A_iS_i \\ 0 & 0 & 1 & D_i \\ 0 & 0 & 0 & 1 \end{bmatrix} \qquad ^{i-1}A_i = \begin{bmatrix} C_i & -S_i & 0 & 0 \\ S_i & C_i & 0 & 0 \\ 0 & 0 & 1 & D_i \\ 0 & 0 & 0 & 1 \end{bmatrix}$$

where C_i = cos θ_i, S_i = sin θ_i, C_{ij} = $C_i * C_j - S_i * S_j$, and S_{ij} = $S_i * C_j + C_i * S_j$.
By substituting the link/joint parameters summarized above, into the link/joint transformation matrices defined for the revolute and prismatic joints, the link transformation matrices were obtained.

The inverse kinematics solution for the SCARA manipulator is used to determine the joint variables for a desired position and orientation of the end effector with reference to the base frame. A geometric approach was used to break down the spatial geometry of the manipulator into several plane geometry problems. This is a simple operation if $\alpha_1 = 0$. By using the link/joint geometric parameters as well as the equations determined, an inverse kinematics solution can be obtained.

$$\theta_2 = \text{Tan}^{-1} \frac{\pm \sqrt{(2A_1A_2)^2 - (Px^2+Py^2)^2 - (A_1^2+A_2^2)^2}}{(Px^2 + Py^2) - (A_1^2 + A_2^2)}$$

$$\theta_1 = \text{Tan}^{-1} \left[\frac{-A_2S_2\ Px + (A_1 + A_2C_2)Py}{(A_1 + A_2C_2)Px + A_2S_2\ Py} \right] \quad \text{and} \quad D_3 = Pz - D_2.$$

Trajectory Planning

To determine a path to be used by the SCARA, a joint-interpolated trajectory was employed using a 4-3-4 joint trajectory [3]. The trajectory of the initial position to the lift-off position was obtained using a fourth degree polynomial equation. For the trajectory of the lift-off to the set-down positions, a third degree polynomial equation was used. The set-down to final position trajectory was obtained using another fourth degree polynomial equation. A simple rule of thumb was applied in the lift-off and set-down positions of the robotic arm, by taking them to be 25% of D parameter distance plus any required rotation. If it is assumed that there are no path constraints or obstacles, then it is only necessary to verify that the path stays within the workspace. To verify these constraints the following section was added.

Workspace Check

The determination of workspace constraints was obtained using the specifications and dimensions mentioned above. The workspace constraints are violated for either the X, Y, or the Z Axis if one of the following conditions in Table 3 are met.

Table 3

I.	If $R_{xy} < 290$ or $R_{xy} > 1030$
II.	If $\theta_1 < 0$ or $\theta_1 > 200$
III.	If $\theta_{2min} < 0$ or $\theta_{2max} > 160$
IV.	If $Pz < -350$
V.	If $R_{min} > R_{xy}$ or $R_{max} < R_{xy}$
VI.	If $Pz > D_2$
VII.	If $Pz < D_2 + D_{3max}$

Kinematic Error Model of a Four Link Manipulator

The kinematic position error for the SCARA robot was determined from the Error Model of an N-Link Manipulator [3]. This error model determines the total error at the end of the manipulator due to link/joint parameter errors. These errors are physical in nature, primarily due to tolerances in the manufacturing of the links, or the speed reduction train. The errors may also be due to aging of the belts, or the thermal effects of the environment, as well as local heating caused by the motors and friction. Therefore using the specifications for the link/joint parameters listed, as well as the $^{i-1}A_i$ matrices determined above, the total positioning error at the end of the manipulator due to the effect of θ_i errors are:

$$Pe = \sum_{i=1}^{4} \{(^{i-1}R_4)^T * [Q_{\theta i} * (^{i-1}P_4 - {}^{i-1}P_i) + V_{\theta i}] * \Delta\theta_i\}.$$

The positioning error due to the effect of α_i errors are:

$$Pe = \sum_{i=1}^{4} \{(^{i-1}R_4)^T * [Q_{\alpha i} * (^{i-1}P_4 - {}^{i-1}P_i)] * \Delta\alpha_i\}.$$

The positioning error due to the effect of A_i errors are:

$$Pe = \sum_{i=1}^{4} \{(^{i-1}R_N)^T * [V_{Ai} * \Delta A_i]\}.$$

Finally, the positioning error due to the effect of D_i errors are:

$$Pe = \sum_{i=1}^{4} \{(^{i-1}R_4)^T * [V_{Di} * \Delta D_i]\}.$$

Where $Q_{\theta i} = \begin{bmatrix} 0 & -1 & 0 \\ 1 & 0 & 0 \\ 0 & 0 & 0 \end{bmatrix}$ and $Q_{\alpha i} = \begin{bmatrix} 0 & 0 & Si \\ 0 & 0 & -Ci \\ -Si & Ci & 0 \end{bmatrix}$

Also $V_{\theta i} = \begin{bmatrix} -a_i Si \\ a_i Ci \\ 0 \end{bmatrix}$ $V_{Di} = \begin{bmatrix} 0 \\ 0 \\ 1 \end{bmatrix}$ and $V_{Ai} = \begin{bmatrix} Ci \\ Si \\ 0 \end{bmatrix}$

Therefore by using the Error Model of an N-Link Manipulator, the calculated errors due to $\Delta\theta_i$, $\Delta\alpha_i$, ΔD_i, and ΔA_i were determined. These errors manifest themselves as $\theta_i + \Delta\theta_i$, $\alpha_i + \Delta\alpha_i$, $D_i + \Delta D_i$, and $A_i + \Delta A_i$. The parameter errors used in the calculations are listed in Table 4.

Table 4

Errors	Joint 1	Joint 2	Joint 3	Joint 4
θ Errors	$\theta_1 = 1°$	$\theta_2 = 1°$	θ_3 = N/A*	θ_4 = N/A*
α Errors	$\alpha_1 = .01°$	$\alpha_2 = .01°$	α_3 = N/A*	α_4 = N/A*
D Error	$D_1 = .5$ mm	$D_2 = .5$ mm	$D_3 = .5$ mm	$D_4 = 0$ mm
A Error	$A_1 = .5$ mm	$A_2 = .5$ mm	$A_3 = .5$ mm	$A_4 = 0$ mm

* == Signifies that these errors are inherently zero due to link/joint parameters.

Nineteen different paths in the workspace were used as well as eleven different error series. The different error series consisted of setting certain errors to zero then determining the total error at the end of the manipulator by adding the calculated errors together for each different path in the workspace.

Error Computation Program
The Kinematic Error Computation Program was written using the kinematics for the SCARA robot developed above. The joint-interpolated trajectories were used to obtain the paths for the

IBM 7540 Robot, using its reduced work-area dimensions. The specifications and tolerances for the IBM 7540 Robot were used by the program to calculate the kinematic errors derived from the kinematic error model developed above. The program performs these error computations for the position of the end-effector with respect to the workplace. The program then outputs the actual (errant) and desired (inerrant) kinematic parameters as calculated, into files that can easily be accessed by graphics programs. The actual Kinematic Error Computation Program was written using QuickBASIC 4.50 © by Microsoft™. The program code is available by contacting the authors.

Computer Generated Plots

The Kinematic Error Compensation Program calculates the actual (errant) and desired (inerrant) kinematic parameters. These computed parameters have been manipulated such that they can be represented graphically. Graphs illustrating the errors in the workspace for select paths (errant vs inerrant paths), as well as the entire recommended workspace, can be obtained. Graphs representing the individual errors (direction and magnitude) at various points of a selected path can also be obtained. Finally, graphs representing the errors in the joint variables have been derived. From these graphs, it was observed that the subset of the workspace with highest accuracy are positions closest to the center of the θ_1 axis (0, 0). The positioning errors increase greatly as one moves farther from the center of the θ_1 axis.

Error Budgeting Output

The error budgeting part of the Kinematic Error Program is in essence a reduced version of a program developed by Dr. Alex Slocum at MIT. This part of the program takes the total calculated position error, and separates the individual errors, namely errors due to θ_1, α_1, D_1, and A_1. These individual errors are then added to obtain the linear errors and the angular errors. This output is useful to determine which errors need to be changed and to what extent to obtain high positioning accuracy of the end effector for each individual path.

Comparison of Angular Errors Versus Linear Errors

It is quite obvious from the various outputs and graphs that the control of the angular parameter errors will achieve the highest positioning accuracy, in comparison to the linear parameter errors. The control of the θ_1 errors is the most critical in obtaining positioning accuracy, in comparison to the other errors. Control of the D_1 errors is of secondary importance in obtaining positioning accuracy. Of all the joints, the variables of joint 1 should be the most tightly controlled. Of secondary concern is the control of joint 2 variables.

Components Affecting Positioning Accuracy to Highest Degree

The link/joint parameters that need to be more tightly controlled to increase positioning accuracy depend on the path, and more specifically the direction of the path. If a path moving chiefly in the X direction or the Y direction is of concern for accuracy, then the θ_i errors should be most tightly controlled. Control of D_i errors or A_i errors are of secondary concern. The control of the α_i errors has very little effect on the positioning accuracy of a path moving primarily in the X or Y direction. If a path moving primarily in the Z direction is of concern for accuracy, then the D_i errors should be most tightly controlled. Control of θ_i errors or A_i errors are of secondary concern. Again, the control of the α_i errors has very little effect on positioning accuracy of a path moving primarily in the Z direction.

Enhanced Assembly Using the SCARA Robot

The application of the SCARA robot manipulator for assembly tasks is enhanced by this study through improved understanding of its end effector positioning accuracy. Certain link/joint parameters have been identified that have a little effect on the positioning accuracy of the end-effector. For a task requiring less accuracy, then these parameters can be neglected, and ease the computational requirements of the controller and computer system. For tasks that require high accuracy, and if the individual kinematic errors from the link-joint parameters can not be controlled then the workpiece or work station should be moved as close to the center of the θ_i axis as possible. If the work station is fixed relative to the robotic manipulator, then the reduction of key link/joint parameter errors must be accomplished to increase the positioning accuracy. This assessment of positioning accuracy affords many more productive uses for the SCARA robot to be accomplished.

BIBLIOGRAPHY

1. IBM Maintenance Information and Hardware Library For The 7535/7540 Manufacturing System. The IBM Corp., 1983.
2. IBM Users Guide For The 7535/7540 Manufacturing System. The IBM Corp., 1983.
3. Fu, K. S., R. C. Gonzalez, and C. S. G. Lee. Robotics: Control, Sensing, Vision, and Intelligence. New York: McGraw-Hill, Inc., 1987.
4. Zeldman, Maurice. What Every Engineer Should Know About Robots. New York: Marcel Dekker, Inc., 1984.
5. Craig, John. Introduction To Robotics: Mechanics and Control. Reading: Addison-Wesley Publishing Co. Inc., 1986.
6. Hulburt, Dave. "The SCARA: Kinematics, Dynamics and Control." MAE 386 Report. WVU, 1989.
7. Paul, Richard. Robot Manipulators: Mathematics, Programming and Control. Cambridge: MIT Press, 1981.
8. Garg, Rajat. "Error Calculations in Kinematic Parameters of Stanford Robotic Arm." MAE 386 Report. WVU, 1989.
9. Lamb, Tony. "Tool/Workpiece Positional Accuracy Using A Computer Error Budgeting Program." MAE 386 Report. WVU, 1989.
10. Slocum, Alex. MIT, Personal Communications. 1989.

Demonstrating Robot Calibration in a Manufacturing Environment

KEN PFEIFFER and LOUIS J. EVERETT
Department of Mechanical Engineering
Texas A&M Unversity
College Station, TX

ABSTRACT

A problem encountered in the implementation of multiple robotic manipulators in a light machining work cell is the requirement to individually teach the part programs to each robot used in the work area. This process is extremely time-consuming and tedious for the operators of this type of system. When equipment within the work cell (i.e..rotary tables, conveyors, etc...) must be taken apart for maintenance or repair, the locations of the fixtures within the work cell are changed and new programs must be developed, or old programs modified by hand, to incorporate these changes. The main reason that each robot must be taught the programs individually is the differences between the robot work cells in terms of tool and fixture locations, as well as the uniqueness of each individual robot. If exact duplicates of manipulator arms and work cells could be made then this problem would be eliminated. However, this is not a practical or realistic approach to the problem. The calibration research at Texas A&M University is directed towards the solution of this problem by means of an efficient, and relatively inexpensive method. This calibration method uses a *lump sum* approach to correct the tolerance buildup within the manipulator arm and the location of the tooling fixtures.

INTRODUCTION

The method used by Texas A&M University researchers utilizes a patented infra-red sensory device mounted on the end of the manipulator arm to locate tooling balls within the work area. Multiple measurements are taken on the tooling balls to provide enough information to develop an accurate and useable, mathematical model of the robot (forward kinematic model). The sensor may also be used to locate tooling points within the work cell to identify the mathematical transformations between surfaces on tooling fixtures. This transformation is necessary information when trying to use a single part description to direct the required motion of the manipulator arm over a tooling fixture that contains several parts. It should be noted that this sensor is not the only means available to identify the corrected forward kinematic model of the manipulator and the locations of tooling fixtures. Other devices such as CMM's, laser interferometers, and force sensors may be used in a similar manner to obtain position data and generate a corrected forward kinematic model.

The combined result of an accurate forward kinematic model and knowledge of locations

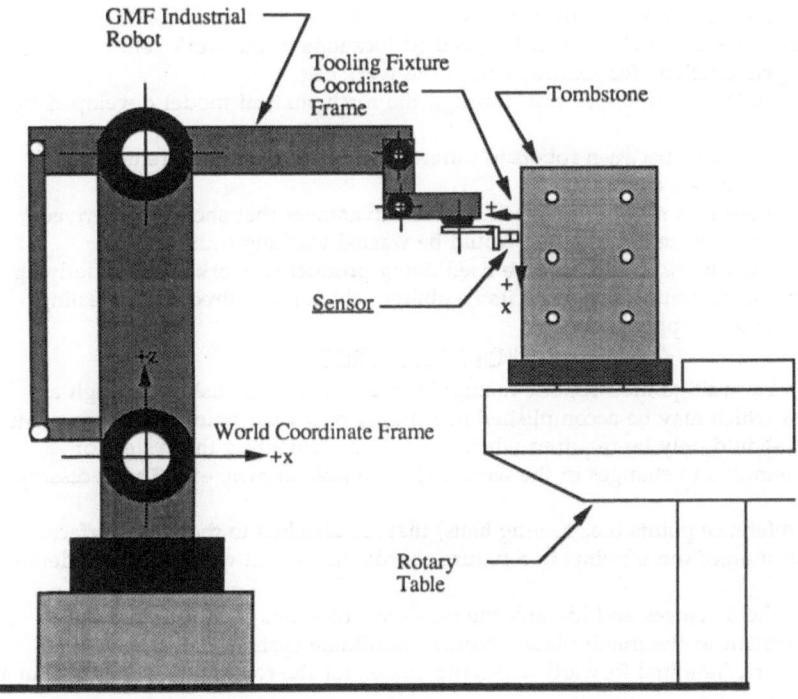

GMF Industrial Robot

Tooling Fixture Coordinate Frame

Tombstone

+z

Sensor

+x

World Coordinate Frame
+x

Rotary Table

Schematic Diagram of Test Site

of work surfaces within the cell is an ability to;

1) *Accurately* direct the manipulator arm to specified locations in the work cell.

2) *Identifiy* changed locations for fixtures within the work cell.

3) *Off-line* program the manipulator arm through the mathematical model developed by the calibration scheme.

4) *Translate* part programs between robots in different work cells, and different types of robots altogether after accurate calibration.

With these capabilities in a robotic work cell the full advantages that should be derived from automation may be realized. Less time would be wasted teaching routines, and modifying programs, and more time would be used doing productive work. The underlying principle of this type of calibration scheme is its flexibility, which is required when dealing with a wide range of robotic applications.

CALIBRATION PROCESS

The calibration of a multi-jointed robotic manipulator arm is accomplished through a series of seven steps which may be accomplished in a matter of several hours. This process is easily repeated, yet should only be required when initially implementing the system or recalibrating the sytem due to changes in the work cell, manipulator arm, etc. The necessary steps are;

1) Measure (3) reference points (i.e.. tooling balls) that are attached to the work surface.

2) Define the location of these points in a fixture coordinate system with one point defined as the origin.

3) Approximate the distances and identify the necessary rotations to move from the fixture coordinate system to the manipulator 'World' coordinate system.

4) Develop a generic/nominal forward kinematic model, for the robot to be calibrated, in a fortran-based data file.

5) Locate the reference points with a sensor and record the joint angles at each point; sensor is mounted on the end of the manipulator arm. [Note: For a 6-joint manipulator arm, (3) measurements at different orientations relative to each reference point is required; (9) total measurements, minimum.]

6) Run the iterative calibration program with the location of the reference points, generic forward kinematic model, and joint angles as input data.

7) Obtain the corrected forward kinematic model.

This calibration scheme works on a closed-loop principle of moving a coordinate system in space. A tooling point coordinate system that is originally defined on one of the tooling balls, is transformed or moved from the tooling fixture to the world coordinate system of the manipulator arm. Then, this coordinate system is moved through each joint in the manipulator arm via the nominal forward kinematics of the robot. Finally, the coordinate system is moved back to the starting point on the tooling fixture when the sensor on the manipulator is located at the point on the tooling fixture thereby closing the kinematic loop.

This movement of a coordinate system, called transformation, is accomplished by using a (4X4) transformation matrix (a mathematical expression of a coordinate system movement in space). Joint angles are taken when the kinematic loop is closed, for all of the tooling balls.

This joint angle information is entered into the calibration program along with the nominal forward kinematic model of the robot, and the location of the tooling balls in the tooling reference frame. Ideally, when the joint angles are entered into the nominal forward kinematic model, the result would be that the robot manipulator is located precisely at the tooling point. However, due to the tolerance build-up of errors in the location of the manipulator arm relative to the fixture, the inaccuracies in the manipulator arm manufacture, etc., this is not the case. The iterative calibration program simply compares the actual location of the manipulator arm (at a tooling point) to where the nominal forward kinematic model suggests that the arm is located. An error is computed and minimized by iteratively modifying the forward kinematic model of the manipulator arm. This iterative process of modifying the forward kinematic model accounts for all of the errors within the work cell in a *lump sum* manner.

Some of these calibration steps have been automated in the laboratory to speed-up data collection time and reduce the required operator attention to the process. The automation of data collection requires a PC, some type of feedback sensor (i.e..three beam infra-red position sensor), and the necessary sofware to make the PC act as a typing emulator to direct the motion of the robot.

DESCRIPTION OF TESTING AT TEXAS INSTRUMENTS, TRINITY MILLS SITE

Several initial tests were conducted to determine an appropriate demonstration to verify the ability of the calibration process to;

1) Provide a correct forward kinematic model.

2) Identify tooling points on a work fixture.

3) Translate program points taught in one reference frame to another reference frame.

4) Direct the robot manipulator to operate in the shifted reference frame with some degree of accuracy.

The manipulator arm used in this demonstration was a GMF industrial robot at Texas Instruments automated manufacturing facility (See Schematic Diagram of Test Site). This robot was found to be very accurate when shifting points in a single plane; an indication that the forward kinematic model used in the controller worked well for some applications. However, the robot was not capable of accurately shifting a part program from the 'teach' plane to another plane within the work cell and would have to be individually taught the points (See schematic diagram of Tombstone). Also, if the equipment in the work cell was disassembled for maintenance or repair work, the entire database of part programs would have to be retaught by hand or changed by a time-consuming process of measurements to identify the changes in the work cell. The local calibration process provides a simpler solution to this problem.

The test performed at Texas Instruments consisted of several steps as follows;

1) Calibrate the GMF robot using the generic forward kinematic calibration software developed at Texas A&M University.

2) Move the tooling ball plate to another surface on the tombstone.

3) Use the sensor to determine the new position of the tooling balls relative to the

manipulator arm.

4) Calculate the transformation between tombstone surfaces.

5) Teach a part program (Draw a box) on one of the surfaces.

6) Mathematically shift the part program to another surface and re-draw the part (a box).

RESULTS OF CALIBRATION TEST

The test was somewhat sucessfully completed in one trip to the Texas Instruments facility. The calculated/shifted boxes are not in the exact location of the original box. The error in the shift was approximately .25 inches maximum. There are several possible reasons for the error in the shift calculations. Some of the more obvious reasons were the 'play' in the location of the tooling ball fixture on the tombstone and the location of the pen attached to the end of the manipulator arm. The calibration procedure assumes that the location of the pen tip is at the location of the trip point on the sensor. These errors could be corrected easily by more precise machining of the test apparatus. However, some other sources of error were also noticed.

Primarily, the orientation of the pen relative to the tombstone was found to be altered during the calculated shift of the part. This orientation error is not particularly crucial to the position of the end effector yet, would be very important for a deburring process where the tool orientation must be maintained to properly work on a part. Initially, the manipulator arm was programmed to have an orientation normal to the tombstone surface. However, when the part was shifted this orientation was found to be several degrees in error. It is speculated that this error occurred because of 'round-off' errors in the calculation of sines and cosines of angles near 90 and/or 0 degrees. For example, the difference between the cosine of 0° and 1° is only 1.52×10^{-4}. Yet, a 1° difference in angles with an arc length of approximately 1000mm (the approx. length of the manipulator arm) is very important because an error of 17.5 mm would result. Double precision programming of variables in the calibration code is used for this reason, however, this may not be enough to compensate for these minute numerical computational errors. Therefore a sensitivity analysis of the system to errors should be made to improve the calibration performance. This sensitivity analysis would allow for a scaling and weighting system to be implemented with the calibration process.

One method of analyzing the sensitivity of the robot kinematics is to calculate the Jacobian matrix for the system. This matrix is used to determine the sensitivity of a change in distance for the tool coordinate system due to a change in the angles on each link. Or inversly, the sensitivity of a change in joint angles needed to change a distance in the tool frame. If the components of the Jacobian matrix are analyzed, a method to minimize these calibration errors and weight the critical kinematic parameters may be found. The Jacobian matrix was calculated for the tested manipulator arm which provided some interesting results. This calculation showed that the sensitivity of the position error is several orders of magnitude different for the different joints in the manipulator arm. Obviously, the sensitivity of the position error to errors in the 'waist' and 'shoulder' joints is most critical and therefore, should be weighted heavily. Whereas, error in the 'wrist' joints are not as crucial to position errors, yet have a greater impact on the orientation of the tool relative to the work surface.

CALIBRATION AND OFF-LINE PROGRAMMING

Ideally, the calibration process would reach its maximum usefulness when implemented in a system which utilized off-line programming of parts to be machined. A CAD system would provide the necessary position information for the part by the assignment of locations relative to a part coordinate frame. This information would necessarily include the type of tools needed for specific operations (i.e..drilling, cleaning) and the orientation of the tool relative to the part for operations such as deburring. When the part is placed in a fixture, the manipulator arm could be equipped with a sensor to locate the part relative to the fixture. Then, the calibrated inverse kinematic solution of the robot arm would be used to generate the required joint angles and trajectories needed to machine the part.

Several important issues must be addressed before this type of system may become feasible for any type of robot in a variety of work cells. The fundamental issue is the calibration of the manipulator arm which compensates for the errors in the system. If the calibration scheme is modified to account for the errors found in the initial testing of the system, then the potential for off-line programming of robotic manipulators and the transferrence of part programs between different robots may be realized. Continued efforts in calibration system research by Texas A&M University would be beneficial in solving this important manufacturing problem.

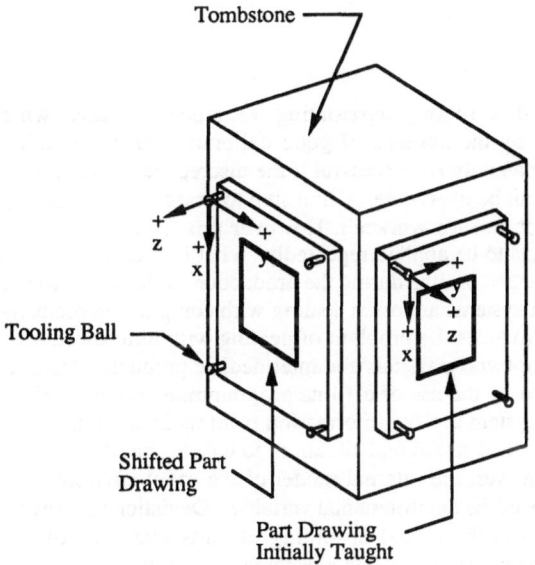

Schematic Diagram of Tombstone and Testing Conditions

On-Line Robot Calibration

F. TUIJNMAN and G.R. MEIJER
Computer Science Department
University of Amsterdam
Amsterdam, The Netherlands

Abstract: The application of off-line robot programming techniques for tasks which require a high precision is hampered by the absence on-line quality control techniques. An essential component for quality control is a low cost recalibration system which can be applied on-line during the production cycles of the robot. In this paper we present a low-cost calibration system for partial recalibration after repair or failure and for preventive maintenance. The system consists of a measuring device and software to perform parameter estimation. The system uses an non-tactile optical sensor. A suitable error model is used so that only a relatively small number of parameters has to be estimated to update the kinematic model of the robot.

Keywords: Calibration, Error models, Robots, Kinematic models, Exception handling, Autonomous systems, Optical sensor, Quality control

1 Introduction

The application of off-line robot programming techniques for tasks which require a high precision is hampered by the absence of good calibration techniques. The use of off-line programming systems can only be successful if the discrepancies between modelled positions and the real positions can be overcome. A first step to bridge this gap is to perform extensive calibration on the robot and it's workcell. However this is an expensive solution which is unacceptable when it has to be applied repeatedly. What is needed is a low cost recalibration system which can be applied on-line during the production cycles of the robot.

Modern production systems are often dealing with complicated products manufactured in small batches. This introduces the problem of dealing with unforeseen situations, errors and unexpected variations of basic materials and intermediate products. One of the problems to be dealt with, is introduced by the use of off-line programming systems. All the modules of the off-line programming system use the information from its internal model. The model should therefore closely correspond to the real situation, to ensure that the resulting robot programs function correctly. However, the internal model of the robot environment often only partly reflects the actual status of the environmental variables. Deviations of part sizes, collision with objects not represented in the model or sliding of parts may lead to a breakdown of the production process. The internal model contains three categories:

- representation of the robot kinematics,
- representation of part dimensions,
- location of part in respect to the robot.

In this paper we address the deviation of the robot kinematics representation and as a step in reaching autonomy of shopfloor systems, we present a low-cost calibration system for partial

The work on partial robot calibration is supported by Esprit I project 623: "Operational control for robot system integration into CIM", Esprit II project 2202 PLATO: "PLAnning TOolbox for CIM-systems" and Esprit II project 5220 CAR: "Calibration Applied to Quality Control and Maintenance in Robot Production, starting late 1990.

recalibration after repair, for preventive maintenance and for the installation and orientation of a calibrated robot in the workcell.

2 Robot Calibration

The calibration of robot manipulators has been a topic of research for considerable time. Early work started by studying the deviations of absolute position as a result from kinematic error sources. The most popular method for describing the kinematics of a robot manipulator is the use of the Denavit-Hartenberg model. Other sources of errors in the absolute position accuracy need be taken into account as well. An important source of errors originates from the gearing systems of the robot joints. Most robots are actuated by a motor and a gear with a high gear-down ratio. The encoders are usually located on the motor shafts and not directly on the joint. Joint compliance, gear eccentricity and backlash [9] are identified as so called "non-geometric" errors and are included in the modelling process [3]. New technologies in robot design and manufacturing introduce leight-weight and flexible robots in the production area. These systems are sensible to structural deformation under dynamic control. Modeling of these effects is essential to estimate the total effect of calibration errors [2].

Taking the error sources discussed above all together, the transformation T of joint encoder values θ to the tool center point position (TCP) given by X, can be represented by a chain of transformations:

$$X = T(\theta); \qquad T: Tg \times Tk \times Td$$

where Tg is the gear train transmission model, Tk the kinematic model and Td the model describing the dynamic effects as elasticity. T is characterized by a set of parameters P. The calibration problem is now defined as the determination of a set of parameters P* such that the difference of the of the real and calculated TCP poses is minimized for all poses in the working area of the robot. The inverse transformation from cartesian TCP position to joint encoder values can be derived with an iterative method [3].

Estimating all the parameters of the set P* for the whole working area of the robot is called global calibration. For most of the commercial available robot systems, a global calibration is carried out by the robot manufacturer. The most widespread technique is the use of theodolite measuring systems. The costs of these systems (typically in the order of 250 KECU), prohibit their application on a wide scale. Until recently calibration of robots was restricted to end-products control or after major overhaul of internal components.

Recent research efforts study the use of restricted calibration techniques. The major motivation of this work is that during the lifetime of the robot only a limited and predictable subset of all the parameters from P need to be recalibrated. Instead of determining P* as a whole, only a selection of the parameters is calibrated. This is called partial calibration. Mooring and Pack recognized the need for partial calibration in the case of robot component failures and routine maintenance [6]. For partial calibration simplified measuring device can be used which only operates in a limited range of the working space. Because the of the low costs (round 2 KECU), the calibration system can be permanently installed at the robot site providing recalibration possibilities as often as needed.

3 On-line Quality Control

In this section our approach for incorporating partial calibration in the on-line robot control is outlined. We believe that partial calibration needs to be an integral part of a production supervision system and that a calibration should take place whenever the quality of production demands so. Concepts were developed for reaching system autonomy on the shopfloor by introducing exception handling techniques in the robot controller [1]. A robot control system for quality monitoring and supervision is developed [4, 5] called the *High Level Interpreter* (HLI) and contains in addition to the traditional control modules such as servo- and logic-control, modules for *monitoring, diagnostics* and *exception handling*.

To start, the robot control system must monitor the actions of the robot. The monitoring runs parallel to the execution of the off-line generated robot program. Different classes of exceptions can be detected by thresholding on the relevant sensor output. For an assembly robot system, five groups of exceptions are identified:

- exceptions resulting from a collision with an obstacle,
- exceptions resulting from the loss of an object,
- exceptions from extensive force feedback,
- exceptions due to a handling error of an object,
- exceptions due to calibration and position errors.

Once a fault condition is detected, a diagnosis is performed to gain additional information on the nature of the exception and to update the internal model of the robot environment. A successful classification of the exception opens the way for the recovery planning function to plan corrective actions. The aim of these actions is to restore the environment to such an extent that the pre-planned robot program can be continued. Recovery planning can be performed by rescheduling of robot tasks or by generating and executing a partial recovery plan [5,7].

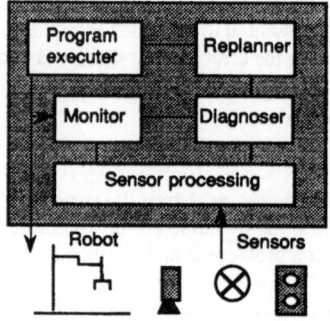

Fig. 1. High Level Interpreter

Figure 1 represents the model of the HLI. During the operation of the robot exceptions can occur, due to some malfunction in the robot itself, a fault in the part that is handled by the robot, or a misplaced object in the environment. Such exceptions, especially if they involve a collision, can cause a small change in the kinematic structure of the robot. Due to an erroneous kinematic model the robot will move to slightly different positions. If it exceeds the tolerances of the manufacturing process, either new exceptions will occur or product quality will degrade. A second cause of changes in the kinematic structure are maintenance operations on the robot. Whether preventive or after a breakdown, maintenance on the robot can require a partial overhaul of mechanical components. In particular the replacement of gears in the drive trains of the joints, or the replacement of shaft encoders, will affect the kinematic behaviour of the robot. In the first case a partial calibration of the drive train is needed whereas in the second case recalibration of the joint home position needs to be carried out.

4 Method

Figure 2 shows the a flow chart of the partial calibration procedure. The procedure is initiated either by a request for regular routine calibration, calibration of specific parameters after maintenance or after an exception. The first step of the procedure is to determine the parameters of the transformation model which need to be updated. At this point a priori knowledge of the exception or maintenance operation can be used to make the parameter selection. Based on the selected parameters, a selection of the available measurement techniques is made and the measurement area is determined.

The second step is the measurement and estimation of the selected parameters. The estimation procedure uses a linearized model for the relations between the measured sensor quantities and the parameters of the kinematic model. The procedure is carried out as follows. The robot is moved first to a position which should give measurements that are somewhere in the middle of the range in which the sensor works. A first set of measurements is made that should give a first correction to the kinematic model. If the position difference is substantial, non-linearities in the measuring device may degrade the accuracy of the re-calibration, and some measurements may even be entirely out of the range of the sensors. In this case the calibration procedure returns to step 1.

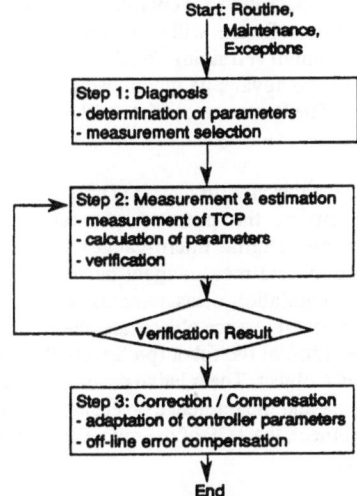

Figure 2: Partial calibration procedure.

After the first calibration, the new kinematic model is loaded into the robot. Then a second calibration is carried out. This one is to verify the results of the previous calibration, but with a new point set .

The last step is the correction or compensation step. Basically two options exists here. In the first option the new parameter values are downloaded into the working memory of the robot controller and thus directly used in the control of the robot. However most state of the art robot controllers either do not allow to change the parameters, or do not use a transformation model containing the parameters measured. In this case the positions used in the robot program have to be updated. This technique is called off-line error compensation [2]. All position values used in the off-line generated robot program are changed to such an extend that when the robot controller executes the robot program, the original position are physically reached by the robot. With Ti the ideal transformation model and Trc the transformation model used by the robot controller, each position X is transformed to a compensation position Xp which is fed to the robot controller:

$$X \rightarrow Ti^{-1} \rightarrow \theta \rightarrow Trc \rightarrow Xp$$

The compensation can be performed prior to robot program execution, but can also be taken up as a filter program in the robot controller itself. A record of the measurements and the new kinematic model is also transferred to a database for later use in statistical analysis and for monitoring possible structural problems in the production process.

5 Realization

An important requirement for the realization of the partial calibration system is that it can be applied on a wide scale and can be 'brought to the robot' instead of the other way round. A second requirement is that it is a low costs system. To comply with these requirements, the control of the partial calibration system which is embedded in the High Level Interpreter is realized with a personal computer or small workstation. Decision making, exception diagnostics and error compensation are controlled from this system. It is envisaged that the HLI, including the calibration software, will be located physically in the same housing as the robot controller. This would entail refraining from the use of a window environment as a user interface, but would have the advantage that no separate computer equipment needs to be installed on the shop-floor. The hardware consists of the sensor interface, and a processor on which the HLI is executed

Tool center point measurements (TCP) determine the cartesian position of the robot, without measurement of the pose or orientation. Several techniques are available for TCP measurements. Digital micro-gauges give a good performance but require direct contact between robot and measurement device. We are using non-tactile measurements based on distance triangulation measurements on a reference object. Depending on the robot system studied, several reference objects are placed on the base of the robot. The sensors are located in the tool position of the robot (preferably through a gripper exchange system) and moved over the reference object. The relative position (X, Y and Z) in robot coordinate system between the sensor frame and the reference object are calculated. Combined with the known position of the reference object, the TCP of the robot is given. With this technique an accuracy of 0.01 mm can be reached.

Another measurement technique applied is based on the direct calculation of the *robot pose* or orientation relative to a reference object. A structured light sensor has been develop which consists of a 2D CCD camera that observes a collection of light stripes cast over an object of fixed location and geometry [8].

From the measured distance of the lines D and the tangent, the pose of the robot can be expressed as the direction of the normal on the reference surface. Given the equation of a surface:

$$a.x + b.y + c.z = d \iff$$
$$1.x + (b/a).y + (c/a).z = d/a$$

The vector $(1, b/a, c/a)^t$ is normal to the surface. The quotient b/a and c/a are given by:

$$b/a = \sin(\alpha)/\text{tangent}$$

$$c/a = +/- D * \sqrt{(1+\text{tangent}-2)}/\delta d$$
$$-\cos(\alpha)/\text{tangent}$$

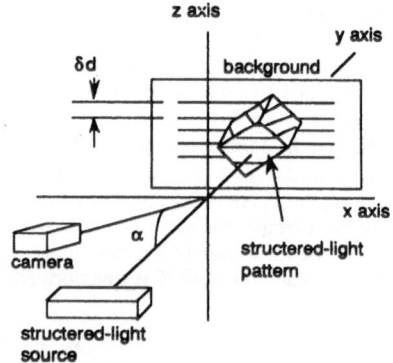

Figure 3: Pose estimation with structured light sensor.

The *test environment* that has been set up consists of a Bosch Scara arm, where the camera and light source are placed at the end effector. The Bosch controller is connected to a SUN workstation, on which the robot program is generated. This SUN workstation is also used by the HLI, and the calibration software. The sensor processing hard- and soft-ware is located in a VME crate, allowing easy exchange of components without rebooting the entire system.

6 Outlook

In the paper we argue that one of the important capabilities of on-line process monitoring, diagnostics and exception handling is to perform partial calibration procedures. After maintenance, exceptions or just routine inspection, partial recalibration of selected parameters of the robot transformation model is needed. We showed that with new sensor techniques and calibration procedures, on-line partial calibration can be an afordable element in the qualtity ensurance functions.

Further work in this area will be focused on the further development and test of the measurement systems and performance testing with the experimental set-up. The work is carried out in close cooperation with both robot manufacturing industry and robot users.

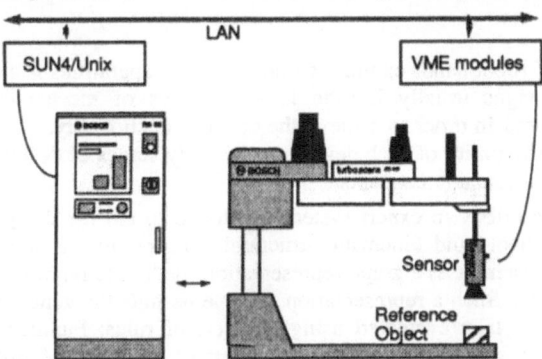

Figure 4: Realization of calibration system with industrial robot.

References

[1] Camarinha, L, Negretto, U, Meijer, G.R, "Information Integration for Assembly Cell Programming and Monitoring in CIM", Proceedings of 21th International Symposium on Automotive Technology and Automation, Wiesbaden 6-10 November 1989.

[2] Duelen, G.; Schröer, K.; Praktische resultate der roboter kalibration (German). Zeitschrift für Wirtschaftlische Fertigung 85-2, pp113. 1990.

[3] Judd, R.P.; Knasinski, A.B.; A technique to calibrate industrial robots with experimental verification. IEEE transactions on Robotics and Automation, Vol 6, No 1, pp20. February 1990.

[4] Meijer G.R., Weller, G.A. Groen, Hertzberger,L.O, "Sensor based control for autonomous robots", Proceedings of IEEE international conference on control and applications, Jeruzalem April 1989, WP-3-4.

[5] Meijer,G.R, Hertzberger,L.O.H. Exception handling for robot manufacturing process control. Proceedings of CIM Europe conference, Madrid may 18-20 1988, IFS publications.

[6] Mooring, B.W.; Pack, T.J.; Calibration procedure for an industrial robot, Proceedings IEEE Int. Conf. on Robotics and Automation. pp786. 1988.

[7] Tuijnman, F, Meijer, G.R., Hertzberger, L.O, (1989), "Data modelling for a Robot Query Language", Proceedings of Conference on Intelligent Autonomous Systems 2 (IAS-2), Amsterdam 11-14 december 1989, Elsevier Science Publishers

[8] Weller, G.A.; Choudry, A.; Hertzberger, L.O.; Meijer, G.R.; Recognition of polyhedral objects under structured illumination. Proccedings of the International Conference on Intelligent Autonomous Systems (IAS), eds. L.O. Hertzberger and F.C.A. Groen, pp 283, Amsterdam, 1986.

[9] Whitney, D.E.; Lozinski, C.A.; Rourke, J.M.; Industrial robot forward calibration method and results. Proceedings ASME conference on Computers & Engineering, pp92-100. 1984.

Expert System for Robot Hand Design Using Graph Representation

M. CHEW, G.F. ISSA and S.N.T. SHEN

Old Dominion University
Norfolk, VA

Summary

The design of robot hands exhibiting manipulative capabilities is a complex and difficult process. Such designs usually include large numbers of mechanical components each specifically structured in order to achieve the desired functionality. It is therefore necessary to investigate the feasibility of such devices at the early stages of design (Conceptual Design Stage) to arrive at acceptable concepts.

This paper describes an expert system developed to aid the designer in evaluating the most appropriate robot-hand kinematic structural concept from a large number of possible alternatives. The system uses a graph representation method to represent the kinematic structures of robot hands. Such a representation scheme permits the generation of large numbers of graphs which are then evaluated using two sets of rules: Fundamental knowledge and heuristic knowledge rule sets. Graphs which satisfy both sets of rules are considered to represent acceptable conceptual designs.

1. INTRODUCTION

This paper describes the design automation of the internal mechanisms of robot hands. It is a sub-system of a larger knowledge base that deals with the design automation of both robot hands and variable-stroke engine designs†. The system automates the preliminary structural design of two-fingered robot hands using graph representation of the kinematic structure of the hands discussed in [1]. This use of graphs as a representation method, enables the system to enumerate large numbers of different graphs which are then evaluated using the two sets of rules.

The overall expert system is made up of two major components. One component is a knowledge base consisting of a set of evaluation rules implemented using an expert system shell called EXSYS [2], presently being restructured for implementation in a more sophisticated expert system shell called GURU [5]. The second component is a set of programs implemented in PASCAL to overcome many of the limitations inherited in an expert system shell. This includes: a customized user interface, incorporation of several types of data

---

† Accepted for publication in the ASME Journal of Mechanisms, Transmission, and Automation in Design. It will also be presented at the September 1990 ASME Conference.

structures to record the facts about test subjects, graphics to support the robot hand representation, and procedures to perform complicated calculations such as the determinant of a matrix and the characteristic equation of a graph.

2. THE DESIGN OF ROBOT HANDS

This section presents the methodology used in the design of robot hands. It describes the approach of design using graphs and presents the evaluation rules used in the resulting expert system.

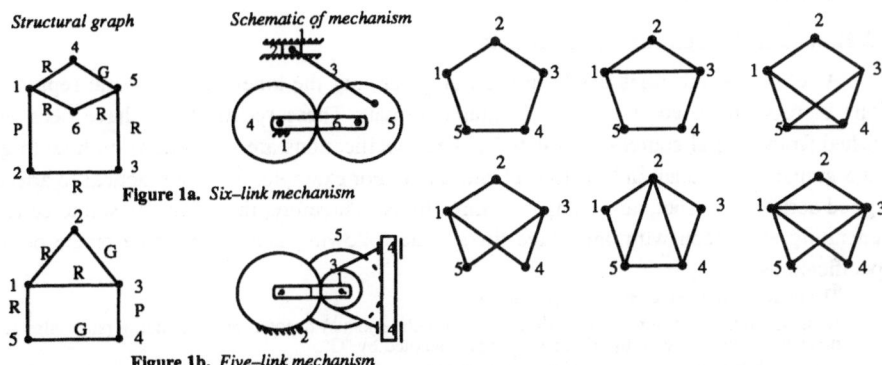

Figure 1a. *Six–link mechanism*

Figure 1b. *Five–link mechanism*

Figure 1. Graph representation of robot–hand design   **Figure 2.** Menu containing possible graph templates for five–link graphs

2.1 Graph Representation And Design Methodology

A mechanical system consists of a number of fundamental joint types (gear pairs, sliding pairs, and turning pairs) connecting various components (cranks, pistons, gear racks) to form a complete mechanical system. A mechanical hand for instance, must have moving parts to be able to manipulate objects. These moving or rotating parts should be free of high friction, vibration, excessive side thrust, and other negative side effects. The analysis of the graph representation of the structures of mechanical hand designs provides sufficient information as to whether such side effects do exist in the designs and/or if there are any components connected together in a way that is not physically practical.

According to the research reported by [1, 3, 4], kinematic structures can be defined precisely with the aid of graph representations, which can be used for the creation of mechanisms in a relatively systematic manner. This means that the structural aspects of a robot hand design can be directly represented by a distinct graph containing sufficient information for evaluating its conceptual feasibility.

The schematic diagrams and their corresponding graph representations of the internal mechanisms of two different robot hands are shown in Figure 1. The figure represents only one finger of the hands since both fingers are assumed identical. The graph representation is directly related to the structure of the schematic diagram. It is a simple representation in which graph vertices correspond to *links* and edges correspond to *joints*. The edge labels represent the type of joints being used.

Figure 1 shows the schematics of the mechanisms which are made up of links each of which is given a distinct number and is connected together by joints. There are several types of links shown in the diagram. In Figure 1a, for example, $link_1$ shown with slashes represents a common ground. $Link_2$ is connected to the ground by a sliding pair or joint and is considered a piston; $links_{4,5}$ form a pair of gears; $link_6$ can be a crank, with other links representing other mechanical components. There are three types of joints in the diagram: turning pairs (R) represented by dots, sliding pairs (P) shown in the figure as the connection between $link_1$ and $link_2$, and gear pairs (G) represented by the contact between the two circles representing $link_4$ and $link_5$.

2.2 Kinematic Structural Specifications

A set of specifications has been used in generating the kinematic structural representations of robot hand configurations in graph format. These specifications have been established from several sources [1, 3, 4, 6, 7]. The specifications are restrictions on how to generate graphs rather than on how to evaluate them. For example, it is not practical to arrive at a good design of a robot hand using only three links. Therefore, there is no reason to consider generating any graph with only three links. The following is a list of some of the obvious specifications:

1. The search is limited to only plane mechanisms.
2. Joints are limited to turning pairs (also called revolute joints) denoted by "R", sliding pairs (also called prismatic joints) denoted by "P", and gear pairs denoted by "G".
3. The search is limited to only four, five, and six-link mechanisms.
4. Maximum number of "P"s or "G"s must not exceed three.

In addition to the above kinematic structural rules, there are rules for evaluation of the resulting graphs that represent different configurations of possible mechanism concepts. Such rules are the evaluation rules which will be described next.

2.3 Evaluation Rules

The knowledge base of this expert system consists of two sets of rules. The first set is the fundamental rule set. Knowledge represented in it consists of basic information in kinematics without regard to the application. Once a graph is tested with this rule set it is passed to a second rule set consisting of application-oriented rules based on heuristics and experience in robot hands design. Such knowledge may differ from one designer or from one application to another. A graph is considered to represent a feasible hand design if it passes both of the rule sets. The following is a list of a few of the rules used in this expert system:

Fundamental rule set:
1. The number of vertices exceeds the number of edges representing turning pairs and sliding pairs by one.
2. The number of turning-pair and sliding-pair edges exceeds the number of geared edges by the degree of freedom of the mechanism.
3. The number of Fundamental circuits (loops) equals the number of geared edges.

Heuristic rule set:
1. Six-link mechanisms are not permitted to contain a gear-rack, while five-link mechanisms can contain only one gear-rack.
2. Consider only mechanisms with six links and seven joints, and mechanisms with five links and six joints.
3. Only one prismatic-pair is permitted to provide the required translation manipulation. Reject configurations with no sliding-pairs.
4. Avoid configurations that contains a slider-crank mechanism.

3. SYSTEM IMPLEMENTATION

The approach used in the design of this expert system follows in the same spirit as that of our previous variable-stroke engine design expert system [6, 7] in which a large number of design alternatives, represented by graphs, were generated and then evaluated using a number of evaluation rules. Following the evaluation, the system must then produce a comprehensive analysis of the design configuration, state whether that configuration is accepted or rejected, and finally show all the steps which contributed to that decision.

The expert system described in this paper is implemented in the EXSYS expert system shell to work with the knowledge base, and Turbo-Pascal external programs for user interface, graph representation, and complex calculations.

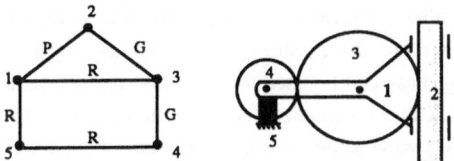 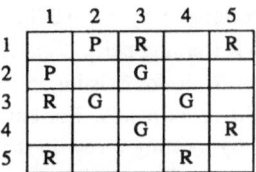

	1	2	3	4	5
1		P	R		R
2	P		G		
3	R	G		G	
4			G		R
5	R			R	

Figure 3. A labeled graph with its corresponding design

Figure 4. Matrix representation of Figure 3.

3.1 Generation of Graphs

The generation process uses the set of specifications described in Section 2.2 to create graph templates which are ready for use. According to these specifications, only four, five, and six-link structures are thought to be appropriate for study. For the four-link structures, two possible templates with four and five joints are generated. For the five-link structure, six possible templates are generated with the number of joints ranging from five to seven (Figure 2). Labeling a template in all the possible ways produces a chain of graphs. As previously mentioned, a chain is a set of graphs having the same structural configurations but differing in the way the edges are labeled with allowable joint types (in our case R's turning-pairs, P's sliding-pairs, and G's gear-pairs). The importance of using a set of specifications or constraints is to limit the number of templates generated and thus eliminate hundreds of meaningless configurations which would otherwise dramatically burden the design effort. In addition, the use of the characteristic equation for detecting redundant graphs and for deleting them, further reduces the search space considerably.

The labeling process could be done automatically by the system upon the user's request. The system generates all the possible permutations of labels, deletes redundant graphs, and displays the graphs for the user one at a time.

The graph shown in Figure 3 is an example of using the labeling process described above. At this point, this graph actually represents the structural design of a robot hand. The duty of the external program is to perform an initial analysis (described in next section) to understand this graph and extract the information regarding the kinematic structure. Appropriate information is then transferred to EXSYS to determine the feasibility of such a preliminary design configuration.

3.2 Analysis of Graph Representation

The purpose of this stage is to extract the necessary knowledge from the graphs and to identify the components that make up the mechanism represented by a graph. Results from this stage are passed to the evaluation stage (Sec. 3.3).

The analysis of a graph is performed on an NxN adjacency matrix that represents the topology of the graph, with N being the number of links or vertices. In Figure 3 the graph has five links and six joints and is represented by a 5 x 5 matrix shown in Figure 4, where the row and column numbers correspond to the link numbers of the graph. The non-empty locations of the matrix correspond to connecting joints and are assigned values of P's, R's, or G's. In this example $link_1$ (shown in the schematic diagram as a laid down funnel) is a ternary link connected to three other links by three joints as follows: it is connected to $link_2$ (shown as an up right rectangle) by a P (sliding pair). It is also connected to $link_3$ (shown as a large circle) by an R (turning pair). Finally, it is connected to $link_5$ (shown as dashed lines "ground") by an other R. Rules existing in the knowledge base can be used to determine the component types of the robot hands. For example, a crank is a mechanical component which could be identified by locating a row in the matrix with two R joints, one of them being connected to the ground link. The two joints ensure that the crank is binary.

values based on 0-10 system		value
1- PASSED ALL RULES, ACCEPT		10
2- Number of Links in Mechanism =	5	
3- Number of Joints in Mechanism =	6	
4- Number of Fundamental Circuits =	2	
5- Number of Turning Pairs =	3	
6- Number of Sliding Pairs =	1	
7- Number of Gear pairs =	2	
8- Degree Of Freedom for Mechanism =	2	
9- First Fundamental Circuit is	1231	
10-Number of Gears in First F-Circuit =	1	
11-Second Fundamental Circuit is	13451	
12-Number of Gears in Second F-circuit =	1	

Figure 5. Results of evaluating Fig. 3

3.3 Evaluation Stage

In this stage the graphs generated are evaluated according to the rules presented in Sec. 2.3. These rules are implemented in the form of IF-THEN-ELSE rules. A rule in EXSYS is made up of a list of IF conditions containing English sentences or algebraic expressions and a list of THEN and ELSE consequences which may contain more statements, probability of a particular choice, or maybe a request to run an external program to perform tasks such as calculations, reading or updating data files, and displaying graphics.

Each design tested will be classified as *Accepted* if it passes all rules, *Marginally-Rejected* if it passes only fundamental rules, or *Absolutely-Rejected* if it fails fundamental rules. The results are followed by a certainty factor which shows the degree of confidence of the rules used. Once a design configuration is evaluated, the user would be in an interactive mode where he can select from a menu to further test the design by changing some attributes, to verify the rules contributing to that decision, or to compare the results of several design alternatives. Figure 5 shows the results of the analysis and evaluation stages of the mechanism presented in Figure 3.

The result of the expert system shows that the design configuration in Figure 3 is feasible since it passes both the fundamental and heuristic rule sets. The designer can verify the results by viewing the rules contributed to this decision.

CONCLUSION

The work presented in this paper shows the applicability of existing expert system techniques, such as rule-based systems, to the conceptual design automation of mechanisms. By using an efficient knowledge representation in the form of graphs, the system was capable of elucidating the structure of each mechanism, and was capable of evaluating such structures using its heuristic reasoning. The augmentation of an expert system shell (EXSYS) with a procedural language (Pascal) permitted the system to handle the overall conceptual design process. This included a customized user interface, graph manipulation, complex calculations, and heuristic reasoning. While the example presented in this paper demonstrated the automation of one application (Robot-Hands), the overall system is also capable of automating other mechanical systems as well.

References

1. Datseris, P. and Palm, W., "Principles on the Development of Mechanical Hands Which can Manipulate Objects by Means of Active Control," *Journal of Mechanisms, Transmissions, and Automation in Design*, pp. 1-9, 1984.

2. EXSYS, Inc., *EXSYS Expert System Development Package.*, EXSYS inc., P.O. Box 75158 Contr. Sta. 14, Albuquerque, NM 87194, 1985.

3. Freudenstein, F. and Maki, E. R., "The Creation of Mechanisms According to Kinematic Structure and Function," *Enviroment and Planning B*, vol. 6, pp. 375-391, Sep 1979.

4. Freudenstein, F. and Maki, E. R., "Development of an Optimum Variable-Stroke Internal Combustion Engine Mechanism From the Viewpoint of Kinematic Structure," *ASME Journal of Mechanisms, Transmissions, and Automation in Design*, vol. 105, pp. 259-268, June 1983.

5. MDBS, Inc., *GURU*, Micro Data Base Systems, Inc., Lafayette, IN 47902, 1987.

6. Shen, S. N. T., Chew, M., and Issa, G. F., "Expert System Approach For Evaluating Engine Design Alternatives," *SPIE proceedings on Applications of Artificial Intelligence VII*, vol. 1095, pp. 533-543, March 1989.

7. Shen, S. N. T., Chew, M., and Issa, G. F., "Expert System Approach Using Graph Representation and Analysis for Variable-Stroke Internal-Combustion Engine Design," *To be published in the International Journal on Artificial Intelligence and Pattern Recognition.*, vol. 4, no. 3, Spet. 1990.

PreGrasp Pose Estimation of Objects Using Local Sensors on Dexterous Hands

V.H. PINTO, L.J. EVERETT and M. DRIELS
Department of Mechanical Engineering
Texas A&M University
College Station, TX

Abstract

In order to successfully grasp an arbitrary object, one must know the pre-grasp position and orientation (pose) of the targeted object. This knowledge may help propose a grasping strategy which will ensure grasping, increase the stability of the grasp, and provide a grasp which will either not require further manipulation of the object or will make manipulation of the object easier. The *pose estimation problem* is the determination of the pose of an object based on sensor data. The sensor data is generated by simulating the response of noncontact proximity sensors having spherical detection ranges. The pose of the object is solved using the concept of triangulation to derive the equations and a nonlinear least-squares technique to solve for the unknown sensor parameters and object frame location relative to the reference frame. For simplicity, the object is modeled as spherical.

1. Introduction

1.1 Motivation

Current work at NASA's Johnson Space Center is developing an autonomous robot to perform tasks such as object rescue and retrieval, space shuttle experimentation, satellite repair, and space station construction. Dexterous end effectors with sensors can help the robot perform this large variety of tasks.

Of the tasks previously mentioned, object rescue and retrieval is considered the most challenging and one of the primary reasons for developing an autonomous robot [1]. Object rescue and retrieval requires locating and tracking the object, movement towards the object, and grasping and maintaining a stable grasp of the object [1]. Stable grasp must be ensured to avoid having the object escape and drift into space. The EVA Retriever, supplied with dexterous end effectors, is currently under development to perform this task.

If the object is obscured from view by the vision sensors during the grasp phase due to the robot arm or other objects, the retriever must rely on localized noncontact sensors to determine the pre-grasp position and orientation (pose) of the object. An added benefit from knowing the object's pose has to do with the manipulation of the object. Using the data, the object can be grasped in such a way which will either not require further manipulation of the object or will make the manipulation of the object easier.

1.2 Objectives

The objective of this paper is to present a method of calculating the pre-grasp position and orientation of an object using data from local, noncontact sensors. The equations will be developed using the concept of triangulation. Software based on a nonlinear least-squares algorithm will then solve for the pose of the object. As an example, we develop the algorithm for the University of Minnesota hand using both simplified proximity sensors and object models.

2. Previous Research

The problem of grasping objects by dexterous hands has been widely considered [2,3,4,5]. In most studies, the object is well defined by vision sensors and/or by tactile sensors and the grasp strategy uses the data from these sensors. The use of vision sensors may be limited by the object being hidden from view by the robot arms or other objects. In the space environment, tactile sensors cannot be used because the object cannot be touched prior to grasping or the object will float away. This investigation is primarily concerned with grasping objects based on noncontact local sensors; that is, using proximity sensors mounted on an end effector to define the pre-grasp pose of the object.

A method to determine the position and orientation of an object's surface using proximity sensors has been studied in [6]. Limited research has been performed on using proximity sensors mounted on a dexterous end effector to determine the position and orientation of an object [7].

3. Pose Estimation

The pose estimation problem will be solved using a triangulation scheme to develop the equations and software based on a nonlinear least-squares algorithm to calculate the pose of the object. We assume that we are given sensor data, knowledge of the finger joint angles, location of the sensors, and the manipulator arm has positioned the end effector close enough to the object to trigger all of the sensors.

3.1. Equation Development

Using a scheme based on triangulation, which is a common method used to optically locate points in space [6], we can develop the necessary equations. For simplicity, it is assumed that both the proximity sensor characteristics and the object type are spherical. From Figure 1 the following transformation equations can be written

$$T_{hb}^{object} = T_{hb}^{sensor_1} T_{sensor_1}^{object} = T_{hb}^{sensor_2} T_{sensor_2}^{object} = T_{hb}^{sensor_j} T_{sensor_j}^{object}$$

for j sensors. It then follows that

$$T_{hb}^{sensor_1} T_{sensor_1}^{object} = T_{hb}^{sensor_2} T_{sensor_2}^{object}$$

$$T_{hb}^{sensor_1} T_{sensor_1}^{object} = T_{hb}^{sensor_3} T_{sensor_3}^{object} \tag{1}$$

$$T_{hb}^{sensor_1} T_{sensor_1}^{object} = T_{hb}^{sensor_j} T_{sensor_j}^{object}$$

where

$$T_{hb}^{sensor} = T_{hb}^{fb} T_{fb}^{link} T_{link}^{sensor} \tag{2}$$

T_{hb}^{object} is the transformation from the hand-base coordinate frame to the object coordinate frame, T_{hb}^{fb} is the transformation from the hand-base coordinate frame to the finger-base coordinate frame, T_{fb}^{link} is the transformation from the finger-base coordinate frame to the link coordinate frame, and T_{link}^{sensor} is the transformation from the link frame to the sensor coordinate frame. T_{sensor}^{object} is given as

$$T_{sensor}^{object} = \mathbf{R} \cdot \mathrm{Sph}(\gamma, \beta, r + R) \tag{3}$$

where [8]

$$\mathrm{Sph}(\gamma, \beta, r + R) = Rot(X_S, \gamma)\ Rot(Y_S, \beta)\ Trans(Z_S, r + R)$$

$$\mathrm{Sph}(\gamma, \beta, r + R) = \begin{bmatrix} \cos\gamma\cos\beta & -\sin\gamma & \cos\gamma\sin\beta & (r+R)\cos\gamma\sin\beta \\ \sin\gamma\cos\beta & \cos\gamma & \sin\gamma\sin\beta & (r+R)\sin\gamma\sin\beta \\ -\sin\beta & 0 & \cos\beta & (r+R)\cos\beta \\ 0 & 0 & 0 & 1 \end{bmatrix} \tag{4}$$

From Figure 2, the value $r + R$ is the radial distance from the sensor to the object's coordinate frame where r is the sensor-to-trigger point distance and R is radius of the sphere, γ is the offset angle in the X_S-Y_S plane, and β is the offset angle from the Z_S axis. The R matrix is used to align the x, y, z axes of the sensor to the defined X_S, Y_S, Z_S axes, and is in general

$$\mathbf{R} = Rot(x, \theta_x)\ Rot(y', \theta_y)\ Rot(z'', \theta_z)$$

where θ_x is a rotation about the x-axis, θ_y is a rotation about the the new y-axis, y', and θ_z is a rotation about the new z-axis, z''.

If we look at the individual matrices and determine the known and unknowns: the T_{hb}^{sensor} matrices are known because the inverse kinematics are known, r (sensor-to-trigger point distance) for each sensor is known because the sensors have been triggered, but both γ and β are unknowns in the T_{sensor}^{object} matrices. Thus for each equation in (1), there are four unknowns (γ and β are unknowns on each side). Therefore *two equations must be available* (i.e.; three individual sensors must trigger) to have six equations with six unknowns.

3.2. Example

Consider Figure 3 which shows a sketch of two fingers of the University of Minnesota hand surrounding a circular object. The three sensors we will use to demonstrate the procedure is one sensor from the first link

474

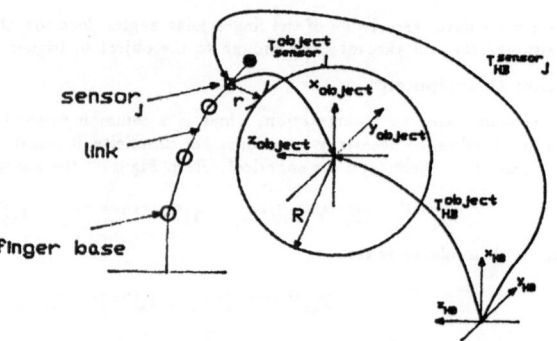

Figure 1. Transformations from hand base frame to object frame.

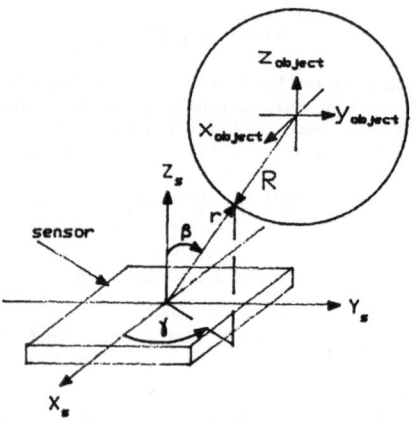

Figure 2. Parameters for a spherically modeled sensor.

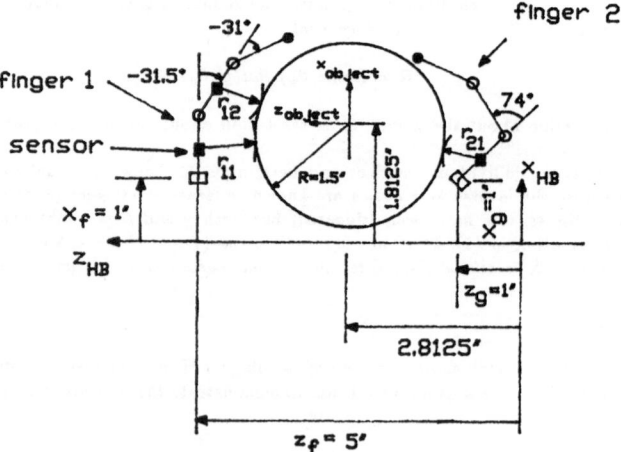

Figure 3. Pose estimation simulation.

sensor parameters	forward solution (degrees)	inverse solution (degrees)	% difference
r_{11}	0.7097087	-	-
γ_{11}	≈ 0	0.38080	-
β_{11}	8.1301	8.1302	≈ 0
r_{12}	0.5216860		-
γ_{12}	≈ 0	0.24633	-
β_{12}	13.8251	13.8251	0
r_{21}	0.7141319	-	-
γ_{21}	≈ 0	≈ 0	-
β_{21}	33.0371	33.0373	≈ 0

Figure 4. Estimated sensor parameters.

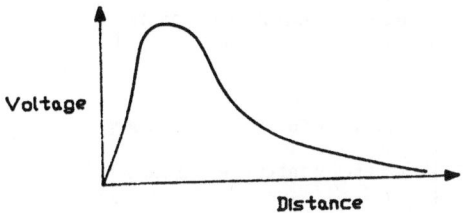

Figure 5. Proximity sensor characteristics.

and one sensor from the second link of finger 1 and one sensor from the first link of finger 2. Thus equation 1 for this example becomes

$$T_{hb}^{sensor_{11}} T_{sensor_{11}}^{object} = T_{hb}^{sensor_{12}} T_{sensor_{12}}^{object}$$

$$T_{hb}^{sensor_{11}} T_{sensor_{11}}^{object} = T_{hb}^{sensor_{21}} T_{sensor_{21}}^{object} \qquad (5)$$

where the first subscript stands for the finger number and the second stands for the link number. Then the following matrices can be found: for the first link of finger 1, $T_{hb}^{sensor_{11}}$ is

$$T_{hb}^{sensor_{11}} = \begin{bmatrix} 1 & 0 & 0 & 1.5 \\ 0 & 0 & -1 & 0 \\ 0 & 1 & 0 & 5 \\ 0 & 0 & 0 & 1 \end{bmatrix}$$

For the second link of finger 1, $T_{hb}^{sensor_{12}}$ is

$$T_{hb}^{sensor_{12}} = \begin{bmatrix} 0.85264 & 0.5225 & 0 & 2.42632 \\ 0 & 0 & -1 & 0 \\ -0.5225 & 0.85264 & 0 & 4.73875 \\ 0 & 0 & 0 & 1 \end{bmatrix}$$

For the first link of finger 2, $T_{hb}^{sensor_{21}}$ is

$$T_{hb}^{sensor_{21}} = \begin{bmatrix} 0.70711 & 0.70711 & 0 & 1.35356 \\ 0 & 0 & -1 & 0 \\ -0.70711 & 0.70711 & 0 & 0.64645 \\ 0 & 0 & 0 & 1 \end{bmatrix}$$

The matrix $\mathbf{R}$ for finger 1 is

$$\mathbf{R}_1 = Rot(x, 90°) = \begin{bmatrix} 1 & 0 & 0 & 0 \\ 0 & 0 & -1 & 0 \\ 0 & 1 & 0 & 0 \\ 0 & 0 & 0 & 1 \end{bmatrix}$$

and for finger 2, $\mathbf{R}$ is

$$\mathbf{R}_2 = Rot(x, -90°)\, Rot(z', 180°) = \begin{bmatrix} -1 & 0 & 0 & 0 \\ 0 & 0 & 1 & 0 \\ 0 & 1 & 0 & 0 \\ 0 & 0 & 0 & 1 \end{bmatrix}$$

Then using equation 3 yields

$$T_{sensor_{1i}}^{object} = \begin{bmatrix} \cos\gamma_{1i}\,\cos\beta_{1i} & -\sin\gamma_{1i} & \cos\gamma_{1i}\,\sin\beta_{1i} & (r_{1i}+R)\cos\gamma_{1i}\,\sin\beta_{1i} \\ \sin\beta_{1i} & 0 & -\cos\beta_{1i} & -(r_{1i}+R)\cos\beta_{1i} \\ \sin\gamma_{1i}\,\cos\beta_{1i} & \cos\gamma_{1i} & \sin\gamma_{1i}\,\sin\beta_{1i} & (r_{1i}+R)\sin\gamma_{1i}\,\sin\beta_{1i} \\ 0 & 0 & 0 & 1 \end{bmatrix}$$

for $i = 1, 2$, and

$$T_{sensor_{21}}^{object} = \begin{bmatrix} -\cos\gamma_{21}\,\cos\beta_{21} & \sin\gamma_{21} & -\cos\gamma_{21}\,\sin\beta_{21} & -(r_{21}+R)\cos\gamma_{21}\,\sin\beta_{21} \\ -\sin\beta_{21} & 0 & \cos\beta_{21} & (r_{21}+R)\cos\beta_{21} \\ \sin\gamma_{21}\,\cos\beta_{21} & \cos\gamma_{21} & \sin\gamma_{21}\,\sin\beta_{21} & (r_{21}+R)\sin\gamma_{21}\,\sin\beta_{21} \\ 0 & 0 & 0 & 1 \end{bmatrix}$$

Taking the products as given by equation 5 and looking at only the last column of each, yields

$$T_{hb}^{sensor_{11}} T_{sensor_{11}}^{object} \begin{bmatrix} 0 \\ 0 \\ 0 \\ 1 \end{bmatrix} = \begin{bmatrix} (r_{11}+R)\cos\gamma_{11}\,\sin\beta_{11} + 1.5 \\ -(r_{11}+R)\sin\gamma_{11}\,\sin\beta_{11} \\ -(r_{11}+R)\cos\beta_{11} + 5 \\ 1 \end{bmatrix} \qquad (6)$$

$$
T_{hb}^{sensor_{12}} T_{sensor_{12}}^{object}
\begin{bmatrix} 0 \\ 0 \\ 0 \\ 1 \end{bmatrix}
=
\begin{bmatrix}
0.853(r_{12} + R)\cos\gamma_{12}\ \sin\beta_{12} - 0.523(r_{12} + R)\cos\beta_{12} + 2.426 \\
-(r_{12} + R)\sin\gamma_{12}\ \sin\beta_{12} \\
-0.523(r_{12} + R)\cos\gamma_{12}\ \sin\beta_{12} - 0.853(r_{12} + R)\cos\beta_{12} + 4.739 \\
1
\end{bmatrix}
$$

$$
T_{hb}^{sensor_{21}} T_{sensor_{21}}^{object}
\begin{bmatrix} 0 \\ 0 \\ 0 \\ 1 \end{bmatrix}
=
\begin{bmatrix}
-0.707(r_{21} + R)\cos\gamma_{21}\ \sin\beta_{21} + 0.707(r_{21} + R)\cos\beta_{21} + 1.354 \\
-(r_{21} + R)\sin\gamma_{21}\ \sin\beta_{21} \\
0.707(r_{21} + R)\cos\gamma_{21}\ \sin\beta_{21} + 0.707(r_{21} + R)\cos\beta_{21} + 0.646 \\
1
\end{bmatrix}
$$

Also, from equation 5, the three equations above must all be equal. Therefore if we equate them, we have six equations with six unknowns ($\gamma_{11}, \beta_{11}, \gamma_{12}, \beta_{12}, \gamma_{21}, \beta_{21}$). A program using a nonlinear least-squares technique was written to solve for the unknowns. The resulting sensor parameters using the radial distances obtained from the forward solution, in which the position of the object coordinate frame is known but r, γ, and β for each sensor are unknown, are given in Figure 4. As one can see, the parameter results match very closely since the percent differences are mostly negligible.

Once obtaining γ and β for any sensor, the object's coordinate frame position is just the last column of the product $T_{hb}^{sensor} T_{sensor}^{object}$ from any sensor. For example, using values obtained from Figure 4 for the sensor on link 1 of finger 1, and using equation 6, the object's coordinate frame position is

$$
T_{hb}^{sensor_{11}} T_{sensor_{11}}^{object} =
\begin{bmatrix}
1.8125 \\
-2 \times 10^{-5} \\
2.8125 \\
1
\end{bmatrix}
$$

which agrees with the input values to the forward solution.

The overall procedure can be programed using the nonlinear least-squares technique to solve a general three-dimensional problem.

3.3. Discussion

There are two problems in the algorithm previously discussed: the fact that only the position is calculated versus both the position and orientation, and that the object and proximity sensor characteristics are modeled as being spherical. Also, there is a limitation in the sensor frame orientations that will allow the least-squares program to converge, which is a result of developing the equations using triangulation.

In order to find the orientation of the object, it will require a total of six more equations if we specify the object orientation by a set of Euler angles ϕ, θ, and ψ.

The algorithm was based on using a spherical model for the grasped object. Other types of objects can be used, but they will require more computation to model them. One possible way is to use Fourier Descriptors, which is a common method used in imaging processes to represent a shape. The problem with using Fourier Descriptors is that the computation time required may make the program run too slow to make it useful for real-time processing. An alternative would be to treat objects as geometric primitives and develop the inverse solution algorithm to handle those specific primitives.

The algorithms were also based on using a spherical model for the proximity sensor characteristics. A more realistic model of the proximity sensor is as shown in Figure 5. This shape must be modeled mathematically but can easily be included as a subroutine converting the sensor voltages or currents to a distance as required by the program.

Another problem with the inverse solution algorithm has to do with the sensor frame orientations. In order for the least-squares program to converge and to solve for the sensor parameters, two successive sensor frames used in the computations cannot be parallel. Physically, this would mean that the lines of site for the two sensors in question do not intersect and the triangulation scheme would not work. This could occur if the relative angle between the two sensors is zero (the finger joint between the two sensors has not been tilted), or successive frames between two fingers are coplanar (the two fingers have equal joint angles). This can be overcome by tilting each joint in the finger or having each finger have different joint angles.

Conclusion

The main focus of this paper was on the pose estimation problem, which was to determine the position and orientation of an object based on sensor data. Using simplified sensor and object models, an algorithm based on a triangulation scheme was developed. The position of the object coordinate frame was found using a nonlinear least-squares technique. Some of the problems with the software are the fact that only the position of the object is calculated versus both the position and orientation of the object, and that both the object to be grasped and the proximity sensor characteristics are modeled as spherical. This was done to make the equation development, and thus programming, simpler.

References

[1] D. McFalls and E. Franke, *A Robotic Assistant for Space Station Freedom*, **Robotics Today**, vol. 2, no. 2, Second Quarter 1989.

[2] H. Kobayashi, *Grasping and Manipulation of Objects by Articulated Hands*, **1986 IEEE Int'l Conf on Robotics and Automation**, vol. 3, pp. 1514-1519.

[3] G. B. Dunn and J. Segen, *Automatic Discovery of Robotic Grasp Configurations*, **1988 IEEE Int'l Conf on Robotics and Automation**, vol. 1, pp. 396-401.

[4] K. Rao, G. Medioni, H. Liu, and G. A. Bekey, *Robot Hand-Eye Coordination: Shape Description and Grasping*, **1988 IEEE Int'l Conf on Robotics and Automation**, vol. 1, pp. 407-411.

[5] S. A. Stansfield, *Robotic Grasping of Unknown Objects: A Knowledge-Based Approach*, **Sandia Report SAND-1087-UC-32**, Sandia National Laboratories, Albuquerque, New Mexico, June, 1989.

[6] M. Furhman and T. Kanade, *Optical Proximity Sensor Using Multiple Cones of Light for Measuring Surface Shape*, **Optical Engineering**, vol. 23, no. 5, Sept/Oct 1984, pp. 546-553.

[7] A. Romiti and T. Raparelli, *Dynamic Six Component Measurement of Robot Precision*, **Proceedings of the 2nd Int'l Conf on Robotics and Factories of the Future '87**, Springer-Verlag, pp. 497-502.

[8] G. B. Thomas and R. L. Finney, **Calculus and Analytic Geometry**, **6th Edition**, Addison-Wesley, Reading, MA, 1984, pp. 717-718.

[9] R. P. Paul, **Robot Manipulators: Mathematics, Programming, and Control**, The MIT Press, Cambridge, MA, 1981.

Chapter VIII

Motion Specification

Introduction

In the first paper, an approximate path planning algorithm first navigates a point robot in polygonal terrains and returns paths with suboptimal lengths. An innovative four-step trajectory planning algorithm is developed in the second paper for a 3-axis articulated robot in which precise joint-level control can be easily achieved, major mechanical vibration can be avoided, and a well tuned speed can be used to maintain the efficiency. Inverse and forward kinematics for the Stewart Platform-based Manipulator are presented in the third paper.

Minimum-time polynomial manipulator trajectory algorithms based on constrained objective optimization with goal programming are developed in the fourth paper. Bezier curves are used to fit the control points along the manipulator path. An efficient discrete time algorithmic search method is proposed in the fifth paper to find a locally minimum time trajectory for the motion of coordinated robots.

A non-heuristic algorithm is used in the sixth paper to plan collision-free paths for two planar robots working coordinately in an unknown environment. A computational model that incorporates the vision process conformably with the motion control of a robot moving as a dynamic environment is explored in the seventh paper. In the eighth paper, a collision detection system has been implemented that plans a collision free path for a space-based robot manipulator through an environment of moving obstacles. The system has been developed to permit selective and/or non-selective collision checking.

In the ninth paper, a model of free space by multivalue coding allows numerical comparisons to be made at its various grid locations that are helpful for robot path planning. The free space model uses a switching function or a tree representation to which boolean algebra rules and mathematical operations are applied. A data base management system for handling the geometric data encountered for modeling a robot and its environment for off-line programming and path planning is considered in the tenth paper. Workcell information handling and storage are recognized as the key factors in reaching integration between robot motion planning and programming.

Approximate and Hierarchical Path Planning

NAGESWARA S.V. RAO, WENCHENG WU and PAI-SHAN LEE

Department of Computer Science
Old Dominion University
Norfolk, VA

Summary

We consider the problem of planning a path for a point robot from a source point s to a destination point d so as to avoid a set of polygonal obstacles in plane. We are interested in planning paths faster at the cost of settling for sub-optimal paths in terms of the distance traversed. We are also interested in the performance of well-known heuristic path planning algorithms. Using known methods, an optimal path, in terms of the path length, can be computed with a time complexity of $O(n^2)$ where n is the total number of obstacle vertices. We present approximate path planning algorithms based on Voronoi diagrams, trapezoidal decomposition and triangulation, which compute an approximate path in $O(n\sqrt{\log n})$ time with preprocessing costs of $O(n\log n)$, $O(n^2)$ and $O(n\log n)$ respectively. We also show that well-known algorithms of unknown terrains can adapted to the present problem to run in $O(n\log n)$ time. For all our algorithms, we estimate the upperbounds on the lengths of the generated paths as functions of the length of a shortest path, and parameters such as maximum clearance, perimeters of the obstacles, etc. Then we present a hierarchical path planning method which has a worst-case complexity of $O(n\sqrt{\log n})$ for path planning after preprocessing in $O(n\log n)$ time. This algorithm has a good average-case performance in that simple paths can be planned faster than complicated ones.

1. Introduction

Motion planning is one of the vital aspects of mobile robots. Several formulations of this problem have been investigated by a number of researchers [11-14,16-21]. One of the simplest of these formulations involves navigating a point robot in a terrain populated by polygonal obstacles in plane. Despite the simplicity, this problem is generic in that the problem of navigating a translating polygon amidst polygonal obstacles can be reduced to this problem [11]. We assume that the terrain is finite-sized and is populated by a finite number of polygonal obstacles. Each obstacle has a finite number of vertices and is indexed by i, $i=1,2, \cdots ,m$ A shortest path for a point robot from a source position s to a destination position d is known to consist of straight line segments whose end points are obstacle vertices (except for s and d which are the start and the end points of the path). This path can be computed by precomputing the visibility graph of the terrain and invoking a graph shortest path algorithm [11]. The time complexity of computing the visibility graph is $O(n^2)$, where n is the total number of obstacle vertices [20]. Also the current best complexity of shortest path algorithm contains $O(e)$ term, where e is the total number of edges of the graph [1,7]. But, the visibility graph can have $O(n^2)$ edges. Approximate algorithms that have lesser time complexities have been studied by a number of researchers [2,3,5,8,17]. In this paper, we study algorithms that have lesser complexity, typically $O(n\log n)$ and $O(n\sqrt{\log n})$, but compromise on the length of the path. We present a summary of our results in this paper, and a detailed treatment including the proofs of various bounds can be found in our forthcoming report.

We present an approximate algorithm, based on the Voronoi diagram of the terrain, which is a variant of the popular retraction algorithm [14]. We then present algorithms that operate on dual graphs based on trapezoidal decomposition and triangulation of the free-space. The dual graphs used in these cases are planar with $O(n)$ nodes, and a shortest path on such graphs can be computed in $O(n\sqrt{\log n})$ time [6]. We use the notion of "growing" the obstacles to estimate bounds on the lengths of the paths generated by these algorithms. For example, the length of the path generated using the Voronoi diagram is upper bounded by $P^* + (\Pi p + q)\delta_{max} - r\delta_{min}$, where P^* is the length of a shortest path ξ, p is the number of vertices of ξ, r is the number of obstacle edges of ξ, and q is the number of Voronoi cells that ξ runs through. And δ_{max} and δ_{min} are the maximum and minimum clearances respectively (the precise definitions are given in Section 2).

We then consider the adaptations of two simple navigation algorithms of [12] originally proposed for navigating a robot with touch sensing through terrains whose model are not known. Let S_1 be the set of obstacles that intersect the line segment joining s and d and $|S_1| = m_1$. For one algorithm we show that the required path can be computed in $O(m_1 \log m_1 + n)$ time. The length of the computed path is upperbounded by $D + 1/2 \sum_{i \in S_1} p_i$, where p_i is the perimeter of the obstacle i, and D is the straight line distance between s and d. We show a similar result for the second algorithm also.

A number of hierarchical path planning algorithm have been studied in literature [3,8,21]. However the time complexities of these algorithms have not been thoroughly investigated. We propose a path planning algorithm that operates on hierarchical triangulation of the terrain, and computes an approximate path in $O(n\sqrt{\log n})$ time after preprocessing in $O(n\log n)$ time. This algorithm has a good average-case complexity, i.e. simple paths can be computed faster than complicted ones. In Section 2, we discuss the approximate path planning algorithms, and in Section 3, we discuss the hierarchical path planning algorithms.

2. Approximate Path Planning

We first consider the path planning algorithms based on the retraction method based on the Voronoi diagram [14]. Second, we consider path planning based on the dual graphs of triangulation and trapezoidal decomposition of free-space [4,15]. For these cases we propose a method of "growing" obstacles to estimate the upper bounds of on the length of the generated path.

2.1. Retraction-Based Algorithms

Let Ω denote the free-space which is the complement of the set of obstacles. For $x \in \Omega$, we define $Near(x)$ as the set of points that belong to the boundaries of obstacles and are closest to x. The Voronoi diagram, $Vor(O)$, of the terrain O is the set of points $\{x \in \Omega \mid Near(x)$ contains more than one point $\}$. In this case, $Vor(O)$ is a union of n straight lines and parabolic arcs [9,14], which can be specified as a combinatorial graph. Now consider the convex hull $C(O)$ of union of all obstacle vertices of the terrain O. Let $E(O)$ denote the polygonal region obtained by pushing the edges of $C(O)$ outwards by a certain distance s. Let us define $Vor_1(O) = (Vor(\Omega) \cap E(O)) \cup \partial E(O)$, where $\partial E(O)$ is the boundary of $E(O)$ [15]. $Vor_1(O)$ can be computed in $O(n\log n)$ time [9].

A retraction mapping $Im: \Omega \rightarrow Vor(O)$ is defined as follows: If $x \in Vor(O)$ then $Im(x) = x$. If not $Im(x)$ is the intersection point obtained by extending a ray from $Near(x)$ through x until it intersects $Vor(O)$ [14]. A navigtaion path from s to d is computed by computing a path on the graph $Vor(O)$ from $Im(s)$ to $Im(d)$ by employing a depth-first-search algorithm [14]. This path, although computable in $O(n)$ time, could yield the longest path from $Im(s)$ to $Im(d)$ on $Vor(O)$. With a nominally higher cost of $O(n\sqrt{\log n})$ we can obtaina shortest path on $Vor_1(O)$ (it is direct to show that $Vor_1(O)$ is a planer graph).

Let $\delta_{min}=\min\limits_{x\in Vor(O)}\{Clearance(x)\}$ and $\delta_{max}=\max\limits_{x\in Vor(O)}\{Clearance(x)\}$, where $Clearance(x)$ is the distance between x and a point in $Near(x)$. The length of the path generated by the proposed method is upper-bounded as: $P\leq P^*+(\Pi p+q)\delta_{max}-r\delta_{min}$, where P^* is the length of a shortest path ξ, p is the number of vertices of ξ, q is the number of Voronoi cells that ξ runs through, and r is the number of obstacle edges that ξ runs through. This shows that this algorithm yields very good paths in terrains that have lesser clearances inbetween the obstacles. For the special case of rectilinear barriers such that the "free-space corridors" are of the same width, we have a better bound given by $P\leq P^*+\Pi/2p\,\delta_{max}$.

The method used to estimate the bound on P is explained as follows. We first choose a shortest path ξ in the original terrain, and then imagine that each obstacle is "expanded" until its boundary touches the Voronoi diagram (but does not cross it). Then we track the expanded version of ξ and obtain an upper bound on the length of this expanded path, which is definitely an upper bound on the path returned by our algorithm. The same method is used in the next section also and the details of the derivation of these bounds will be given in a forthcoming report.

Let $l$min be the length of a shortest obstacle edge. Then proceeding along the above lines we can show that $P\leq\Pi p\,\delta_{max}+P^*\sqrt{1+\left[\dfrac{l_{min}}{\delta_{max}-\delta_{min}}\right]^2}$. For the case $l_{min}\geq\delta_{max}-\delta_{min}$, we have $P\leq p\,\Pi\delta_{max}+1.414P^*$.

2.2. Decomposition-Based Algorithms

We partition the free-space into polygons with disjoint interiors. We define a dual-graph G, where each polygon is represented by a node and two nodes are connected by an edge if and only if the corresponding polygons share an edge. Note that such a graph is planar.

The trapezoidal decomposition of free space can be computed in $O(n^2)$ time [15] using plane sweep methods. We assume that the sweep line is horizontal. The dual graph can also be constructed during the sweep. It can be shown that the number of trapezoids is $O(n)$ and the dual graph has $O(n)$ edges and vertices. Let δ be an upper bound on the width of the trapeziods and also on the distance between any two obstacle boundary points that are joined by a horizontal segment that lies in free-space. Now we can show that $P\leq P^*+p\sqrt{5}\delta+1/2(q-p)\delta$ where p is the number of vertices, of a shortest path ξ, that support a horizontal tangent and q is the number of trapeziods that ξ runs through. We can also bound P by a function of p, q and δ only as $P\leq[\sqrt{5}/2-1]q\,\delta+[\sqrt{5}+1]p\,\delta$. For the example of rectilinear obstacles with a uniform corridor width of δ, we have $P\leq P^*+1/\sqrt{2}q\,\delta$.

Now consider the dual graphs based on triangulation. Here each triangle contains at most two free-space edges. Each node of the dual graph denotes the mid point of a free-space edge of a triangle. The two nodes corresponding to a traingle are connected by an edge. This dual graph contains $O(n)$ vertices and edge. The bound on the path generated by using this dual graph is given by $P\leq P^*+1/2q\,\delta$, where q is the number of triangles that a shortest path intersects (an intersection could be at a vertex), and δ is an upper bound on the distance between any two vertices that are connected by a straight line that lies entirely inside free-space.

We can also consider a dual such that each node corresponds to the centroid of a triangle and two nodes are connected by a dual edge if and only if the corresponding triangles share a free-space edge. The bound on the length of the path generated by this method is given by $P\leq P^*+\delta(n+q)$, where q is the number of traingles that a shortest path intersects (an intersection could be at a vertex), and δ is an upper bound on the length of a dual graph edge.

We can obtain the constrained triangulation based on the Voronoi diagram in $O(n\log n)$ time [15]. We can also use the other types of decompositions to yield algorithms that are similar to the ones of this section. A survey of algorithms to obtain several types of decompositions can be found in [4].

2.3. Unknown Terrains Algorithms

Two algorithms BUG1 and BUG2 have been proposed in [12] to navigate a point robot in terrains whose model is not a priori known. The automaton uses touch sensing. Note that in our problem the terrain model is known. Let M-line be the line segment joining s and d.

We first consider the algorithm BUG1 which can be described as follows: The automaton or robot R moves along the M-line until it reaches an obstacle at point p. It moves along the boundary until it meets M-line again at a point closer to d than p, then it starts moving along the M-line towards d. This algorithm can be adapted to our case as follows: we compute all the intersection points of the boundary of the obstacles with M-line in $O(n)$ time. Let S_1 be the set of obstacles that intersect the M-line, and $|S_1|=m_1$. Then we sort these points on the M-line in $O(m_1\log m_1)$ time. Then consider the intersection point H_i closest to s and navigate around the boundary of the polygon i to obtain an intersection point L_i closest to d. Then of the two paths from H_i to L_i along the boundary of the polygon i, choose the shorter. Repeat the procedure to navigate from L_i to d. The navigation path is obtained by concatenating the chosen paths around the polygons of S_1 in $O(n)$ time. The total time complexity of this algorithm is given by $O(m_1\log m_1+n)$. The bound on the distance traversed by the robot can be easily seen to be $P\leq P^*+1/2\sum_{i\in S_1}p_i$, where D is the length of the line segment joining s and d, and p_i is the length of the perimeter of the obstacle i.

Consider the second algorithm. For each obstacle we compute a vertex nearest to d, and then proceed sequentially from s. We first obtain the nearest (to s) point of intersection of M-line and an obstacle boundary. Then we compute the shorter path around the obstacle to the point L_i on the obstacle boundary nearest to d. The same procedure is repeated treating L_i as new source point. To support this algorithm we preprocess the terrain such that for each vertex we identify the angular range in which a ray emanating from that vertex into free space intersects the same obstacle (we take the nearest intersection point). We store this information in a balanced binary tree for each vertex. The preprocessing can be performed in $O(n^2)$ time by modifying the algorithm of [20]. Let S_2 be the set of obstacles that the robot meets during the navigation, and $|S_2|=m_2$. Given s and d, for each obstacle the next obstacle can be found in $O(\log n)$ time. By proceeding sequentially from s, the required path can be computed in $O(m_2\log n+n_2)$ time, where n_2 is the total number of vertices of obstacles of S_2. It is straight forward to establish that $P\leq D+1/2\sum_{i\in S_2}p_i$.

3. Hierarchical Path Planning

We first grow the terrain by an infinitesimally small amount such that the connectivity of the free-space is retained. Then we triangulate the free-space into *free* triangles and *obstacle* triangles. Then using this triangulation G we construct a hierarchical structure (similar to that of [10]) as follows. We construct a sequence of triangulations $S_1, S_2, \cdots, S_{h(n)}$, where $S_1=G$, and S_i is obtained from S_{i-1} by (a) removing a set of independent (i.e. nonadjacent) vertices of S_{i-1} and their incident edges; (b) retriangulating the polygons arising from the removal of vertices and edges. After the operation (b) the resultant triangles can partially contain obstacles and such triangles are called *mixed*. For each S_i we store a dual graph whose nodes represent triangles and an edge between two traingles indicates that these two triangles share share an edge that partially lies in free-space. Let n_i be the number of vertices of S_i and the following properties are shown in [10].

Property 1: $n_i = \alpha_i n_{i-1}$ with $\alpha_i \leq \alpha < 1$ for $i = 2, \cdots, h(n)$.

Property 2: Each triangle $R_j \in s_i$ intersects at most H triangles in S_{i-1}, and vice versa.

By property 1, it follows that $h(n) \leq \lceil \log_{1/\alpha} n \rceil = O(\log n)$. And properties 1 and 2 jointly imply that $O(n)$ storage suffices for storing all dual graphs. The hierarchical path planning algorithm tries to plan a shortest path through free triangles at the highest level. If such a path exists, it is returned. If not a shortest path through mixed triangles is found at the current level and it is recursively refined by going deeper into the structure. The correctness of this approach is proved from the following theorem.

Theorem 1: *(i) Two triangles Δ_1 and Δ_2 at level i that share a boundary will have non-empty subset of the boundary that lies in free-space Ω. (ii) In any mixed triangle Δ at level i any two points in $\Delta \cap \Omega$ are connected through a path that runs inside $\Delta \cap \Omega$.*

The time complexity of this algorithm can be shown to be $\dfrac{2n\sqrt{\log n}}{1-\alpha} = O(n\sqrt{\log n})$, which is the same as the time complexity of planning a path at the lowest level. However our algorithm runs in lesser time on the average for the paths that are found at the higher levels. The hierarchical structure can be constructed in $O(n \log n)$ time.

4. Conclusions

We present approximate path planning algorithms for navigating a point robot from point s to d in polygonal terrains. These algorithms have running times of $O(n \log n)$ and $O(n\sqrt{\log n})$ and return paths with suboptimal lengths. Our algorithms are based on Voronoi diagrams, trapezoidal decompositions, traingulations and unknown terrains algorithms. We obtain the upper bounds on the lengths of the paths generated by these algorithms. We then present a herarchical algorithm that plans a path in $O(n\sqrt{\log n})$ time with a preprocessing cost of $O(n \log n)$. This algorithm exhibits good average-case behavior.

References

[1]   R.K. Ahuja, K. Mehlhorn, J.B. Orlin, R.E. Tarjan, Faster algorithms for the shortest path problem, *J. Asso. Comput. Mach.*, 1990.

[2]   A. Basu, J. Aloimonos, Approximate constrained motion planning, *Proc. 1990 IEEE Int. Conf. Robotics and Automation*, 1990, 1833-1838.

[3]   R.A. Brooks, T. Lozano-Perez, A subdivision algorithm in configuration space and findpath with rotation, AI Memo 684, AIlab, MIT, 1983.

[4]   B. Chazelle, Approximation and decomposition of shapes, in *Algorithmic and Geometric Aspects of Robotics*, Eds. J.T. Schwartz and C. Yap, Lawrence Erlbaum Associates, Hillsdale, NJ, 1987, 145-185.

[5]   K.L. Clarkson, Approximation algorithms for shortest path motion planning, *proc. 19th Ann. Symp. on Theory of Computing*, 1987, 56-65.

[6]   G.N. Frederickson, Fast algorithms for shortest paths in planar graphs with applications, *SIAM J. Computing*, 1987, 1004-1022.

[7]   M.L. Fredman, R.E. Tarjan, Fibonacci heaphs and thier uses in improved network optimization algorithms, *Proc. 25th Ann. Symp. Foundations of Computer Science*, 1984, 338-346.

[8]   S. Kambhampati, L.S. Davis, Multiresolution path planning for mobile robots, *IEEE J. Robotics and Automation*, vol. 2, 1986, 135-145.

486

[9]   D.G. Kirkpatrick, Efficient computation of continuous skeletons, *Proc. 20th Ann. Symp. Foundations of Computer Science*, 1979, 18-27.

[10]   D.G. Kirkpatrick, Optimal search in planar subdivisions, *SIAM J. Computing*, vol. 12, 1983, 28-35.

[11]   T. Lozano-Perez, M.A. Wesley, An algorithm for planning collision-free paths among polyhedral obstacles, *Commun. ACM*, vol. 22, 1979, 560-570.

[12]   V.J. Lumelsky, A.A. Stepanov, Path-planning strategies for a point mobile automaton moving amidst unknown obstacles of arbitrary shape, *Algorithmica*, vol. 2, 1987, 403-430.

[13]   J.S.B. Mitchell, An algorithmic approach to some problems in terrain navigation, *Artificial Intelligence*, vol. 37, 1988, 171-201.

[14]   C. O'Dunlaing, C.K. Yap, A 'Retraction' method for planning the motion of a disc, *J. Algorithms*, vol. 6, 1985, pp. 104-111.

[15]   F.P. Preparata, M.I. Shamos, *Computational Geometry: An Introduction*, Springer-Verlag, 1985.

[16]   N.S.V. Rao, An algorithmic framework for navigation in unknown terrains, *IEEE Computer*, June 1989, pp. 37-43.

[17]   N.S.V. Rao, S.S. Iyengar, C.C. Jorgensen, C.R. Weisbin, Robot navigation in an unexplored terrain, *J. Robotics Systems*, vol. 3, 1986, 389-407.

[18]   J.T. Schwartz, J.E. Hopcroft, M. Sharir (Eds), *Planning, Geometry and Complexity of Robot Motion*, ABlex Pub. Co., Norwood, NJ, 1986.

[19]   M. Shair, Algorithmic motion planning in robotics, *IEEE Computer*, March 1989, 9-20.

[20]   E. Welzl, Constructing the visibility graph for n-line segments in $O(n^2)$ time, *Inform. Process. Lett.*, vol.20, 1985, 167-171.

[21]   D.Zhu, J.C. Latombe, Constraint reformulation in a hierarchical path planner, *Proc. 1990 IEEE Int. Conf. Robotics and Automation*, 1990, 1918-1923.

Bandlimited Trajectory Planning for Continuous Path Industrial Robots

J.T. HUANG

University of South Western Louisiana
Lafayette, LA

Summary

This paper provides a trajectory planning algorithm for high speed, high precision continuous path robots. The basic idea is to tune the speed along the path according to the curvature, so the generated velocity and acceleration commands are bounded and continuous. Frequency range of the velocity and acceleration commands is also controllable.

This proposed planning algorithm can be summarized in four steps. The first step is to fit the Cartesian trajectory by polynomials. In step two, the radius of curvature along the path is obtained, and a maximum velocity along with a cut-off frequency are also chosen. In the third step, a digital filter corresponding to the cut-off frequency is applied to smooth the velocity commands. Finally the defined speed is converted into Cartesian position commands in step four. A physical simulation using 3-axes robot is conducted to show the advantages of this method.

Introduction

In recent years the articulated continuous path robots have been increasing used in automated manufacturing. A typical application is the seam sealing process for car assembly. Currently most common approach in path planning is to use simple trajectories, such as a straight line or a circular arc. The motion of the manipulator arm is controlled by constant acceleration and deceleration (hereafter only acceleration is mentioned) at starting and destination points, and a constant travel speed is assigned in between [1]. This method limits the maximum acceleration as well as the maximum speed in Cartesian space. It has demonstrated reliable performance for simple trajectories. The problem usually happens in using this method is that the suddenly applied acceleration force may result in unexpected structural vibration. For an arbitrary path, this type of robot should be kept at a very low speed through the entire path in order to avoid significant tracking error, especially in the abrupt changing portion of the path. Therefore, it is too inefficient to be applied in continuous path planning.

To improve the efficiency, Shin and McKay [2] developed a minimum-time control criterion to achieve the maximum efficiency by applying full acceleration capability along the entire path. In their study, the minimization of traveling time is the only concern. Several important factors, such as the nonlinear stiffness of the transmission mechanism and the structural dynamics of the manipulator arm, are neglected. Without considering these factors, the suddenly changed acceleration may cause serious mechanical vibration, and the nonlinear response of the transmission will make the system characteristics unpredictable. Achieving a precise joint control then becomes very difficult [3].

The objective of this paper is to develop an innovative trajectory planning algorithm in which a precise control in joint-level can be easily achieved, major mechanical vibration can be avoided, and a well tuned speed can be used to maintain the efficiency. In order to meet these conditions, the following preliminary design criteria are derived.

- Velocity is continuous and bounded in Cartesian space.
- Acceleration is continuous and bounded in Cartesian space.
- Frequency ranges of velocity and acceleration commands are bounded and the cut-off frequencies are less than the major resonance frequency of the structure.

This method is based on the 3-axes articulated robot. The reason to simplify the structure of a 6-axes articulated robot into a 3-axes is that the first three axes dominate the dynamic behavior of the entire robot. The last three axes can be easily added to this configuration to reach the desired orientation. Figure 1 shows its configuration and the arbitrary path.

Theoretical Background

The following theories provide the basis of the proposed trajectory generation method. They are also proven in this section.

- Theory #1 : A bandlimited trajectory in Cartesian space is also bandlimited in joint space.

- Theory #2 : A digital filter provides the same frequency range no matter it is applied forward or backward.

Proof of Theory #1 :

The velocity vectors in the joint space trajectory $\underline{\Delta\theta}$ and in Cartesian space $\underline{\Delta P}$ are related by a Jacobian matrix [J].

$$\underline{\Delta P} = [J] \ \underline{\Delta\theta} \tag{1}$$

Each element of [J] matrix relates a joint velocity vector to a Cartesian velocity vector. It is a

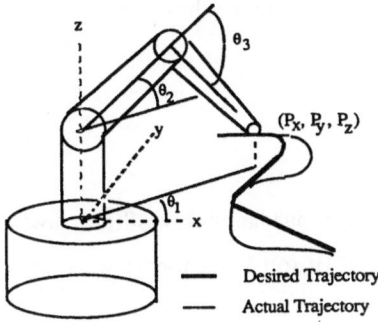

Figure 1 Configuration of the 3-Axes
Articulated Robot

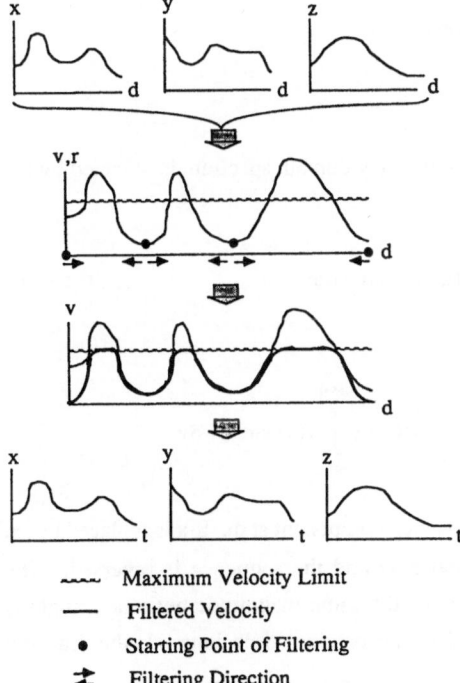

Maximum Velocity Limit
Filtered Velocity
● Starting Point of Filtering
⇕ Filtering Direction

Figure 2 Signal Processing of the Algorithm

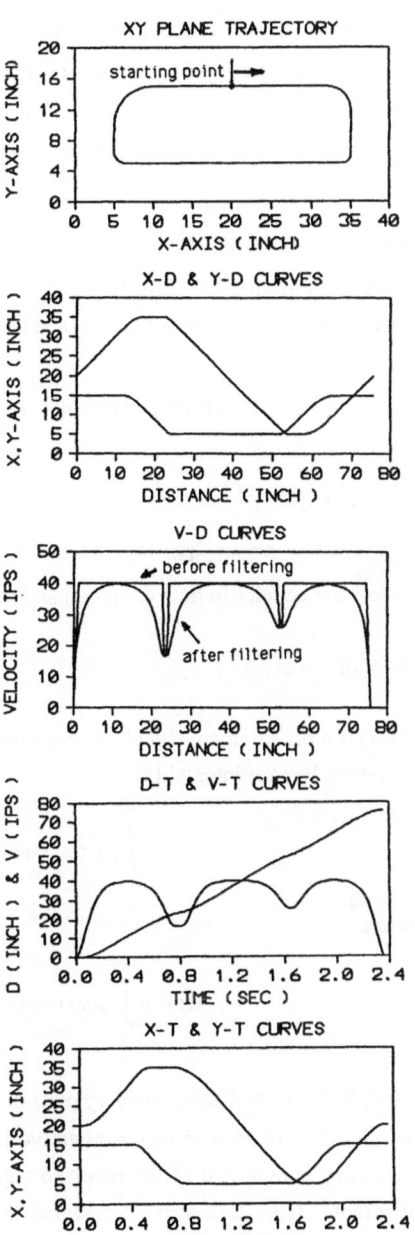

Figure 3 Results of the Example

linear time domain function and can be converted into S domain by LaPlace transform. In S domain, if S is replaced by $j\omega$ then (1) is expressed as

$$\underline{\Delta P(j\omega)} = [J(j\omega)] \; \underline{\Delta\theta(j\omega)} \tag{2}$$

Since the robot discussed in this paper does not include any redundant link, a $\underline{\Delta\theta(j\omega)}$ always maps to an unique $\underline{\Delta P(j\omega)}$. It proves the nonsingularity of $[J(j\omega)]$ in the robot's work space. Therefore, the inverse matrix exists and (2) can be written as

$$\underline{\Delta\theta(j\omega)} = [J(j\omega)]^{-1} \; \underline{\Delta P(j\omega)} \tag{3}$$

Because $[J(j\omega)]^{-1}$ is not singular, a zero joint velocity vector $\underline{\Delta\theta(j\omega_0)}$ at a particular frequency ω_0 gives a zero Cartesian velocity vector $\underline{\Delta P(j\omega_0)}$. This result means that a robot has the same cut-off frequencies in both Cartesian and joint spaces.

Proof of Theory #2 :

If $f(x)$ is a time domain signal, the corresponding frequency domain spectrum is obtained by the following Fourier integral [4] :

$$f(x) = \int_0^\infty [\, A(\omega) \cos \omega x + B(\omega) \sin \omega x \,] \, d\omega \tag{4}$$

where

$$A(\omega) = \int_{-\infty}^{+\infty} f(v) \cos \omega v \, dv, \qquad B(\omega) = \int_{-\infty}^{+\infty} f(v) \sin \omega v \, dv$$

$A(\omega)$, $B(\omega)$ are real and imaginary parts of the frequency component at ω. If x is replaced by -x, the signal is mirrowed to the negative horizontal axis and the sequence is reversed. The frequency component of the reversed signal remains the same magnitude but phase angle is different. Then this mirrowed signal is shifted to the original time interval, the real and imaginary parts $A(\omega)$ and $B(\omega)$ are not affected at this step. This proves Theory #2.

Proposed Method

The proposed method is divided into four steps. The implementation of each step is discussed in this section. Figure 2 shows the data processing in each step.

Step #1 : Sample the desired trajectory in Cartesian space. The x, y and z position components are recorded in every distance increment Δd. The x-d, y-d and z-d curves are obtained at this step. An on-line or off-line curve fitting [5] is applied to compress the size of the data representing the three curves so that memory size for data storage can be reduced.

Step #2 : Find the radius of curvature r for each point of the trajectory to obtain the r-d curve. Define the cut-off frequency then convert the r-d curve into velocity vs. distance (v-d) curve. Search the starting points and define their directions for data filtering.

Step #3 : Apply a digital filter to smooth the v-d curve in step #2.

Step #4 : Based on the v-d curve, compute the traveled distance d at time t. Find the corresponding Cartesian coordinates x, y and z for each distance d then construct x-t, y-t and z-t curves.

In step #2, the maximum allowed velocity in Cartesian space is limited by the maximum angular velocities of the three joint actuators. Obviously the cut-off frequency should be bounded by the resonance frequency. Since the maximum velocity and a cut-off frequency can also define the maximum acceleration, a lower cut-off frequency should be selected if joint actuators cannot reach that acceleration.

The purpose of defining v-d curve by r-d curve is to tune the speed according to the radius of curvature. When a robot moves along a circle at constant speed, the frequency is decided by the radius of the circle. If the radius is greater than a certain value, the full speed can be assigned while the frequency range is still maintained. Based on this concept, a r-d curve is converted proportionally into a v-d curve by finding a corresponding maximum radius of curvature r_{max}. The converted velocity curve is then limited by the maximum velocity v_{max}. This v-d curve so far is not bandlimited in time domain. To remove high frequency components, a digital filter is needed. In order to apply the digital filter, this curve is divided into several segments. Each segment starts from one of the minimum points of the curve and ends at an adjacent maximum point or the straight line defining the maximum velocity. The signal filtering direction is always from the minimum to the maximum point. Thus the processing sequence of the segment with negative slopes should be reversed.

Because the velocity curve now gives the envelope of the bandlimited velocity curve, the filtered velocity should be bounded by this curve. Therefore, the digital filter applied in step #3 should be overdamped to avoid any overshoot [6]. Usually a digital filter with higher attenuation beyond cut-off frequency is applied to limit the acceleration frequency range. Once the bandlimited velocity curve is obtained, the traveled distance vs. time (d-t) curve can be constructed. The distance in d-t curve is mapped to the x-d, y-d and z-d curves in step #1 to obtain the position commands x-t, y-t and z-t curves.

Case Study

An xy plane trajectory in Figure 3 is taken as the example and analyzed to show the advantages of this method. The rectangular trajectory has four corners with radius 3", 1", 1.5" and 5" respectively. A robot with a maximum speed 40 ips (inch per second) and cut-off frequency 3 Hz is commanded to follow this trajectory. The total trajectory length is 75.5 inches. To meet the 3 Hz frequency requirement, a speed less than 18.8 ips should be set for the 1" radius corner. Including the time for acceleration and deceleration at starting and destination ends, the total time consumption is more than 4 seconds if robot with a constant speed is applied.

The proposed method is then applied in this case. The x-d and y-d curves in Figure 3 are constructed at the first step. The velocity boundary defined by the maximum speed and the radius of curvature of the path is processed by an overdamped filter. Its filtered velocity is shown in the v-d curve. The d-t and v-t curves show the smooth of the robot's position and velocity commands. Finally, the x-t and y-t curves defined the trajectory commands along the two directions. The total time consumption is 2.13 seconds which almost twice as faster as the constant speed algorithm. Most importantly, the frequency range of the velocity commands is limited.

References

1.   John J. Craig, Introduction to Robotics -- Mechanics & Control, Addison-Wesley, 1986.

2.   Kang G. Shin & Neil. D. McKay," Minimum-Time Control of Robotic Manipulators with Geometric Path Constraints ," IEEE Transaction Automatic Control, June 1985.

3.   J. T. Huang, On-Line Self-Tuning Adaptive Control for Industrial Robots, Ph.D. Thesis, University of Wisconsin-Madison, 1987.

4.   Erwin Kreyszig, Advanced Engineering Mathematics, 3rd Ed., John Wiley & Sons,1972.

5.   Curtis F. Gerald, Applied Numerical Analysis, 2nd Ed., Addison-Wesley, 1978.

6.   G. F. Franklin and J. D. Powell, Digital Control of Dynamic System, Addison-Wesley, 1980.

Trajectory Planning and Kinematic Control of a Stewart Platform-Based Manipulator

CHARLES C. NGUYEN, SAMI S. ANTRAZI and ZHEN-LEI ZHOU

Robotics and Control Laboratory
Department of Electrical Engineering
Catholic University of America
Washington, D.C.

Summary

This paper deals with the trajectory planning and control of a robot manipulator that has 6 degrees of freedom and was designed based on the mechanism of the Stewart Platform. First the main components of the manipulator will be described and its operation will be explained. We then briefly present the solutions for the forward and inverse kinematics of the manipulator. After that, two trajectory planning schemes will be developed using the manipulator inverse kinematics to track straight lines and circular paths. Finally experiments conducted to study the performance of the developed planning schemes in tracking a straight line and a circle will be presented and discussed.

1 Introduction

Successful robotic assembly of parts requires that the robot end-effector be able to perform very precise motion. As a result, research effort has been enormously spent [1,12] to study the application of closed-kinematic chain (**CKC**) mechanism in the design of robot manipulators performing high precision motion because CKC mechanism generally provides better positioning capability than open-kinematic chain (**OKC**) mechanism. CKC mechanism was first implemented in the design of the Stewart Platform [1] originally intended for simulating aircraft motion. Later, the Stewart platform attracted considerable interest of robotic researchers and was proposed in the design of several robot manipulators and end-effectors [2,8]. Hunt [2,3] developed numerous structural designs for robot manipulators using the mechanism of the Stewart Platform which initiated intensive research in [4,5]. Sugimoto and Duffy [6] employed the theory of linear algebra and screw systems to obtain a general method which describes the instantaneous link motion of a single closed-loop mechanism. Yang and Lee [7] studied the feasibility of manipulators designed based on the Stewart Platform whose inverse dynamics and kinematics were later investigated by Do and Yang [8] and Sugimoto [9]. Nguyen and Pooran applied the Lagrangian formulation to derive dynamical equations for a six-degree-of-freedom (**DOF**) CKC manipulator [10] whose kinematics was also developed in [11]. Later, they derived a learning-based control scheme for the above manipulator to perform repetitive tasks [12].

A Stewart platform-based manipulator was designed and built at the Goddard Space Flight Center (**GSFC**) [13] to serve as a testbed for evaluating the feasibility of autonomous assembly of parts in space. In this paper, we first describe the main components of the manipulator and explain its operation. We then briefly review the forward and inverse kinematics of the manipulator. After that, a Cartesian trajectory planning scheme is developed and employed with a trajectory control scheme using the manipulator inverse kinematics. Finally experiments conducted to study the performance of the developed trajectory planning and control scheme are presented and discussed.

2 The Robot Manipulator

This section is devoted to briefly describe the main components of the manipulator and present the solutions for its inverse and forward kinematics.

Figure 1: The robot manipulator.

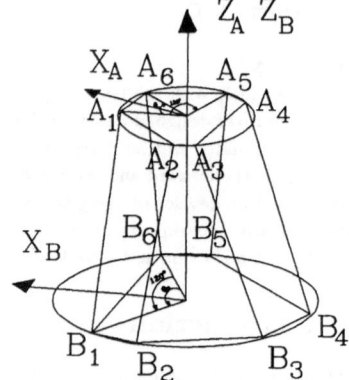

Figure 2: Frame assignment for platforms.

2.1 Hardware Description

Figures 1 and 2 show the robot manipulator which was designed based on the Stewart Platform mechanism and mainly consists of a lower base platform, an upper payload platform, a compliant platform, a gripper and six linear actuators. The movable payload platform is supported above the stationary base platform by six axially extensible rods with ballnuts and ballscrews providing the extensibility. Stepper motors were selected to drive the ballscrews to extend or shorten the actuator lengths whose variations will in turn produce the motion of the payload platform and consequently the motion of a gripper attached to the compliant platform. Each end of the actuator links is mounted to the platforms by 2 rotary joints whose axes intersect and are perpendicular to each other. Passive compliance is provided through the compliant platform, which is suspended from the payload platform by six spring-loaded pistons arranged in the Stewart Platform mechanism. The compliance is passively provided by permitting strain on two opposing springs acting in the pistons. Thus the pistons are compressed and extended when resistive and gravitational forces are applied on the gripper. The rotation of each stepper motor is controlled by sending out proper commands to an indexer which then transmits proper pulse sequences to the stepper motor drive. Therefore, precise gripper motion can be produced by properly controlling the motions of six manipulator legs. Figure 2 presents the planning and control scheme employed to control the motion of the manipulator gripper. A Cartesian path specified with desired starting and ending velocities and accelerations is converted into 6 Cartesian trajectories using a Cartesian trajectory planning scheme. Then based upon the desired Cartesian trajectories, joint-space trajectories will be determined by a planner which sends proper commands through the RS232 port of a personal computer to the indexers. The indexer will then transmit pulses to the stepper motor drives where microstepping permits each revolution (360^0) of the stepper motor to be equivalent to 25,000 steps. Therefore the drive rotates the stepper motor one angular increment of $\frac{360^0}{25,000} = 0.0144^o$, each time it receives one step pulse. Furthermore, through the linear motion converter system consisting of the ballnut and the ballscrew, each angular increment (=1step) is converted into 8 μ-inches of linear translation of the manipulator leg.

2.2   Inverse Kinematic Solution

The lengths of the manipulator legs are selected as joint variables. To define the Cartesian variables we proceed to assign coordinate frame {A} to the movable payload platform and {B} to the base platform as shown in Figure 2. The position and orientation of Frame {A} with respect to Frame {B} are selected as the Cartesian variables in the sense that the position of Frame {A} is the position of its origin with respect to Frame {B}. Now denoting the angle between AA_i and x_A by λ_i, and the angle between BB_i and x_B by Λ_i for i=1,2,...,6, we have obtain $\Lambda_i = 60(i-1)^\circ$; $\lambda_i = 60(i-1)^\circ$ for i=1,3,5 and $\Lambda_i = \Lambda_{i-1} + \theta_B$; $\lambda_i = \lambda_{i-1} + \theta_A$ for i =2,4,6. Furthermore, if Vector $^A\mathbf{a}_i = (a_{ix}\ a_{iy}\ a_{iz})^T$ describes the position of the attachment point A_i with respect to Frame {A}, and Vector $^B\mathbf{b}_i = (b_{ix}\ b_{iy}\ b_{iz})^T$ the position of the attachment point B_i with respect to Frame {B}, then they can be written as $^A\mathbf{a}_i = [\ r_A cos(\lambda_i)\ \ r_A sin(\lambda_i)\ \ 0\ ]^T$ and $^B\mathbf{b}_i = [\ r_B cos(\Lambda_i)\ \ r_B sin(\Lambda_i)\ \ 0\ ]^T$ for i=1,2,...,6 where r_A represents the radius of the payload platform and r_B that of the base platform.

We proceed to consider the vector diagram for an ith actuator given in Figure 4. The length vector $^B\mathbf{q}_i = (q_{ix}\ q_{iy}\ q_{iz})^T$, expressed with respect to Frame {B} can be computed by

$$^B\mathbf{q}_i = {}^B\mathbf{a}_i - {}^B\mathbf{b}_i \tag{1}$$

where Vector $^B\mathbf{a}_i$ and Vector $^B\mathbf{b}_i$ describe the positions of A_i and B_i, respectively both in terms of Frame {B}. However, $^B\mathbf{a}_i$ can be computed by

$$^B\mathbf{a}_i = {}^B_A\mathbf{R}\ {}^A\mathbf{a}_i + {}^B\mathbf{d} \tag{2}$$

where $^B_A\mathbf{R}$ is the matrix representing the orientation of Frame {A} with respect to Frame {B} and Vector $^B\mathbf{d}$ contains the Cartesian coordinates x, y, z of the origin, A of Frame {A} with respect to Frame {B} such that $^B\mathbf{d} = [\ x\ \ y\ \ z\ ]^T$.

Now substituting (2) into (1) and using $l_i = \sqrt{q_{ix}^2 + q_{iy}^2 + q_{iz}^2}$, we obtain after intensive simplifications:

$$
\begin{aligned}
l_i^2 &= x^2 + y^2 + z^2 + r_A^2 + r_B^2 + 2(r_{11}a_{ix} + r_{12}a_{iy})(x - b_{ix}) \\
&\quad + 2(r_{21}a_{ix} + r_{22}a_{iy})(y - b_{iy}) + 2(r_{31}a_{ix} + r_{32}a_{iy})z - 2(xb_{ix}),
\end{aligned} \tag{3}
$$

for i=1,2,...,6 where r_{ij} for i,j=1,2,3 are the elements of the matrix $^B_A\mathbf{R}$.

Equation (3) represents the solution to the inverse kinematic problem in the sense that for a given Cartesian configuration, composed of the position and orientation specified by $^B\mathbf{b}_i$ and $^B_A\mathbf{R}$, respectively, the actuator lengths l_i for i=1,2,...,6, can be computed using (3).

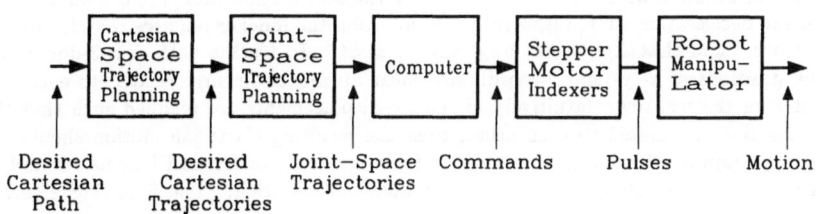

Figure 3: Trajectory planning and control of the robot manipulator.

2.3   Forward Kinematic Solution

The forward kinematics concerns with the determination of the Cartesian position and orientation of the payload platform when the actuator lengths l_i for i=1,2,...,6, are given. Consequently, the forward kinematic problem can be formulated as to find x, y, z (position) and r_{ij}

496

(orientation) for i=j=1,2,3 to satisfy Equation (3) for a given set of l_i for i=1,2,...,6. Thus the problem is reduced to solving 6 highly nonlinear simultaneous equations with 6 unknowns, 3 of which represent the orientation. A closed-form solution does not generally exist, which leads us to seek an iterative numerical method to solve the above set of nonlinear equations. One widely used technique for solving nonlinear equations is the Newton-Raphson method, which is employed in this paper to solve the forward kinematic problem.

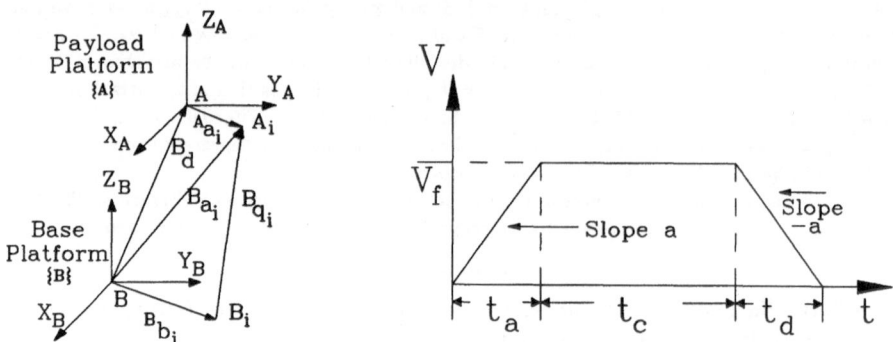

Figure 4: Vector diagram for the ith leg.   Figure 5: Trapezoidal velocity profile.

3 Trajectory Planning and Kinematic Control

Two trajectory planning schemes developed to control the motion of the robot manipulator gripper are presented in this section. The first scheme is designed for tracking straight lines while the second for arbitrary paths.

3.1 Straight-Line Trajectory Planning Scheme

The indexer has two main modes of operation: the *normal mode* and the *continuous mode*. In the normal mode, based on the information about the velocity v_f, acceleration a, and the distance to be traveled Δ_l, the indexer will determine the appropriate leg velocity profiles which are either a trapezoid or a triangle depending on the relationship between the given information. For simplicity, the straight-line planning scheme will only utilize the trapezoidal profile which is illustrated in Figure 5 where t_a, t_c, and t_d denote the acceleration time, the constant velocity time, and the deceleration time, respectively. In addition the indexer requires that $t_a = t_d$. By inspection, we found that $\Delta_l = v_f(t_c + t_d)$ and $v_f = at_a$. To track a path in a 3-dimensional space, the positions of x- y- and z-coordinates must always be linearly related to each other anytime during the tracking. Intuitively, if the leg displacements are planned such that their velocities are linearly related to each other, then the resulting Cartesian motion should be a linear path. Computer simulation which utilized the manipulator forward kinematics and was performed to verify the above fact has agreed with our intuition. The following algorithm facilitates the trajectory planning for straight lines.

Algorithm 1: Straight Line Trajectory Planning

1. Use the manipulator inverse kinematics to compute the leg lengths corresponding to the starting point P_s and the final point P_f of the straight line, namely l_{is} and l_{if} for i=1,2,...,6.

2. Compute $\Delta_{li} = l_{if} - l_{is}$ for i=1,2,...,6 and find Δ_{lk} whose absolute value is the biggest.

3. Select a_k and v_{fk} for the k-th leg such that $a_k \leq a_{max}$; and $v_{fk} \leq v_{max}$; $v_{fk} \leq a_k\sqrt{|\Delta_{lk}|}$ to ensure trapezoidal profile where a_{max} and v_{max} denote the maximum acceleration and velocity of the stepper motor, and then compute $t_a = \frac{v_{fk}}{a_k} = t_d$ and $t_c = \frac{\Delta_{lk}}{v_{fk}} - t_a$.

4. For $i \neq k$; i=1,2,..,6 compute $a_i = \frac{\Delta_{li}}{t_a(t_a+t_c)}$ and $v_{fi} = t_a a_i$.

3.2 Trajectory Planning Scheme For Arbitrary Curves

In the *continuous mode* of operation, the indexer requires the acceleration a, final velocity v_f and the direction of the rotation (direction of the linear displacement) of the stepper motor. The stepper motor will accelerate to the velocity v_f and continue to run at this velocity until new velocity and acceleration are given in the same direction of rotation. The current planning mainly consists of dividing a space curve into n segments and then planning the velocity profiles of the manipulator legs in the continuous mode so that each segment will be reached within a specified time. The following algorithm is proposed to plan for arbitrary space curves.

Algorithm 2: Arbitrary Curve Trajectory Planning

1. Divide the space curve into n segments.

2. Use the manipulator inverse kinematics to compute the leg lengths corresponding to each segment point on the curve, namely l_{ij} for i=1,2,...,6 (leg number) and j=1,2,...,n+1 (segment point number)

3. Compute $\Delta_{ij} = l_{i,j+1} - l_{i,j}$ for i=1,2,...,6 and j=1,2,...n.

4. For each segment, select an appropriate travel time t_j for j=1,2,...,n, and compute the corresponding acceleration and final velocity at the end of each segment.

In general, the travel times for the segments are constant and equal to each other during the tracking of curves which do not require the change of leg direction. However when direction of any leg has to change, the travel time can be selected efficiently using the *look ahead method*. Using this method, the algorithm looks at the next segment point and determine if any change in leg direction is necessary. For example if the direction of a leg requires direction change, then its travel time will be recomputed to ensure that the velocity at the end of the segment will be zero to allow direction change. After that, the recomputed travel time will be set for the remaining legs for the next two segments. Finally the travel time of all legs will be set back to the old value before the leg direction change occurs. The above process can be repeated any time a leg direction change is necessary.

4 Experiments

This section presents the results of experiments conducted to evaluate the performance of the developed trajectory planning schemes. In particular, Algorithm 1 was used to track a straight line in the x-y plane of the base frame {B}, which is described by $y(t) = x(t)$. The required profiles of the manipulator legs required to track the straight line were determined before the tracking using Algorithm 1 and are illustrated in Figure 6. The path that the manipulator gripper actually tracked, is presented in Figure 7 together with the desired path. Figure 7 shows excellent tracking capability with insignificant errors. Algorithm 2 was applied to compute the leg velocity profiles shown in Figure 8 to track a circle in the x-y plane of the {B} with a radius of 0.8 inches. Figure 9 shows the desired path plotted together with the path the manipulator actually tracked. The average errors for x- y- and z-coordinates were computed to be 1.096×10^{-4} inches, 2.326×10^{-4} inches, and 2.974×10^{-5} inches, respectively.

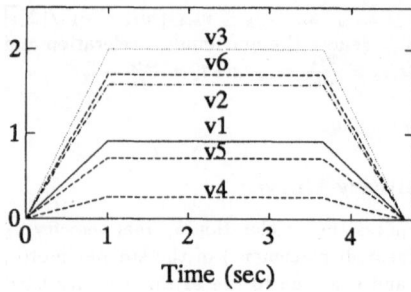

Figure 6: Leg velocity profiles (Alg. 1).
vertical axis=velocity (rev/sec)

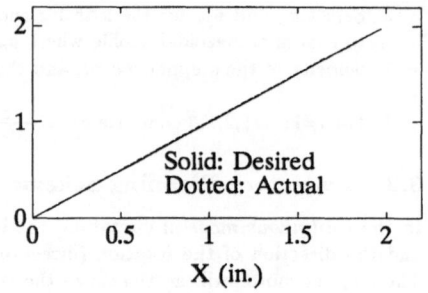

Figure 7: Tracking a straight line.
vertical axis=y-axis (inch)

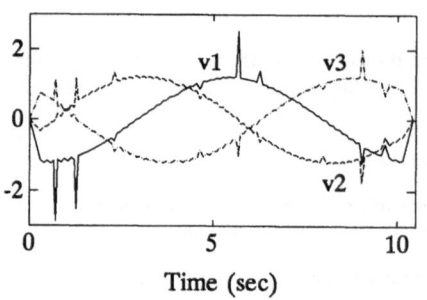

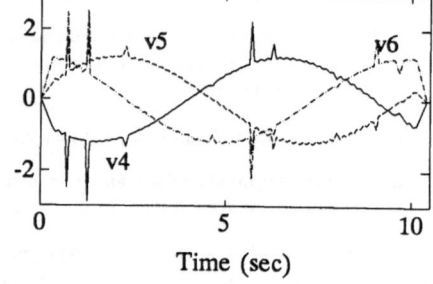

Figure 8: Velocity profiles for 6 legs (Algorithm 2).
vertical axis=velocity (revolution/sec)

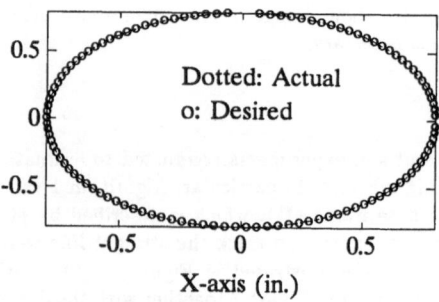

Figure 9: Tracking a circular path.

5 Conclusion

The problem of trajectory planning for a Stewart Platform-based manipulator was addressed in this paper. We first described the main components of the manipulator and briefly presented the development of its inverse and forward kinematics. We then employed the normal and continuous modes of operation of the stepper motor indexer to derive two trajectory planning schemes for tracking straight paths and circular paths, respectively. Experiments performed to evaluate the developed algorithms showed excellent tracking results. Future research will focus on investigating the application of the algorithms to carry out assembly tasks such as mating space connectors with or without force and position feedback. In addition, development of a 6 DOF force sensor system based on the mechanism of the Stewart Platform is under way.

6 Acknowledgment

The research presented in this paper was conducted under a research grant, NAG 5-780 funded by the Goddard Space Flight Center. The authors are grateful to Dr. Charles E. Campbell, Jr. at Goddard (NASA) for helpful suggestions for the optimization of the leg velocity profiles.

References

[1] Stewart, D., "A Platform with Six Degrees of Freedom," *Proc. Institute of Mechanical Engineering*, vol. 180, part 1, No. 5, pp. 371-386, 1965-1966.

[2] Hunt, K. H., *Kinematic Geometry of Mechanisms*, Oxford University, London 1978.

[3] Hunt, K. H., "Structural Kinematics of in-parallel-actuated Robot Arms," *Trans. ASME, J. Mech., Transmis., Automa. in Des.*, Vol. 105, pp. 705-712, 1983.

[4] Fichter, E.F. and MacDowell, E.D., "A Novel Design for a Robot Arm," *ASME Int. Computer Technology Conference*, San Francisco, pp. 250-256, 1980

[5] Fichter, E.F., "A Stewart Platform-Based Manipulator: General Theory and Practical Construction," *Int. Journal of Robotics Research*, pp. 157-182, Summer 1986

[6] Sugimoto, K. and Duffy, J., "Application of Linear Algebra to Screw Systems," *Mech. Mach. Theory*, Vol. 17, No. 1, pp. 73-83, 1982.

[7] Yang, D. C. and Lee, T. W., "Feasibility Study of a Platform Type of Robotic Manipulators from a Kinematic Viewpoint," *Trans. ASME Journal of Mechanisms, Transmissions, and Automation in Design*, Vol. 106. pp. 191-198, June 1984.

[8] Do, W.Q.D. and Yang, D.C.H., "Inverse Dynamics of a Platform Type of Manipulating Structure," *ASME Design Engineering Technical Conference*, Columbus, Ohio, pp. 1-9, 1986.

[9] Sugimoto, K., "Kinematic and Dynamic Analysis of Parallel Manipulators by Means of Motor Algebra," *ASME Journal of Mechanisms, Transmissions, and Automation in Design*, pp. 1-5, Dec. 1986.

[10] Nguyen, C.C., Pooran, F.J., "Dynamic Analysis of a 6 DOF CKCM Robot End-Effector for Dual-Arm Telerobot Systems," *Journal of Robotics and Autonomous Systems*, Vol. 5, pp. 377-394, 1989.

[11] Nguyen, C.C., Pooran, F.J., "Kinematic Analysis and Workspace Determination of a 6 DOF CKCM Robot End-Effector" *Journal of Mechanical Technology*, Vol. 20, pp. 283-294, 1989.

[12] Nguyen, C. C., Pooran, F.J., "Learning-Based Control of a Closed-Kinematic Chain Robot End-Effector Performing Repetitive Tasks", to appear in *Journal of Microcomputer Applications*, Vol. 8, No. 3, 1989.

[13] Premack, Timothy et al, "Design and Implementation of a Compliant Robot with Force Feedback and Strategy Planning Software," *NASA Technical Memorandum 86111*, 1984.

Planning and Execution of Polynomial Manipulator Trajectories

FRED BAREZ

Mechanical Engineering Department
San Jose State University
San Jose, CA

Summary

A method is presented to plan and executed minimum-time manipulator
trajectories for predefined Cartesian end-effector path in a workspace
containing obstacles. The method employs a polynomial path
configuration function along with a dynamic search approach with goal
programming. The Cartesian path of the manipulator is represented
with Bezier Polynomials connecting the initial and the final positions
of the end-effector. The minimum-time trajectory planning problem is
formulated as the problem of minimizing the total travel time subject
to constraints on joint positions, velocities, accelerations, jerks,
torques and end-effector accelerations. An efficient algorithm is
used to determine the minimum-time path close to the given path. A
numerical example for a two-degree of freedom manipulator is presented
to demonstrate the application of this method.

Introduction

Trajectory planning is an important off-line process which is
concerned with the generation of a time history of a manipulator's
joint position, velocity, acceleration, jerk, and input torque.
Manipulator trajectory defines the time sequence of intermediate
configurations in a desired motion along a path. An executable
trajectory should require the end-effector of a manipulator to move to
a point inside its workspace or move with a velocity, acceleration or
joint torque that is physically possible. The trajectory optimization
is to find a trajectory function which satisfies the given
constraints and minimizes the measure of the total traveling time.
Therefore, the minimum-time path planning should be considered since
infinitely many collision-free paths are usually possible and the
shortest distance path does not necessarily produce the shortest
traveling time.

A number of techniques have been developed for planning minimum-time trajectories of robotics manipulators[1], [2], which usually include complex computations. Some of these methods make fairly specific assumptions about the form of the joint torque/force constraints, thereby limiting their applicability [3], [4].

More recently many researchers have addressed the minimum-time problem of moving a manipulator along a prescribed path in a Cartesian coordinate. Bobrow, Dubowsky, and Gibson[5] used a parametric function to define the geometric path and applied the phase-plane technique to solve the problem. Sahar and Hollerbach [6] employed the time optimal path search method by using a dynamic time scaling and joint tessellations to successfully analyze a two-degree of freedom manipulator. Lin and Chang[7], and Rajan[8] used kinematic approach along with cubic spline functions to develop a time-scheduling technique to minimize the traveling time of a manipulator subject to physical constraints on velocity, acceleration, and jerk for each point. Bobrow[9], and Barez[10] used approximation curve techniques to execute minimum-time manipulator trajectories in a Cartesian coordinate. Chen and Peikari[11] developed an algorithm based on parametrization method to reduce the dimensionality of the problem while manipulating the end-effector along a uniform Beta spline.

In this paper the design of minimum-time polynomial trajectories for robotic manipulators in Cartesian coordinates is discussed. The geometric path is defined by means of Bezier curves and the manipulator is subject to various constraints including joint positions, velocities, accelerations, jerks, and torques, and the constraint on end-effector acceleration.

Dynamic Model of the Manipulator

The dynamic model of the manipulator is obtained using the Newton-Euler dynamic modeling formulation. This method is independent of the type of manipulator configuration. When Newton-Euler equations are applied, they yield and equation of motions for the manipulator which can be written as [12]:

$$T = M(\theta)\ddot{\theta} + H(\theta,\dot{\theta}) + G(\theta) + F(\theta,\dot{\theta}) \qquad (1)$$

Where T is the input torque vector, θ, $\dot{\theta}$, and $\ddot{\theta}$ are the joint displacement, velocity and acceleration of the manipulator, respectively. Mass matrix, $M(\theta)$ and the gravitational force vector, $H(\theta,\dot{\theta})$, and the friction matrix, $F(\theta,\dot{\theta})$ are complex functions of θ and $\dot{\theta}$. Forward and backward recursive equations are applied to propagate kinematics information from the inertial frame to end-effector frame and to propagate the forces and moments exerted on each link from the end-effector frame to the manipulator base frame. It is assumed that each link of a manipulator as a rigid body.

Polynomial Path Planning and Trajectory Generation
The manipulator path is generated using a Bezier curve. A Bezier curve is associated with the vertices, also called control points, of a characteristic polygon which define the curve shape. The Bezier curve is defined by:

$$P(u) = \sum_{i=o}^{n} P_i B_{i,n}(u) \qquad (2)$$

Where P_i represent the n+1 vertices of a characteristic polygon and $B_{i,n}(u)$ is the Bernstein polynomial blending function:

$$B_{i,n}(u) = C(n,i)\, u^i (1-u)^{n-i} \qquad (3)$$

and where $C(n,i)$ is the binomial coefficient:

$$C(n,i) = \frac{n!}{i!\,(n-i)!} \qquad (4)$$

The Bezier curves are variation-diminishing and never oscillate wildly away from their defining control points. However, moving any control point will change the shape of the curve.

To find a joint trajectory that approximates the desired path closely, the Cartesian path points are transformed into N sets of joint displacements, with one set for each joint. Application of Bezier polynomial will provide trajectories that are smooth and have small overshoot of angular displacement between two adjacent knot points. The continuity conditions for joint displacement, velocity, and acceleration must be satisfied on the entire trajectory for the Cartesian robot path.

Optimization Procedure

Minimum-time joint trajectory is a constrained non-linear optimization problem with a single objective function. The optimization procedure used in this work is the non-linear optimization search method with goal programming based on the Modified Hooke and Jeeves Direct Search Method [13].

The cost function in this optimization, the total traveling time for the robotic manipulator, should be minimized to maximized the speed of operation. Therefore, the optimization problem is to adjust the time intervals subject to the constraints on the joint positions, velocities, accelerations, jerks, torques, and end-effector acceleration within each of the segments of the polynomial joint trajectory to minimize the total traveling time. N joints have to be considered simultaneously since the optimization problem is in the joint space coordinates to minimize the manipulator path for the traveling time.

The problem is therefore expressed as to minimize an objective function subject to certain constraints as follows:

$$t_f - \sum_{i=1}^{n-1} S_i - \sum_{i=1}^{n-1} (t_{i+1} - t_i) \qquad (5)$$

Subject to :

$$\theta_{min} \leq \theta \leq \theta_{max} , \qquad \ddot{\theta}_{min} \leq \ddot{\theta} \leq \ddot{\theta}_{Max} , \qquad T_{min} \leq T \leq T_{max}$$

$$\dot{\theta}_{min} \leq \dot{\theta} \leq \dot{\theta}_{max} , \qquad \dddot{\theta}_{min} \leq \dddot{\theta} \leq \dddot{\theta}_{max} , \qquad E_{\alpha_{min}} \leq E_{\alpha} \leq E_{\alpha_{max}}$$

where θ, $\dot{\theta}$, $\ddot{\theta}$, and $\dddot{\theta}$ are the joint displacement, velocity, acceleration, and jerk, respectively. T is the torque and E_{α} is the end-effector acceleration.

Numerical Example

The technique is applied to study a two-degree of freedom planar manipulator with two links and joint rotation capability about the z-axis as shown in Figure 1. Each link is assumed to be 20 inches long and to have a mass of 7.4 lbm. The two-link manipulator parameters are shown in Table 1. The Cartesian end-effector path is defined by five control points P_0 (starting point) through P_4

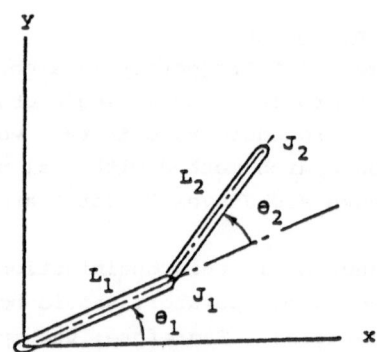

Figure 1. Two-Link Planar Manipulator

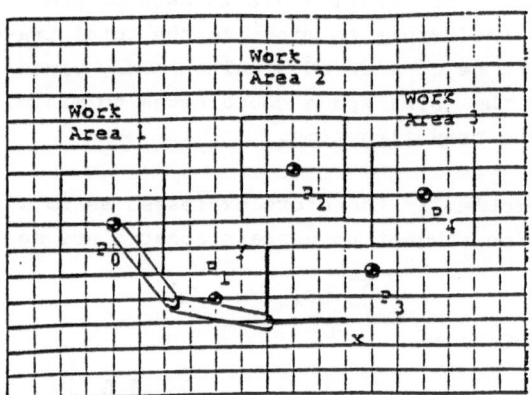

Figure 2: Manipulator and Control Points

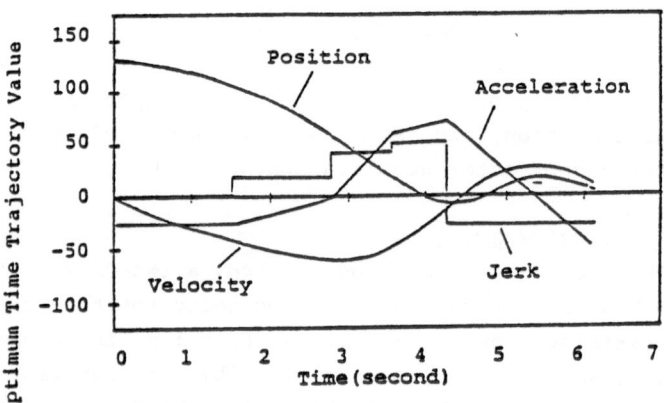

Figure 3: Minimum-Time Trajectories of Joint 1

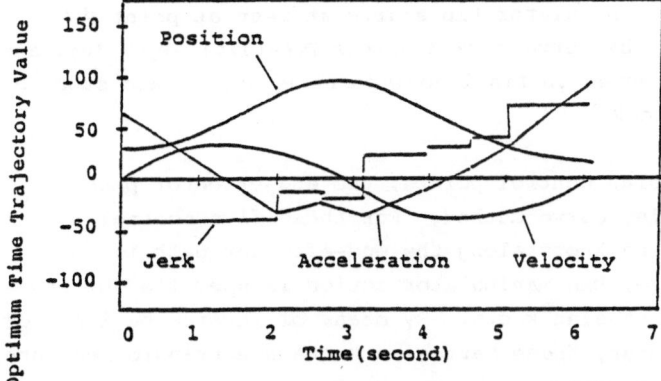

Figure 4: Minimum-Time Trajectories of Joint 2

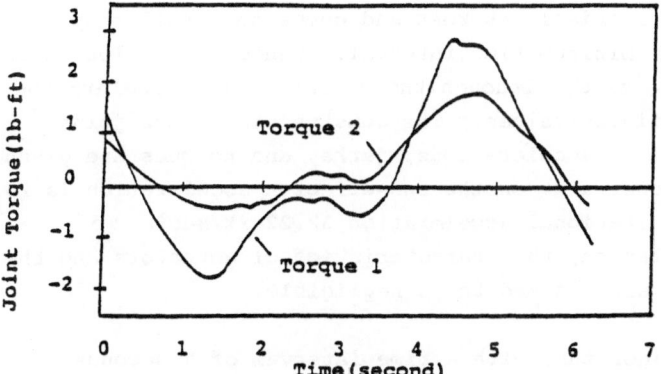

Figure 5: Optimum Joint Torque

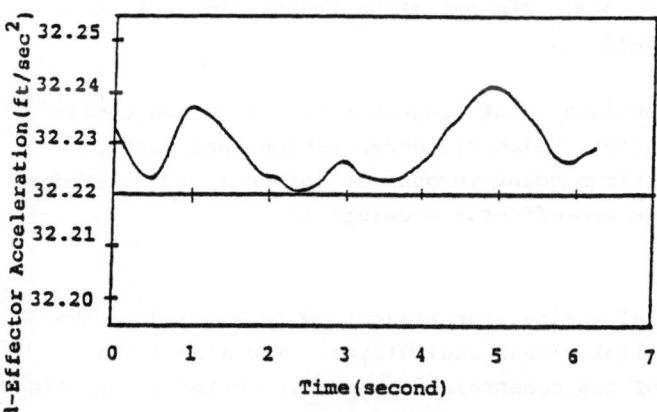

Figure. 6: Optimum End-Effector Acceleration.

(final point). The manipulator tip starts at rest at point P_0(-30,20), moves along the curve toward points P_1(-10,5), P_2(5,30), and P_3(20,10). Then it moves to final goal point P_4(30,25) and stops at P_4 as shown in Figure 2.

For the given Cartesian control points, the end-effector path is generated using Bezier curve method. For these five control points, nine Cartesian knots along the end-effector path is generated. Therefore, the manipulator motion is specified by the sequence of these Cartesian knots. By means of inverse manipulator kinematics and Jacobian, these Cartesian knots are transformed into sets of joint displacements.

The manipulator is initially at rest and comes to a full stop at the end of the last minimum-time interval. Hence, the velocities at t_1 and t_n must be zero, although the corresponding acceleration and jerk may have finite values. The constraints on the joint positions, velocities, accelerations, jerks, and torques are given in Table 2. The constraint on the end-effector acceleration is 2g, where g is the gravitational acceleration 32.22 ft/sec^2. To simplify the calculation, the characteristics of actuators and the frictional effects are assumed to be negligible.

The optimization algorithm, with a time interval of 5 seconds yields a minimum total traveling time of 6.38 seconds. This is the total travel time along the eight segments of the Bezier curve defined by the knot points. The optimized travel time for each segment is shown in Table 3.

Figure 3 and 4 show optimum joint trajectories, using the Bezier curve, including position, velocity, acceleration, and jerk. Figure 5 shows the optimum joint torques at joints 1 and 2. Figure 6 presents the optimum end-effector acceleration.

Conclusion

Minimum-time polynomial manipulator trajectory is applied to the case of a simple two-link planar manipulator. The algorithms developed are those of the constrained objective optimization with goal programming.

The manipulator path is generated using Bezier curve to fit the control points along this path. Constraints are applied to joint positions, velocities, accelerations, jerks, torques, and the end-effector acceleration. Incorporation of the joint jerk constraint and the end-effector acceleration into the model are important considerations. In an actual manipulator path, large jerks may result in excessive mechanical wear in the joints and the addition of the end-effector acceleration constraint will improve the manipulator tracking capabilities.

References

1. Luh, J. Y. S. and Walker, M. W., 'Minimum-Time Along the Path for a Mechanical Arm,' IEEE Proc. of 16th Conf. on Decision and Control, 1977, pp. 755-759.

2. Lozano-Perez, T. and Wesley, M. A., 'An Algorithm for Planning'Collision-Free Path among Polyhedral Obstacles, Communications of the ACM, Vol. 22, 1979, pp. 560-570.

3. Luh, J. Y. S. and Lin, C. S., 'Optimum Path Planning for Mechanical ManipuContr., Vol. 102, 1981, pp. 142-151.

4. Lin, C. S., Chang, P. R. and Luh, J. Y. S., 'Formulation and Optimization of Cubic Polynomial Joint Trajectories for Industrial Robots,' IEEE Trans. Auto. Contr., Vol AC-28, No. 12, 1983, pp. 1066-1074.

5. Bobrow, J. E., Dubowsky, S. and Gibson, J. S., 'Time-Optimal Control of Robotic Manipulators Along Specified Paths,' Int. J. Robotics Res., Vol. 4, No. 3, 1985, pp. 3-17.

6. Sahar, G. and Hollerbach, J. M., 'Planning of Minimum-Time Trajectories for Robot Arms, Int. J.Robotics Res., Vol. 5, No. 3, 1986, pp. 90-100.

7. Lin, C. S. and Chang, P. R., 'Approximate Optimum Path ofRobot Manipulators Under Realistic Physical Constraints, IEEE Int. Conf. on Robotics and Automation, 1985, pp. 737-742.

8. Rajan, V. T., 'Minimum Time Trajectory Planning,' IEEE Conf. on Robotics and Automation, 1985, pp. 759-764.

9. Bobrow, J. E., 'Optimal Robot Path Planning Using the Minimum-Time Criterion,' IEEE J. of Robotics and Automation, Vol. 4, No. 4, 1988, pp. 443-450.

10. Barez, F., 'Simulation of Minimum-Time Robotic Manipulator Trajectories Using Parabolic Blending Technique,' Proceedings of International Symposium on Robotics and Manufacturing, ACTA Press, 1989, pp. 44-48.

11. Chen, C. M. and Peikari, B., 'Minimum Time Robot Path Planning,' Proceedings of International Symposium on Robotics and

Manufacturing, ACTA Press, 1989, pp. 187-191.

12. Craig, C., Introduction to Robotics: Mechanic and Control, Addison-Wesley Publishing, Reading, MA, 1986

13. Hooke, R. and Jeeves, T. A., 'Direct Search Solution of Numerical andstatical Problems,' J. of the Assn. of Computing Machinery, Vol. 8, 1961, pp. 212-229.

TABLE 1: Two-Link Manipulator Parameters

Link Number	l_1	l_2
Length(in.)	20	20
Mass(lb_m)	7.4	7.4
I_{xx} (lb-ft-s^2)	2.9E-3	2.9E-3
I_{yy} (lb-ft-s^2)	.8E-3	.8E-3
I_{zz} (lb-ft-s^2)	2.9E-3	2.9E-3
$I_{xy}=I_{xz}=I_{yz}$	0	0

TABLE 2: Joint Constraints

Joint Number	Joint 1		Joint 2	
Bounds	lower	upper	lower	upper
Position(deg)	-200	200	-250	250
Velocity(deg/sec)	-150	150	-120	120
Acceleration(deg/sec^2)	-85	85	-60	60
Jerks(deg/sec^3)	-50	50	-45	45
Torque(lb-ft)	-15	15	-8	8

TABLE 3: Optimum Time of Travel

Segment	Optimum Time(sec)
1	1.13
2	0.71
3	0.68
4	0.79
5	0.75
6	0.72
7	0.65
8	0.95
Total Time	6.38 seconds

Efficient Trajectory Planning Algorithm for Coordinately Operating Multiple Robots

YUNG-PING CHIEN and QING XUE

Department of Electrical Engineering
Purdue University School of Engineering and Technology at Indianapolis
Indianapolis, IN

Abstract

An efficient trajectory planning algorithm that uses the search method to generate locally minimum time trajectories for coordinately operating multiple robots is proposed. The algorithm can be applied to non-redundant robots which operating coordinately in a three-dimensional workspace. The control sampling period of the robots is assumed to be constant. The path of the rigid object carried by the robots is specified. The generated robot trajectories satisfy the joint velocity, acceleration and torque constraints as well as the path constraints.

1. Introduction

The generation of the minimum time path-following trajectory for a single robot manipulator has attracted attention in the robotics research for long time. The minimum time solution for a robot motion along a prescribed path was described in [1,2]. The joint torque constraints and velocity constraints are included in the problem formulation. Exhaustive search is involved in finding the admissible region of the s-ṡ plane. The planning of the straight line trajectory for a robot is proposed in [3] by considering the bounds on the joint position, velocity, acceleration, jerk and torque. The minimum time trajectory planning is formalized in [4] by specifying the number of sampling intervals for a path and then minimizing the length of the sampling intervals.

In this paper, the problem of finding a locally minimum-time path-following trajectory for the motion of the coordinated multiple robots carring a rigid object is discussed (Figure 1). The continuous geometric path of the carried object is specified in the world coordinate system since it is easily visualized by the path designer. The given object path is parameterized as in [1,2] but the exhaustive search in finding the admissible region in the phase plane is avoided. An algorithmic search method is used in searching the trajectory. The resulting trajectory satisfies the joint velocity, acceleration and torque constraints of the robots.

2. System Model and Motion Constraints

It is assumed that there are n non-redundant robots in a common work cell. The task of the robots is to cooperatively move an object from an initial position and orientation to a final position and orientation along a given path. The velocities of the object at the starting and final positions are assumed zero.

For convenience, the following nomenclature is used throughout the paper:

Δt -- sampling period.

$\lambda(k)$ -- normalized scalar parameter that describes the position and orientation of the rigid object at time $k\Delta t$ on the given object path in the world coordinate system, where $k=0,1,2,\cdots$. $\lambda(\cdot)=0$ at the starting position and orientation of the path and $\lambda(\cdot)=1$ at the final position and orientation of the path.

p, Φ -- 3x1 vectors of position and orientation, respectively.

The left superscript of p and Φ can be o or i which denotes the rigid object or robot i, respectively. The left subscript w or o denotes the world coordinates or object coordinates, respectively. The right subscript i or f denotes the initial or final value, respectively. For example,

$_o^i p, _o^i \Phi$ -- grasping position and orientation of robot i relative to the object coordinate frame, respectively.

$_w^i p(k), _w^i \Phi(k)$ --grasping position and orientation of robot i relative to the world coordinate system at kth sampling instant, respectively.

$_w^o p(k), _w^o \Phi(k)$ --the position of the reference position and the orientation of the rigid object, respectively, relative to world coordinates at the sampling instant $k\Delta t$.

$_w^o p_i, _w^o \Phi_i$ -- initial position and orientation of the rigid object respectively relative to world coordinates.

$_w^o p_f, _w^o \Phi_f$ -- final position and orientation of the rigid object respectively relative to world coordinates.

$^i q, ^i \dot{q}, ^i \ddot{q}, ^i \tau$ -- 6x1 column vectors of the position, velocity, acceleration of joint angles of robot i, and the joint torque of robot i, respectively.

The right subscript j of q and τ refers to the joint j, j=1, $\cdots$, 6, e.g., $^i q_j$ is the position of the joint j of robot i.

M --   The mass of the rigid object.

The task of the robots is to move the rigid object from $[{}^o_w p_i, {}^o_w \Phi_i]^T$ starting at zero velocity to $[{}^o_w p_f, {}^o_w \Phi_f]^T$ finishing at zero velocity along a given path in the world coordinate system, where the superscript T denotes transposition. It is assumed that the position and orientation of the object on the given path can be expressed by a function F in a parametric form:

$$[{}^o_w p(k), {}^o_w \Phi(k)]^T = F[\lambda(k), [{}^o_w p_i, {}^o_w \Phi_i]^T, [{}^o_w p_f, {}^o_w \Phi_f]^T], \tag{1}$$

The grasp position and orientation of robot i, $[{}^i_o p, {}^i_o \Phi]$, i=1,···, n, relative to the object coordinate system are fixed during the motion. They can be expressed in the world coordinate system as the function

$$[{}^i_w p(k), {}^i_w \Phi(k)]^T = G({}^i_o p, {}^i_o \Phi, {}^o_w p(k), {}^o_w \Phi(k)), \tag{2}$$

where G is the transformation from the object coordinate system to the world coordinate system.

Let ${}^i A(\cdot)$ be a nonlinear transformation from the joint space of robot i to the world coordinate system. Assuming that ${}^i A(\cdot)$ is non-singular, ${}^i q(k)$ can be calculated by the inverse kinematic equation:

$$^i q(k) = {}^i A^{-1}([{}^i_w p(k), {}^i_w \Phi(k)]^T). \tag{3}$$

Equation (3) may give several ${}^i q(k)$. A feasible ${}^i q(k)$ can be selected which is adjacent to ${}^i q(k-1)$. The velocity and acceleration of the joints of robot i at time $k\Delta t$ can be approximated, respectively, by

$$\dot{q}(k) = (q(k) - q(k-1))/\Delta t, \quad \ddot{q}(k) = (q(k) - 2q(k-1) + 2q(k-2))/\Delta t^2. \tag{4}$$

Assuming the position, velocity, and acceleration of joint j of robot i, are constrained, respectively, by

$$^i q_{j,min} \le {}^i q_j(k) \le {}^i q_{j,max}, \quad \left| {}^i \dot{q}_j(k) \right| \le {}^i \dot{q}_{j,max}, \quad \left| {}^i \ddot{q}_j(k) \right| \le {}^i \ddot{q}_{j,max}. \tag{5}$$

given the position, velocity, and acceleration of joint j of robot i at time $(k-1)\Delta t$, the minimum and maximum feasible values of joint j of robot i at time $k\Delta t$ can be calculated, respectively, by

$$^i q_{j,min}(k) = \max\{^i q_{j,min}, [^i q_j(k-1) + \Delta t(\max\{-^i \dot{q}_{j,max}, [^i \dot{q}_j(k-1) + \Delta t(-^i \ddot{q}_{j,max})]\})]\}, \tag{6}$$

$$^i q_{j,max}(k) = \min\{^i q_{j,min}, [^i q_j(k-1) + \Delta t(\min\{^i \dot{q}_{j,max}, [^i \dot{q}_j(k-1) + \Delta t(^i \ddot{q}_{j,max})]\})]\}. \tag{7}$$

By the Lagrange-Euler equation of motion, the required torques ${}^i \tau(k)$ of the joint actuators of robot i can be calculated [4]. The required torque of joint j of robot i must satisfy

$$^i \tau_{j,min}(k) < {}^i \tau_j(k) < {}^i \tau_{j,max}(k) \tag{8}$$

where ${}^i \tau_{j,max}(k)$, ${}^i \tau_{j,min}(k)$, are the maximum and minimum feasible driving torques respectively. At each sampling instant, ${}^i \tau_{j,max}(k)$, ${}^i \tau_{j,min}(k)$, can be expressed as functions of the joint velocities at time instant k, the gear ratio, the maximum and minimum actuator driving voltage and other actuator characteristics [4].

Hence, the trajectory planning problem is presented as maximize $\Delta\lambda(k) = \lambda(k) - \lambda(k-1)$ subject to

$$^i q_{j,min}(k) \le {}^i q_j(k) \le {}^i q_{j,max}(k) \tag{9}$$

$$^i \tau_{j,min}(k) < {}^i \tau_j(k) < {}^i \tau_{j,max}(k) \tag{10}$$

$$^i \dot{q}_j(0) = 0 \tag{11}$$

$$^i \dot{q}_j(f) = 0 \tag{12}$$

for all i and j at each sampling instant $k\Delta t$, where f is the final sampling instant.

3. Locally Time Minimum Trajectory Planning by a Search Method

The problem is solved in two steps. In the first step, the constraint equation (12) is ignored during the search of the locally minimum time trajectory. In the second step, the constraint equation (12) is incorporated.

3.1. Plan a locally minimum time trajectory without considering final speed

In this section, the constraint of stopping the rigid object at the final position and orientation is temporarily released. Since the total motion distance of the rigid object is known, the locally minimum time trajectory implies a locally maximum speed trajectory.

3.1.1. Non-zero final velocity trajectory search algorithm

When the object is at the position $\lambda(k-1)$ on the path at time instant k-1, a feasible $\lambda(k)$ may or may not be found depends on the velocity and acceleration of the rigid object at $(k-1)\delta t$. The strategy adopted in the search algorithm is (1) if a feasible range of $\lambda(k)$ can be found, the maximum feasible $\lambda(k)$ is used; (2) if a feasible range of $\lambda(k)$ cannot be found, set the currently maximum reachable $\lambda_{max} = \lambda(k-1)$ and use the maximum deceleration (using minimum feasible λ) of the rigid object from time k-a on until λ_{max} is passed, where a is the minimum integer required to achieve the goal of (2).

Non-zero Final Velocity Trajectory Search Algorithm:

1. [Initialization] Set k=0, $^i\dot{q}_j(k)=0$, and $^i\ddot{q}_j(k_c)=0$ for all i and j, $\lambda(k)=0$, and $\lambda_{max}=0$.

2. [Check criterion for stop] If $\lambda(k) >= 1$, stop.

3. [Find the maximum feasible $\lambda(k)$] k=k+1. Call the maximum feasible $\lambda(k)$ search algorithm (to be described in Section 3.1.2). If the return value is not "no feasible $\lambda(k)$", then λ_{max}=maximum feasible $\lambda(k)$, $\lambda(k)$=maximum feasible $\lambda(k)$, and goto step 2.

4. [Determine the beginning time of deceleration] Find the latest sampling period k_a, where k_a<k, in which maximum feasible $\lambda(k_a)$ is used. Set k=k_a.

5. [Find the minimum feasible $\lambda(k)$] Call the minimum feasible $\lambda(k)$ search algorithm (to be described in Section 3.1.4). If the return value is "no feasible $\lambda(k)$", then goto step 4; else set $\lambda(k)$=minimum feasible $\lambda(k)$. If the minimum $\lambda(k) \geq \lambda_{max}$, or $\lambda(k)-\lambda(k-1)<\delta_1$ (velocity of the object close to zero), then goto step 2; else, k=k+1 and goto step 5.

The output of this algorithm is a list of locally minimum time $\lambda(k)$, k=0,1, $\cdots$, without considering the constraint of having zero velocity at the final position and orientation.

3.1.2. Find the maximum feasible $\lambda(k)$

In order to search the maximum feasible $\lambda(k)$, the lower and upper boundary of the search region $\lambda_{low}(k)$ and $\lambda_{high}(k)$, respectively, are required. Since $\lambda(k) \geq \lambda(k-1)$, the initial estimated lower boundary $\lambda_{low}(k)$ is $\lambda(k-1)$. The upper boundary $\lambda_{high}(k)$ is first estimated as $\lambda(k-1)$ plus a small real value δ_1, which is the maximum acceptable search error. The correct search direction is determined by comparing the relative position and torque values of the robot joints which correspond to the two latest estimated $\lambda(k)$s, and the joint positional and torque limits at time k. Once $\lambda_{low}(k)$ and $\lambda_{high}(k)$ are determined, the binary search method is adopted to find the maximum feasible $\lambda(k)$ between $\lambda_{high}(k)$ and $\lambda_{low}(k)$.

The required joint torques of a robot are functions of the mass of the carried rigid object. Therefore, the proper load distribution among the coordinately operating robots reduces the chances that required joint torques exceed the joint torque limits. Since the required joint torques of a robot are also functions of the required position, velocity and acceleration of the carried rigid object, the load distribution should be changed during the motion in order to reduces the chances that required joint torques exceeds the joint torque limits. Instead of finding the best load distribution, the maximum load that each robot can carry at the given joint position, velocity and acceleration is searched. If the sum of the maximum feasible load of all robots is less than the mass of the rigid object, the torque constraints of the robots can be satisfied; otherwise, the required motion is not feasible.

Maximum Feasible $\lambda(k)$ Search Algorithm:

1. [Initialize and loop] Set m=0, highflag=false, feasibleflag=false, $\lambda_{low}(k)=\lambda(k-1)$, $\lambda_{high}(k)=1$, $\lambda^m(k)=\lambda_{low}(k)$.

2. [Compute the joint and torque boundaries of robot at time k] Compute $^iq_{j,min}(k)$ and $^iq_{j,max}(k)$ by equations (10) and (11), respectively. Compute $^i\tau_{j,min}(k)$ and $^i\tau_{j,max}(k)$, respectively (described in Section 2.4).

3. [Determine new $\lambda^m(k)$] m=m+1. If highflag=false ($\lambda_{high}(k)$ has not been found), $\lambda^m(k)=\lambda^{m-1}(k)+2^{m-1}\delta_1$; else $\lambda^m(k)=0.5(\lambda_{low}(k)+\lambda_{high}(k))$

4. [Check stopping condition] If $\lambda^m(k)-\lambda^{m-1}(k)<\delta_1$ or $\lambda_{low}(k)>\lambda_{high}(k)$, goto step 11.

5. [Check if the estimated upper boundary of $\lambda(k)$ is greater than 1] If $\lambda^m(k)>1$, then set $\lambda_{high}(k)=1$, highflag=true.

6. [Find grasp positions and orientations of all robots and inverse kinematics solution of all robots; find the maximum feasible load of each robot] That is, find $^i_wp(\lambda^m(k))$ and $^i_w\Phi(\lambda^m(k))$ for all i. Then find the joint value $^iq_j(\lambda^m(k))$ for all i and j. Find the joint velocity and acceleration by equation (4). Call maximum load search algorithm (Section 3.1.3) to obtain the maximum feasible load of the robot i at time (k), $^iM^m(k)$, at the given position $^iq^m(k)$, velocity $^i\dot{q}^m(k)$, and acceleration $^i\ddot{q}^m(k)$.

7. [Check if $\lambda^m(k)$ is feasible] If $^iq_{j,min}(k)<^iq_j(\lambda^m(k))<^iq_{j,max}(k)$ for all i and j, and $\sum_{i=1}^n {}^iM^m(k)\geq M$, then set feasibleflag=true, $\lambda_{low}(k)=\lambda^m(k)$ and goto step 3.

8. [Check constraints and reset estimation boundary of $\lambda(k)$] If (a) $^iq_j^m(k)-^iq_{j,min}(k)$, $^iq_j^m(k)-^iq_{j,max}(k)$, $^iq_j^{m-1}(k)-^iq_{j,min}(k)$, and $^iq_j^{m-1}(k)-^iq_{j,max}(k)$, have the same sign and $|^iq_j^m(k)-^iq_{j,min}(k)| > |^iq_j^{m-1}(k)-^iq_{j,min}(k)|$ (or $|^iq_j^m(k)-^iq_{j,max}(k)| > |^iq_j^{m-1}(k)-^iq_{j,max}(k)|$), for any i and j, or (b) $\sum_{i=1}^n {}^iM^m(k)>\sum_{i=1}^n {}^iM^{m-1}(k)>M$, the search is in the wrong direction, goto step 9; else the search is in the

right direction, goto step 10.

9. [Reestimate boundary of $\lambda(k)$] If $\lambda^m(k) > \lambda^{m-1}(k)$ then $\lambda_{high}(k) = \lambda^{m-1}(k)$, highflag=true; else $\lambda_{low}(k) = \lambda^{m-1}(k)$. Goto step 3.

10. [Reestimate boundary of $\lambda(k)$] If highflag=false and $\lambda^m(k) > \lambda^{m-1}(k)$, then $\lambda_{low}(k) = \lambda^{m-1}(k)$;
    else if highflag=false and $\lambda^m(k) \leq \lambda^{m-1}(k)$, then $\lambda_{high}(k) = \lambda^{m-1}(k)$; else if ${}^iq_j^m(k) - {}^iq_{j,min}(k)$, ${}^iq_j^m(k) - {}^iq_{j,max}(k)$, ${}^iq_j^{m-1}(k) - {}^iq_{j,min}(k)$, and ${}^iq_j^{m-1}(k) - {}^iq_{j,max}(k)$ have the same sign for any i and j or $\sum_{i=1}^n {}^iM^{m-1}(k) > \sum_{i=1}^n {}^iM^m(k) > M$, and $\lambda^{m-1}(k) > \lambda^m(k)$, then set $\lambda_{high} = \lambda^m(k)$;
    else if ${}^iq_j^m(k) - {}^iq_{j,min}(k)$, ${}^iq_j^m(k) - {}^iq_{j,max}(k)$, ${}^iq_j^{m-1}(k) - {}^iq_{j,min}(k)$, and ${}^iq_j^{m-1}(k) - {}^iq_{j,max}(k)$ have the same sign for any i and j or $\sum_{i=1}^n {}^iM^{m-1}(k) > \sum_{i=1}^n {}^iM^m(k) > M$, and $\lambda^{m-1}(k) < \lambda^m(k)$, then set $\lambda_{low}(k) = \lambda^m(k)$;
    else if $\lambda^m(k) > \lambda^{m-1}(k)$, then set $\lambda_{high} = \lambda^m(k)$;
    else if $\lambda^m(k) < \lambda^{m-1}(k)$, then set $\lambda_{low} = \lambda^m(k)$. Goto step 3.

11. [Send searching results] If feasibleflag=false, send message "no feasible $\lambda(k)$"; else, send "maximum feasible $\lambda(k) = \lambda^{m-1}(k)$." Stop.

The superscript m is the number of estimations. The highflag is a variable to indicate if a high boundary of the search range has been found. The feasibleflag is a variable to indicate if a feasible $\lambda(k)$ has been found. $\lambda_{low}(k)$ and $\lambda_{high}(k)$ are the lower and upper bounds for searching the maximum $\lambda(k)$.

3.1.3. Find the maximum feasible load

The input of this algorithm is the required joint position, velocity, and acceleration of the robots at time k, the torque limits of robot joints at time k and the mass of the carried rigid object. The outputs of the algorithm are the maximum feasible load of robot i, ${}^iM(k)$, where i=1, ⋯, n, at the given joint position, velocity and acceleration of the robot. It is assumed that the minimum load of robot i at time k is zero. The maximum load of robot i is the mass of the rigid object M.

Maximum Load Search Algorithm:

1. [Start algorithm] For i=1 to n do the following:

2. [Initialize the search boundary] ${}^iM_{low}=0$, ${}^iM_{high}=M$, p=0. ${}^iM^p=M$,

3. [Find joint torque] Assume that the load of robot i is ${}^iM^p$ and the required joint position, velocity and acceleration of robot i at time k are, respectively, ${}^iq_j(k)$, ${}^i\dot{q}_j(k)$, ${}^i\ddot{q}_j(k)$, find the required joint torque vector ${}^i\tau^p(k)$ by equation (12).

4. [Check constraints] If ${}^i\tau_{j,min}(k) < {}^i\tau_j^p(k) < {}^i\tau_{j,max}(k)$ for all j, ${}^iM_{low}={}^iM^p$; else, ${}^iM_{high}={}^iM^p$;

5. [Check stop condition] If ${}^iM_{high} - {}^iM_{low} > \delta_2$, then set p=p+1, ${}^iM^p=0.5({}^iM_{high} + {}^iM_{low})$, and goto step 3; else set ${}^iM(k) = {}^iM_{low}$, and p=0.

6. Stop.

The variable p is a loop counter and δ_2 is the maximum acceptable error.

3.1.4. Find the minimum feasible $\lambda(k)$

The procedure for finding the minimum $\lambda(k)$ is similar to that of finding the maximum $\lambda(k)$. The only difference between them is the way to update the upper and lower boundaries for searching the minimum feasible $\lambda(k)$.

3.2. Locally Minimum Time Trajectory Planning

In this section, the problem is to incorporate the zero final velocity constraint to the planning algorithm. In order to rest the rigid object at the final position and orientation, the maximum deceleration should start before the rigid object arrives the final position and orientation. Since the system is not linear, it is difficult to pinpoint at where the rigid object will stop when the maximum deceleration starts. Therefore, the rest position search algorithm is designed in Section 3.2.1 to find the rest position of the carried object when the maximum deceleration is started at give positions, velocities and accelerations of all robot joints.

Since the sampling time period is a constant, it is usually not possible for the rigid object to reach the final position and orientation at zero velocity by using a sequence of maximum deceleration periods right after a maximum acceleration period. Two adjacent sampling instants k_b and k_b+1 can be found in the sequence of the trajectory points planned by using the non-zero final velocity trajectory search algorithm such that when starting the maximum deceleration at k_b, the object can be brought to rest before reaching the final position and orientation, and when starting the maximum deceleration at k_b+1, the object cannot be brought to rest before reach the final position and orientation. The time instant k_b is denoted as the break point which is searched by the algorithm described in Section 3.2.2.

From the definition of the break point, there must be a feasible $\lambda(k_b+1)$ such that the rigid object reaches the final position and orientation at zero velocity. The fine tuning algorithm described in Section 3.2.3 determines the value of $\lambda(k_b+1)$ such that when the object stop maximum acceleration motion at time k_b and start maximum deceleration motion at time k_b+1, the object will have zero velocity at the final position and orientation.

3.2.1. Rest position search algorithm

The input of the algorithm is the current time k_c, the current position and orientation of the object in terms of the parameter $\lambda(k_c)$, the values of the current joint position ${}^iq_j(k_c)$, velocity ${}^i\dot{q}_j(k_c)$, and acceleration ${}^i\ddot{q}_j(k_c)$. The output of the algorithm is a sequence of trajectory points that leads the object to a position and an orientation at which the velocity of the object is zero.

1. [Initialization] $k=k_c$. ${}^iq_j(k)={}^iq_j(k_c)$, ${}^i\dot{q}_j(k)={}^i\dot{q}_j(k_c)$, and ${}^i\ddot{q}_j(k_c)={}^i\ddot{q}_j(k_c)$.

2. [Find the minimum feasible $\lambda(k)$] Call the Minimum Feasible $\lambda(k)$ Search Algorithm.

3. [Check the stop criterion of the carried object] If the return message is "velocity can be zero", then send message "stop at $\lambda(k)$" and stop; else, set $k=k+1$ and go to 2.

3.2.2. Break point search algorithm

The input of the break point search algorithm is a sequence of trajectory point generated in the Maximum Feasible $\lambda(k)$ Search Algorithm. Let k_f be the number of sampling periods required in the trajectory planned by the non-zero final velocity trajectory planning algorithm.
Break Point Search Algorithm:

1. [Initialize the lower and upper search boundaries of k_b] Set $k_{b,low}=1$, $k_{b,high}=k_f$, and $p=0$ (p denotes the number of estimation of k_b).

2. [Check stop criterion] If $k_{b,high}=k_{b,low}+1$, then $k_b=k_{b,low}$ is the break point and stop. else, $p=p+1$, continue to 3.

3. [Estimate the starting time of deceleration] Set $k_b^p= \left\lceil \dfrac{k_{b,low}+k_{b,high}}{2} \right\rceil$. From the output of the maximum $\lambda(k)$ search algorithm, the value of $\lambda(k_b^p)$ and the corresponding joint positions, velocities, and accelerations can be found.

4. [Find the value of λ_r^p at which of the carried object rests] Call the rest position search algorithm by using the k_b^p obtained in step 3 as the input. Obtain the trajectories of the robots and λ_r^p from the output of the rest position search algorithm.

5. [Update the boundary of the estimation of k_b] If $\lambda_r^p \geq 1$, $k_{b,high}=k_b^p$; else, $k_{b,low}=k_b^p$. Go to 2.

If the maximum deceleration is used from time k_b on, the object can have zero velocity before reaching the final position and orientation. If the maximum acceleration stops at k_b+1 and the maximum deceleration starts at k_b+1, the object cannot have zero velocity before passing the final position and orientation. As depicted in Figure 4, k_b is 22 and k_b+1 is 23. Therefore, there should be a value of $\Delta\lambda(k_b)=\lambda(k_b+1)-\lambda(k_b)$ in the time period (k_b, k_b+1) such that the object can have zero velocity at the final position and orientation by stopping the maximum acceleration at k_b and starting the maximum deceleration at k_b+1 and afterwards. The value of $\Delta\lambda(k_b)$ is determined by the fine tuning algorithm described in the next section.

3.2.3. Fine tuning algorithm

The input of this algorithm is the output of the non-zero final velocity trajectory search algorithm, and k_b obtained from the break point search algorithm. The output of the algorithm is the trajectories of all robots.
Fine Tuning Algorithm:

1. [Initialize loop counter] Set $p=1$, $\lambda_{high}(k_b+1)$ equals the maximum feasible $\lambda(k_b+1)$, $\lambda_{low}(k_b+1)$ equals the minimum feasible $\lambda(k_b+1)$ calculated based on given $\lambda(k_b)$.

2. [Estimate $\lambda^p(k_b+1)$ and find corresponding joint values of robots] $\lambda^p(k_b+1)=0.5(\lambda_{low}(k_b+1)+\lambda_{high}(k_b+1))$. Find the position and orientation of the object at $\lambda^m(k_b+1)$. Find the joint positions, velocities, and accelerations of each robot corresponds to $\lambda^p(k_b+1)$.

3. [Find the value of λ_r^p at which of the carried object rests] Call the Rest Position Search Algorithm by using the values found in step 2 as the input. Obtain the trajectories of the robots and λ_r^p from the output of the rest position search algorithm.

4. [Update the boundaries of the estimation of $\lambda^p(k_b+1)$] If $\lambda_r^p \geq 1$, $\lambda_{high}(k_b+1)=\lambda^p(k_b+1)$; else, $\lambda_{low}(k_b+1)=\lambda^p(k_b+1)$.

5. [Check stopping criterion] If $\left| \lambda_r^p-1 \right| < \delta_3$, then go to 6; else, $p=p+1$, go to 2.

514

6.   Stop.

The δ_3 in step 5 is a small constant indicating the maximum acceptable search error. The locally minimum time trajectory is composed by part of the sequence of the trajectory points generated by the non-zero final velocity trajectory search algorithm up to time k_b, followed by the sequence of trajectory points obtained in this algorithm. As shown in Figure 4, the trajectory point 23 at time 23 is selected by using the maximum feasible lambda(23), and the trajectory point 23a at time 23 is selected by using the minimum feasible lambda(23). After using the fine tuning algorithm (Figure 5), the trajectory point 23b at time 23 is selected by using a feasible $\lambda(23)$ between the maximum $\lambda(23)$ and the minimum $\lambda(23)$.

4. Discussion and Conclusion

The search method can be applied to any kind non-redundant robot manipulators. The computational complexity of the planning algorithm is $O(k^2(\log\frac{1}{\delta_1})(n\log\frac{M}{\delta_2}))$, where k is the minimum number of sampling period required by the trajectory, and δ_1 and δ_2 are constants. It is assumed that a trajectory requires at least two sampling periods.

In this paper, an efficient discrete time algorithmic search method for solving the problem of finding a locally minimum time trajectory for the motion of coordinated multiple robots without violating the joint velocity, acceleration and torque constraints of the robots is proposed. The task of the robots is to move a rigid object along a given path in the world coordinate system. The shape of the path is arbitrary. The position and orientation of the carried rigid object along a given path is expressed in a parametric form in terms of a normalized scalar parameter λ. Unlike other approaches to the same problem, which also parameterize the object path, the proposed method does not use the exhaustive search to find admissible region in the parameter phase plane, thus providing an efficient trajectory planning.

References

1.   J. E. Bobrow, S. Dubowsky, and J. S. Gibson, "Time-Optimal Control of Robotic Manipulators along Specified Paths," *International Journal of Robotic Research*, Vol. 4, No. 3, 1985, pp. 3-17.

2.   K. G. Shin and N. D. Mckay, "A Dynamic Programming approach to Trajectory Planning of Robotic Manipulators," *IEEE Transactions on Automatic Control*, Vol. AC-31, No. 6, June 1986, pp. 491-500.

3.   B. H. Lee, "An Approach to Motion Planning and Motion Control of Two Robots in a Common Workspace," Ph.D. Dissertation, College of Engineering, The University of Michigan, 1985.

4.   H. H. Tan and R. B. Potts, "A Discrete Trajectory Planner for Robotic Arms with Six Degrees of Freedom," *IEEE Transection on Robotics and Automation*, Vol. 5, No. 5, October 1989, pp. 681-690.

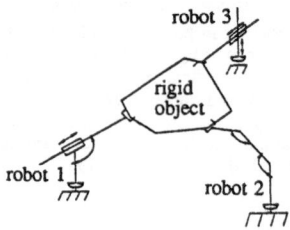

Figure 1. Multiple robots move an object.

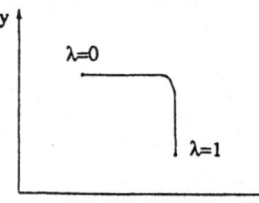

Figure 2. Desired path of object in xy-plain.

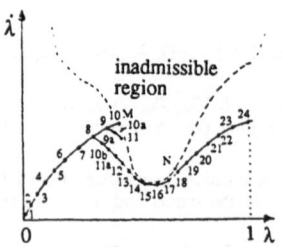

Figure 3. Trajectory generated by the non-zero final velocity search algorithm.

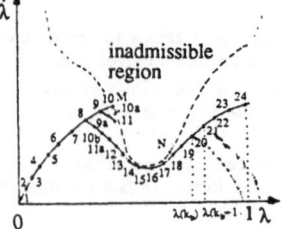

Figure 4. Break point on the trajectory.

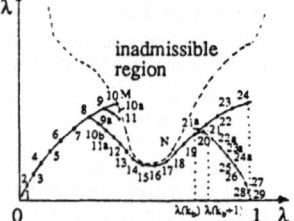

Figure 5. The final trajectiory is (0-8,9a,10b,11a,12-20, 21a-24a,25-29).

Path Planning for Coordinated Planar Robot Arms Moving in Unknown Environment

YUNG-PING CHIEN and QING XUE

Department of Electrical Engineering
Purdue University
School of Engineering and Technology at Indianapolis
Indianapolis, IN

Abstract

This paper investigates the problem of planning a collision-free path for two planar robot manipulators which coordinately carry a small object from a start position to a target position in a two dimensional (2-D) workspace which is filled with unknown stationary obstacles. The shapes of the obstacles are arbitrary. The positions of the obstacles are not known a priori. In order to obtain the obstacle information, sensors are used to provide the immediate surroundings of robot arms. It is assumed that any point on the robots is subject to collision. It is shown that the proposed non-heuristic algorithm guarantees the robots to reach the target position or to conclude that no path exists.

1. Introduction

Planning a collision-free path for robot arms which move in an environment with obstacles is to find the continuous joint motion so that the arm can move from the given start position to a given target position without colliding with any obstacle. The collision-free path generation for the motion of robots in a completely known environment has been studied extensively. The well known approaches are the Piano Mover's problem [1] and Configuration Space method [2]. The path planning of a 2-D planar robot with incomplete information about the stationary environment has been proposed in [3]. In this approach, the path of a planar robot is planned continuously, based on the arm's target position, current position, and the sensory feedback.

This paper presents a collision-free path planning algorithm for two coordinately operating planar robot arms each of which has two revolute joints and two links. The robots move an object coordinately from one position to another position in a 2-D environment which is filled with unknown obstacles. The links of the robots and the line segment connecting the bases of robots are modeled as 5-link closed chain. It is assumed that the robot arms know their current positions and target position, and any point of the robot body is capable of feeling a contact or at a distance with obstacles. This situation is similar to a person hold a small object with two arms. When the arms touch an obstacle, the points of contact are known. There are no restrictions on the shape of the obstacles and the arm links. It is assumed that robots can move along the boundaries of the obstacles. How robots sense the obstacles and move along the boundaries of the obstacle are not in the scope of this paper. Interested reader can refer to [4].

The proposed algorithm is non-heuristic ("exact") -that is, this procedure guarantees either reaching the target point or concluding that the target point cannot be reached. The use of the algorithm may be extended in certain robot hand applications and bipad walking robots.

2. Model

The links of robots and the straight line segment connecting the bases of the robots together can be modeled as a 5-link closed chain (Figure 1). The links and the joints of the closed chain are denoted, respectively, as ℓ_i and θ_i, $i=1,2,\cdots,5$. $\theta_1=0$ when ℓ_1 and ℓ_5 overlap. $\theta_2=0$ when ℓ_1 and ℓ_2 aline but not overlap. $\theta_5=0$ when ℓ_4 and ℓ_5 aline but not overlap. $\theta_4=0$ when ℓ_4 and ℓ_3 aline but not overlap. $\theta_3=0$ when ℓ_2 and ℓ_3 overlap. θ_i increases in counterclockwise direction. Each joint is assumed to have its minimum and maximum motion limits, $\theta_{i,min}$ and $\theta_{i,max}$, respectively. In order to be able to physically move the carried object, the motion of robot arms must also satisfy the following constraints:

(1) Collision-free Constraint: all the links except the virtual link between the bases of robots must not collide with obstacles in workspace.

(2) Closed Chain Constraint: the adjacent links should always be connected to each other.

(3) Joint Limit Constraint: all joints move within their motion limits.

Although the 5-link closed chain has five joint variables, it has only 2-degrees of freedom of motion [5]. Therefore, two robot joints can be considered as the independent joint variables and the other three joints are dependent joint variables. We arbitrarily assign joint variables θ_4 and θ_5 as independent variables and θ_1, θ_2, and θ_3 as dependent joint variables.

Definition 1: A *Configuration* of the 5-link closed chain is defined as a tuple $(\theta_2, \theta_4, \theta_5)$. The *Configuration* at which $\theta_2 \leq 0°$ is called the *Large Configuration* (LC) and the *Configuration* at which $\theta_2 \geq 0°$ is called the *Small Configuration* (SC). □

Definition 2: The space of θ_4 and θ_5 which corresponds to all large configurations is denoted as the *Large Configuration Space* (LC-space). The space of θ_4 and θ_5 which corresponds to all small configurations is denoted as the *Small Configuration Space* (SC-space). The space of θ_4 and θ_5 which corresponds to all configurations of the 5-link closed chain is denoted as the *Configuration Space* (C-space). The *Configuration space* (C-space) consists LC-space and SC-space. In case that LC-space and SC-space do not need to be distinguished, LC-space or SC-space is denoted as a 2D-subspace. □

The 5-link closed chain is presented as a point in the C-space. For each point in the C-space, there is an unique configuration of the 5-link closed chain.

Definition 3: All the adjacent points in C-space that do not satisfy the Collision-free Constraint are denoted as a *real-obstacle*, all the adjacent points in C-space that do not satisfy the Closed Chain Constraint are denoted as a *closed-chain-obstacle*, and all the adjacent points in C-space that do not satisfy the Joint Limit Constraint are denoted as a *joint-limit-obstacle*. A C-space obstacle can be either a *real-obstacle*, a *closed-chain-obstacle*, or a *joint-limit-obstacle*. □

Figure 2 shows an example of C-space of the 5-link closed chain. Each 2D-subspace is represented by a rectangle. The boundaries of the rectangle represent the joint-limit-obstacles. The closed-chain-obstacles are presented by the darkly shadded areas. The shape and the number of the closed-chain-obstacles are the same in both LC-space and SC-space. The shape and the number of the real-obstacles in LC-space and SC-space may be different (lightly shadded regions). Each connected free-space in LC-space or SC-space not occupied by C-space obstacles is denoted as a free-region. For example, there are two free-regions in Figure 2b. The current robot position in C-space is denoted as C-point. The problem of collision-free path planning for a 5-link closed chain is then transformed to finding a collision-free path of C-point in C-space.

Definition 4: The *Configuration* of the 5-link closed chain which satisfies $\theta_2 = 0°$ and the closed chain constraint is denoted as *Transition Configuration* (TC) (Figure 3). All TC's in LC-space (or SC-space) form a curve, denoted as TC-curve.

The 5-link closed chain can change its configuration from LC-space to SC-space, or vise versa, only through TCs (at which links ℓ_1 and ℓ_2 are alined but not overlapped). TC can be considered as both LC and SC. By the definition of C-space obstacle, TC-curve is actually the boundary of the *closed-chain-obstacle*.

3. Strategy for path planning

The path planning algorithm divides the problem according to the following two cases.

Case (1) S and T are in the same 2D-subspace.

Without loss of generality, the cases that both the start and target configurations are in LC-space is discussed. When the closed chain is restricted to be in LC all the time, the algorithm Bug2 [3] can be used to find a collision-free path within each free-region separately. Unless otherwise specified in the text, M-line is assumed as a straight line segment between S and T in a 2D-subspace. In complicated situations, the C-point may need to move between LC-space and SC-space several times before reaching the target point. For example, each line in Figure 4 connecting two free-regions represents all the TC's common to the two free-regions. When the C-point is to move from S to T, C-point has to move through regions 1 to 5.

Whether there is a "door" from the current free-region to another free-region in the other 2D-subspace needs to be remembered during the collision-free path searching. If a collision-free path in the current free-region cannot be found and a "door" to the other free-region from the current region cannot be found, it can be concluded that there is no collision-free path. If it is concluded that there is no collision-free path in the current path but there is a "door" to an unsearched free-region, there may be a path residing partially in SC-space and partially in another free-region of LC-space. Then, the C-point moves through the "doors" to unsearched free-regions to find a collision-free path.

Case (2) S and T are in different 2D-subspaces.

Without loss of generality, it is assumed that the start position is in SC-space and target position is in LC-space. The C-point has to go LC-space through a TC-point. Once the C-point moves to LC-space through a TC, the strategy described in case (1) can be used.

4. Analysis

Since the work environment is unknown, the problem is that how to find all the free-regions that the C-point can reach. To solve this problem, TC-curve segments which represents "doors" seen from the searched free-regions in LC-space and that in SC-space are remembered in the path searching process.

4.1. Properties of TC-curve

Since a path may pass through several free-regions, all the free-regions reachable by the C-point have to be searched before claiming no path exists. Hence, once a free-region is searched, the free-region should be remembered. In this subsection, the properties of TC-curve are discussed. It will be clear that only a finite number of TC's need to be remembered in order to find free-regions that are reachable by the C-point. Due to the length limitation of the paper, the proofs of the lemmas and theorem are omitted.

Lemma 1: The position of TC-curve in LC-space is the same as that in SC-space.

Lemma 2: If the C-point meets the boundary of a real-obstacle at a point which is on a TC-curve in a 2D-subspace, the C-point must meet the boundary of a real-obstacle at the same point in the other 2D-subspace.

Lemma 3: If there are no joint limits for the robots and no real-obstacles, there is at most one continuous TC-curve in both LC-space and SC-space.

If the independent joint variables have limits but there are not real-obstacles in C-space, TC-curve can be a closed curve or several open curves depends on the imposed robot joint limits and/or lengths of the robot links. When there are real-obstacles, it may divide TC-curve into finite number of segments. A segment can be either located on the boundary of a free-region or inside real-obstacles. A segment located on the boundary of the free-region is a collision-free segment. The C-point can move from one 2D-subspace to the other 2D-subspace only through the collision-free segments of a TC-curve.

Lemma 4: If the C-point can move from a TC point p on a collision-free segment to a given point in a 2D-subspace, the C-point can move from any TC-point on the same collision-free segment to the given point.

During path planning, the C-point moves along either M-line or the boundary of obstacles. If TC-curve does not intersect with real-obstacles and joint-limit-obstacles, by lemma 3, there is one and only one collision-free segments on TC-curve. If TC-curve intersects real-obstacles and/or joint-limit-obstacles, by lemma 4 and lemma 2, the intersections can be used to represent the collision-free TC-curve segments.

4.2. Convergence conditions

During the collision-free path planning, after a free-region is searched and no collision-free path can be found, the other free-regions reachable by the C-point, if there is any, need to be searched.

Theorem 1: Assuming C-point moves to a new free-region, if possible, only after all the intersections between TC-curve and joint-limit- and real-obstacles are found in the current free-region. If the numbers and the positions of the intersections between real-obstacles and TC-curve and between joint-limit-obstacles and TC-curve that have been reached by the C-point in LC-space and that have been reached by the C-point in SC-space are the same, all the free-regions that are reachable by the C-point have been found; otherwise, at least one free-region reachable by the C-point has not been searched.

Corollary 1: If both S and T are in the same 2D-subspace but not in the same free-region, and TC-curve does not intersect the boundaries of joint-limit- and real-obstacles, there is no collision-free path in the C-space.

5. Algorithm

The procedure SAME_SPACE is used to plan a collision-free path for the 5-link closed chain whose starting and target configurations are both in LC-space. If both the start position and target position are in SC-space, the strategies discussed in this procedure can still be used.

Procedure SAME_SPACE:

Input: Start and target configurations.

Output: collision-free path or the conclusion that collision-free path does not exist.

0. [Initialization] Set $n=m=1$, $LCTC_m = SCTC_n = \varnothing$, TRAN_LIST=nil, S=start configuration, T=target configuration.

1. [Search for a collision-free path] Start from S, while keeping the closed chain in LC-space, move to T by calling procedure LOCAL. If T can be reached, a path is found and stop; If T cannot be reached and TC_flag=0, there is no collision-free path and stop; otherwise, continue.

2. [Find all intersections between real-obstacles and TC-curve, and joint-limit-obstacles and TC-curve in free-region m] Call procedure FIND_TC. Save all the returned intersection points in TC_QUEUE to the set $LCTC_m$. Continue.

3. [Check stopping condition in LC-space] One of the following cases occurs at this point:

   case 1. If $\bigcup_{p=1}^{m} LCTC_p = \bigcup_{q=1}^{n} SCTC_q$, there is no collision-free path. Stop.

   case 2. If $LCTC_m \subset \bigcup_{q=1}^{n} SCTC_q$, set $SCTC_q = SCTC_q - LCTC_m$ for all $1 \leq q \leq n$. Then set $LCTC_m = \varnothing$, Go back to SC-space through the last point stored in TRAN_LIST. Delete the point from TRAN_LIST. Go to Step 5.

   case 3. Otherwise, set $LCTC_q^1 = LCTC_m$, for all $q=1$ to n, set $TEMP_q = LCTC_m \cap SCTC_q$ and $LCTC_q^{q+1} = LCTC_q^q - TEMP_q$ and $SCTC_q = SCTC_q - TEMP_q$. Set $LCTC_m = LCTC_q^{q+1}$, Move to a free-region in SC-space through a TC-point in $LCTC_m$. Append the TC-point in the list TRAN_LIST. Assign $m=m+1$, $LCTC_m = \varnothing$, and go to Step 4.

4. [Find all reachable intersections between real- and joint-limit-obstacles and TC-curve in a free-region in SC-space] Call the algorithm FIND_TC. Save all the returned intersection points in TC_QUEUE to the set $SCTC_n$. Continue.

5. [Check stopping condition in SC-space] One of the following cases occurs at this point:

   case 1. If $\bigcup_{p=1}^{m} LCTC_p = \bigcup_{q=1}^{n} SCTC_q$, there is no collision-free path. Stop.

   case 2. If $SCTC_n \subset \bigcup_{p=1}^{m} LCTC_p$, for $p=1$ to m, $LCTC_p = LCTC_p - SCTC_n$, and $SCTC_n = \varnothing$. Go back to LC-space through the last point on TRAN_LIST. Delete the last element in TRAN_LIST. Go to Step 3.

   case 3. Otherwise, set $SCTC_p^1 = SCTC_n$, for $p=1$ to m, set $TEMP_p = SCTC_p^p \cap LCTC_p$ and $SCTC_p^{p+1} = SCTC_p^p - TEMP_p$ and $LCTC_p = LCTC_p - TEMP_p$. Set $SCTC_n = SCTC_p^{p+1}$. Move to a free-region in LC-space through a TC-point in $SCTC_n$. Append the TC-point in the list TRAN_LIST. Set the TC-point as a new start point S, $n=n+1$, $SCTC_n = \varnothing$, and go to Step 1.

Step 0 is initialization. $LCTC_m$ and $SCTC_n$ are the sets of all intersections between TC-curve and the boundaries of joint-limit-obstacles and TC-curve and between TC-curve and real-obstacles met so far in free-region m in LC-space and in free-region n in SC-space, respectively, where m and n are integers. TRAN_LIST is a list of TC points at which C-point moves between LC-space and SC-space. In Step 1, corollary 1 is used to check stopping condition. In case that TC-curve coincide with the boundary of real-obstacles and joint-limit-obstacles, by lemma 4, only the two end points of the coincide curve need to be remembered. Step 2 calls the procedure FIND_TC to find all the possible 'doors' to move from free-region m in LC-space to SC-space. In step 3, stopping conditions are checked according to theorem 1. Steps 4 and 5 are similar to steps 2 and 3, respectively, except C-point is in SC-space.

Procedure LOCAL

Input: start and target points of the path.

output: (1) claim whether a collision-free path can be found in the current free-region. (2) TC_flag (TC_flag=1 if a TC-point is met in path search)

0. [Initialization] Set $j=0$, S=start point, T=target point, TC_flag=0. Start at S. Go to Step 1.

1. [Motion along M-line] Move along M-line toward T until one of the following occurs:

   case 1. T is reached. A collision-free path is found. Return.

   case 2. An obstacle is encountered. Define the point as hit point, H_j. Go to Step 2.

2. [Motion along the boundary of the obstacle] Move along the boundary of the obstacle at left direction until one of the following occurs:

   case 1. TC_flag=0 and a TC point is met. Set TC_flag=1. Go to Step 2.

   case 2. T is reached. A collision-free path is found. Return.

   case 3. M-line is met again. If the distance between the current position and T is shorter than that between the hit point H_j and T, define the current point as the leave point L_j, update $j=j+1$ and go to Step 1. If the distance between the current position and T is longer than that between the hit point H_j and T, go to Step 2. If the point is the last hit point H_j before a companion leave point L_j is defined, there is no collision-free path in the current free-region. Return.

The procedure LOCAL is the same as algorithm Bug2 [3] except it remembers if a TC on the boundary of an obstacle is met during path searching.

Procedure FIND_TC

Input: current position of the C-point.

Output: TC_QUEUE which contains all the intersections of TC-curve and real-obstacles and that of TC-curve and joint-limit-obstacles in the currently resided free-region of a 2D-subspace.

0. [Initialization] Set the current point as S, TS and TT. TC_QUEUE=∅, queue SEARCH=∅.

1. [Find all intersections of TC-curve and the boundaries of joint-limit- and real-obstacles] Move along the boundary of the obstacle at the left direction until TT is met. Append all the intersection points between TC-curve and boundaries of real-obstacle and joint-limit-obstacle met during the motion to the queue SEARCH. If the queue SEARCH is empty, go back to S and return. Otherwise, go to Step 2.

2. [Check return condition] Move to the last intersection point in the queue SEARCH by calling the procedure LOCAL. Append this intersection point to the queue TC_QUEUE. Delete this intersection point from the queue SEARCH. If the queue SEARCH is empty, return; otherwise, go to Step 3.

3. [Check if there are TC points on the boundary of another real-obstacle or joint-limit-obstacle] Move along TC-curve in the direction away from the real-obstacle or joint-limit-obstacle until reach another intersection of TC-curve and the boundary of a real-obstacle or joint-limit-obstacle. If the intersection point is in TC_QUEUE, go to Step 2; If the intersection point is in the queue SEARCH, move the point to TC_QUEUE; go to Step 2; otherwise, put it in the queue SEARCH. Set the current point as TS and TT. Go to Step 1.

This procedure is called only when no collision-free path can be found in the current free-region by the procedure LOCAL. At this moment, the C-point is on the boundary of either a joint-limit-obstacle or a real obstacle. Step 0 is initialization. In Step 1, the C-point moves along the boundary of the obstacle for a complete circle and remember all the intersections between TC-curve and the boundary of the real-obstacles and joint-limit-obstacle in a set SEARCH. If TC-curve and the boundaries of the real-obstacles or joint-limit-obstacles are coincide, by lemma 4, only the two end points of the coincided curves need to be remembered. In Step 2, the C-point moves to the last intersection point remembered in the queue SEARCH. If the set SEARCH is empty, all the intersections in the current region have been found by step 1. Since there may be intersections between TC-curve and boundaries of other real-obstacles and joint-limit-obstacles, in Step 3, other real-obstacle or joint-limit-obstacles in the region are searched by checking if other end of a collision-free TC-curve segment is met before. If the other end of a collision-free TC-curve segment is met before, the obstacle has been met before; otherwise, a new obstacle in the current free-region is met and the intersections of TC-curve and boundary of the new obstacle should be found.

If the start position is at SC-space and the target position is at LC-space the procedure DIFF_SPACE is followed. If the start position is at LC-space and target position is in SC-space, the strategies discussed in this procedure can still be used.

Procedure DIFF_SPACE:

0-2. Same as the procedure LOCAL except that the target point is an arbitrary TC-point.

3. [Path Planning by procedure SAME_SPACE] Move to LC-space through the TC-point. Assign the TC point as the new start point so that both the start and target points are in LC-space. Then the procedure SAME_SPACE is called to find the collision-free path of the C-point.

6. Conclusion

The described algorithm is the first to use non-heuristic method to plan collision-free path for two planar robots working coordinately in an unknown environment. As described in the paper, the robots and the straight line segments connecting the bases of the robots are modeled as a five link closed chain. The algorithm is based on separating the configuration space of the 5-link closed chain into two 2D-subspaces, i.e., large configuration space and small configuration space. The algorithm converges and is applicable to farely large robot population in the current robot world, such as SCARA type robots.

520

References

1. J. T. Schwartz and M. Sharir, "On the Piano Movers' Problem: III. Coordinating the Motion of Several Independent Bodies: The Special Case of Circular Bodies Moving Amidst Polygonal Barriers," *The International Journal Of Robotics Research*, vol. 2, No. 3, pp. 46-75, Fall 1983.

2. Tomas Lozano-Perez, "Spatial Planning: a Configuration Space Approach," *IEEE Transactions on Computers*, vol. C-32, No. 2, pp. 26-38, February 1983.

3. V. J. Lumelsky and A. A. Stepanov "Dynamic Path Planning for a Mobile Automaton with Limited Information on the Environment," *IEEE Transactions on Automation Control*, Vol. AC-31,No.11, November 1986.

4. K. Sun and V. J. Lumelsky, "Motion Planning with Uncertainty for a 3d Cartesian Robot arm," *5th International Symposium on Robotics Research*, Tokyo, Japan, August 1989.

5. H. Mabie, *Mechanisms and Dynamics of Machinery*, (Fourth Edition), John Wiley & Sons.

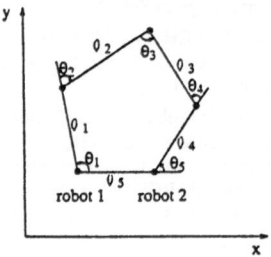

Figure 1. 5-link closed chain.

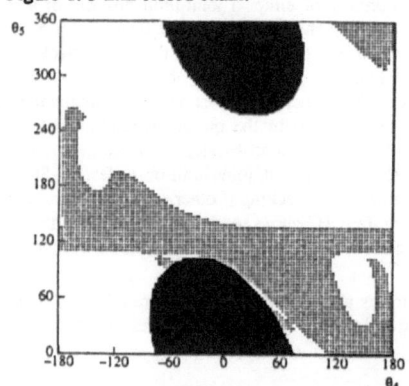

Figure 2(a). LC-space of the 5-link closed chain.

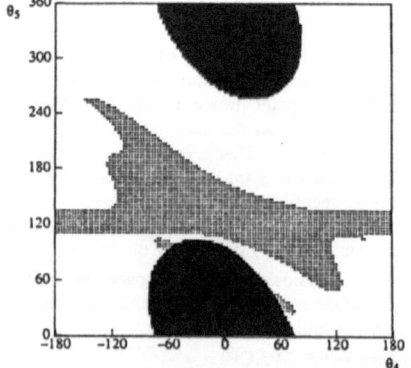

Figure 2(b). SC-space of the 5-link closed chain.

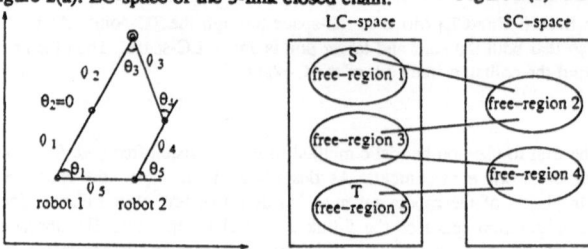

Figure 3. Transition configuration.   Figure 4. A path between S and T.

A Theory of Collision Avoidance on Visual Guidance of Robot Motion in Dynamic Environment

Q. ZHU
Department of Mathematics and Computer Science
University of Nebraska at Omaha
Omaha, NE

J. LIU
School of Electrical and Electronic Engineering
Nanyang Technological Institute
Singapore

ABSTRACT
A theoretic study and computer simulation of model and control strategy for visual guidance of robot motion in dynamic environment is presented. Characteristics that make the visual guidance and motion control in dynamic environment distinct from that in static environment are discussed. The paper inspects the system attributes that must be taken account of and explores the inter-relations between these attributes. A conformable model of vision processing with the motion control of robot is developed. Computer simulation verifies the model and provides valuable insight toward the optimization of the process.

1. INTRODUCTION

Real-time visual guidance and control of robot motion is an important and growing area in robotics and factory automation [1 - 6]. It is a challenging problem because of the complexity of the unknown environment to be encountered by the mobile robot. The environment complexity includes the variations of physical appearances of obstacles, their kinematic and dynamic behaviors, and the planned or unplanned perturbations of their motion. This is particularly true for the collision avoidance where the obstacles (may be other robots) in the environment are also moving at relatively high speed [4 - 6].

Many research work has been done on the visual guidance of mobile robot (or autonomous vehicle) moving in a static environment where the positions of obstacles do not change during the period of time the robot is operating [1 - 4]. One large and fruitful area is called "motion planning" where given a description of the environment and desired initial and final positions, a geometrical motion-planning algorithm computes and figures out a collision-free trajectory [1 - 3]. However, the visual guidance and control of robot motion in dynamic environment where the obstacles are also moving is less systematically studied.

Motion control of robot in a dynamic environment has two processes: (1) global path planning, which is aimed on the accomplishing of entire mission to be carried out by the robot, (2) local path planning, which is focused on finding a collision-free path within the sight of the robot. A global planning usually specifies an ultimate goal of the motion or a schematic specification of directions or routes of the motion. Such planning is conducted according to the task requirement and the overall static environment, therefore is sometimes called *task-level planning*. The process does not care very much about the details of the motion path and does not foresee any dynamics of the environment.

Moving obstacle detection and collision avoidance are dealt with in the local path planning. Such process emphasizes on the safety issue of the motion. It includes a sequence of operations consisting of obstacle detection and local trajectory planning[5]. The vision system is responsible for acquiring necessary information of the scene and determines whether there is any object, static or in motion, which may interfere with the planned trajectory of the robot. If the obstacle in the environment is also in motion, the system has to determine its velocity and acceleration, predict its moving trajectory, and check for possible interference. If a possible collision is detected, the trajectory planning process must be activated to identify a collision-free path for the robot to proceed. The reliability and effectiveness of the path planning largely depend upon the correctness of the detection of obstacle motion and the timely control of robot maneuver. It is the main concern of this paper.

2. MODEL OF VISUAL GUIDANCE OF ROBOT MOTION

A. Motion State Description

The motion state of a mobile robot at time instance t is described by:

$$(\mathbf{p_r}(t), \mathbf{v_r}(t), \mathbf{a_r}(t)),$$

where $\mathbf{p_r}(t)$ denotes the position of the robot in Cartesian coordinate system, $\mathbf{v_r}(t)$ the velocity, and $\mathbf{a_r}(t)$ the acceleration.

We use $\mathbf{s_r}(t)$ to denote the moving distance of the robot in time interval $[t, t+\Delta t]$, where Δt is the control cycle. From the motion rule of rigid body object, we have

$$\mathbf{s_r}(t) = \mathbf{p_r}(t+\Delta t) - \mathbf{p_r}(t) = \int_t^{t+\Delta t} \mathbf{v_r}(\tau) \, d\tau \tag{2.1}$$

where

$$\mathbf{v_r}(\tau) = \mathbf{v_r}(t) + \int_t^\tau \mathbf{a_r}(t) \, dt. \tag{2.2}$$

We use $T_c(t)$ to denote the *vision and control processing time* of the system. $T_c(t)$ includes the time for processing the scene image, detecting moving obstacles, figuring out a collision-free path, and issuing a maneuver command. Basically, we should have

$$T_c(t) < \Delta t. \tag{2.3}$$

In an asynchronous system Δt is a function of t, denoted as $\Delta t(t)$. We use $T_d(t)$ to denote the time delay incurred due to the kinematics of the robot. 2.3 thus becomes:

$$\Delta t(t) \geq T_c(t) + T_d(t). \tag{2.4}$$

Taking account of 2.4 with 2.1, we then have

$$\int_t^{t+T_d(t)\,+T_d(t)} v_r(\tau)\ d\tau \le s_r(t). \tag{2.5}$$

Equation 2.5 can have two interpretations: one is that the robot should only proceeds in a speed that allows sufficient time for visual and control processing to determine the next motion plan before reaching to the current destination; the other is that the vision and control system of the robot should have the motion trajectory of next time interval determined within the planned moving range $s_r(t)$ of current time interval.

B. Safety Motion Principle

The central problem of local path planning is to explore the value of $v_r(t)$ (or $a_r(t)$), which determines the motion trajectory of robot from time t to $t+\Delta t(t)$. Main consideration here is the safety issue of the motion. To represent the safety attribute of robot motion in dynamic environment, we use $s_a(t)$ which is defined according to following safe motion principle:

A robot should always keep $s_a(t)$ distance away from any obstacle whose motion behavior is uncertain.

$s_a(t)$ is the range within which the robot has sufficient time to go through a number of sampling and control cycles. By these cycles, the motion states of obstacles can be detected and maneuvers can be made by the robot to have any possible collisions avoided. Note that in static environment, this $s_a(t)$ can be very small. In dynamic environment it is a function of the velocity of robot and the velocities of obstacles.

Let $v_{O_i}(t)$ denote the velocity of an object O_i. $s_{a_i}(t)$ is the safety range of robot with respect to O_i. $s_{a_i}(t)$ is obtained by:

$$s_{a_i}(t) = \alpha \int_t^{t+\Delta t} v_{O_i}(\tau)\ d\tau + \beta \int_t^{t+\Delta t} v_r(\tau)\ d\tau + \gamma_0, \tag{2.6a}$$

where α and β are two coefficients, $\alpha, \beta \ge 0$. γ_0 is a constant in which the kinematics of the robot is embedded. Simply assuming $v_{O_i}(t)$ and $v_r(t)$ be constants during the time interval, above equation becomes:

$$s_{a_i}(t) = \alpha\, s_{O_i}(t)\ \Delta t(t) + \beta\, s_r(t)\ \Delta t(t) + \gamma_0. \tag{2.6b}$$

$s_a(t)$ can be obtained by taking the maximum of the $s_{a_i}(t)$'s:

$$s_a(t) = MAX\{s_{a_i}(t)\}. \tag{2.7}$$

A *trajectory-planning range* $s_t(t)$ is the distance within which the local trajectory planning of robot motion is made. From above discussion we see that:

$$s_t(t) \ge s_r(t) + s_a(t) \tag{2.8}$$

We call 2.8 the *motion planning constraint equation*.

C. Visual and Control Equation

A *field of view* $s_v(t)$ is the scope of scene taken by the robot vision system. Obviously, we should have

$$s_v(t) \geq s_t(t) \tag{2.9}$$

Let $s_o(t)$ denote the *observing range* in which the motions of obstacles are detected. It allows the motion planning be made with respect to the presence of obstacles in the scene. There is

$$s_v(t) \geq s_t(t) + s_o(t) \tag{2.10}$$

We call 2.10 the *visual guidance constraint equation*.

From 2.8, we have

$$s_r(t) \leq s_v(t) - s_o(t) - s_a(t). \tag{2.11}$$

Replacing $s_r(t)$ of 2.5, we obtain the following functional description of visual guidance and control equation of robot motion:

$$\int_t^{t+T_\delta(t)+T_\delta(t)} v_r(\tau)\, d\tau \leq s_v(t) - s_o(t) - s_a(t). \tag{2.12}$$

or

$$\int_t^{t+T_\delta(t)+T_\delta(t)} v_r(\tau)\, d\tau \leq s_t(t) - s_a(t). \tag{2.13}$$

D. Optimization

Equation 2.12 shows that the speed of robot and motion performance is mainly constrained by (1) $T_c(t)$, (2) $s_v(t)$, and (3) $s_a(t)$. While the roles of $s_v(t)$ and $s_a(t)$ cannot be neglected, $T_c(t)$ is the most critical factor. Other factor involved in the visual guidance and control of robot motion in dynamic environment is the *number of objects* within $s_v(t)$, denoted as $n_{obj}(t)$. Both $s_v(t)$ and $n_{obj}(t)$ effect $T_c(t)$, which in fact can be represented as:

$$T_c(t) \propto (s_v(t),\ n_{obj}(t),\ P_c), \tag{2.14}$$

where P_c represents the computational capability of the robot system, which depends on the hardware and software used for the visual and control processing. Note that should other environment complexities be considered, they also effect $T_c(t)$ directly or indirectly.

An optimal control state of robot motion in dynamic environment can be derived from above discussion:

$$\int_t^{t+T_\delta(t)+T_\delta(t)} v_r(\tau)\, d\tau = s_r(t), \tag{2.15}$$

and an optimal system attributes setting should be:

$$s_r(t) = s_t(t) - s_a(t) = s_v(t) - s_0(t) - s_a(t). \tag{2.16}$$

3. COMPUTER SIMULATION

The above robot motion model is simulated with different combinations of $s_v(t)$, $s_t(t)$, and $s_a(t)$ with respect to $n_{obj}(t)$, and $v_{oi}(t)$. The simulation applies a trajectory guided path planning approach[6]. A global goal is pursued by the robot during the simulation. The system is evaluated according to following measurements: (1) Average $T_c(t)$; (2) Collision Rate; (3) Average Deviation of the trajectory away from a global path; (4) Average robot motion speed. Simulation results are illustrated in Fig. 3.1.to 3.4, respectively.

It is observed from the simulation results that the increase of $s_v(t)$ has only very limited effect to the improvement of motion performance. The reason is that collision avoidance in dynamic environment is basically a local operation. The moving state of far away object has very little influence to the current trajectory of the robot. On the other hand, the increase of $s_v(t)$ increases the $T_c(t)$, thus degrades the system performance.

4. CONCLUSION

The problem of visual guidance and collision avoidance of robot motion in dynamic environment has been investigated. Basic vision and robot control models are studied. When a robot moves in dynamic environment, it senses the environment and plans a collision-free path which is consent to a global goal specified at the task level. A computational model that incorporates the vision process conformably with the motion control of robot is explored. Simulation results provide valuable insight into the optimization of system attributes involved in the visual guidance and control process.

REFERENCES
[1] A. Chattergy, "Some Heuristics for the Motion of a Robot", *The International Journal of Robotics Research,* Vol. 4, No. 1, pp. 59-66, Spring 1985.
[2] J. L. Crowley, "Motion for an Intelligent Mobile Robot", *The First Conference on Artificial Intelligence Application*, IEEE Computer Society, Denver CO. pp. 51-56, December 1984.
[3] O. Khatib, " Real-time Obstacle Avoidance for Manipulators and Mobile Robots", *The International Journal of Robotics Research*, Vol.5, No.1, pp.90-98, Spring 1986
[4] C. R. Weisbin, et al, "Autonomous Mobile Robot Navigation and Learning", *IEEE COMPUTER*, pp. 29-35, June 1989
[5] Q. Zhu, and N. K. Loh, "Modeling and Control Strategies for Dynamic Obstacle Avoidance By Mobile Robots", *Proceedings of 1989 Korean Automatic Control Conference*, Seoul, Korea, October 1989.
[6] Q. Zhu, "A Stochastic Algorithm for Visual Guidance of Robot Motion in a Dynamic Environment", *IEEE International Conference on Systems Engineering*, Pittsburgh, PA, August 9-11, 1990.

526

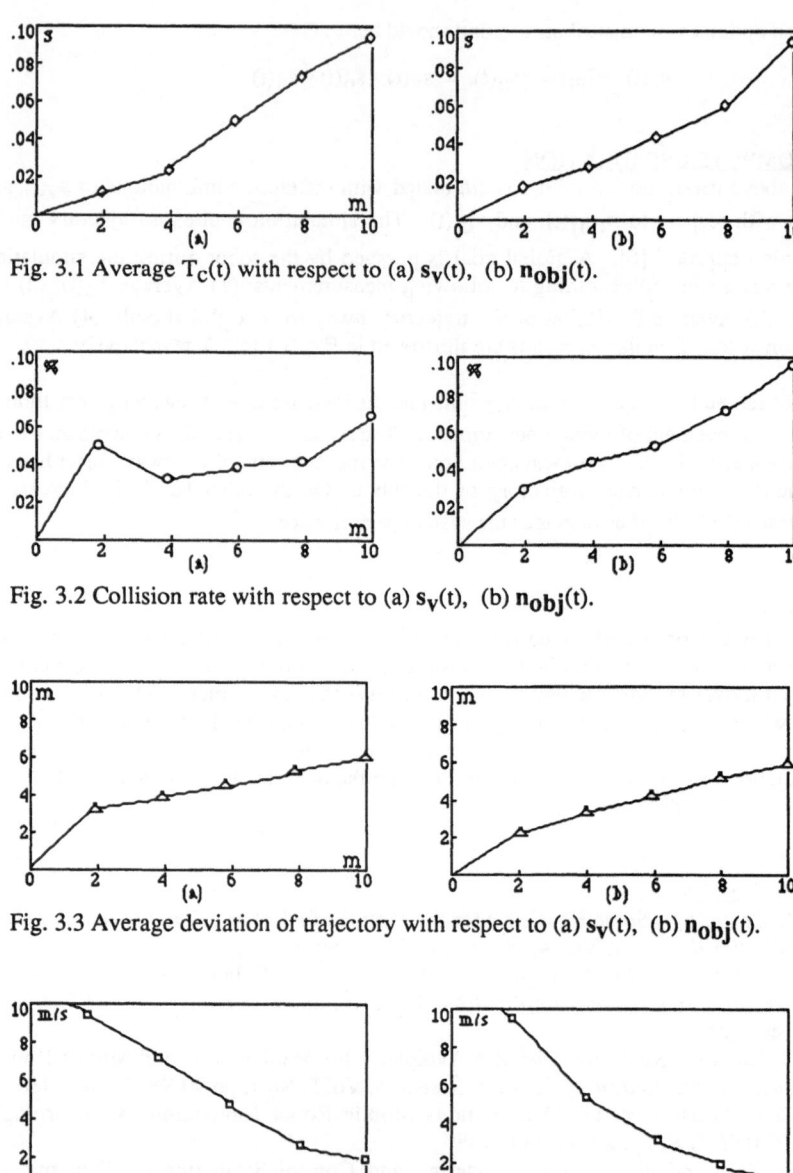

Fig. 3.1 Average $T_C(t)$ with respect to (a) $s_v(t)$, (b) $n_{obj}(t)$.

Fig. 3.2 Collision rate with respect to (a) $s_v(t)$, (b) $n_{obj}(t)$.

Fig. 3.3 Average deviation of trajectory with respect to (a) $s_v(t)$, (b) $n_{obj}(t)$.

Fig. 3.4 Average motion speed with respect to (a) $s_v(t)$, (b) $n_{obj}(t)$.

A Collision Prediction System for a Robotic Environment

NANCY SLIWA
NASA Ames Research Center
Palo Alto, CA

WILLIAM BYNUM and CHARLES WATLAND
Department of Computer Science
College of William and Mary
Williamsburg, VA

Abstract

A number of papers in the past have dealt with the formidable problem of planning a collision-free path for a robot or robotic manipulator through an environment of obstacles [see 1, 2, 3, 4].

This paper describes a system designed to perform the more tractable task of modelling the objects in a robotic laboratory environment and warning when collision between two or more of them appears imminent. The collision detection system has been implemented in the Intelligent Systems Research Laboratory of the Automation Technology Branch at NASA Langley Research Center. Acceptable performance has been demonstrated with as many as 30 objects, including two six-joint PUMA 560 manipulators. The system has both "selective" and "non-selective" collision checking modes. In the selective collision checking mode, collision checks between objects that can collide is triggered when the cumulative distance change between the two objects has exceeded a threshold. In the non-selective mode, collision checks between objects that can collide are always performed. The system has been tested under various degrees of clutter in the environment.

Introduction

Telerobotic operations in space have unique collision avoidance requirements. Many collision avoidance systems that have been developed in recent years [1, 4] can work in real-time in a fixed environment. That is, they assume that the objects in the environment are not changing with each movement of the manipulator. In the space environment, however, the environment is constantly changing as work is being performed by the manipulator system; in fact, that is the whole objective of telerobotic activity – to change the environment. A collision avoidance system must

Research was supported in part by NASA Grants NAG-1-775 and NAG-1-906.
Direct correspondence to second author at:
Department of Computer Science,
College of William and Mary,
Williamsburg, VA 23185,
(804) 221-3456, e-mail: bynum@cs.wm.edu

also be able to handle unexpected objects coming into the range of the manipulators, such as space debris, loose-floating tools, etc. It would be unacceptable to wait a substantial amount of time to update an environmental database at each system cycle.

In space operations, the areas that a manipulator must avoid hitting are not necessarily just the boundaries of solid objects. Zones of avoidance also include areas of unacceptably high thermal and cosmic radiation, areas of interference with scientific and communications receivers and transmitters, areas of thruster plumes, etc. These must be able to be modelled and avoided by the system.

Many collision avoidance systems cannot work in real-time in three dimensions, or with more than a single manipulator, particularly in a sufficiently real-time manner. These aspects are all important to space telerobotic activities.

Perhaps one of the most difficult aspects of space telerobotic operations to deal with is the presence of the human operator as a controller of the manipulator system. In many cases, the operator's attention will be focused on one portion of the operation relevant to the task at hand, while a potential collision with another part of the manipulator is unperceived. These collisions could be cause by unexpected objects coming into range, or by the operator's inputs themselves. A collision avoidance system must handle both inadvertent operator inputs, and purposeful but potentially hazardous inputs. It must do this in a way that does not hinder the mission of the operator. Most activities that telerobotic systems will be performing will require "controlled" collisions, that is, connecting parts together and grasping objects. A collision avoidance system should be able to determine from the task context when a specific "collision" is acceptable or not.

The collision avoidance task can be divided into four categories: *collision detection, collision prediction, path planning*, and *trajectory planning*. A *collision detection* system will detect when a collision has occurred. This is usually too late to help, particularly in space telerobotic systems where one is dealing with exceptionally expensive equipment which is difficult to access for repair, and whose operation is usually critical to mission success. A collision detection system could at most initiate safing procedures to prevent catastrophic failure as the result of a collision.

A *collision prediction* system can give sufficient advance warning to an operator to prevent the actual collision. This is much preferable, but more difficult to do in a timely manner. *Path planning* involves actual precalculation of a collision free path. *Trajectory planning* is similar to path planning, but also involves precomputing a path over a specific time sequence.

This paper deals with a collision prediction system. This system was developed to research techniques for real-time collision prediction in a complex, dynamic space environment with shared computer/human control of a manipulator system.

This collision prediction research effort proposed a number of techniques to allow rapid collision prediction suitable for space operations. These are described in detail below:

Simple geometric modelling

Objects are represented as spheres (and lists of spheres) to allow simple and rapid collision checks between objects. Planes were also introduced as a modelling type since it is very difficult to approximate a planar object with any precision with a group of spheres. Capped cylinders are also being considered as a modelling type, to provide better modelling accuracy. There is a continuing trade-off between the rapidity of collision checks and the accuracy of the environmental model.

Unlimited object motion

Objects have equations of motion associated with them to allow predictive tracking. This is relatively easy to do with manipulators and the objects they grasp (through kinematic equations) and, of course, stationary objects. It would be more difficult to infer equations of motion for unexpected objects that enter the environment. These would presumably be perceived by a sensor such as a camera, and tracked through several frames in order to determine future expected motion.

Unlimited collision potential

The system will allow any object to potentially collide with any other object. This can be very useful for tracking whether a manipulator is about to collide with itself, or with unexpected objects. It also allows collision checks between objects and sections of the manipulator other than the tip (end-effector), such as the shoulder or elbow.

Heuristic limits to collision checks

In order to limit the amount of computation required at each cycle to check for potential collisions between all objects in the environment, this project examined several techniques for eliminating checks.

One is a static check elimination: if two objects can never collide (e.g., the separate fixed bases of two manipulators) then these objects can be declared "mutually exclusive" in the database, and collision checks will never be run between these two objects. Actually, a large number of objects in a typical environment can be eliminated in this way, given the geometry of most mechanisms.

Another method of eliminating collision checks was proposed but not implemented in this initial collision prediction system: elimination by the dynamic position of objects in the environment. Some objects can move in and out of range of other objects. For example, a space manipulator on a moving platform may move from one position on the Space Station, where it must avoid certain antennae, to another position where the antennae are out of reach, but the solar arrays are now a concern. In dynamic collision elimination, objects in the antennae cluster would be removed from the collision-check list of the manipulator, while the solar array objects would be added. Using this approach, major sections of the geometric database could be removed from consideration, greatly reducing the required calculations at each cycle.

Another approach that was examined was the use of heuristics to eliminate continuous collision checks. For example, a simple heuristic would be to assume that two objects cannot collide until the sum of the distances travelled by both objects was greater than their original distance apart. Operationally, this would mean initializing the distances between each pair of objects, and keeping track of how much each had moved. Once the sum of their movements reach a certain threshold approaching the original distance, a collision check would be done, and their mutual distances re-initialized. The question arises as to whether the time saved by eliminating collision checks would be offset by the added bookkeeping overhead required by this approach. This study attempted to acquire data concerning this issue.

Description of the Collision Detection System

The collision detection system is implemented on a Symbolics 3620 in Symbolics Common Lisp. Objects in the environment are modelled using spherelists [see 5]. The system provides the user with an OBJECT type for handling spherelists. The type contains "current" and "next" spherelists, a name used to identify the collection of spherelists, a mobility type ("mobile" or "fixed"), and a location of the "center" of the collection of spherelists to be used in the calculation of the approximate distance travelled by the spherelist over the last cycle. The "next" spherelist is used to hold the anticipated position of the OBJECT at the beginning of the next update cycle. In addition to the OBJECT type, the system provides a PLANE type for specifying bounding planes in the environment such as walls, floor, and ceiling.

The system can handle any arrangement of objects and planes. No details of the environment are built into the system. This sort of generality makes it possible to accommodate equipment location changes in the laboratory without requiring extensive changes to the system software.

To introduce an object into the environment, the object is specified by means of a file of LISP functions that is arranged into three sections. The first section of the file defines the object as a collection of elements of the system OBJECT type.

The second section of the file defines a "collision-init" function that names the objects and surfaces in the environment with which the object can collide. For example, the collision-init function for LINK5-BLACK-PUMA might declare that it can possibly collide with LINK3-WHITE-PUMA, LINK4-WHITE-PUMA, and LINK5-WHITE-PUMA. If these names are not known, an OBJECT can declare ALL, which means that it can possibly collide with everything in the environment. The system needs these declarations because it maintains a list (in the form of a lower-triangular array) of pairs of OBJECTs that must be checked for collision. Each bounding surface in the environment owns a list into which an OBJECT type can be inserted if it can collide with the surface.

The third section of the file consists of a LISP function that implements the kinematics of the object. This section of the file is expected to contain the data structures needed to keep track of the position and velocity of the object.

The system maintains two lists of LISP functions, the "collision-init" list and the "move" list that contain the functions defined in the second and third section of each file for each object in the environment. At system initialization, the functions on the "collision-init" list are executed after all files have been loaded. Objects can be introduced into the environment dynamically at run-time not just at system initialization time. However, as a conservative measure, the "collision-init" function of such an object is expected to declare the possibility of collision with all existing objects in the environment.

Then, the system connects into the laboratory communication network to receive data from the PUMA manipulators and finally enters the collision detection phase in which it continuously cycles through the move list, executing each function in turn and checking for potential collisions.

The PUMA manipulators are currently modelled with five links, each of which is a separate OBJECT. Any number of manipulators can be instantiated. To model the actual conditions in the laboratory, only two are used at the present time.

The collision detection system has both a "selective" and a "non-selective" collision checking mode. In the selective collision checking mode, the cumulative distance travelled since the last collision check is maintained for each pair of OBJECTs that could possibly collide. When the distance travelled for the two OBJECTs becomes greater than the distance between them, the two OBJECTs are checked for possible collision; otherwise no collision check is performed. In the non-selective mode, a collision check is performed for each pair of OBJECTs that could possibly collide, regardless of the distance travelled by the two OBJECTs. The user can switch between the two modes by means of a mouse-activated pop-up menu.

The selective strategy was implemented in an attempt to minimize heuristically the amount of computation required per cycle, increasing the speed of the system for real-time operation. The non-selective strategy was retained as a means of quantitatively measuring the effectiveness of the selective strategy.

A number of debugging facilities have been implemented to enhance the system and to check that it is working correctly. A pop-up menu allows the user to toggle each debugging feature on or off (see Figure 1). The foremost of these facilities is a graphical representation of the system's view of the environment. This graphical representation is computationally expensive and is not needed for operation of the system because the Symbolics 3620 that hosts the system is located with a clear view of the laboratory environment. On the other hand, the graphic representation of the environment was an indispensable debugging tool in the development of the system.

The major events of the system can be timed using the microsecond timer provided by the Symbolics. The source of the joint velocities of the manipulators can be varied between being determined locally (assuming that the time interval between each position is constant) or determined from the microVAX controlling the manipulators. Various parts of the collision detection calculations can be displayed.

Integration with the Intelligent Systems Research Laboratory

The collision avoidance system was developed and tested with the graphics interface, and then integrated with the manipulators in the Intelligent Systems Research Laboratory (ISRL) at NASA Langley Research Center. The ISRL houses two Unimation PUMA 560 six-joint robotic manipulators with vision/image acquisition, graphics, and sensor systems supported by network of computers, consisting

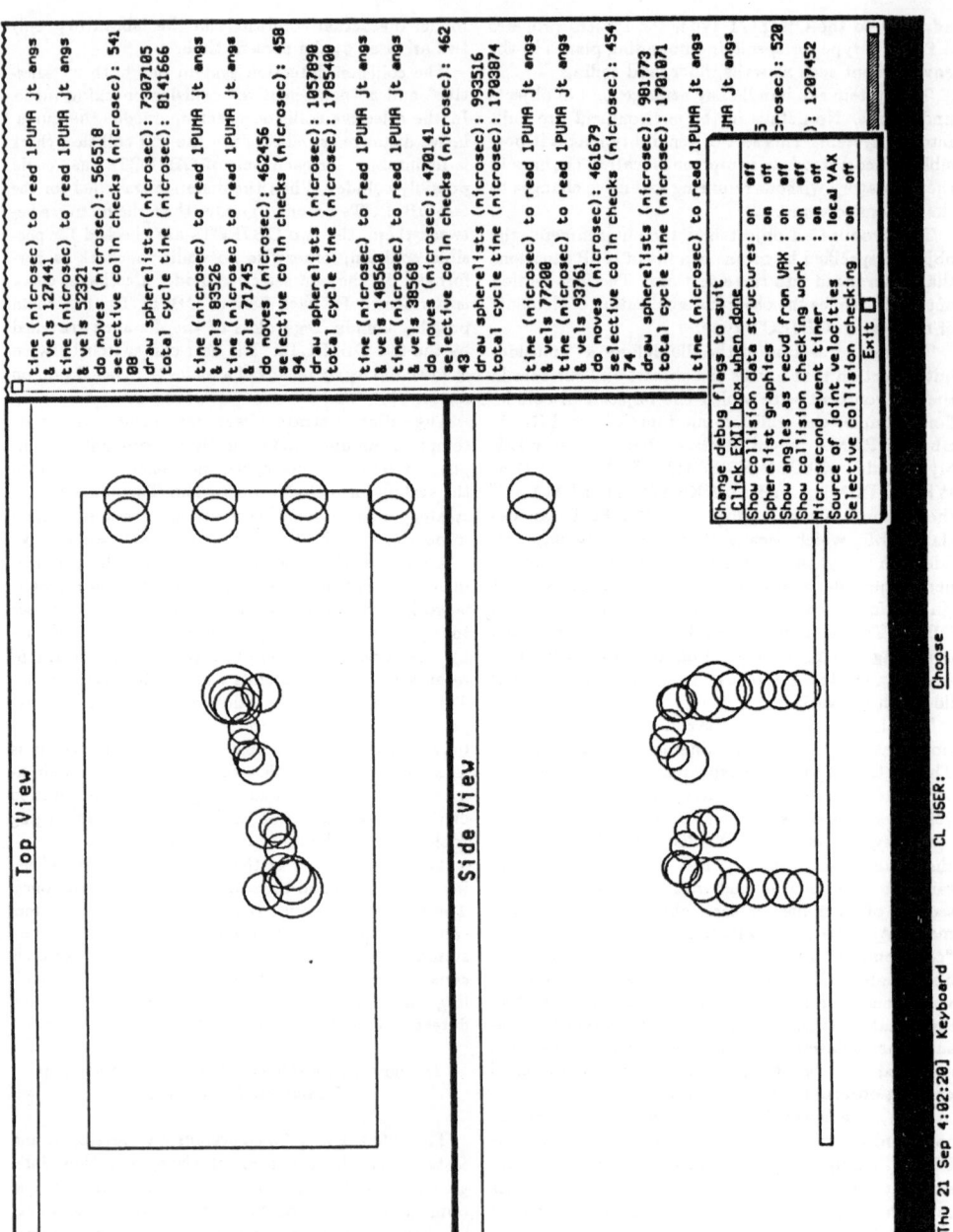

of a VAX, microVAX workstations, and Symbolic machines. The laboratory is used as a testbed for research into topics relevant to the automation of remotely controlled space systems, such as manipulator dynamics and control, man-machine interface of telerobotic systems, operating systems and languages for distributed and networked computers, and applications of artificial intelligence to telerobotics.

The descriptions of the current states of the two manipulators are kept in blackboard-type data structures owned by the controlling program running on a microVAX. The collision prediction software running on the Symbolics communicates via a DECnet connection to a stub running on the microVAX to obtain current joint angles and velocities for both manipulators.

To predict the future positions of the manipulators, the system can either use the sampled joint angles and velocities or the joint velocities can be calculated locally through the joint angles of two successive joint angle samples. The user can select between these two alternatives at any time by means of a pop-up menu. This flexibility is useful for debugging purposes in determining which of the two velocity calculation methods will most accurately predict the actual positions of the manipulators.

Experimental Results

This collision prediction system has performed successfully with the hardware of the ISRL during initial testing and evaluation. A number of measurements were taken.

For an uncluttered environment (just the two manipulators), the system can complete a cycle of selective collision checks in an average time of 0.58 seconds, with the graphical representation of the environment disabled. The corresponding time for non-selective collision checking is 0.60 seconds, so these preliminary figures indicate that the selective collision checking is slightly more advantageous. The graphical representation of the environment essentially added 1.5 to 2.0 seconds to these cycle times.

Using the OBJECT definition facilities of the system, a file declaring a varying number of wandering mobile objects was developed for inclusion into the environment at startup time. These OB-JECTs were declared to be able to collide with all other OBJECTs in the environment to impose the maximum computational burden on the system. For selective collision checking, there was little variation in total cycle times between an uncluttered environment (just the two manipulators), and two, four, eight, or sixteen wandering mobile objects. The average total cycle time 0.57 seconds for two wandering objects, 0.56 seconds for four, 0.56 seconds for eight, and 0.61 seconds for sixteen.

The variation of these figures is due to the effects of ephemeral garbage collection on the Symbolics with the shorter times occurring when less garbage collection was required. The increase in average cycle time from the uncluttered environment to sixteen wandering objects is much smaller than expected.

The possibility of hosting the collision detection system on a VAX was explored. The Symbolics Common LISP version of the system was ported to VAX Common LISP and tested on a VAX 750. The average total cycle times were roughly comparable. The VAX-hosted system was not satisfactory, however, because of the stop-and-copy garbage collection used in VAX Common LISP. When dynamic memory is exhausted in VAX Common LISP, all processing is suspended until the copy operation to the free half of dynamic memory is completed. This suspension can last for several seconds. In a collision detection system, this sort of delay is simply not acceptable.

The preliminary timing results obtained have demonstrated that the system can function effectively in detecting collisions with the moderate object velocities that are typical in the laboratory.

Future Work

The collision prediction system described in this paper was developed as the first component of an integrated modular collision avoidance system, which was to have included path planning and trajectory planning modules as well. However, a different approach to collision detection became available to the ISRL as a result of new graphics technology. The solid geometry modelling system used by the Lab to model and display the task environment to the telerobotic operator has collision detection capabilities built in at the hardware level. It was decided to use this capability as the foundation for the Lab's collision avoidance work. The hardware approach offered much finer resolution, thus better accuracy, of the boundaries of objects, with real-time response over a very large number of objects. Future development of the collision avoidance system described herein has therefore been suspended.

However, this approach would still be valid in the event that such sophisticated hardware capability were not available, as is currently the case for space operations.

Conclusions

On the basis of preliminary testing, the collision detection system described here has demonstrated its usefulness and effectiveness in warning of expected collisions in the laboratory environment. The system provides a general, extensible collision detection framework into which new objects can easily be introduced.

References

1. Brooks, Rodney A. Solving the Find-Path Problem by Good Representation of Free Space. *IEEE Trans. Systems, Man, Cybernetics*, v. SMC-13, no. 3 (Mar. - Apr. 1983), pp. 190 - 197.

2. Freund, E.; Hoyer, H. Real-time Path-finding in Multirobot Systems Including Obstacle Avoidance. *International J. Robotics Research*, v. 7, no. 1 (Feb. 1988), pp. 42 - 70.

3. Khatib, Oussama. Real-time Obstacle Avoidance for Manipulators & Mobile Robots. *International J. Robotics Research*, v. 5, no. 1 (Spring 1986), pp. 90 - 98.

4. Lozano-Pérez, Tomás. A Simple Motion- planning Algorithm for General Robotic Manipulators. *IEEE J. Robotics and Automation*, v. RA-3, no. 3 (June 1987), pp. 224 - 238.

5. Orlando, Nancy. An Intelligent Robotics Control Scheme. *American Controls Conference*, San Diego, CA, June 6 - 8, 1984, 6 pp.

Multivalue Coding: Application to the Autonomous Robots

A. PRUSKI

Laboratoire d'Automatique et d'Electronique Industrielles
University of Metz
Metz, France

Summary

This paper describes a free space modeling method by multivalue coding. Each code defines some numerical values representing a set of grid cells. The idea consists of using the grid as a Karnaugh board which rows and columns are binary coded rather Gray coded. This operating method allows to define, for each code, its grid location and allows numerical comparison in order to locate a code relatively to an other. This aspect is helpfull for path planning. The free space model is with a switching function or a tree represented to which boolean algebra rules and mathematic operations are applied. We describe an application to mobile robot path planning

INTRODUCTION

An environment model divides the space in accessible or non accessible areas. The dividing is by two methods performed. The first one describes the free space either with particular areas which imposes shapes and which geometric parameters represents the model [1] or uses obstacle bounderies to form the biggest areas which have particular characteristics[2][7]. The second method lies on the whole space grid which meshes are orthogonal or not and which strips size is either defined in a constant manner or to the environment adapted [5][11]. The model may be from a grid derived.[8] proposes to use the quadtrees to represent a binary grid. [12] propose a model based on crossing points determined by the obstacles vertices and of the robot orientation dependant. The points are connected by a graph called visibility graph defining the path possibilities. An other method uses highways defined with the bounderies of the Voronoï diagram[10]. The environment dividing aim is twice:create homogeneous areas which have some characteristics and define the links between them. The link description is generaly performed with a graph where the nodes represents an environment area defined either with the model directly or determined with a treatment of some areas by the model described.

We propose a grid based model homogeneous or not composed with a set of codes representing blocs of elementary cells. The used code is such it defines the size of an environment region, the links with the other codes and its grid location.

PRINCIPLE

The method lies on a set of binary coded numerical values. This coding that we call multivalue coding is considered, according to the case, either as a numerical value or as a set of numerical values. In the first case we apply to it mathematical operations (addition, substraction...) and logical operations (and, or...) in the second case.

DEFINITION

Let us consider the space of numbers which elements takes three states 0, 1 or X. The X value represents simultaneously the values 0 and 1. Let us build the set of multiple numbers.

$$T = a_n \, a_{n-1} \, ... \, a_1 \, a_0$$

with ai one of the three states above defined. In decimal basis we have

$$(N)_{10} = a_n 2^n + a_{n-1} 2^{n-1} + ... + a_1 2^1 + a_0 2^0 \qquad \text{with } a_i = 1, 0 \text{ or } X.$$

We see that one value gives more values N if T contains X elements. For example:T=1X11X => $(N)_{10}$=(22,23,30,31).

In that paper and in the application to path planning we consider only a sub-space of multivalue codes: the continuous multivalue codes defined in following manner:$a_0 = X$ and $a_j = X$ only if $a_i = X$ with $j=i+1$.

It may be possible to apply mathematical and logical operations. In this paper we are focused only on the application that modelises an environment and performs the path planning for a mobile robot.

APPLICATION TO THE MOBILE ROBOTS

1-The environment modeling

The environment modeling consists of creating in memory a space map or a picture in which the robot moves. The model describes either the whole space ,to say accessible or non accessible areas , or only the free space. The multivalue number environment representation is performed with a 2D or a 3D grid. In the following , we use the two dimension model to explain the principle. In two dimensions the environment is dividing in rows and columns which intersection forms the elementary cells which are set to "1" if it is accessible and reset to "0" if it is part of an obstacle. Starting from the binary grid we group the "1" states to create accessible environment regions representing free space. This operation is performed in manner of a Karnaugh's board. The principle of Karnaugh's logical equations simplification consists of erasing variables where logic state is modified if going from a cell to the neighbor. The row and column coding with a GRAY code allows only one variable to change state at a time.The code's disadvantage constitutes the difficulty to relate the code and its location in the board. We propose to use the natural binary code allowing to link the numerical code value to the group location without altering the grouping. In monodimensional case,the natural binary code allows to group 2^k cells with $1 < k < n$ and 2^n the maximal board size. In case of logical equation simplification, the modified variable is erased. In the computer we work with a fixed format, thus the modified variable (bit) is replaced with a third state X. In practical case, a group is coded with q bits, n for the row code and m for the column code. Each third state needs adding a code determining the state type:explicit (1 or 0) or non explicit (X). Thus a group is defined with 2q bits:

- q bit for the code
- q bits the validation or type of bit

For example, a 32 bit word represents any code of a 256*256 cells grid. A true state is with a "1" represented,a false state with a "0" and a non explicit state with any value ("0" or "1").

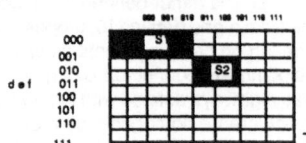

The coding of grid in Figure 1 gives:

Fig1: Coding example.

$$S = S1 + S2 = \bar{a}\ \bar{d}\ \bar{e} + a\ \bar{b}\ \bar{d}\ e \quad \text{represent the logical equation.}$$

In validation code, the bit corresponding to the code bit is set to "1" if the code is explicit (or valued) and reset to "0" if its state is non explicit. Let us take the 8 row and 8 column map of fig. 1 corresponding to 3 variables (a, b, c) in row and 3 in column coded in following manner:

$$\bar{a}\ \bar{d}\ \bar{e} = \underset{\substack{abc\ def\ abc\ def \\ code \quad\quad validation}}{\underbrace{0XX0XX}|\underbrace{100\ 110}} \qquad a\ \bar{b}\ \bar{d}\ e = \underset{\substack{abc\ def\ abc\ def \\ code \quad\quad validation}}{\underbrace{10X01X}|\underbrace{110110}}$$

In the computer the X states are replaced with "0" corresponding to the lowest value of the set of values. Following this example we see that the freespace is described with a continuous multivalue number on which may be applied mathematical and logical equations. Note the following property: the validation code of a continuous multivalue code represents the two's complement of the set cardinal.

2-Coding algorithm

The grid is in memory defined as a list of m words (or rows of the map) of n variables (bits) in 2D and a set of lists in 3D. The coding of the monodimensional case is performed always on a word WORD (or a set of words if the size of the grid is greater than the computer word size) with the following algorithm.

```
j=0;q=1;
DO
        i=1;Compute the maximum mx of i at position j;
        DO
                Create the mask M(j,i) of n bits where i consecutives bits starting at the jst position from
                left side are set to '1';
                R = WORD AND M(j,i);
                if R = M    then {/* it exists a code who groups minimum i bits*/i=2i;}
                            else
```

if i=1   then {/*it don't exist a code*/i=mx;}

         else   /* coding*/

           {code(q) = j ; valid(q) =~ (i/2-1);/* ~=one's comple-

ment*/

         q=q+1;j=j+i/2; i=mx;}

       WHILE (i<mx);

   WHILE (j<n);

The time complexity of this algorithm is O(n). In the second dimension case we have a list of m words of n variables (bits) representing the bit map. In the first time, the coding consists of performing a set of (m-1) logical AND between the words and in second time of performing the coding operation of each resulting word as seen in monodimensional case.

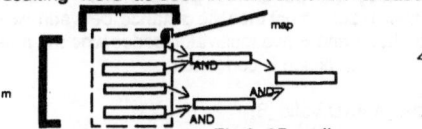

Fig.2: 2D coding.   Fig.3:3D coding

The second dimension (Fig.2) (code and validation) is determined with the level of the AND operation. The codes are defined with the containing of the words before the AND operation. A code is validated if it don't be in a word resulting of an logical AND. The algorithm is performed with a time complexity of O(nm). Like in 2D,the 3D case coding (Fig.3) is realised after performing AND operations between the binary words.The logical operations are performed in first time with all 2D grids (level I) and in second time between the level II. A code is validated if it don't be in the neighbor level. The number of AND operations for a m words of n variables and depth p grid is (m-1)(2p-1) and the complexity is O(nmp).In all dimensions the codes are unique and minimal .

3- Creating a Multivalue Code Tree :MCT

The obtained list is itself sufficient.Each code contains enough informations to determine the free space size and it's location.Code transformations are directly usable as we can see in the following. Nevertheless neighbor codes fusion operation or a code search in the list involves an important search time (O(n^4) corresponding to all code analysis in 2D case). To limit the computing time we propose to create a tree. The proposed coding allows a fast tree building. We consider in first time a monodimensional code having the two parts: code and validation noted as in the following example:

ex: C = 00000000/1111110   A   depth i

        code/validation   V = 0   V = 1

We consider a set of three nodes as following: B   C   depth i+1

    - A is a parent node of B and C with depth i

    - B and C are sons of A with depth i+1.

A boolean variable V is associated to each node. We assume the convention that if the left branch going from A to B is active than V=0 and if the right branch going from A to C is active than V=1.

The Multivalue Code Tree (MCT)building consists of giving each node the value of the code binary variable from left to right starting from the source of the tree. A branch is active if the validation variable corresponding to the code bit is set to "1".

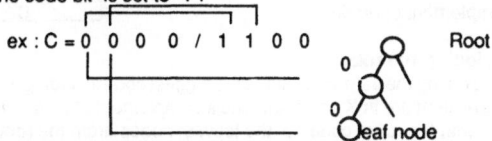

Each lief node corresponds to a multivalue code. This tree correspond to a 2'tree (bitree).In the second dimension each bitree node may be the origin of an other bitree. To create such tree we start with a primary tree corresponding to one dimension code and a second bitree (or secondary tree) for the second dimension of the code.

ex: C = 10010/11110   10000/11100

       X   valid_x   Y valid_y

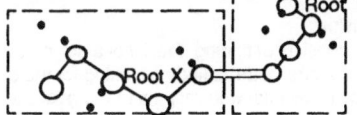

Each tree may be performed taking the code X for primary tree and Y for secondary tree or the inverse. In 3D or more dimension we repeat the same procedure. The secondary tree nodes turns origin of a third tree and so on.

PATH PLANNING
The path planning consists of finding the list of codes including the source point to a code including the goal point. Before describing the algorithm principle, we define some operations used in 2D case.

1 Distance computing
The path planning proposed algorithm is an heuristic algorithm using the distance between the actual computing code and the goal code to choose the better code that perform the goal approach in fastly manner. If an intersection in X or Y exists then the distance is zero. If not then the distance between two codes in X or in Y is performed with two operations. Let A and B two multivalue codes who the non explicit variables are replaced with 0. The existance of an intersection is computed:
* either on codes :
inter = (Code_A XOR CODE_B) AND Valid_A AND Valid_B;
if (inter ≤ 0) then intersection exist
* or on MCT :
Two codes are intersecting if the way followed from the root to the leave node is the same for the two codes.
The distance is given by:
$$Disq(Aq, Bq) = Min (|\sim Valid\_Aq + (Code\_Aq-Code\_Bq)|, |\sim (\sim Valid\_Bq) + (Code\_Aq-Code\_Bq)|)$$

with q= x or y. $\boxed{Dis = Dis_x + Dis_y}$

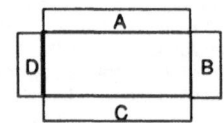

example: we consider an 8 * 8 grid and two codes A and B:

```
        X       Y
A = 000/000 000/111
B = 011/111 100/110
```
- computing of intersection existance:
inter_X = (000 XOR 011) AND 000 AND 111 = 0   => intersection on X and distance = 0
inter_Y = (000 XOR 100) AND 111 AND 110 = 100 => no intersection and the distance is equal to

$$Dis_y(A, B) = Min(|0 +(0-4)|, |-1 + (0-4)|)=Min(4,5) = 4$$

2 Bounderies computing
The bounderies computing consists of computing the codes representing the four surrounding codes having one cell thickness. Let us consider a multivalue code composed with CODE_X, Valid_X, CODE_Y, Valid_Y:
* A = { CODE_X, Valid_X, CODE_Y-1, ~0 }
* B = { CODE_X + ~Valid_X + 1, ~0, CODE_Y, Valid_Y }
* C = { CODE_X, Valid_X, Code_Y + ~Valid_Y + 1, ~0 }
* D = { CODE_X-1, ~0, CODE_Y, Valid_Y }

with ~ the one's complement operation.

3 Computing the code including the robot
The first code in the list representing the path constitutes the biggest code including the robot. The source point is a code where the validation are in X and Y the one's complement of 0. It consists to follow on a MCT the path represented with the robot code to the leaves nodes from the root node. The choice of the better code is performed according to the biggest size computed as following:
size = (~Valid_X + 1)*(~Valid_Y + 1)
The complexity of the operation is $O((\log n)^2)$.

4 The algorithm
The principle consists of finding the list of a set of contiguous codes from the source code to the goal code. The contiguous code is finding looking for the code that intersect one of the four bounderies code which distance is minimal with respect to the goal code.

4.1 Finding the contiguous code of CODE

This algorithm consists to find the better code contiguous of CODE according to the least distance to goal code. For example ,we assume that CODE is a D type surrounding code:

CODE = { CODE_X,Valid_X,CODE_Y,Valid_Y }

```
{      Node_X = CODE_X;Work_Valid_X = Valid_X;
       Node_Y = Code_Y;Work_Valid_Y = ~0;
       Max_Y = CODE_Y OR ~Valid_Y;
       Distance = Big_Value;
       DO{
       Following the way on MCT from root to {Node_Y,Work_Valid_Y} on primary tree;
       Verify for all crossing node on primary tree if way to {Node_X, Work_Valid_X} on secondary tree
       exists.
       if exists then compute the distance to goal and memorise it if lower than Distance;
       if no more node on the way then Node_Y= Node_Y+ 2^((log n)-depth) (detph of leaf node in primary tree-
       the root has the depth 0);
           }WHILE (Node_Y <= Max_Y)
}
```

The complexity of this algorithm depends of the cardinality m of the code. The complexity is $O(m \log n)$.

4.2 Path planning algorithm

The path planning algorithm consists of finding the sequence of contiguous codes from source point to goal point in a depth first manner. A breath first search is also possible if we want cross big areas to remain far from the obstacles. We have chosen the depth first manner in order to find the best path in term of minimal distance. The algorithm is given below.

1- i=1;

2- Create a list P of codes initially containing in P(1) a code including the source point;

3- Prune the MCT of code included in P(i). i=i+1;

4- Find the best (with respect to the distance) contiguous code of P(i-1);

5- if don't exists then two cases occurs :
* either all bounderies codes are treated and jump in 7;
* or choice another boundery code and jump in 4;

if exist then include it in P(i);

6- if P(i) includes goal point then END;

else jump in 3;

7- i=i-1 ;if i=0 then NO PATH.END.

else jump in 4;

The complexity of this algorithm is $O(n^2 \log n)$.

5.1 The path drawing.

The code list defining the path decribes the sequence of environment blocs which the robot must cross to go from a source point to goal point. We may reduce all blocs to a point which is constitued of the center of the recoverments of twocodes .Two case occurs:
- either two consecutive codes present a common area,
- or the codes are joined but don't be covered.

In the first case, it exists a code intersection in X and Y. The cross point computing consists to determine the intersection area center. In the second case the lack of recoverment dont't allow to compute the intersection between the codes X and Y, but it obligatory exists an intersection with the part X or Y of the code following the principle of which the path was find. It consists of taking , according to the coordinates having no intersection, the maximum of the lowest multivalue code and the center of the segment representing the code bounderies having an intersection.

0 1 2 3 4 5 6 7

Let us take an example.

Let two multivalue codes:

```
CODE_A     0000  0000        1100  1111
           CODE_X CODE_Y     VALID_X VALID_Y
CODE_B     0100  0000        1111  1100
```

We are in the case with only one recoverment in Y.

in X (CODEA_X XOR CODEB_X) AND VALIDA_X AND VALIDB_X = 0

(0000   XOR 0100) AND 1100 AND 1111 = 0100 => no intersection

in Y (CODEA_Y XOR CODEB_Y) AND VALIDA_Y AND VALIDB_Y = 0

(0000   XOR 0000) AND 1111 AND 1100 = 0000 => intersection

Intersection computing:

INTER_CODE_Y= (CODEA_Y OR ~VALIDA_Y) AND (CODEB_Y OR ~VALIDA_Y)

= >(0000 OR 0000) AND (0000 OR 0011) =0000

INTER_VALID_Y= VALIDA_Y OR VALIDB_Y

=> 1111 OR 1100 = 1111

The Y bound is finding with INTER_CODE_Y + (~INTER_VALID_Y + 1)/2 (in the cell size unit).

The CODE_X don't have an intersection. The lowest code is CODE_X=0000 and the greatest code value of the multivalue code CODEA_X is

CODEA_X_MAX = CODEA_X OR ~VALIDA_X + 1= 0100

5.2 Path cost computing

The path cost computing is performed according to the distance. Let $Px(j)$ and $Py(j)$ the cross points coordinates, the cost is computed with

$$C = \sum_{j=1}^{n-1} [\; \|(Px(j+1) - Px(j))\| + \|(Py(j+1) - Py(j))\|\;]$$

with n the number of codes of the path and $\|\;\|$ the square power or the absolute value.

5.3 Path smoothing

The swiftly approach criterion have the disadvantage, in some cases, to find a longer path. The use of an algorithm like A* /NIL 80/ allows to give a path who cost is lower. It is computed performing the sum of the path from source point to code p and an heuristic cost from point p to the goal . This method find an admissible path but it may have a high computing cost if the number of crossing points is large. Another technique is employed to find a smoothed path swiftly. It may longer in terms of distance . We consider the initial path as a set of points noted P ={P1,P2,... Pu ...,Pn}. We try to join the points Pi to Pj (j>i) with a line with an linear interpolation algorithm. Each generated point must be in the free space. This is verified in O(log(n)) time on the MCT. If it is not verified Pi is joined to P(j-1).

CONCLUSION

The described modeling method allows an environment representation under a code form by dividing the space. The code contains informations related to the representing area size and its grid location. This informations allows to find the path between a point to another. The size of the robot was not considered. It may be considered verifying if the crossing area can contain the mobile. The dimension of the multivalue code is not limitative to say that more informations , non obligatory geometric, may be added allowing performing path planning under constrainsts /PRU90/.

The following example needs for a grid of 256 *256, cells 595 codes, 3096 nodes , 5sec. modeling time and 1 sec. path planning time.

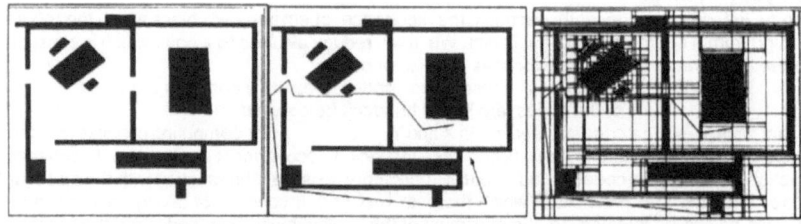

Information Management for Off-Line Robot Programming

H. AFSARMANESH, G.R. MEIJER and F. TUIJNMAN

Computer Science Department
University of Amsterdam
Amsterdam, The Netherlands

abstract

This paper describes an approach to the specification and modeling of information required for off-line robot programming. The approach is characterized by the combined structural and operational descriptions of the data. Database modeling requirements for robot programming applications are considered and techniques to model specific concepts in this application are described. The data management environment used for this is EDEN, an experimental database management system for industrial automation, which is currently under development. EDEN is based on the 3DIS, an extensible object-oriented information management framework. In this paper we show some data modeling problems encountered in off-line robot programming systems and how the EDEN system addresses them.

1. Introduction

Off-line robot programming is the generation of robot programs at a location different from where the robot is. A major research effort aimed at the development of an off-line robot programming environment has been carried out in the Esprit I project "Operational Control for Robot System Integration into CIM; systems planning, implicit-programming and explicit-programming". After an initial period of developing stand-alone planning and programming tools, the need for advanced information support became apparent. In particular the use of a common system for information handling and storage was recognized as the key factor in reaching integration between planning and programming functions of a CIM system. In the past database support for CIM has concentrated on Computer Aided Design (CAD). In this area the shortcomings of traditional solutions developed for business applications became clear. Although considerable effort has been put in the development of data conversion between CAD systems, the exchange of data between other CIM modules as listed above was an open question. An attempt was made to design an information support system for the CIM modules developed in the project. As a first task to reach this objective, the information and the operations applied on them by the different CIM modules were analyzed and brought into chart.[1]

These include geometric data for the parts, the robot and its environment, typically derived from CAD systems, process parameters (for example for spot or arc welding applications) determined by a production engineer, and data describing peripheral devices, such as feeders, conveyors and other transport mechanisms. A user of an off-line robot programming system requires easy access to all this data. This is only possible if the interface to this data is provided by a single database system. However, current database technology is not suitable to store, represent and support access to various views of such a wide variety of data in an efficient and elegant way.[2]

The EDEN system (see [3] for a detailed description), currently under development, is

540

(3DIS),[4] an extensible object-oriented database model, which has also been used for VLSI/CAD application environments (see[5] and[6]). The 3DIS provides an approach in which data and the descriptive information about the data are handled in a uniform framework. It is especially suited to represent information objects of various levels of abstraction and modality, and supports various views of the same data which is essential for this type of engineering database. In addition to that, the EDEN system supports a number of novel modeling constructs, one of which makes it possible to interpret parts of the data as graphs and to express retrievals, constraints and updates in terms of graph operations.

2. An Overview of EDEN

The EDEN database management environment is based on the 3DIS data model, to which the following extensions have been added.

(1) A graph interpretation of the database and a graph query language

(2) A library of operations to manipulate the data modeled by the schema. This library includes operations on standard mathematical data types (e.g. matrix multiplication)

In figure 1, the users model of EDEN is graphically represented. The users model of the EDEN system basically consists of an object-oriented database management system that gives the user access to some predefined information concepts and supports manipulation of these concepts through a limited set of standard operations. For instance, a modeling construct for graphs and an operation to find the shortest path in a graph is supported by EDEN. Because the standard set of operations is a part of the database environment, the user has the tools to manipulate the data in the database without using an application program.

The 3DIS data modeling constructs support: (1) information objects of various levels of abstraction and modalities, (2) dynamic descriptive and structural information (meta-data), and (3) user design, manipulation, and evolution of databases. Data and meta-data are handled uniformly.

3DIS databases contain collections of interrelated objects which represent any identifiable information fact. For example, a part **screw10** , a parts attribute Weight, a string of characters **SC342P** , a part's type (meta-data) **SCREW** , and a procedure **Add-To-Inventory** , are all modeled uniformly as objects. What distinguishes different kinds of objects is the set of structural (meta-data) and non-structural (data) relationships defined on them.

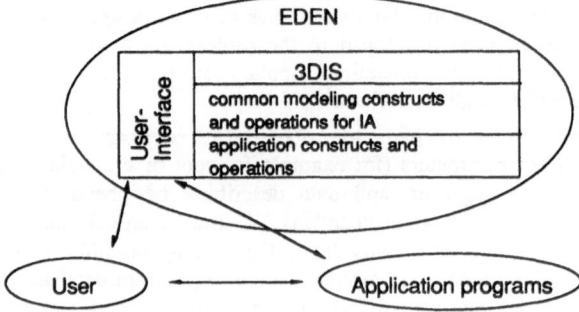

Figure 1: The EDEN system

3DIS supports *atomic, composite* and *type* objects. Atomic objects represent symbolic constants which cannot be decomposed. Strings, numbers, text, video objects, and behavioral (procedural) objects are examples of *atomic* objects. *Composite* objects describe non-atomic entities of application environments. The information content of these objects can be interpreted through their decomposition into atomic objects. *Mapping* objects are a special kind of composite object that describe relationships between other objects. Mapping objects may be decomposed into their domain and range type objects and their inverse mapping objects. *Type* objects contain the descriptive and classification information of a database. Every type object is a structural specification of a group of atomic or composite objects.

Basic relationships among objects are defined with the three fundamental abstraction mechanisms, *generalization, classification,* and *aggregation.* Generalization represents type/supertype relationships by relating a type object, e.g. **FIXTURE**, to a more general type object, e.g. **SOLID**. Classification represents instance/type relationships by relating an atomic or composite object, e.g. **screw10**, to its generic type object(s), e.g. **SCREW**. Aggregation represents mapping/type relationships by relating a mapping object, e.g. **Weight**, to a type object, e.g. **SCREW**. The 3DIS model also supports other kinds of useful abstractions like the *generic recursion abstraction* for recursive definition of concepts and entities.

3. Database representation of application domain concepts

In this section we discuss in some detail how two concepts from the off-line robot programming environment have been represented in an EDEN database. The two examples are: (1) the production task knowledge, and (2) the position information.

EXAMPLE 1: REPRESENTING THE PRODUCTION TASK KNOWLEDGE

Production tasks, such as part welding or assembly, consist of a number of subtasks on which certain dependency (precedence) relations are defined by the application.[7] The precedence relations among production tasks can be represented by a Directed Acyclic Graph (DAG). A precedence graph therefore, shows the decomposition of an assembly task into a number of subtasks and their order dependencies. In this graph, the tasks are represented by nodes and the dependency of two tasks is represented by an arc, connecting the two nodes. The dependency arc indicates that the node which is at the top end of the arc needs to be successfully performed before the node at the lower end can be executed.

In Figure 2, a precedence graph for a simple mechanical assembly kit, the 'Cranfield Benchmark', is given. The assembly kit consists of two side plates with a pendulum like structure in between. In this precedence graph the sub-assembly is represented by a separate node named *spacer. A second precedence graph represents the decomposition of *spacer into its subtasks. In [2] a number of operations are defined for precedence graphs, which make it possible to do advanced queries. As an example, it is possible to ask what all the next possible operations are after some assembly operations have been carried out.

A second important piece of task knowledge is the information on the transfer of global task descriptions into their executable instructions, called the task structure in this paper. Each global production task is first decomposed into its subtasks. Then, the process is repeated until the subtasks are detailed enough to be directly expressed in a sequence of explicit instructions or elementary operations, that can be carried out by a robot. Task structures can be represented by a tree structure. Figure 3 represents the task structure for the side-plate1 assembly task of the 'Cranfield Benchmark' example.

542

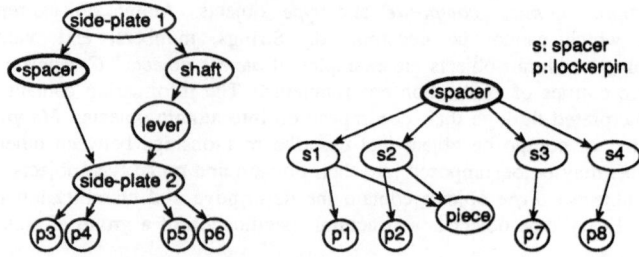

Figure 2: A precedence graph for the Cranfield benchmark

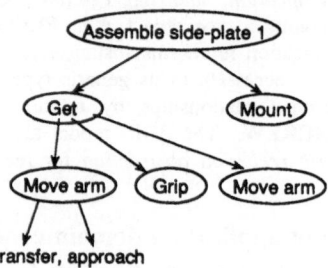

Figure 3: Task structure for 'Assemble side-plate 1'

The database schema designed in EDEN to represent both the precedence graph and the task structure uses the generic recursion abstraction technique. The recursion abstraction technique provides a tool to model recursive entities and concepts of application environments, such as lists, ordered multisets, binary trees, and design components. The general format for this abstraction consists of four specific type objects and certain subtype/supertype relationships defined among them. Figures 4 and 5 show the schema representing the precedence graphs for assembly tasks, and the task structures for the list of robot operations performing each assembly operation respectively.

In Figure 4, the three subtypes NIL, SINGLE-ASSEMBLE-OPERATION, and RECURSIVE-ASSEMBLE-OPERATION disjointly partition the type ASSEMBLE-OPERATION-SET-COMPONENT. Therefore, any member of ASSEMBLE-

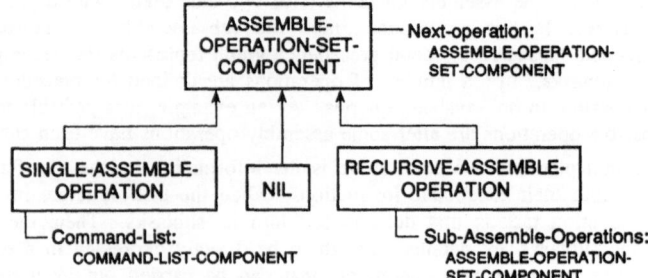

Figure 4: Precedence graph schema for assembly tasks

OPERATION-SET-COMPONENT (a node in the precedence graph) is a member of one of these three subtypes. Members of the SINGLE-ASSEMBLE-OPERATION are atomic objects (nodes that are not recursively defined). Members of RECURSIVE-ASSEMBLE-OPERATION are composite objects (composite nodes in the graph), further defined in terms of their constituents. The type NIL, has exactly one member, nil. Every atomic and composite object ASSEMBLE-OPERATION object points to its next operation(s), through the Next-Operations mapping (attribute).

The RECURSIVE-ASSEMBLE-OPERATION objects also point to their sub-assemble-operations that appear in the first level of their definition subtree, through the Assemble-Operation-Set-Component mapping. In Figure 2, *spacer is a composite node. Therefore, *spacer will be represented in the database as a recursive-assembly-operation for which the sub-assemble operations are those operations directly connected to it in the subgraph defining it. The Single-Assemble-Operation objects point to a list of commands organized in the task structure, through the Command-List mapping.

The system automatically generates the inverses of all mappings introduced in the database. However, for simplicity, the schemas in Figure 4 and 5 do not represent the inverse mappings for all mappings defined between types. For example, for the mapping Next-Operations defined on ASSEMBLE-OPERATION-SET-COMPONENT, there is an inverse mapping (Preceding-Operations) that gives all preceding assemble-operations for each operation. Whenever an operation has no next operation (a leaf node in the precedence graph) the Next-operations mapping points to nil.

EXAMPLE 2: REPRESENTING POSITION INFORMATION

In robot programming, positions are often expressed relative to local coordinate frame One reason is that it makes it easier to carry out changes. For example, if a robot has to pick several items from a pallet, it is best to express the positions in the coordinate frame of the pallet. If the pallet's position is changed, only the pallet's new position has to be entered, and the pick positions relative to the pallet can remain unchanged.

Positions are represented by coordinates relative to a coordinate frames, and homogeneous 4*4 matrix transformations determine the relation between coordinate frames. The schema defined to store relations between coordinate frames allow any two coordinate frames to be related to each other by a coordinate transformation. If this structure is interpreted as a graph, any cycle represents redundant information. Such cycles can arise as a consequence of storing both design data and measured data (calibrations). To check the consistency of this redundant information, the following procedure has been adopted. As long as their are no cycles in the graph, the information stored is not redundant. So when a transformation between two coordinate frames is added, it is necessary to check whether there already existed a path between these two frames. If no such path exists, the new transformation can be stored immediately. If it exists the consistency of the new transformation with the existing information has to be checked.

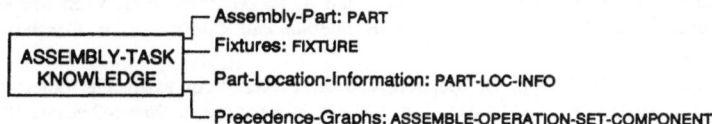

Figure 5: Task knowledge schema

544

To support these operations the EDEN system offers also a graph based query language. This query language views the database as a graph structure, where the nodes are formed by (non-mapping) objects, and arcs by mappings between objects. The arcs are labeled by mappings. EDEN supports a number of special graph queries that correspond directly to efficient graph search algorithms. Two specific operations defined on graphs that are used for the position information are:

(1) Find a (shortest) path; This returns a list of edges e_i

(2) Calculate a generic distance along a path of the following form, where g and h have to be specified by the user $d_i = g(d_{i-1}, h(e_i))$ and $d_1 = h(e_1)$ If h returns the length attribute of an edge and g is summation, d will be the distance along a path.

The consistency procedure mentioned above can be easily defined in terms of the general graph operations. To find a transformation between two not directly connected frames a path is found first, and than a generic distance is calculated, where h returns the matrix and g performs matrix multiplication. The consistency check consists of comparing the result with the matrix that has to be added to the database.

4. Conclusions

We have discussed some information concepts that make it possible to capture the semantics and to represent in an elegant way the complex structured data that is used in an off-line robot programming environment. This supports the development of a single, unified database in which all the information required for off-line robot programming can be stored. We have shown how the 3DIS modeling constructs can be used to represent task knowledge, and how the recursion abstraction is used. The availability to view the database as a graph makes it possible to formulate more complex queries and constraints, the use of which has been demonstrated for the modeling of position information.

References

1. L.M. Camarinha-Matos, U. Negretto, G.R. Meijer, J. Moura-Pires, and R. Rabelo, "Information integration for assembly cell programming and monitoring in CIM," in *Proceedings of 21st ISATA, Wiesbaden, Germany*, (November 1989).

2. F. Tuijnman, G.R. Meijer, and L.O. Hertzberger, "Data Modeling for a Robot Query Language," pp. 208-218 in *Proceedings of the Conference on Intelligent Autonomous Systems 2 (IAS-2)*, , Amsterdam (1989).

3. H. Afsarmanesh, G.R. Meijer, and F. Tuijnman, "EDEN: an Experimental Database management Environment for iNdustrial automation," Technical Report, University of Amsterdam (December 1990).

4. H. Afsarmanesh and D. McLeod, "The 3DIS: An Extensible Object-Oriented Information Management Environment," *ACM Transactions on Information Systems* 7(4) pp. 339-377 (Oct. 1989).

5. H. Afsarmanesh, E. Brotoatmodjo, E. Byeon, and A. Parker, "The EVE VLSI Management Environment," in *Proceedings of the IEEE International Conference on Computer-Aided Design*, (1989).

6. H. Afsarmanesh, H. Knapp, D. McLeod, and A. Parker, "An extensible, object-oriented approach to database for VLSI/CAD," pp. 607-618 in *Readings in Object-Oriented Database Systems*, ed. S. Zdonik and D. Maier,Morgan Kaufmann, Palo Alto (1990).

7. G.R. Meijer and L.O. Hertzberger, "Exception Handling for Robot Manufacturing Process Control," in *Proceedings of the CIM Europe conference*, , Madrid (1988).

Chapter IX

Manipulator Mechanics

Introduction

A simple and comprehensive menu-driven computer-based method for trajectory planning and force analysis in a planar robot is developed in the first paper. The robot designer is able to vary parameters and study their effect on the robot performance. In the second paper, a simple method to analyze the effect of torque and force on the first three links of a PUMA robot has been determined. Minimum time trajectory and bang-bang control with discontinuity points and knot points smoothed by parabolic blend are used. The workspace of a robotic arm using the Articulated Total Body model is calculated in the third paper. Computation of the workspace of the end effector is important in determining the effectiveness of a robot.

Flexible robot dynamics are analyzed in the fourth paper using a velocity transformation matrix method that transforms Cartesian equations into complex joint space equations. The dynamic response of a serial link manipulator with flexibility in both its links and joints is developed in the fifth paper based on a modal approach for the Lagrangian formulation of the equations of motion. Considerations are given for developing control algorithms for the motion of flexible robotic arms using compliance. In the sixth paper, the kineto-elastodynamic acceleration, based on an approximate method of the coupler of a four-bar linkage due to elastic motion of the coupler, is compared with rigid body acceleration. Positional inaccuracies in a robot arm due to deflections of the joints and links are sensed by an unloaded mating link in a parallel with each link.

A scheme for active positional correction of robot arms is described in the seventh paper. A methodology that selects Kalman filter to estimate unmeasured states and applies the separation theorem to design a feedback controller for a general n-link robot is developed in the eighth paper. A multi-layer neural net for recognition of sinusoidal, saw tooth, and square signals is discussed in the ninth paper. The multi-layer perception program shows that the learning is very fast and the recognition results are excellent. In the tenth paper, the current state of the system, linked to previous measurement data through a state space model and estimated by use of least squares techniques, is developed as a simple filter for the state estimation of linear systems under unknown noises. Close relations between the Kalman filter and the least-squares filter are found.

Computer Aided Analysis of a Planar Robot

SHAILESH SHAH and YOGESHWAR HARI

Department of Mechanical Engineering and Engineering Science
University of North Carolina at Charlotte
Charlotte, NC

ABSTRACT

This paper presents the computer aided analysis of a planar robot.
Forward Kinematics, inverse kinematics, velocity analysis, force
analysis and trajectory planning are considered in this work.
Forward kinematics involves the process of solving for the cartesian configuration of the robot in terms of its articulated links.
The inverse kinematics solution gives a particular link trajectory
and provides the desired co-ordinates. The trajectory is composed
of a constant velocity linear segment to which parobolic segments
are blended or merged at appropriate times. Equations for the
above functions were programmed to determine the velocity and
forces on each link. The program is written in such a way that the
designer can vary parameters and study the effect on the links.

INTRODUCTION

The objective of this paper is to develop a simple and comprehensive method for trajectory planning and force analysis in a planar
robot, as seen in Figure 1(page2).

NOMENCLATURE

d	= Length of link 1
h	= Length of link 2
f	= Length of link 3
θ	= Orientation angle of link 1
ψ	= Orientation angle of link 3
α	= Orientation angle of tool
P_x, P_y	= Co-ordinates of the end effector
v	= Velocity
a	= Acceleration
$X(t)$	= Cubic polynomial function of time
t_0	= Initial tome
t_f	= Final time
X_0	= Position at X_0
X_f	= Position at X_f
I_q	= Inertia in a frame
F_k	= Force Matrix
r	= Co-ordinate transformation matrix
$D_k{}^q$	= Displacement vector
J	= Jacobian Matrix

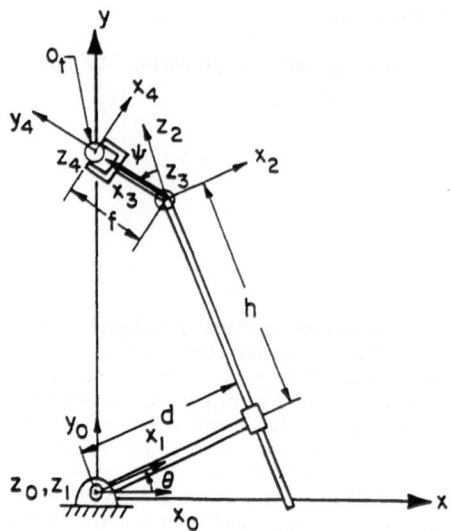

Figure 1

The <u>forward</u> <u>Kinematics</u> of the robot is used to develop the equations to determine the position and orientation of the end effector relative to the base as a function of link values. The Denavit-Hartenberg (D-H) Co-ordinate transformation matrix for determining the position is given by:

$$
T_{i-1} = \begin{bmatrix}
C\theta_i & S\theta_i & 0 & b_i \\
C\alpha_i S\theta_i & C\alpha_i C\theta_i & -S\alpha_i & -d_i S\alpha_i \\
S\alpha_i S\theta_i & S\alpha_i C\theta_i & C\alpha_i & -d_i C\alpha_i \\
0 & 0 & 0 & 1
\end{bmatrix}
$$

The positional vectors are given by

$$P_x = dC\theta - hS\theta - fS(\theta+\psi),$$

$$P_y = dS\theta - hC\theta - fC(\theta+\psi).$$

<u>Inverse</u> <u>Kinematics</u> involves determination of the particular set of values that will produce a known and desired effector configuration. The equations for the parameter are

$$h = \sqrt{P_x^2 + P_y^2 + f^2 - d^2 + 2f[P_x S\alpha - P_y C\alpha]},$$

$$\theta = ATan\left[\frac{d(P_y - fC\alpha) - h(P_x + fS\alpha)}{d(P_x - fS\alpha) - h(P_y + fC\alpha)}\right],$$

$$\psi = (\alpha-\theta).$$

In order to achieve the desired cartesian motion, the equations associated with the manipulator are solved.

$$X = \begin{bmatrix} P_x \\ P_y \\ \propto \end{bmatrix} = \begin{bmatrix} dC\theta - hS\theta - fS(\theta+\psi) \\ dS\theta - hC\theta - fC(\theta+\psi) \\ \theta+\propto \end{bmatrix}$$

Velocity $(\dot{X})$ $= J\dot{\theta}$.

$$\begin{bmatrix} P\dot{X} \\ P\dot{Y} \\ \propto \end{bmatrix} = \begin{bmatrix} (-dS\theta - hC\theta - fC(\theta+\psi)) & -S\theta & fC(\theta+\psi) \\ (-dC\theta - hS\theta - fS(\theta+\psi)) & C\theta & fS(\theta+\psi) \\ 1 & 0 & 1 \end{bmatrix} \begin{bmatrix} \dot{\theta} \\ \dot{h} \\ \dot{\psi} \end{bmatrix}$$

Acceleration $(\ddot{X})$ $= \dot{J}\dot{\theta} + J\ddot{\theta}$.

The forces in the planar robot are determined with respect to the base frame. The equations are

$$F_k = R_k{}^q F_q,$$
$$M_k = R_k{}^q + (D_k{}^q (R_k{}^q F_q)),$$
$$\tau = J^T F,$$

$$\begin{bmatrix} \tau_1 \\ f_2 \\ \tau_2 \end{bmatrix} = J^T \begin{bmatrix} f_{xt} \\ f_{yt} \\ m_{zt} \end{bmatrix} .$$

The trajectory plays a main role in the motion of the end effector. It is a specification of the desired time dependent paths in either cartesian or link space. Linear Segment Parabolic Blend (LSPB) trajectory, composed of constant velocity segments and parabolic segments is used.

The equations are:

for $0 < t < tb$; $X(t) = X_0 + V(t-t_0)^2 / tb,$

$X(t) = V(t-t_0) / tb = \text{Velocity},$

$X(t) = V / tb = \text{Acceleration},$

$tb = (X_0 - X_f + V(t_f-t_0)) / V,$

550

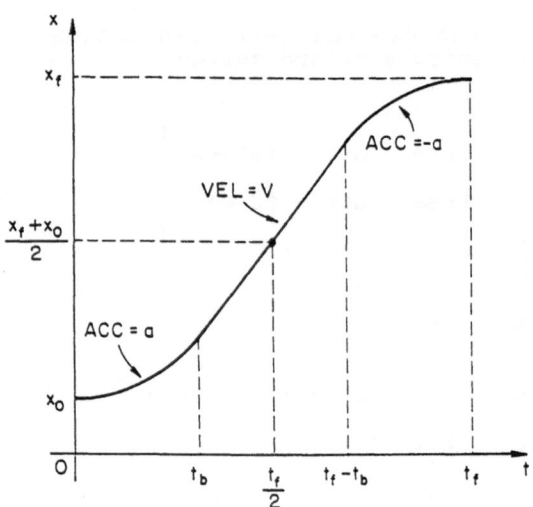

Figure 2

for $t < 0 < (t_f - t_0)$; $X(t) = (X_f + X_0 - V(t_f - t_0) + V(t - t_0)) / 2$

for $(t_f - t_0) < t < t_f$;

$x(t) = X_f - (a(t_f - t_0)^2) / 2 + a(t_f - t_0) (t - t_0) - (a(t - t_0)^2 / 2$

EXAMPLES

Forward Kinematics, Inverse Kinematics, Velocity analysis, Force analysis and Trajectory planning was done for a planar robot having the following dimensions:

P_x = 6
P_y = 7
f = 2
d = 5
$\propto$ = 30 degrees
X_0 = 4
X_f = 3.

Using inverse Kinematics the parameters θ, h and ψ are calculated. These and the physical dimensions are used to calculate velocity and angular acceleration of the links. The results for the velocity analysis are shown in Fig.3. The force analysis and trajectory plannung results are shown in Figure.4 and Figure.5.

REFERENCES

Wolovich, William A., Robotics: Basic Analysis and Design., CBS College Publishing., NY, NY, 1987.

```
THESE ARE THE RESULTS OF VELOCITY ANALYSIS
PX      H          THETA       PSI        THETADOT   HDOT       PSIDOT
6       7.193836   0           30          15.92914  -1.390079  -15.92914
6.5     7.681229  -21.85446   1282.169    -14.77925   1.017205   14.77925
7       8.170145  -25.1694    1472.101     14.01624  -1.296438  -14.01624
7.5     8.660328  -28.21131   1646.39     -13.20552   1.026429   13.20552
8       9.151572  -31.00823   1806.642     11.49524  -.210164   -11.49524
8.5     9.643717  -33.58525   1954.295    -6.694932  -1.068089   6.694932
9      10.13663   -35.96469   2090.626    -1.841526   2.133989   1.841526
9.5    10.63021   -38.16636   2216.773     9.624327  -1.740747  -9.624327
10     11.12435   -40.20785   2333.742    -8.30612   -.4580504   8.30612
10.5   11.619     -42.10471   2442.424    -2.978467   2.166535   2.978467
11     12.11409   -43.87075   2543.611     9.400578  -.5977234  -9.400578
```

Figure 3

```
           THESE ARE RESULTS OF FORCE ANALYSIS

     FORCES IN LINK 3
     F3(1)= 9.193836 F3(2)=-2 F3(3)= 0
     FORCES IN LINK 2
     F2(1)= .3511839 F2(2)= 0 F2(3)= 9.402303
     FORCES IN LINK 1
     F1(1)= .3511839 F1(2)= 9.402303 F1(3)= 0
     MOMENTS IN LINK 3
     M3(1)= 0 M3(2)= 0 M3(3)= 2
     MOMENTS IN LINK 2
     M2(1)= 0 M2(2)=-2 M2(3)= 0
     AS LINK 1 IS REVOLUTE IN PLANAR ROBOT NO RESULTANT
     TORQUE CAN BE TRANSMITTED TO LINK 1 BY EXTERNAL
     FORCES AND MOMENT
     THESE ARE THE TORQUES
     T1=-4.824863 T2= 7.967305 T3=-9.192997
```

DO YOU WANT TO GO TO THE MENU (Y/N)? ▮

Figure 4

```
        THESE ARE THE RESULTS OF TRAJECTORY PLANNING
   FOR 1 <=T<= .5  X(T)= 4 +  2 * (T - 1 )^2
   FOR .5 <=T<= 4.5  X(T)= 3.5 + -2 * (T - 1 )
   FOR 4.5 <=T<= 4  X(T)=-21 +  12 * (T - 1 )-  2 * (T - 1 )^2
```

DO YOU WANT TO GO BACK TO THE MENU (Y/N)? ▮

Figure 5

Computer Aided Analysis of the First Three Links in a Puma Robot

DURAISWAMI PALANIVELU and YOGESHWAR HARI

Department of Mechanical Engineering and Engineering Science
University of North Carolina at Charlotte
Charlotte, NC

ABSTRACT

This paper presents an analysis of loading on the drive motors of a Puma robot. The first three links which make up the main structure are considered. The link torques are determined for various time intervals. The dynamic moments are found by differentiating the kinetic energy equations and the torque at the pivot is found by equating the static forces and the dynamic moment. The time dependent variables are derived from the global co-ordinates. The function of the global determinants are specified by the control program of the robot. The trajectory movement between two points can be specified for a minimum time, minimum distance, minimum energy or minimum stress. A trajectory of minimum time is chosen. A Bang-Bang control with discontinuity points and knot points, smoothed by a parabolic blend is used. The above equations were programmed to determine the torque on the links at various conditions. The program is user friendly. The designer can vary design parameters or application conditions to study the effects of torque on the three links.

INTRODUCTION

The Puma robot has six degrees of freedom and has six revolute links to deliver a tool or maintain a position at any point in the space. The point has to lie within the working envelope. The first three links make up the main structure of the robot. The rest make up the wrist, which usually holds the part to be carried (see Fig. 1). The links are powered by DC servo motors. The robot is controlled by a programmable microcomputer.

The objective of the paper is to develop a simple method to analyze the effect of torque and force on the first three links.

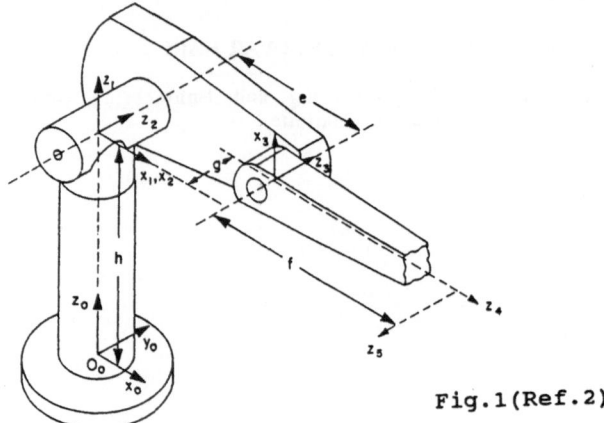

Fig.1(Ref.2)

NOMENCLATURE

m_1 = Mass of link1 (Kg)
r_1 = Radius of link 1 (Cm)
l_1 = Length of link1 (Cm)
m_2 = Mass of link 2 (Kg)
l_2 = Length of link 2 (Cm)
b_2 = Height of link 2 (Cm)
a_2 = Depth of link 2 (Cm)
m_3 = Mass of link 3 (Kg)
l_3 = Length of link 3 (Cm)
b_3 = Height of link 3 (Cm)
a_3 = Depth of link (Cm)
e = Distance between the axis of the first link and the joint of the axis of link 2 & 3 (Cm)
f = Distance between the axis of the wrist and the joint of the axis of link 2 & 3 (Cm)
g = Offset between the base and the wrist (Cm)
h = height of the base which holds link 1
X_n, Y_n, Z_n = Local co-ordinate system for links n = 1,2,3
S_n = Sin(theta$_n$), n= 1,2,3
S_{23} = Sin(theta2 + Theta3)
C_n = Cos(theta$_n$), n = 1,2,3
C_{23} = Cos(theta2 + theta3)

The torque is determined at the base of the link, where the power is applied. The kinetic energy equation was derived for each link in terms of its mass, geometry and angular velocity.

The equations for the kinetic energy are:

$$K_1 = \frac{1}{2}\left[\frac{1}{4}m_1 r_1{}^2 + \frac{1}{12}m_1 l_1{}^2\right]\dot{\theta}_1{}^2,$$

$$K_2 = \frac{m_2}{24}\left[(a_2{}^2+b_2{}^2)\dot{\theta}_1{}^2 s_2{}^2 + (a_2{}^2+4l_2{}^2)\dot{\theta}_1{}^2 c_2{}^2 + (b_2{}^2+4l_2{}^2)\dot{\theta}_2{}^2\right],$$

$$K_3 = \frac{m_3}{2}\left\{\left[\frac{1}{12}S_{23}{}^2(a_3+l_3{}^2) + \frac{f^2}{4}C_{23}{}^2 + efC_2C_{23} + 2fgS_1C_1C_{23} + \right.\right.$$

$$\left. 2egS_1C_1C_2 + e^2C_2{}^2 + g^2\right]\dot\theta_1{}^2 + \left[\frac{(b_3{}^2+l_3{}^2)}{12} + \frac{f^2}{4}\right]\dot\theta_3{}^2 +$$

$$\left[\frac{C_{23}{}^2}{12}(a_3{}^2+b_3{}^2) + \frac{f^2}{4} + efS_2S_{23} + e^2 + efC_{23}C_2 + \right.$$

$$\left.\frac{(b_3{}^2+l_3{}^2)}{12}\right]\dot\theta_2{}^2 + \left[gf(C_1C_{23}{}^2-S_1{}^2S_{23}) + geS_2(C_1{}^2-S_1{}^2)\right]\dot\theta_1\dot\theta_2$$

$$+\left[\frac{(b_3{}^2+l_3{}^2)}{6} + \frac{f^2}{2} + efS_2S_{23} + efC_2C_{23}\right]\dot\theta_2\dot\theta_3 +$$

$$\left.\left[gfS_{23}(C_1{}^2 - S_1{}^2)\right]\dot\theta_1\dot\theta_3\right\}.$$

Therefore the total kinetic energy of the entire manipulator is given by

$$K = K_1 + K_2 + K_3$$

Let L_n denote the generalized moment acting on the n_{th} link in the direction of permissible link motion. From the above definitions given the Lagrangian equation can be seen to be

$$\text{Lagrangian } (L_n) = \frac{d}{dt}\left(\frac{\partial K}{\partial\dot\theta_n}\right) - \left(\frac{\partial K}{\partial\dot\theta_n}\right) \quad \text{for } n = 1,2,3$$

To determine the Lagrangian equation, time dependent variables and link angles are required. These variables are derived from the global co-ordinates, which are a function of time and Jacobian operators.

The dynamic moments are found by differentiating the kinetic energy equations by the Lagrangian operators. The torque at the pivot is found by equating the static forces and the dynamic moment. Analysis of free body diagram for each link yeilds the torque equations.

The torque equations for the pivot point for each link are:

$$\tau_1 = L_1$$

$$\tau_2 = L_2 + gl_2C_2 \left(\frac{m_2}{2} + ml + m_3\right) + gl_3C_{23}(ml+ \frac{m_3}{2})$$

$$\tau_3 = L_3 + \left(\frac{m_3}{2} + ml\right) glC_{23}$$

The trajectory of the robot's movement is specified by minimum time. A Bang-Bang control with knot points smoothed by a parabolic blend is used.

COMPUTER PROGRAM

The procedure described above is programmed in BASIC.The input values are the physical dimensions of the robot, the starting and end points for the movement, the load to be carried by the links and the total time desired for the movement. Once this data is given, the program calculates the constant acceleration associated with the Bang-Bang trajectory (BBPB) and displays it on the screen. If the acceleration is acceptable to the user, it then calculates the torque.

The total time given by the user is divided into 10 equal increments. The program cycles for each time interval.The X, Y, Z positions for an increment time are found using BBPB and using the equations for inverse kinematics. The velocity vectors, Vx, Vy, Vz, and angular velocity are also found using BBPB. Angular accelerations are derived from inverse Jacobian and the time differentiated Jacobian. With the help of the above values the the program calculates the Lagrangian equations which in turn yeilds three torques. An inverse kinematics solution will produce four different set of angles for link 1, 2 and 3. The user can obtain results for the other angles if desired.

RESULTS OF AN EXAMPLE

Torque was determined on the three links for a robot which has the following dimensions:

Base height	= 50Cm
Radius and mass of link 1	= 10Cm,25kg
Elbow length, height, depth & mass	= 25Cm,30Cm,10Cm,15Kg
F-arm length, height, depth & mass	= 30Cm,35Cm,10Cm,20Kg
Elbow Offset	= 5Cm
Starting point of wrist	= 5,5,5
End point of the wrist	= 6,10,19
Load Carried	= 15 Kgs

Time interval of 3 and 30 seconds were used. The displacement of the end effector X, Y, Z and the the torque acting on the three links was plotted with respect to the time interval. The time interval for the first graph is 3 seconds and the second graph is 30 seconds. It can be seen from graphs 1(d) and 2(d) that the torques are erratic when the time interval is very short. For longer time intervals the curves are smoother.

REFERENCES

1. Fu, K.S,. Gonalez, R.C,. Lee, C.S.G,. Robotics Control, Sensing, Vision and Intelligence. McGraw-Hill, Inc,.NY, NY 1987.
2. Wolovich, William A,. Robotics:Basic Analysis and Design CBS Publishing,. NY, NY,. 1987.

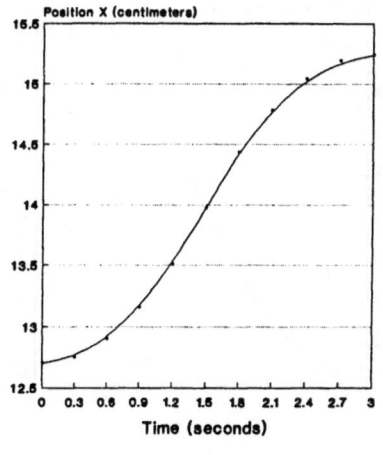

(a)

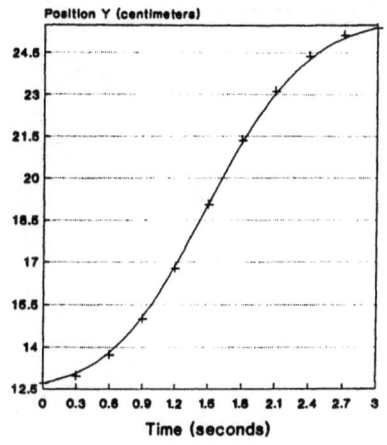

(b)

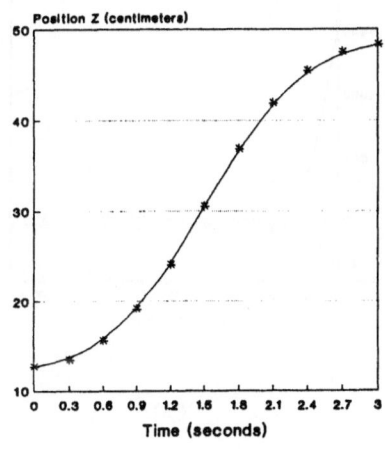

(c)

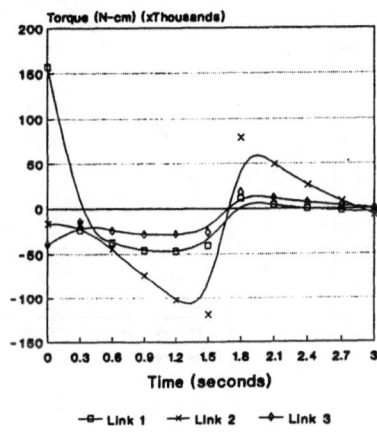

(d)

Graph_1

1(a), 1(b), 1(c)   Displacement Vs Time

1(d)   Torque Vs Time for each link

558

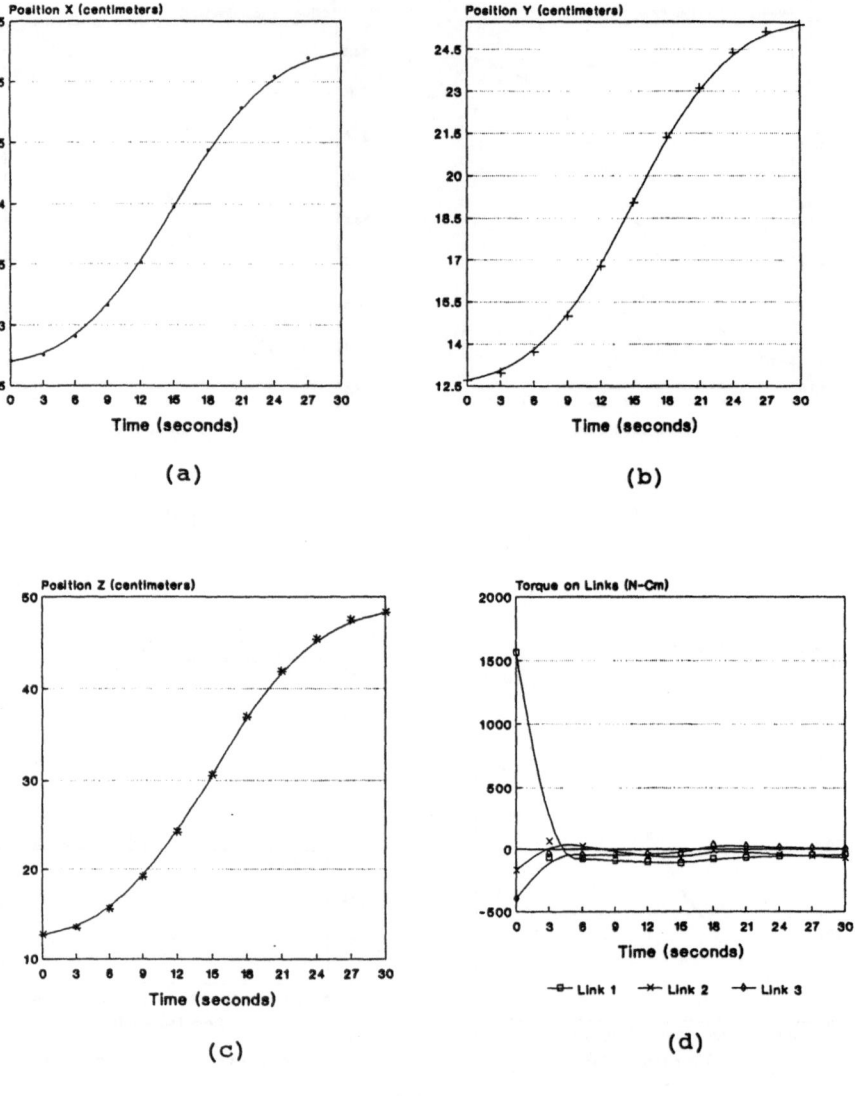

Graph_2

2(a), 2(b), 2(c)   Displacement Vs Time

2(d)   Torque Vs Time for each link

Work-Space Calculation of a Robotic Arm Using the Articulated Total Body Model

XAVIER J. R. AVULA
Department of Mechanical and Aerospace Engineering
University of Missouri-Rolla
Rolla, MO

INTS KALEPS and LOUISE OBERGEFELL
Armstrong Aerospace Medical Research Laboratory
Wright-Patterson Air Force Base
Dayton, OH

ABSTRACT

In this paper the work-space of a robotic arm and the arm positions at various times are calculated using the Articulated Total Body (ATB) Model developed by the Armstrong Aerospace Medical Research Laboratory, Wright-Patterson Air Force Base in an effort to advance the Robotics Telepresence Technology. The behavior of the right arm is simulated on the computer with a force applied to the center of the hand, and the generated work-space is graphically displayed.

I. INTRODUCTION

Some technological advancements, in addition to being beneficial to humans, bring forth risks that threaten personal safety. Several defense and industrial operations are performed in hazardous environments contaminated by nuclear materials and toxic chemicals, and in extreme temperatures. For operations in such hazardous environments, the employment of mobile robots, which are remotely controlled by humans, is a natural alternative. Also, some operations in modern air-and spacecraft place such great demands on the senses that it is humanly impossible to perform optimally without the aid of some electronic and mechanical manipulators. The technology of utilizing and controlling robots by a human in the control loop is called Robotic Telepresence, and lately, much attention is being directed towards its development. In the Robotic Telepresence Program of the Harry G. Armstrong Aerospace Medical Research Laboratory (AAMRL), Wright-Patterson Air Force Base, the Articulated Total Body (ATB) model is being used as a dynamics and feedback control simulation tool. The ATB model is a computer model developed under the sponsorship of the AAMRL to aid in the study of crew member dynamics during ejection from high-speed aircraft.

The ATB model is based on the rigid body dynamics which uses Euler's equations of motion with constraint relations of the type employed in the Lagrange method. The model has been successfully used to study the articulated human body motion under various types of body segment and joint loads. The technology of robotic telepresence will provide remote, closed-loop, human control of mobile robots.

The objective of the present study is to calculate the work-space of a robotic arm using the ATB model. In consideration of similarities between the robotic arm and the human arm, and in view of human-in-the-loop control of remotely located robots through exoskeletal devices, such a calculation will be beneficial.

II. PREVIOUS STUDIES IN ROBOTICS

Kinematics is a fundamental tool in the study of geometry of the end-effector arm motions of a robot. The problem of finding the end-effector position and orientation for given joint displacements is referred to as the direct kinematics problem. The direct kinematics problem of defining the position of the end-effector as a function of the joint angle values can be solved using a straightforward geometry [1]. However, only simple kinematic structures can be handled by this method.

The problem of solving for the joint angle and position when the end-effector is moved to a specified position and orientation is the inverse kinematic problem. Only a few simple kinematic structures yield closed-form solutions for the inverse problem. If the closed-form solution cannot be obtained, numerical methods based on iterative algorithms such as the Newton-Raphson method can be employed. The iterative methods are computationally complex and time consuming. There are numerous alternative kinematic modeling methods to iterative methods. The problem of determining the position and orientation of a rigid body in space is studied by using the method of screw axes and line geometry. Sugimoto and Duffy [2], and Woo and Freudenstein [3] investigated various aspects of rigid body motions in the light of screw theory and line geometry which were found to be more efficient in numerical kinematic analysis.

The design of arm linkages and the determination of workspace volume are important objectives in robotics technology. Roth [4] was the first to address the design problem of finding appropriate kinematic structures and link dimensions to allow the arm to cover a specified workspace. Kumar and Waldron [5], Gupta and Roth [6], and Tsai and Soni [7] also have dealt with work space volume and shape determinations by a variety of analytical and numerical methods.

Let us now turn our attention to the Articulated Total Body model which we have considered as a tool for understanding human articulated motion from robotics point of view.

III. THE ARTICULATED TOTAL BODY (ATB) MODEL

The dynamic behavior of the ATB model is determined by classical methods of analysis. (See Fleck and Butler [8].) The body segments are coupled to form an open chain of interconnected rigid bodies. For an arbitrary segment n, the translational dynamic equation is

$$M_n X_n = \Sigma F_{nk} + \Sigma f_{nj} \tag{1}$$

in which M_n = mass of segment n, X_n = position of the center of gravity of the nth segment, f_{nj} = constraint force at joint j acting on segment n, F_{nk} = kth external force acting on the nth segment.

The angular dynamic equation in the inertial system is

$$\dot{D}_n^{-1} \phi_n \omega_n + D_n^{-1} \dot{\phi}_n \omega_n + D_n^{-1} \phi_n \dot{\omega}_n = D_n^{-1} \Sigma \text{ Torques} \tag{2}$$

where, ω_n = angular velocity vector of segment n, ϕ_n = inertia matrix about the center of gravity of segment n, D_n = the direction cosine matrix associated with segment n.

A vector exponential integrator is used to solve Equations (1) and (2) with appropriate constraint equations. The constraints considered in the ATB program are: (i) linear position constraint, (ii) angular joint constraint (locked joint, pinned joint), (iii) distance constraints (zero distance, fixed distance, rolling and sliding constraints), (iv) force constraints, and (v) torque constraints.

The integration of Equations (1) and (2) yields the position vector components of body segments.

In the ATB model the kinematic variables are determined by the integration of dynamical equations of motion. A recent modification in the ATB program utilizes a direct kinematic method to automatically position a seated occupant before the dynamic loads are applied. With this modification the model computes the position of each body segment in the inertial reference frame provided that the location of the first segment and the angular orientation of each segment are known. This is similar to the direct kinematics method in robotics involving straightforward computations. A computer subroutine based on the relationship

$$Z_i + D_i^T r_{j,i} = z_{j+1} + D_{j+1}^T r_{j,j+1} \tag{3}$$

in which z_k = vector from the origin of the inertial coordinate system to the center of gravity of segment k, D_k = direction cosine matrix for segment k (the superscript T stands for the transpose), $r_{j,k}$ = vector from the center of gravity of segment k to joint j whose components are expressed in the local reference coordinate system of segment k, and i = joint (j).

Equation (3) has been used to solve for the location of each successive segment from that of the previously defined segments. This subroutine specifies the axes abut which the segments are rotated, and the sequence of rotations to achieve the desired orientation.

IV. AN EXAMPLE

In this example the behavior of the right arm of a human male, to which a five-pound force is applied in the vertically upward direction, is simulated. The force remains constant throughout the simulation. For each segment of the arm, the x-y coordinate system is parallel to the cross-section, and the z-axis is oriented along the longitudinal axis of the segment. The following are the segment properties:

Right Upper Arm

Principal Moments of Inertia:

$I_{xx} = 0.102$ lb-sec^2-in

$I_{yy} = 0.099$ lb-sec^2-in

$I_{zz} = 0.011$ lb-sec^2-in

Major and Minor Axes:

a = 1.9 in
b = 1.8 in
c = 6.0 in

Right Lower Arm

Principal Moments of Inertia:

$I_{xx} = 0.12$ lb-sec^2-in

$I_{yy} = 0.12$ lb-sec^2-in

$I_{zz} = 0.007$ lb-sec^2-in

Major and Minor Axes:

a = 1.775 in
b = 1.775 in
c = 5.8 in

562

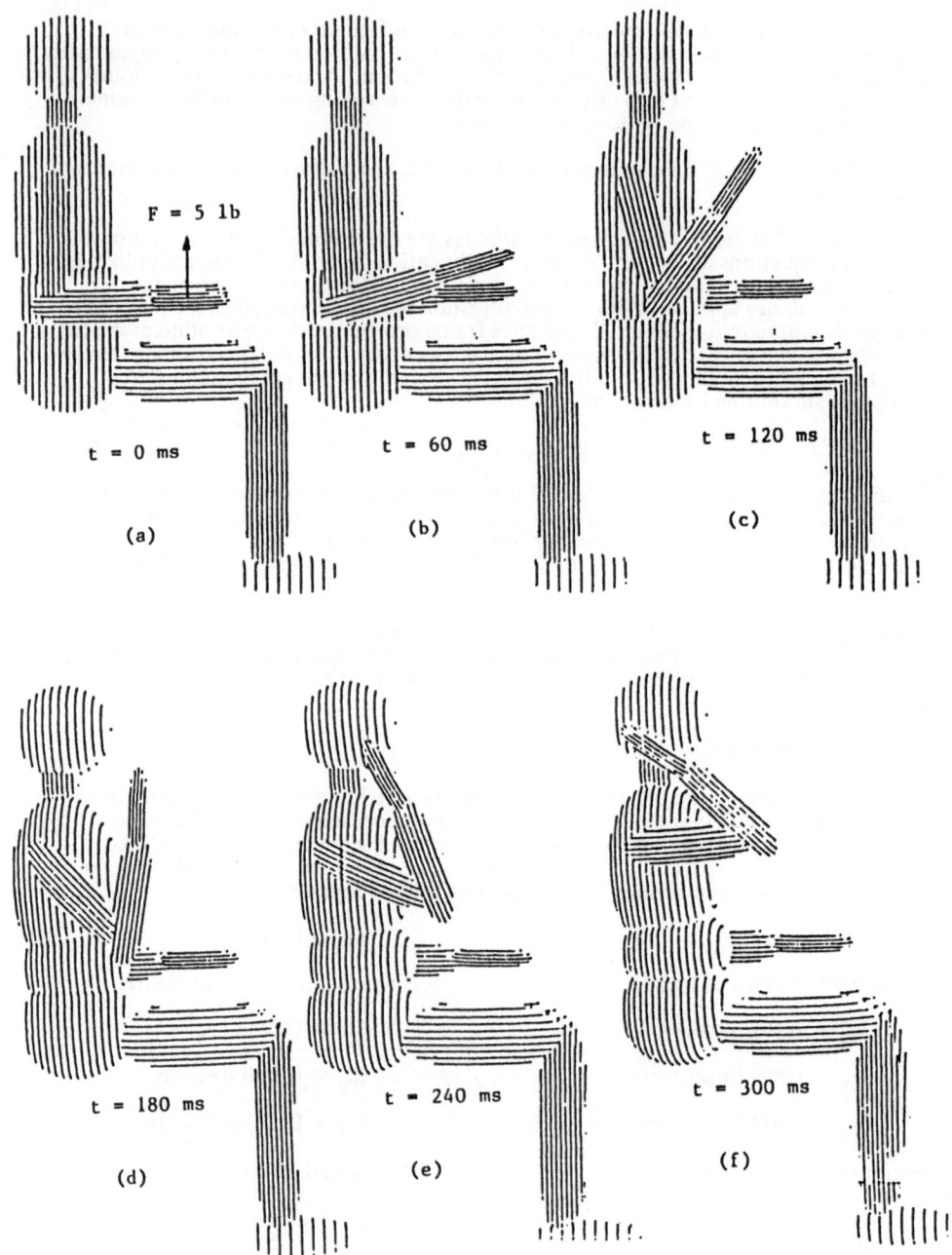

Figure 1. ATB simulation of the right arm.

The seated occupant at time t = 0 sec is shown in Fig. 1a. A force F = 5 lb is applied at the center of gravity of the hand, and the subsequent motion of the arm is simulated. The arm positions which are calculated by the ATB model software are displayed in Fig. 1 (b-f) for times t = 60, 120, 180, 240, and 300 milliseconds. Figure 2 is constructed by superposition of arm motion sequences to indicate the workspace generated under the prescribed conditions. It must be emphasized once again that the segment positions have been calculated by solving the dynamical equations of motion, Equations (1) and (2), and not the kinematic equations.

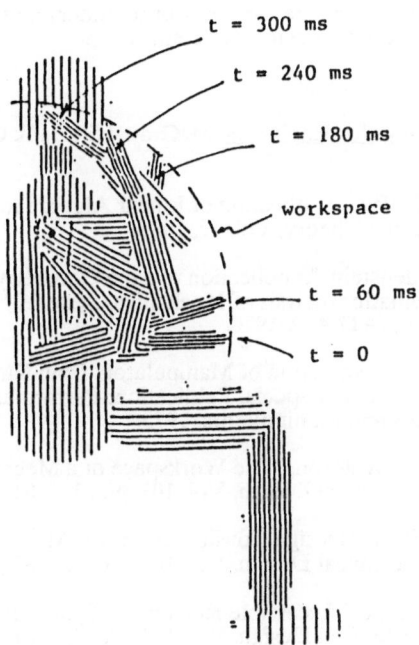

Figure 2. Workspace of the ATB right arm with F = 5 lb(↑)
(Arm positions are superposed)

564

V. DISCUSSION

There exists a vast body of knowledge on the three-dimensional kinematic analysis methods addressing the behavior of robot manipulators. Numerous methods of analysis have been developed in attempting to reduce the computational complexities associated with kinematic structures representing robotic arms. Using rotation vectors and quaternion algebra, the number of arithmetic operations required for kinematic solutions can be reduced to only three or four parameters to define the relative spatial orientation of a rigid body. Regardless of the method chosen, as we increase the number of links and joints in a kinematic chain the analysis becomes more complex and leads to huge mathematical expressions. This has motivated researchers to seek a unified theory of notations for simplification. In reference to the Robotic Telepresence Program at the Armstrong Aerospace Medical Research Laboratory and the role of the dextrous hand in this program, the kinematics of the robotic arm has been analyzed using the ATB model equations which are less complex.

The computation of workspace of the end-effector is important in determining the effectiveness of a robot. In view of using the ATB model as a simulation tool in the Robotic Telepresence Program the methods of neuromuscular control should be investigated to determine appropriate mechanisms for controlling the ATB hand. The knowledge of these mechanisms can be carried over to understand the dextrous hand behavior and to specify exoskeletal devices for control.

REFERENCES

1. Koren, Yoram, Robotics for Engineers, McGraw-Hill Book Company, New York, 1986.

2. Sugimoto, K. and J. Duffy, "Application of Linear Algebra to Screw Systems," Mechanism and Machine Theory, Vol. 17, 1982.

3. Woo, L. and F. Freudenstein, "Application of Line Geometry to Theoretical Kinematics and Kinematic Analysis of Mechanical Systems," Journal of Mechanisms, Vol. 5, pp. 417-460, 1970.

4. Roth, B., "Performance Evaluation of Manipulators from Kinematics Viewpoint," National Bureau of Standards Special Publication: Performance Evaluation of Programmable Robots and Manipulators, 1975.

5. Kumar, A. K. and K. J. Waldron, "The Workspace of a Mechanical Manipulator," ASME Journal of Mechanical Design, Vol. 103, No. 3, 1981.

6. Gupta, K. C. and B. Roth, "Design Considerations for Manipulator Work Space," ASME Journal of Mechanical Design, Vol. 104, No. 4, 1982.

7. Tsai, Y. C. and A. H. Soni, "Accessible Region and Synthesis of Robot Arms," ASME Journal of Mechanical Design, Vol. 103, No. 4, 1981.

8. Fleck, J. J. and F. E. Butler, "Development of an Improved Computer Model of the Human Body and Extremity Dynamics," Report No. AMRL-TR-75-14, July 1975.

Effect on Flexibility on Manipulator Dynamics

H. ASHRAFIUON and C. NATARAJ

Department of Mechanical Engineering
Villanova University
Villanova, PA

Summary

The variational equations of motion are developed for flexible serial manipulators with rotary joints which account for full coupling between the rigid body motion and link deformation. Velocity and acceleration transformation equations are developed to conveniently transform Cartesian space equations into joint space. Small deformation is assumed such that vibration modal coordinates and mode shapes can represent the elastic motion of the flexible links. Flexibility and mass properties of the links are obtained by finite element method. A case study of an industrial robot is presented to show the effect of bending and torsional vibrations on end-effector motion.

Kinematics of Flexible Manipulators

Using the finite element approach, flexibility of a link i can be modeled by discretizing it into several elements. Lumped mass approximation then allows the element masses to be distributed at the nodes connecting adjacent elements. The displacement field u_i', representing linear and rotational displacements of the link's n_i nodes, may be written as a linear combination of deformation modes:

$$u_i' - \Phi_i a_i \tag{1}$$

where Φ_i is a $6n_i \times m_i$ modal matrix containing the m_i selected deformation modes via Ritz approximation and a_i is the vector of m_i corresponding modal coordinates of link i. The prime notation is used to represent vectors in link reference frame. Separating the linear displacements of a node k, u_{tik}', from its rotation, u_{rik}', one may write (see Fig. 1):

$$u_{tik}' - \mu_{ik} a_i \tag{2a}$$
$$u_{rik}' - \phi_{ik} a_i \tag{2b}$$

μ_{ik} and ϕ_{ik} are $3 \times m_i$ modal submatrices containing those rows of Φ which, respectively, correspond to linear and rotational displacements of node k.

Based on the assumption of small rotations, a 3×3 transformation matrix may be defined to represent the orientation of node k as follows [1]:

$$\mathbf{E}_{ik} - \mathbf{I}_3 + \tilde{\mathbf{u}}'_{rik} \tag{3}$$

where $\mathbf{I}_3$ denotes a 3 x 3 identity matrix and "~" which is normally used in matrix calculus for cross product representation, defines a skew-symmetric matrix; e.g., given $\mathbf{c} - [c_x, c_y, c_z]^T$:

$$\tilde{\mathbf{c}} - \begin{bmatrix} 0 & -c_z & c_y \\ c_z & 0 & -c_x \\ -c_y & c_x & 0 \end{bmatrix} \tag{4}$$

Consider a typical rotary serial manipulator shown in Fig. 2. A set of reference frames are assigned to each link at the joint between that link and the link preceding it. The position and orientation of the reference frame of each link with respect to the base is determined recursively, starting from the base. Position of link i's reference frame o_i with respect to base 0 is given in terms of its preceding link i-1 as:

$$\mathbf{r}_i - \mathbf{r}_{i-1} + \mathbf{A}_{i-1}\mathbf{s}'_{i-1} - \sum_{j-1}^{i-1} \mathbf{A}_j\mathbf{s}'_j , \quad i-1,\ldots,p \tag{5}$$

Here, $\mathbf{A}_j$ is a 3 x 3 transformation matrix representing the orientation of o_j with respect to 0 and $\mathbf{s}_j$ is a vector starting from o_j to o_{j+1}. Using Fig. 1 and Eq. (2), one may write:

$$\mathbf{s}'_j - \mathbf{s}'_j{}^0 + \mu_j\mathbf{a}_j \tag{6}$$

$\mathbf{s}'_j{}^0$ is the undeformed vector and μ_j is the translational modal submatrix of the node where joint i is located on link i-1 henceforth called the joint node. Orientation of reference frame o_i with respect to 0 is also determined recursively as [2]:

$$\mathbf{A}_i - \mathbf{A}_{i-1}\mathbf{E}_{i-1}\mathbf{T}_i - \sum_{j-1}^{i} \mathbf{E}_{j-1}\mathbf{T}_j , \quad i-1,\ldots,p \tag{7}$$

where $\mathbf{T}_j - \mathbf{T}_j(\theta_j)$ is the 3 x 3 joint j transformation matrix and $\mathbf{E}_j$ is the jth link node joint transformation matrix defined in Eq. (3).

The generalized coordinates representing arm kinematics are defined as:

$$\mathbf{q} - [\boldsymbol{\theta}^T, \mathbf{a}^T]^T \tag{8}$$

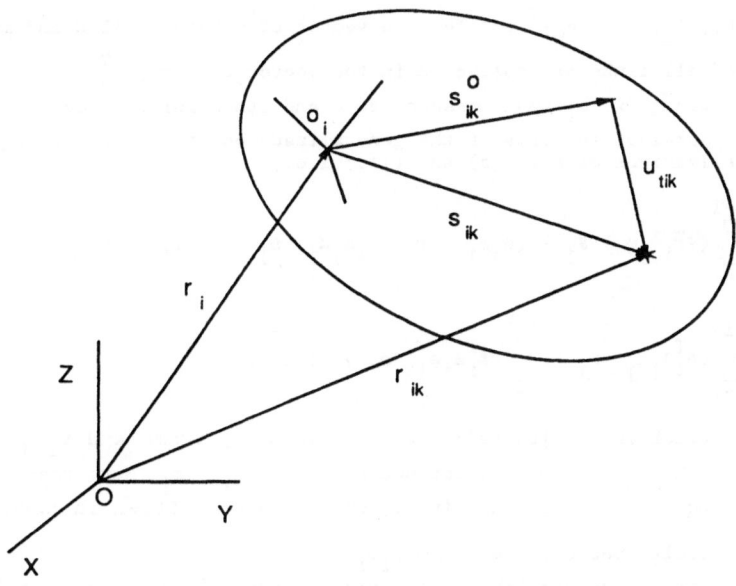

Figure 1.   Kinematics of Flexible Body i

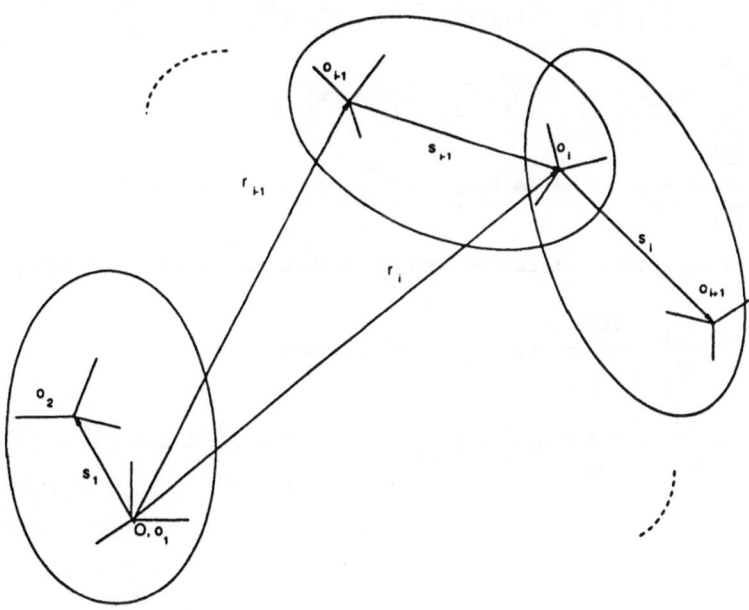

Figure 2.   Kinematic Structure of a Serial Rotary Manipulator

where $\theta = [\theta_1, \theta_2, \ldots, \theta_p]^T$ denotes the vector of p joint angles and modal coordinates of all links are collected in the vector $a = [a_1^T, a_2^T, \ldots, a_p^T]$.

Linear velocity of o_i with respect to 0 and its angular velocity in o_i frame can be expressed in terms of the generalized coordinates by taking the total time derivatives of Eqs. (5) and (7).; i.e.,

$$\dot{r}_i = \sum_{j=1}^{i-1} (-\tilde{r}_{i,j} u_j) \dot{\theta}_j + (A_j \mu_j - \tilde{r}_{i,j+1} A_j \phi_j) \dot{a}_j , \quad i=1,\ldots,p \tag{9a}$$

$$\omega_i' = \sum_{j=1}^{i} (A_i^T A_j u_j') \dot{\theta}_j + \sum_{j=1}^{i-1} (A_i^T A_j \phi_j) \dot{a}_j , \quad i=1,\ldots,p \tag{9b}$$

u_j' is a unit vector in the jth joint direction in o_j frame and $r_{i,j}$ is a vector from joint j to joint i defined as $r_{i,j} = r_i - r_j$. Note that, time derivative of Eq. (7) yields a matrix [3] which can be written in terms of the angular velocity vector; i.e., $\dot{A}_i = A_i \tilde{\omega}_i'$.

Similarly, linear and angular acceleration vectors are derived by taking the total time derivative of Eqs. (9a) and (9b):

$$\mathbf{Y}_i = \sum_{j=1}^{i-1} (-\dot{\tilde{r}}_{i,j} u_j) \ddot{\theta}_j + (-\ddot{\tilde{r}}_{i,j} A_j - \tilde{r}_{i,j} A_j \tilde{\omega}_j') u_j' \dot{\theta}_j + (A_j \mu_j - \tilde{r}_{i,j+1} A_j \phi_j) \ddot{a}_j$$
$$+ [A_j \tilde{\omega}_j' \mu_j - (\dot{\tilde{r}}_{i,j+1} A_j + \tilde{r}_{i,j+1} A_j \tilde{\omega}_j') \phi_j] \dot{a}_j , \quad i=1,\ldots,p \tag{10a}$$

$$\dot{\omega}_i' = \sum_{j=1}^{i} (A_i^T A_j u_j') \ddot{\theta}_j + (A_i^T A_j \tilde{\omega}_j' - \tilde{\omega}_i' A_i^T A_j) u_j' \dot{\theta}_j$$
$$+ \sum_{j=1}^{i-1} (A_i^T A_j \phi_j) \ddot{a}_j + (A_i^T A_j \tilde{\omega}_j' - \tilde{\omega}_i' A_i^T A_j) \phi_j \dot{a}_j , \quad i=1,\ldots,p \tag{10b}$$

From these equations, it can be concluded that for links 1 through p:

$$\mathbf{v}_{r_i,\theta_j} = \frac{\partial \dot{r}_i}{\partial \dot{\theta}_j} = \frac{\partial \mathbf{Y}_i}{\partial \ddot{\theta}_j} = -\tilde{r}_{i,j} u_j , \quad j=1,\ldots,i-1 \tag{11a}$$

$$\mathbf{v}_{r_i,a_j} = \frac{\partial \dot{r}_i}{\partial \dot{a}_j} = \frac{\partial \mathbf{Y}_i}{\partial \ddot{a}_j} = A_j \mu_j - \tilde{r}_{i,j+1} A_j \phi_j , \quad j=1,\ldots,i-1 \tag{11b}$$

$$\mathbf{V}_{\omega_i,\theta_j} = \frac{\partial \omega_i'}{\partial \dot{\theta}_j} = \frac{\partial \dot{\omega}_i'}{\partial \ddot{\theta}_j} = \mathbf{A}_i^T \mathbf{A}_j \mathbf{u}_j' \ , \quad j=1,\ldots,i \tag{11c}$$

$$\mathbf{V}_{\omega_i,a_j} = \frac{\partial \omega_i'}{\partial \dot{a}_j} = \frac{\partial \dot{\omega}_i'}{\partial \ddot{a}_j} = \mathbf{A}_i^T \mathbf{A}_j \phi_j \ , \quad j=1,\ldots,i-1 \tag{11d}$$

and

$$\dot{\mathbf{V}}_{r_i,\theta_j} = \frac{\partial \mathbf{Y}_i}{\partial \dot{\theta}_j} = (-\dot{\tilde{\mathbf{r}}}_{i,j}\mathbf{A}_j - \tilde{\mathbf{r}}_{i,j}\mathbf{A}_j \tilde{\omega}_j')\mathbf{u}_j' \ , \quad j=1,\ldots,i-1 \tag{12a}$$

$$\dot{\mathbf{V}}_{r_i,a_j} = \frac{\partial \mathbf{Y}_i}{\partial \dot{a}_j} = \mathbf{A}_j \tilde{\omega}_j' \mu_j - (\dot{\tilde{\mathbf{r}}}_{i,j+1}\mathbf{A}_j + \tilde{\mathbf{r}}_{i,j+1}\mathbf{A}_j \tilde{\omega}_j')\phi_j \ , \quad j=1,\ldots,i-1 \tag{12b}$$

$$\dot{\mathbf{V}}_{\omega_i,\theta_j} = \frac{\partial \dot{\omega}_i'}{\partial \dot{\theta}_j} = (\mathbf{A}_i^T \mathbf{A}_j \tilde{\omega}_j' - \tilde{\omega}_i' \mathbf{A}_i^T \mathbf{A}_j)\mathbf{u}_j' \ , \quad j=1,\ldots,i \tag{12c}$$

$$\dot{\mathbf{V}}_{\omega_i,a_j} = \frac{\partial \dot{\omega}_i'}{\partial \dot{a}_j} = (\mathbf{A}_i^T \mathbf{A}_j \tilde{\omega}_j' - \tilde{\omega}_i' \mathbf{A}_i^T \mathbf{A}_j)\phi_j \ , \quad j=1,\ldots,i-1 \tag{12d}$$

Such relations have been obtained by Jerkovsky [4] and Kim [5] for rigid mechanical systems. Recently Chang [6] has also determined somewhat similar relations as those of Eq. 11 for flexible mechanical systems. Collecting the joint and modal coordinates as presented in Eq. 8 and using the above notation, Eqs. (9) and (10) may be written in matrix form as:

$$\begin{bmatrix} t_i \\ \omega_i' \end{bmatrix} = \begin{bmatrix} \mathbf{V}_{r_i,\theta} & \mathbf{V}_{r_i,a} \\ \mathbf{V}_{\omega_i,\theta} & \mathbf{V}_{\omega_i,a} \end{bmatrix} \begin{bmatrix} \dot{\theta} \\ \dot{a} \end{bmatrix} \tag{13a}$$

and

$$\begin{bmatrix} \mathbf{Y}_i \\ \dot{\omega}_i' \end{bmatrix} = \begin{bmatrix} \mathbf{V}_{r_i,\theta} & \mathbf{V}_{r_i,a} \\ \mathbf{V}_{\omega_i,\theta} & \mathbf{V}_{\omega_i,a} \end{bmatrix} \begin{bmatrix} \ddot{\theta} \\ \ddot{a} \end{bmatrix} + \begin{bmatrix} \dot{\mathbf{V}}_{r_i,\theta} & \dot{\mathbf{V}}_{r_i,a} \\ \dot{\mathbf{V}}_{\omega_i,\theta} & \dot{\mathbf{V}}_{\omega_i,a} \end{bmatrix} \begin{bmatrix} \dot{\theta} \\ \dot{a} \end{bmatrix} \tag{13b}$$

$\mathbf{V}_{r_i,\theta}$ and $\mathbf{V}_{\omega_i,\theta}$ are 3 x p matrices constructed by appropriate collection of 3 x 1 vectors $\mathbf{V}_{r_i,\theta_j}$ and $\mathbf{V}_{\omega_i,\theta_j}$, respectively. Similarly, $\mathbf{V}_{r_i,a}$ and $\mathbf{V}_{\omega_i,a}$ are 3 x m matrices constructed by 3 x m_i matrices $\mathbf{V}_{r_i,a_j}$ and $\mathbf{V}_{\omega_i,a_j}$; m is the total number of modes of all the links.

Combining each of Eqs. (13a) and (13b) for all links, the velocity and acceleration transformation equations from joint to Cartesian space are derived as:

$$\dot{\mathbf{x}} - \mathbf{V}\,\dot{\mathbf{q}} \tag{14a}$$

$$\ddot{\mathbf{x}} - \mathbf{V}\,\ddot{\mathbf{q}} + \dot{\mathbf{V}}\,\dot{\mathbf{q}} \tag{14b}$$

In the above equations, $\mathbf{V}$ is called velocity transformation matrix, $\dot{\mathbf{q}}$ and $\ddot{\mathbf{q}}$ are the 1st and 2nd time derivatives of $\mathbf{q}$ in Eq. (8), $\dot{\mathbf{x}} - [\dot{\mathbf{t}}^T, \boldsymbol{\omega}'^{,T}, \dot{\mathbf{a}}^T]^T$ is the Cartesian velocity vector of the system, where $\dot{\mathbf{t}} - [\dot{\mathbf{t}}_1^T, \dot{\mathbf{t}}_2^T, \ldots, \dot{\mathbf{t}}_p^T]^T$ and $\boldsymbol{\omega}' - [\boldsymbol{\omega}_1'^{,T}, \boldsymbol{\omega}_2'^{,T}, \ldots, \boldsymbol{\omega}_3'^{,T}]^T$. The velocity transformation matrix, $\mathbf{V}$, and its time derivative, $\dot{\mathbf{V}}$, are defined as:

$$\mathbf{V} - \begin{bmatrix} \mathbf{V}_{r,\theta} & \mathbf{V}_{r,\theta} \\ \mathbf{V}_{\omega,\theta} & \mathbf{V}_{\omega,\theta} \\ \mathbf{0} & \mathbf{I} \end{bmatrix} \quad (15a) \quad \text{and} \quad \dot{\mathbf{V}} - \begin{bmatrix} \dot{\mathbf{V}}_{r,\theta} & \dot{\mathbf{V}}_{r,\theta} \\ \dot{\mathbf{V}}_{\omega,\theta} & \dot{\mathbf{V}}_{\omega,\theta} \\ \mathbf{0} & \mathbf{0} \end{bmatrix} \quad (15b)$$

where $\mathbf{I}$ is an $m \times m$ identity matrix, $\mathbf{V}_{r,\theta}$ and $\mathbf{V}_{\omega_i,\theta}$ are $3p \times p$ and $\mathbf{V}_{r_i,a}$ and $\mathbf{V}_{\omega_i,a}$ are $3p \times m$ matrices which, respectively, consist of submatrices $\mathbf{V}_{r_i,\theta}$, $\mathbf{V}_{\omega_i,\theta}$, $\mathbf{V}_{r_i,a}$ and $\mathbf{V}_{\omega_i,a}$.

Dynamics of Flexible Manipulators

Incorporating D'Alembert's principle into the principle of virtual work and assuming no surface traction, the variational equations of motion of body i is written as [7]:

$$- \int_{\gamma_i} \rho_i \delta \mathbf{r}_{ik}^T \ddot{\mathbf{r}}_{ik} d\gamma_i + \int_{\gamma_i} \rho_i \delta \mathbf{r}_{ik}^T \mathbf{f}_{ik} d\gamma_i - \int_{\gamma_i} \rho_i \delta \boldsymbol{\epsilon}_{ik}^T \mathbf{r}_{ik} d\gamma_i , \quad i-1,\ldots,p \tag{16}$$

In the above equation, ρ_i is material density, γ_i is the volume, $\delta\boldsymbol{\epsilon}_{ik}$ and $\mathbf{r}_{ik}$ are the strain variation and stress vectors containing the stress and strain tensors at node k, $\mathbf{f}_{ik}$ is the body force distribution at node k, $\delta\mathbf{r}_{ik}$ is linear virtual displacement of node k, and $\ddot{\mathbf{r}}_{ik}$ is the linear acceleration vector of node k.

To derive the equations of motion in terms of generalized coordinate, $\mathbf{r}_{ik}$ may be written in the following form, as shown in Fig. 1,

$$\mathbf{r}_{ik} - \mathbf{r}_i + \mathbf{A}_i \mathbf{s}_{ik}' - \mathbf{r}_i + \mathbf{A}_i (\mathbf{s}_{ik}^0 + \boldsymbol{\mu}_{ik}\mathbf{a}_i) \tag{17}$$

Taking the total time derivative of Eq. (17) twice, yields:

$$\mathbf{r}_{ik} - \mathbf{r}_i - A_i \bar{s}'_{ik} \dot{\omega}'_i + A_i \mu_{ik} \ddot{a}_i + A_i \bar{\omega}'_i \bar{\omega}'_i s'_{ik} + 2A_i \bar{\omega}'_i \mu_{ik} \dot{a}_i \tag{18}$$

Substituting from Eqs. (10) and (11) into Eq. 18:

$$\mathbf{r}_{ik} - \sum_{j=1}^{i} \left[(\mathbf{V}_{r_i,\theta_j} - A_i \bar{s}'_{ik} \mathbf{V}_{\omega_i,\theta_j}) \ddot{\theta}_j + (\dot{\mathbf{V}}_{r_i,\theta_j} - A_i \bar{s}'_{ik} \dot{\mathbf{V}}_{\omega_i,\theta_j}) \dot{\theta}_j \right]$$
$$+ \sum_{j=1}^{i-1} \left[(\mathbf{V}_{r_i,a_j} - A_i \bar{s}'_{ik} \mathbf{V}_{\omega_i,a_j}) \ddot{a}_j + (\dot{\mathbf{V}}_{r_i,a_j} - A_i \bar{s}'_{ik} \dot{\mathbf{V}}_{\omega_i,a_j}) \dot{a}_j \right]$$
$$+ A_i \mu_{ik} \ddot{a}_i + A_i \bar{\omega}'_i \bar{\omega}'_i s'_{ik} + 2A_i \bar{\omega}'_i \mu_{ik} \dot{a}_i \tag{19}$$

The first variation of Eq. (17) yields:

$$\delta \mathbf{r}_{ik} - \delta \mathbf{r}_i + \delta A_i s'_{ik} + A_i \mu_{ik} \delta a_i \tag{20}$$

Using Eqs. (5), (7), (11), and the cross product rule $\bar{c}d - -\bar{d}c$,

$$\delta \mathbf{r}_{ik} - \sum_{j=1}^{i} \left[(\mathbf{V}_{r_i,\theta_j} - A_i \bar{s}'_{ik} \mathbf{V}_{\omega_i,\theta_j}) \delta\theta_j \right]$$
$$+ \sum_{j=1}^{i-1} \left[(\mathbf{V}_{r_i,a_j} - A_i \bar{s}'_{ik} \mathbf{V}_{\omega_i,a_j}) \delta a_j \right] + A_i \mu_{ik} \delta a_i \tag{21}$$

Using the stress-strain and strain-displacement relations for linear elastic deformation and using the variational principle, one may write [8]:

$$\int_{\gamma_i} \rho_i \delta \epsilon_{ik}^T \mathbf{r}_{ik} d\gamma - \delta a_i^T G_i \tag{22}$$

where

$$G_i - \int_{\gamma_i} B_i^T D_i B_i d\gamma \; a_i - \Phi_i^T K_i \Phi_i a_i \tag{23}$$

B_i, D_i, K_i, and Φ_i are respectively modal strain, material property, finite element stiffness matrix, and the modal matrices, and G_i is the force due to stain energy of body i.

Substituting Eqs. (19), (21), and (22) and using the generalization of Eqs. (14) - (15), the variational equations of motion of Eq. (16) can be written in terms of the generalized coordinates for the whole system. Since Eq. (16) must hold for all virtual displacements $\delta\theta$ and δa, then equations of motion may be written as:

$$\bar{M} \ddot{q} - Q \tag{24}$$

where

$$\tilde{M} - V^T M \ V + \text{Diag}(J_1, \ldots, J_p) \tag{25}$$

$$Q - V^T(f - h - M V \dot{q}) + [G'^T, G^T]^T \tag{26}$$

$$G' - [d_1 \dot{\theta}_1 + c_1 \text{sign}(\dot{\theta}_1), \ldots, d_p \dot{\theta}_p + c_p \text{sign}(\dot{\theta}_p)]^T \tag{27}$$

In the above equations, J_i is the ith actuator inertia, d_i and c_i are viscous damping and coulomb friction coefficients at joint i, G is comprised of G_i's in Eq. (23), M and f are Cartesian mass matrix and force vector, and h is a Cartesian vector which is quadratic in velocity. Using the subscript notation for rigid-translation (t), rigid-rotation (r), and elastic (e) displacements; M, f, and h are constructed as:

$$M - \begin{bmatrix} M_{tt} & M_{tr} & M_{te} \\ M_{tr}^T & M_{rr} & M_{re} \\ M_{te}^T & M_{re}^T & M_{fe} \end{bmatrix}, \quad f - \begin{bmatrix} f_t \\ f_r \\ f_e \end{bmatrix}, \quad h - \begin{bmatrix} h_t \\ h_r \\ h_e \end{bmatrix} \tag{28}$$

Each Submatrix of M is a block diagonal matrix comprised of the appropriate Cartesian mass matrices of each body. Similarly components of f and h are comprised of the appropriate Cartesian vectors of each body. Individual components of Cartesian mass matrix and force vectors of a flexible body have been presented (in a slightly different form) by Yoo [8].

Case Study

The method presented in this paper was applied to the PUMA 560 shown in Fig. 3. Full description of the kinematic structure and geometry of the robot can be found in reference [9]. End-effector's straight up position is (-0.1491, 0, 1.5925) meters with respect to the base and has a maximum payload of 2.27 kg.

The desired task for the arm is to move a 2.27 kg object at its straight up position to the location (-0.0246, 0.6011, 0.5872) in 1 second. The arm starts from the rest and comes to rest at its final destination. All links including the base are assumed to be rigid except links 3 and 6 which are assumed to be hollow Aluminum tubes and are modeled with 3D beam elements to account for bending and torsion modes.

Four flexible models have been developed. The first model accounts only for the first two bending modes of links 3 and 6. Model 2 also accounts for the 1st torsion mode of link 3 in addition to the bending modes. Models 3 and 4 are developed from models 1 and 2, respectively, except bearings at joints 3, 4, and 6 are also considered to be flexible.

A program was written to calculate the joint angles, velocities, and accelerations, and the corresponding torques [10] required to perform this task based on a rigid body model of the robot. The torque functions were then applied to the flexible models. Figure 4 shows the root mean square error of end-effector position from the desired path for all four models. Figure 5 shows the same plots as Fig. 4 but in this case the desired task is performed in only 0.2 seconds, resulting in much higher speeds. It can be seen from these results that both link and joint flexibility are important in position accuracy but at higher speeds joint flexibility is more significant and torsional vibration also becomes noticeable.

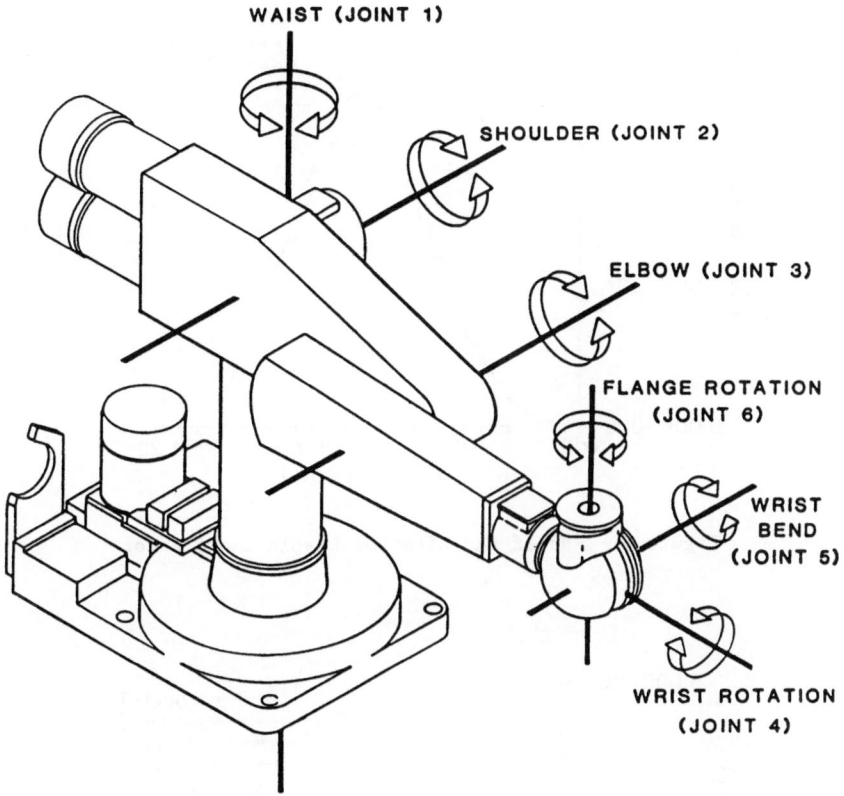

WAIST (JOINT 1)

SHOULDER (JOINT 2)

ELBOW (JOINT 3)

FLANGE ROTATION (JOINT 6)

WRIST BEND (JOINT 5)

WRIST ROTATION (JOINT 4)

Figure 3. PUMA 560 Configuration

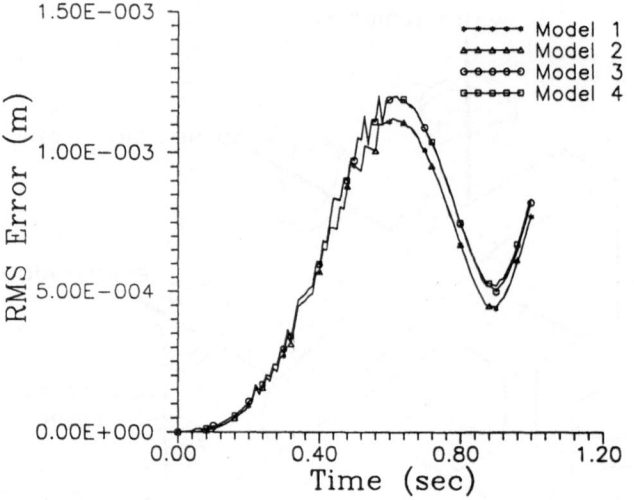

Figure 4. RMS of End-Effector Displacement Error

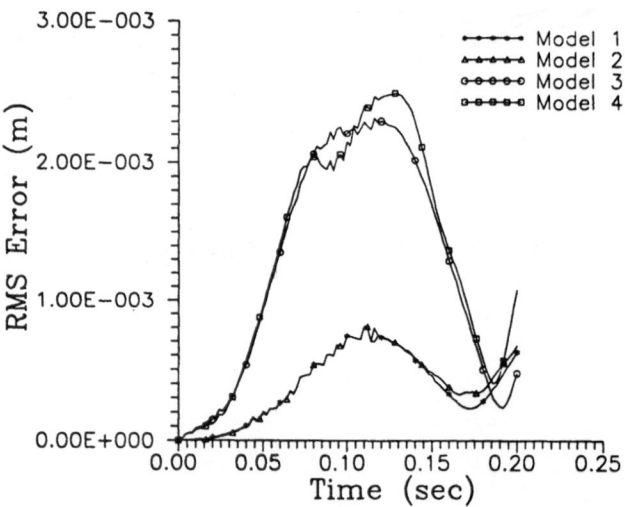

Figure 5. RMS of End-Effector Displacement Error- High Speed Case

Conclusions

A velocity transformation matrix has been presented which conveniently transforms easy to develop Cartesian equations into complex joint space equations for efficient solution of flexible robot dynamics. The method has been applied to an industrial robot to verify the significance of link and joint flexibility on the end-effector motion.

References

1. Book, W.J. Analysis of Massless Elastic Chains With Servo Controlled Joints. Trans. ASME J. Dyn. Syst. Measurement Contr. 101(3) (1979) 187-192.

2. Book, W.J. Recursive Lagrangian Dynamics of Flexible Manipulator Arms. The Int. J. of Robotics Res. 3(3) (1984) 87-101.

3. Sunada, W.; Dubowsky, S. The Application of Finite Element Methods to the Dynamic Analysis of Flexible Spatial and Co-Planar Linkage systems. Trans. ASME J. Mech. Design. 103 (1981) 643-651.

4. Jerkovsky, W. The Structure of Multibody Dynamics Equations. J. Guid. and Contr. 1(3) (1978) 173-182.

5. Kim, S.S.; Vanderploeg, M.J. A general and Efficient Method for Dynamic Analysis of Mechanical Systems Using Velocity Transformations. Trans. ASME J. Mech., Transm., and Autom. in Design. 108 (1986) 176-182.

6. Chang, C.W.; Shabana A.A. Spatial Dynamics of Deformable Multibody Systems with Variable Kinematic Structure: Part 2-Velocity Transformation. Trans. ASME J. Mech., Transm., and Autom. in Design. 112 (1990) 160-167.

7. Shames, I.H.; Dym, C.L. Energy and Finite Element Methods in Structural Mechanics. Hemisphere Publishing Corp.: McGraw-Hill Book Co. 1985.

8. Yoo, W.S.; Haug, E.J. Dynamics of Flexible Mechanical Systems Using Vibration and Static Correction Modes. Trans. ASME J. Mech., Transm., and Autom. in Design. Paper 85-DET-71 (1985).

9. Elgazzar, S. Efficient Kinematic Transformations for the PUMA 560 Robot. IEEE J. of Rob. and Autom. RA-1(3) (1985) 142-151.

10. Luh, J.Y.; Walker, M.; Paul, R.P. On-Line Computational Scheme for Mechanical Manipulators. Trans. ASME J. Dyn. Syst. Measurement Contr. 102 (1980) 69-76.

Dynamics of Flexible Manipulators With Application to Robotic Assembly

E. WEHRLI and P. COIFFET
Office National d'Etudes et de Recherches Aerospatiales (ONERA)
Chatillon, FRANCE

Abstract - The development of high speed and light weight manipulators for complex robotic assembly tasks and remotely operated space programs has motivated research work in the field of dynamic modeling and control. A kinematic and dynamic model of serial link manipulators with flexible links and joints has been developed. Simulation results are presented for the case of a 1 or 2 link planar arm which include compliance and vibration damping concepts.

Introduction

New generations of high speed and light weight robots have created new problems in trajectory tracking and control. It has become necessary to develop models of the mechanical structure which include flexibility of the links and the joints [5], in order to achieve vibration damping, shorter settling times and higher precision. Several models have been developed based on the finite element approach [1], and modal representation [4],[7] and [15]. Simultaneous research work has been conducted in the domain of mechanisms and multibody dynamics of articulated flexible structures [2],[3],[8],[14] with applications in space manipulators and other space station projects. Efficient recursive formulations of the dynamics [11],[13] have been widely developed for application to open chain robotic manipulators and space mechanisms. These methods have been extended to closed chain structures [12]. Different approaches [6],[9],[10] have been presented for open loop and closed loop control of the motion of flexible arms. It is still an area for challenging research work where good modeling of the dynamics and robust controller design are necessary. Factory automated assembly and space operations should benefit greatly from advances in this field.

The first part of this work presents the developments of a simulation tool for the dynamic behavior of a serial link manipulator, based on a modal approach for the Lagrangian formulation of the equations of motion. These equations are presented for a planar arm with 1 or 2 links. Results are given for the motion of a 1 link arm with application to vibration damping and open loop control. Finally simulation results are presented for a simple assembly task with a 2 link arm, with parameterized values of the joint and link stiffness and damping. Some considerations are given for developing control algorithms for the motion of flexible robotic arms using compliance.

Kinematic Model of a Flexible Arm

A kinematic model is developed for the motion of a serial link manipulator with flexibility in both the links and joints. All deformations are small compared to the joint displacements corresponding to the rigid body motion of the structure. The joints are supposed to be of revolute type. This formulation can be extended to closed chain robotic arms, parallel mechanisms and articulated flexible space structures with revolute and prismatic joints. Links or bodies of the multibody system consist of two rigid hubs with flexible midsection (figure 1). To link i we assign two frames R_{i1b} and R_{i2b} bound to the hubs, with their Z axes coinciding with the joints axes. The geometry of the undeformed link is described with three of the Denavit-Hartenberg parameters (length a_i, twist α_i, offset d_{i+1}). The transformation from R_{i1b} to R_{i2b} is represented by a 4×4 matrix T_{ui}. When the midsection of link i is deformed, R_{i2b} moves to a new position R_{i2c}, and the transformation T_{fi} from R_{i2b} to R_{i2c} can be written by decomposition of the deformations on a modal basis :

$$T_{fi} = I_{4\times4} + \sum_{j=1}^{m_i} q_{ij}(t) A_{f,ij} \tag{1}$$

where q_{ij} is the modal coordinate (considered second order quantity), and $A_{f,ij}$ is the time-invariant mode shape transformation matrix for mode j of link i.

The transformation from R_{i2c} to $R_{i+1,1b}$ (the frame bound to the proximal hub of the next link), in the absence of joint deformation, is a rotation of angle θ_{i+1} (the fourth Denavit-Hartenberg parameter) about the Z axis of R_{i2c}, which is represented by the 4×4 matrix $T_{r,i+1}$. Due to the flexibility of the contact surfaces, the elasticity of gears and drive systems and other elements of the joint, the joint rotation $T_{r,i+1}$ brings us to an intermediate frame $R_{i+1,1a}$, and an infinitesimal elastically restrained transformation $T_{a,i+1}$ representing the joint deflection, brings the frame $R_{i+1,1a}$ to $R_{i+1,1b}$ (figure 2). $T_{a,i+1}$ can be written in a summation form similar to that used for T_{fi}, with the q_{ij} 's, $j = 1, ..., m_i$, being replaced by the six infinitesimal deformation variables $p_{i+1,l}(t)$, $l = 1, ..., 6$ for joint $i + 1$ (three rotations and three translations), associated with appropriate constant 4×4 matrices.

The global transformation matrix from the base of joint i to the extremity of link i, from frame $R_{i-1,2c}$ to frame R_{i2c}, is written as :

$$_i^{i-1}T_d = T_{r,i}(\theta_i) \ T_{a,i}(p_{il}) \ T_{u,i} \ T_{f,i}(q_{ij}) \tag{2}$$

A first order approximation for T_d which is linear in the variables p and q is : $T_d = T_r \ T_u + T_r \ T_a \ T_u + T_r \ T_u \ T_f$. The term $T_{r,i} \times T_{u,i}$ is the usual Denavit-Hartenberg transformation matrix representing the rigid body motion. If R_0 is a frame attached to the ground, the transformation from R_0 to $R_{i+1,2c}$ is given by the recursive relationship : $_{i+1}^0T_d = \ _i^0T_d \ _{i+1}^iT_d$.

578

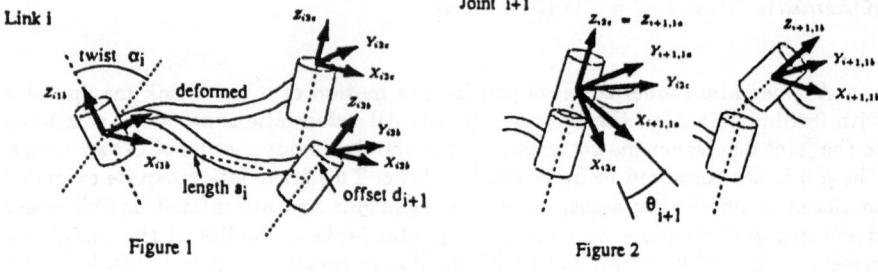

Link i

twist α_i

deformed

z_{i3b}

z_{i3e}

Y_{i3e}

X_{i3e}

z_{i3b}

Y_{i3b}

X_{i3b}

length a_i

offset d_{i+1}

Figure 1

Joint i+1

$z_{i3e} = z_{i+1,1e}$

$Y_{i+1,1e}$

Y_{i3e}

$X_{i+1,1e}$

X_{i3e}

θ_{i+1}

$z_{i+1,1b}$

$Y_{i+1,1b}$

$X_{i+1,1b}$

Figure 2

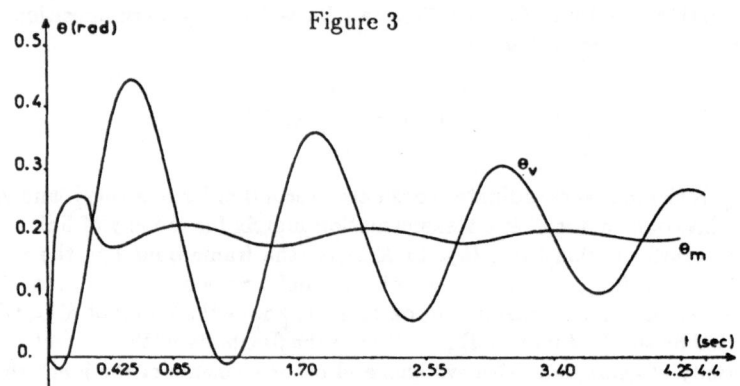

Figure 3

θ_v

θ_m

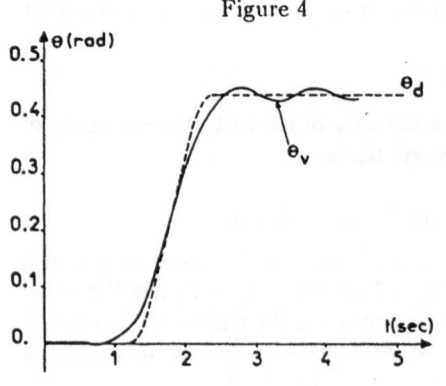

Figure 4

θ_d

θ_v

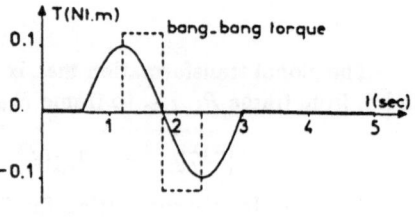

bang-bang torque

Figure 5

The instantaneous position of a material point on link i relative to R_0 is given by : $^0P = {}^0_i T_d\, {}^iP$, where ^iP is the instantaneous position of point relative to R_{i1b}, and is expressed by decomposing the displacements on a mode shape basis.

$$^iP = {}^iP_o\ (undeformed)\ +\ \sum_j q_{ij}(t)\ \Phi_{ij}({}^iP_o) \qquad (3)$$

The kinematic model forms the basis for obtaining the kinetic energy and the potential energy in order to derive the equations of motion. Velocities of material points of link i are obtained from : $^0\dot{P} = {}^0_i\dot{T}_d\, {}^iP + {}^0_i T_d\, {}^i\dot{P}$.

Lagrangian Formulation of the Dynamics

The mode shapes, as presented previously for describing the deformations of the links, can be obtained from a finite element analysis of each of the bodies of the structure. The Lagrangian formalism enables a straightforward derivation of the equations of motion. The Lagrangian can be expressed in terms of nodal displacements and their time derivatives. Here it is formulated as a function of the joint angles θ, the joint deformation variables p and the link modal coordinates q, and their time derivatives, as follows : $L(\theta,p,q,\dot{\theta},\dot{p},\dot{q}) = KE(\theta,p,q,\dot{\theta},\dot{p},\dot{q}) - PE(\theta,p,q)$. For a serial link manipulator with n joints and n links, the equations of motion take the form :

$$\frac{d}{dt}\left(\frac{\partial L}{\partial \dot{\theta}_i}\right) - \frac{\partial L}{\partial \theta_i} = T_i \qquad i=1,...,n\ (T_i\ :\ torque) \qquad (4)$$

$$\frac{d}{dt}\left(\frac{\partial L}{\partial \dot{p}_{il}}\right) - \frac{\partial L}{\partial p_{il}} = 0 \qquad i=1,...,n\ ;\ l=1,...,6 \qquad (5)$$

$$\frac{d}{dt}\left(\frac{\partial L}{\partial \dot{q}_{ij}}\right) - \frac{\partial L}{\partial q_{ij}} = 0 \qquad i=1,...,n\ ;\ j=1,...,m_i \qquad (6)$$

The kinetic energy is written using a lumped mass approach by summing for each link the translation kinetic energy of infinitesimal elements k of mass m_k.

$$KE = \sum_{i=1}^{n} \sum_{elts\ k} \frac{1}{2} m_k Trace({}^0\dot{P}(k)\ {}^0\dot{P}(k)^T) \qquad (7)$$

The utilization of a so-called variational formulation of the problem, in the finite element community, coincides with a Lagrangian formulation when generalized coordinates are used. Moreover, with a discretization of the links in sufficiently small elements this formulation is equivalent to writing the square inertia matrix corresponding to the variables q of link i as : $T_\Phi^T\, M\, T_\Phi$, where M is the finite element mass matrix of link i and T_Φ is the transformation matrix from the basis of the node displacements to the basis of the m_i first modes. The potential energy is composed of the strain energy and a gravitational energy term which is a function of the position vectors of material points of each link, as obtained from relation (3).

Equations for a Planar Flexible Arm

The system of equations (4), (5) and (6) which are second order differential equations, can be set in the following form :

$$I \ddot{Y} + C \dot{Y} + K Y = F(t) + G(\dot{Y}^t \dot{Y}) \qquad (8)$$

Y is the vector of generalized coordinates : θ_i, p_{il} and q_{ij}. I and K are the usual inertia and stiffness matrices. A diagonal damping matrix C has been introduced in the system to represent material damping in the links and the joints. F is the external force vector corresponding to input torques. G is a force vector, function of the squares of derivatives of the generalized coordinates, which correspond to centrifugal and Coriolis coupling terms in the kinetic energy.

The K matrix is diagonal as the link deformations are expressed in terms of the modal components in the basis of the normal modes. Diagonal terms corresponding to joint variables are zero. For a one-link structure, I is a constant matrix and G is a null vector, which results in a linear system of ordinary differential equations with constant coefficients. For the two-link structure where the first joint and link of length l_1 are supposed to be rigid, the second link of length l_2 is flexible and the only flexibility in the second joint is torsional motion corresponding to variable : $p_t = p_{21}$, the inertia matrix is no more constant. It is a function of the Y vector, and the dependance on the generalized coordinates is related to the motion of the link reference frame in the ground based reference frame. The differential equation corresponding to the variable θ_2 has the following form :

$$I_{21} \ddot{\theta}_1 + I_{22} \ddot{\theta}_2 + I_{23} \ddot{p}_t + \sum_{j=1}^{m} I_{2,j} \ddot{q}_j = T_2 + G_2(\dot{Y}^t \dot{Y}) \qquad (9)$$

The I_{21} inertia term depends on θ_2 and p_t as follows :

$$I_{21} = I_r + \sigma_b l_1 cos(\theta_2 + p_t) + m_T [\, l_2^2 + l_1 l_2 cos(\theta_2 + p_t) \,] \qquad (10)$$

where I_r is the sum of the rotational inertias of masses along the second link (rotor of the actuator, transmission, link, tip payload), $\sigma_b = \sum_{elts \, k} m_k x_k$ is an inertia per unit length for the second link, and m_T is the mass of the tip payload. All other inertia terms in this equation are constant.

The nonlinear force term $G_2(\dot{Y}^t \dot{Y})$ arise from the coupling of the motion of the flexible second link with the rotation of the first link, as shown by its expression :

$$G_2(\dot{Y}^t \dot{Y}) = l_1 \dot{\theta}_1 sin(\theta_2 + p_t) \{ (\sigma_b + m_T l_2) (\dot{\theta}_2 + \dot{p}_t) + \sum_{j=1}^{m} (\sigma_j + m_T \Phi_j(l_2)) \dot{q}_j \} \qquad (11)$$

The equations for the one-link arm can be obtained by setting l_1 equal to zero. Also coupling terms appear in the differential equation for variable θ_1, so that for smooth end point motion both torques should be computed simultaneously.

Numerical Applications

The formulation of the dynamics of a flexible manipulator has been applied to a 1-link arm driven by a 30 Nt.m maximum torque actuator through an elastic joint (or transmission). The rotor inertia is $I_m = 0.15 \ kg.m^2$. The joint torsional stiffness is $K_t = 2500 \ Nt.m/rad$. The link is an aluminium beam of cross section dimensions $0.001 \times 0.066 \ m$ and length $L = 1 \ m$. The first 2 natural frequencies of the clamped-free beam representing the link are 0.8 and 5 Hz. These modes have been obtained from a discretization of the link in 6 Euler-Bernoulli beam elements. Material damping ratio is set equal to 0.02.

Figure 3 shows the time response of the rotor angle θ_m and the tip position θ_v over a 4.5 second time span. The position of the tip of the link is expressed as an angle θ_v computed from the formula : $\theta_v \ = \ \theta_m \ + \ p_t \ + \ \sum_j \ q_j \Phi_j(L)/L$. The torque profile is the following : $T = 30. \ from \ t = 0. \ to \ t = 0.03 \ sec$, $T = 0. \ from \ t = 0.03 \ to \ t = 0.29 \ sec$, $T = -30. \ from \ t = 0.29 \ to \ t = 0.30 \ sec$ and $T = 0. \ after \ t = 0.3 \ sec$. The joint is very stiff compared to the link so that the rotation at the base of the link is nearly equal to that of the rotor. The flexibility of the link is modelled with a single mode. Tip response shows that after 2 sec the tip of the link starts oscillating with a 1.23 sec period, which corresponds to the lowest frequency of the system, and that the rotor angle is in quadrature with the tip angle. The same results are obtained with a 2 mode representation of the arm.

For control purpose, the response of the tip of the manipulator is also computed for a different torque profile. Rigid link robotic arms are commonly driven by bang-bang torque profiles resulting in large amplitude oscillations in the case of existence of some elasticity in the structure. With a finite element model of a flexible one-link arm, Bayo [6] integrated the system differential equations via the frequency domain, to obtain a torque profile which achieved the same tip motion that the bang-bang torque with a rigid link. For the same purpose, an anticipative sinusoidal torque profile is applied here to the flexible arm which joint is assumed rigid, and results are presented on figure 4. θ_d represents the desired tip motion obtained by applying a bang-bang torque profile over a 1.2 sec period starting at $t = 1.2 \ sec$ to the same but rigid arm. θ_v represents the tip response with a sinusoidal torque profile with period equal to 2.4 sec, and starting at $t = 0.6 \ sec$, as shown on figure 5. The response of the arm is close to that of the rigid link case and vibration damping at the end of the trajectory is good.

The model of a 2-link planar arm with flexible second joint and link has been developed for dynamic simulation and for testing the effects of compliance in some robotic tasks. For high speed and long reach automated assembly tasks and space transfer operations, the flexibility of some of the members of the manipulator arm cannot be neglected. With this model it can be taken into account as well as the elasticity of the transmissions and the joints. Simulations are run for an arm which total reach is 2 m, with both links of 1 m length. The second link is identical to the one in the previous application, but with a 0.010 *kg* payload mass at the tip and it is represented with 5 modes at respective frequencies 0.798, 4.85, 13.22, 25.16 and 39.53 *Hz*. The mass of the first link is $m_1 = 0.4$ *kg*, while it is of 0.182 *kg* for the second one. The inertias of the rotor at the base, the stator, the rotor and the transmission at the second joint are respectively 0.3, 0.15, 0.15 and 0.05 $kg.m^2$. The stiffness of the second joint is $K_t = 2500$ *Nt.m/rad* and the damping ratio for the link and the joint are set equal to 0.001.

A simple assembly motion such as pin insertion is simulated. Initially the arm is aligned along the Y axis with the second link resting along the first one ($\theta_1 = \Pi/2$, and $\theta_2 = -\Pi$). Torques to be applied at the joints are computed by solving the inverse dynamics for the arm with rigid links, so that the tip of the arm would slide along the X axis with trapezoidal joint angular velocity profiles.

Figure 6 and 7 show the time history of joint angles 1 and 2 for a 4 sec time run. Both angles vary smoothly and the shape of the curves is close to that which would be obtained with the rigid link arm as represented with the dashed line. At low speed the natural frequencies of the system are not highly excited. The trajectory of the tip of the arm is represented on figure 8 for a 2 sec (dotted line) and a 4 sec operation (dashed line), as well as intermediate configurations of the arm. Also, it can be noted that at the end of the task, when the applied torques are small, inertia and flexibility effects bring significant perturbations to the motion of the arm.

The time histories of joint angle 2 (dashed line) and the angle defining the tip position (solid line) for the 2 sec operation are given on figure 9. This second angle is defined in the same manner as θ_v for the 1-link arm. The two curves cross each other at time intervals corresponding to half the first period of oscillation of the link. Therefore the prevailing deformation mode of the second link is its first mode, other modes of vibration are not much excited in this operation. Figure 10 presents joint angle 1 time history for the same 2 sec insertion task while varying structural damping. Values of the damping ratio are of 0.001 for the solid line (previous case) and 0.02 for the dotted line. Increasing damping produces a smoother motion, but higher torques should be used. Passive damping is a first step to realize smooth compliant assembly operations.

The natural frequency of the joint with the present value of K_t is of 35 *Hz*. By tuning K_t to obtain a frequency equal to the first 0.8 *Hz* natural frequency of the flexible link, it might be possible to counteract some inertia and flexibility effects. Stabilization of the tip of the arm at the end of the trajectory should be obtained by introducing joint torques: $T = -k\Delta\theta$ proportional to the joint angle error. Further research will be done on active damping of vibrations and controlling the motion of the tip at high speed, by applying anticipative torques as for the 1-link arm.

Conclusion

A method has been presented for computing the dynamic response of a serial link manipulator with flexibility in both the links and the joints. Robotic applications of this work are vibration control, reduction of cycle time, compliance control for assembly operations. Numerical results were presented for a one-link structure with

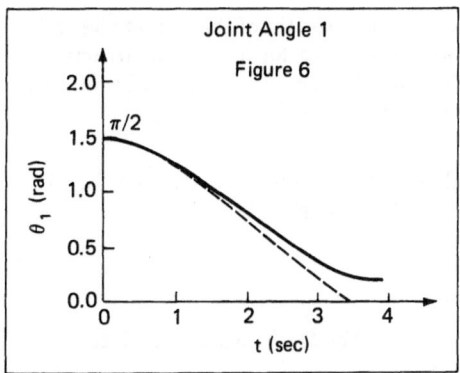

Joint Angle 1

Figure 6

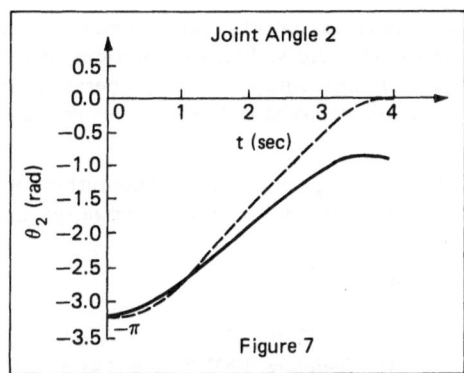

Joint Angle 2

Figure 7

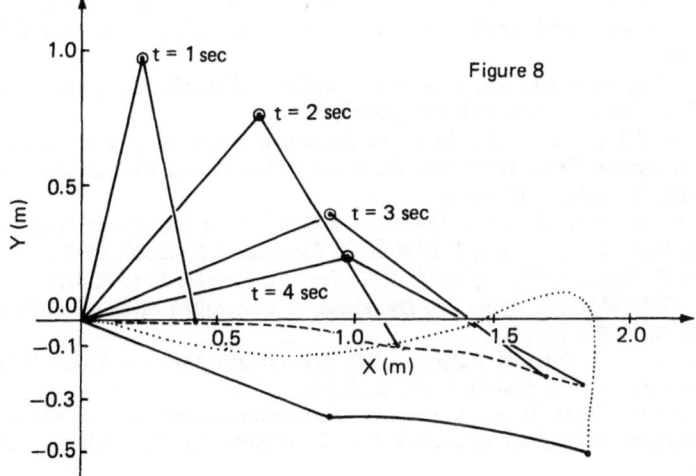

Figure 8

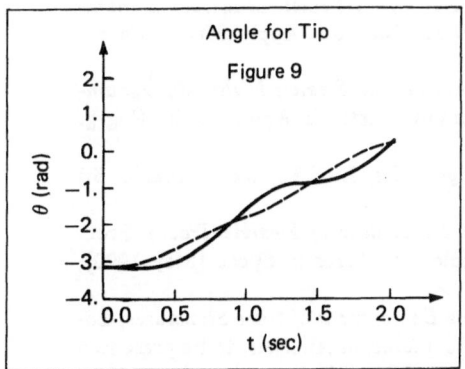

Angle for Tip

Figure 9

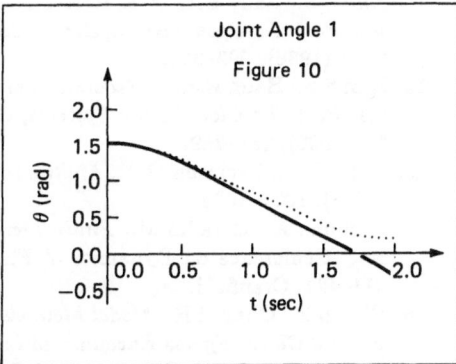

Joint Angle 1

Figure 10

insights into control problems and vibration damping of the tip of the arm at the end of the trajectory. Some simulations results were presented for a two-link structure performing a simple assembly operation, which highlight some characteristics of the dynamic behavior of flexible arms and the effect of passive damping.

Aknowledgements :   The authors would like to thank Dr R. Ohayon for his support through the study of large space structures at O.N.E.R.A.

References

1. Sunada W.H., *Dynamic Analysis of Flexible Spatial Mechanisms and Robot Manipulators*, PhD Thesis (1984), U.C.L.A..
2. Yoo W.S., Haug E.J., *Dynamics of Articulated Structures, Part I : Theory*, J. of Structural Mechanics **14**,1 (1986), 105–126.
3. Yoo W.S., Haug E.J., *Dynamics of Articulated Structures, Part II : Computer Implementation and Applications*, J. of Structural Mechanics **14**,2 (1986), 177–189.
4. Book W.J., *Recursive Lagrangian Dynamics of Flexible Manipulator Arms*, Int. J. of Robotics Research **3**,3 (1984), 87–101.
5. Dado M.H.F., Soni A.H., *Dynamic Response Analysis of 2-R Robot with Flexible Joints*, Proc. IEEE Int. Conf. on Robotics and Automation (1987), 479–483, Raleigh, N. Carolina.
6. Bayo E., *A Finite-Element Approach to Control the End-Point Motion of a Single-Link Flexible Robot*, J. of Robotic Systems **4**,1 (1986), 63–75.
7. Wehrli E., Kokkinis T., *Dynamic Behavior of a Flexible Robotic Manipulator*, Proc. IUTAM/IFAC Symp. on Dynamics of Controlled Mechanical Systems (1988), 309–320, Zurich.
8. Cardona A., *An Integrated Approach to Mechanism Analysis*, Thèse de Dr en Sciences Appliquées (1989), Univ. de Liège.
9. Siciliano B., Book W.J., *A Singular Perturbation Approach to Control of Lightweight Flexible Manipulators*, Int. J. of Robotics Research **7**,4 (1988), 79–90.
10. Cannon R.H.,Jr., Schmitz E., *Initial Experiments on the End-Point Control of a Flexible One-Link Robot*, Int. J. of Robotics Research **3**,3 (1984), 62–75.
11. Kim S.S., Haug E.J., *A Recursive Formulation for flexible Multibody Dynamics, Part I : Open Loop Systems*, Comput. Methods Appl. Mech. Engrg. **71**,3 (1988), 293–314.
12. Kim S.S., Haug E.J., *A Recursive Formulation for Flexible Multibody Dynamics, Part II : Closed Loop Systems*, Comput. Methods Appl. Mech. Engrg. **74** (1989), 251–269.
13. Kane T.R., Levinson D.A., *Multibody Dynamics*, J. of Applied Mechanics **50** (1983), 1071–1078.
14. Cardona A., Geradin M., *Finite Element Modeling of Flexible Tracks*, Proc. Int. Conference on Dynamics of Flexible Structures in Space (May 1990), 411–424, Cranfield.
15. Wehrli E., Ohayon R., *Modal Methods for Deployment of truss Structures, Local and Global Effects Encountered in the Vibration Analysis*, to be presented at Euromech 268 on Dynamics and Control of Flexible Structures (Sept 1990), Munich.

Kineto-Elastodynamic Effect on the Design of Elastic Mechanisms

ECHEMPATI RAGHU
Department of Mechanical Engineering
Washington State University
Pullman, WA

A. BALASUBRAMONIAN
Department of Mechanical Engineering
Regional Engineering College
Calicut, Kerala India

Summary

In this paper an approximate method to find out the kineto-elastodynamic acceleration of an elastic coupler of a four-bar linkage produced due to elastic motion of the coupler is explained. A comparison of rigid body acceleration with the resultant acceleration of the elastic link is also made. The results (in MKS units) clearly indicate the importance of considering KED effects while designing high-speed linkages with elastic links.

Introduction

Most modern high-speed linkages are designed with light-weight links to reduce the overall weight. However, these links may become elastic during the operation of the linkage. So far the design of such linkages are purely based on dynamic effects such as deflection, stress, vibration and stability. The vibration of links gives rise to an additional acceleration field besides the acceleration field produced by the gross rigid body motion of the links. This causes additional inertia force (kineto-elastodynamic inertia force or KED inertia force). The varying forces and moments transmitted to the machine foundation due to the KED inertia force can be substantial depending upon the mass distribution of the link and the acceleration field resulting from the vibration of the moving links.

Procedure Followed

A. Vibratory Response:

Figure 1 shows a Four-Bar Linkage with elastic coupler.

The equation of motion for vibratory response of the elastic coupler of a four-bar linkage has been derived by energy method [1]. The normal mode shape of the coupler was assumed to be that of simply supported beam and the first mode solution was assumed

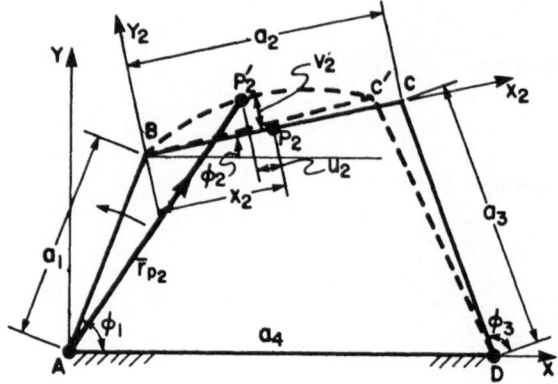

Figure 1. Four-bar linkage with elastic coupler.

to be in the form

$$v_2 = \sin(\pi x_2/a_2)q_2(t) \tag{1}$$

The differential equation of motion obtained using orthogonality conditions and changing the independent variable from time to angular position is

$$\ddot{q}_2 + \frac{C_D \dot{q}_2}{\dot{\phi}_1 m_2} + q_2 \left\{ \frac{E_2 I_2 (\pi/a_2)^4}{m_2 \dot{\phi}_1^2} + \pi^2 \left(\frac{a_1 \cos(\phi_1 - \phi_2)}{2a_2} + \frac{J_3 \ddot{\phi}_3}{m_2 a_3 a_2^2 \sin(\phi_3 - \phi_2)\dot{\phi}_1^2} \right. \right.$$

$$\left. \left. + \frac{\cot(\phi_3 - \phi_2)}{a_3^3 m_2 \dot{\phi}_1^2} \left[J_2 \ddot{\phi}_2 - a_1 \dot{\phi}_1 \sin(\phi_1 - \phi_2)M_2 r_{2c} \right] \right) - 0.395 \ddot{\phi}_2^2/\dot{\phi}_1^2 \right\}$$

$$= 4a_1 \frac{\sin(\phi_1 - \phi_2)}{\pi} - \frac{2\ddot{\phi}_2 a_2}{\pi \dot{\phi}_1^2} \tag{2}$$

where

a_i	=	length of ith link of the four-bar linkage
E_2	=	modulus of elasticity of the coupler material
I_2	=	area moment of link 2
J_2	=	mass moment of inertia of link 2 about B
J_3	=	mass moment of inertia of link 3 about D
M_i	=	mass of link i
m_2	=	mass/unit length of link 2
P_2	=	point on the undeflected coupler
r_{1c}	=	distance of the center of mass of link 1 from A
r_{1c}	=	distance of the center of mass of link 2 from B

r_{1c}	=	distance of the center of mass of link 3 from D
v_2	=	transverse deflection of link 2
$\phi_i, \dot{\phi}_i, \ddot{\phi}_i$	=	angular position, velocity, and acceleration of ith link with respect to the ground
C_D	=	damping coefficient

Runge-Kutta Merson method of numerical integration was used for the solution of this equation.

B. KED Acceleration

The elastic motion in the transverse direction of any point on the coupler as given in Eq. (1) above, is rewritten as

$$v_2 = q_2 \sin(\pi x_2/a_2) \tag{3}$$

The velocity and acceleration of the point due to this motion at any instant are given by

$$\dot{v}_2 = \dot{q}_2 \sin(\pi x_2/a_2) \tag{4}$$

and

$$\ddot{v}_2 = \ddot{q}_2 \sin(\pi x_2/a_2), \tag{5}$$

respectively.

In the program developed for the numerical integration of the equation of motion, $(dq_2/d\phi_1)$ is obtained at every $1°$ interval of the crank position. Since

$$d\phi_1 = \dot{\phi}_1 dt \tag{6}$$

$$dq_2/dt = \dot{q}_2 = \dot{\phi}_1(dq_2/d\phi_1) \tag{7}$$

Hence $\ddot{q}_2$, KED acceleration of the mid-point of the coupler at any instant, can be calculated using the relation

$$\frac{d^2 q_2}{dt^2} = \ddot{q}_2 = \left\{ \frac{(dq_2/d\phi_1)|_{(\phi_1-1)} - (dq_2/d\phi_1)|_{\phi_1}}{(\pi/180\dot{\phi}_1)} \right\} \dot{\phi}_1 \tag{8}$$

where $(dq_2/d\phi_1)$ at $(\phi_1 - 1)$ and $(dq_2/d\phi_1)$ at ϕ_1 refer to the instantaneous previous and the current crank positions respectively and were obtained from the program.

C. Resultant Acceleration and Resultant Shaking Force

The rigid body acceleration of the center of mass of the links are calculated using relations obtained from the dynamic analysis of the linkage with links assumed to be rigid [2]. The KED acceleration was vectorially added to rigid body acceleration of the center of mass of

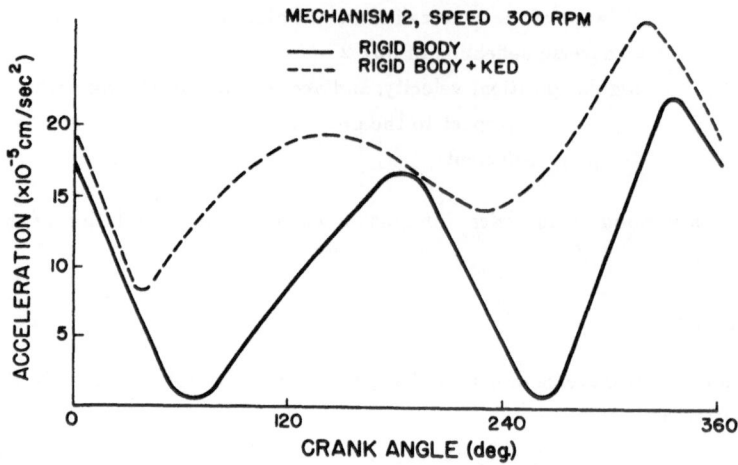

Figure 2. Rigid body and **KED** accelerations of the mid point of the coupler.

the coupler to get the resultant acceleration. The total KED inertia force can be obtained from

$$F_s = - \int_0^{a_2} (m_2 dx_2) \ddot{q}_2 \sin\left(\frac{\pi x_2}{a_2}\right) dx_2 \qquad (9)$$

Therefore, the total shaking force transmitted on to the input/output link pivots will be the vectorial sum of the rigid body plus the KED inertia force.

Results and Discussion

The parameters of the linkage used for this study are given in Table 1. In this linkage only coupler was found to have elastic deflection at the speed considered. Figure 2 shows the comparison between the absolute rigid body and the resultant acceleration of the center of mass of the coupler. In some positions of the crank KED acceleration is considerably more compared to the rigid body acceleration. The resultant acceleration is always found to be higher than the rigid body acceleration. In the present analysis it is assumed that the elastic motion of link (coupler) is limited to first mode only. This was verified from the results obtained for the elastic motion using equation 1.

Conclusions

Using the approximate procedure explained it is possible to determine the KED acceleration due to the elastic motion of links. The results clearly indicated the importance of considering the KED effects while designing high speed linkages with elastic links. This effect will also significantly affect the forces and moments transmitted to the foundation, which are considered while balancing the linkages.

Table 1
Parameters of the linkage

Link	Pivot to Pivot distance cm	Width in plane of the linkage cm	Thickness cm	Mass $(kg\text{-}sec^2/cm)$
Crank	23.0	2.0	2.0	7.034×10^{-4}
Coupler	81.0	1.0	2.0	13.12×10^{-4}
Rocker	81.0	2.0	2.0	26.24×10^{-4}
Ground	122.0			

Type of linkage:	Crank-rocker Four-Bar linkage
Link material:	Mild steel
Mass density of the material:	0.079×10^{-4} kg-sec^2/cm^4
Modulus of elasticity:	$E_2 = 2 \times 10^6$ kg/cm^2
Moment of inertia of coupler:	$I_2 = 1.33 \times 10^{-3}$ cm^4
Damping ratio:	.048

References

1. A. Balasubramonian and E. Raghavacharyulu, " Vibratory Response of an Elastic Coupler of a Four-Bar Linkage," Paper No. 86-DET-194, ASME Design Engineering Technical Conference, Columbus, Ohio, October, 1986.

2. G.G. Lowen, F.R. Tepper and R.S. Berkof, "The Quantitative Influence of Complete Force Balancing on the Forces and Moments of Certain Families of Four-Bar Linkages," Mechanism Mach. Theory, 9, 1974, pp. 299-323.

3. A. Balasubramonian, "Some Studies on the Elastodynamic Behavior of Unbalanced and Counterweighted Four Bar Mechanisms," PhD Thesis, IIT, Delhi, 1986.

Acknowledgements

The first author thankfully acknowledges the facilities provided by the Department of Mechanical and Materials Engineering at Washington State University, Pullman, Washington, in writing this paper. He also thanks Ms. Jo Ann Hicks for final preparation of this manuscript.

Scheme for Active Positional Correction of Robot Arms

SAEED B. NIKU

Mechanical Engineering Department
California Polytechnic State University
San Luis Obispo, CA

ABSTRACT

This paper describes a system which can measure the positional inaccuracies in a robot arm due to deflections of the joints and links, and which can make corrections such as to maintain the stated positional accuracy of the robot arm under different loading conditions. As a result, with the application of this system, the robot arm may be designed much lighter and still carry the same loads with the same accuracy.

In this system, there will be a mating link in parallel with each link. The main link, connected to the actuator, will be the load carrying member while the mating link moves along with the main link unloaded. If there is any deflection in the main link, it will be detected by the sensor on the mating link. The error signal will be used to activate the control system and the software to correct the position of the main link.

Introduction

One of the most important specifications of any robotic system is its positional accuracy. This is a measure of how accurately the robot can position the hand or the end effector compared to what is desired. Accuracy is dependent on many factors, including its control system, software, and rigidity. The final position of the hand is determined by the magnitude of each joint and link parameter as well as backlashes and joint and link deflections. Thus, the positional accuracy of the robot arm is measured at its maximum load capacity and when the arm is fully extended horizontally.

Since a robotic arm is a multi-degree of freedom open kinematic chain, joint and link deflections will not be known to the controller except through external measurements. This will have a direct effect on the positional accuracy of the robot. Since the deflections in the joints and links are functions of the external loads carried by the robot as well as the location and orientation of each link, and therefore, they are time dependent and difficult to calculate, robot arms are designed to have minimal deflections. This is accomplished by making the joints and the links very strong and bulky, increasing their moment of inertia and consequently their weight. In other words, a robot is made many times stronger than needed to carry the specified load in order to increase its rigidity and to keep the deflections to a minimum. Robots can generally carry much bigger loads than specified, but with less accuracy.

In order to fully specify the location and the orientation of a robotic hand in space six degrees of freedom are required. Many robots are designed with six axes, making it possible to accomplish this. Nevertheless, there are a significant number of robot arms in the market that possess fewer number of axes. Out of the six degrees of freedom, three are necessary to position the arm at the desired location, and three to orient it at the desired orientation. Generally, even if fewer degrees of freedom are used in the design of the robot, there will be three axes for positioning. These are either prismatic or revolute joints. Structurally, for most manipulator arms, the three main links which are used to position the end effector at the desired location are relatively long and primary load carrying members, while the remaining orientation links are relatively short. As a result, most deflections and subsequent positional errors are due to the first few links. In fact since the waist of a robot arm which houses the first joint is generally the frame of the robot and a non moving part, its share of the deflection, and consequently, its contribution to the error is minimal. Therefore, in a manipulator, only deflections of the two main arm links are important and should be considered.

It is stipulated that if this error can be measured and compensated, the robot arm links can be made significantly lighter but with the same positional accuracy.

Methodology

This correction scheme is accomplished by introducing a mating link in parallel with each of the two main links of the robot. The two parallel links are driven by the main actuator and move simultaneously toward the desired location. The difference between the two parallel links is that the main link is also the load carrying member while the mating link moves unloaded. As a result, depending on its orientation and location and the magnitude of the load, only the main arm will experience a deflection. The positional error will translate into a distance developed between the tips of the two arms. This distance is actively kept to a minimum by additionally moving the main arm a certain angle while the mating link is moved in the opposite direction an equal amount. The motion of the mating link is provided by a secondary small actuator which operates only on the mating link. Thus, the final positional error will be at the desired robot accuracy.

An error detection system will detect and measure the distance and the error developed between the two links. This information can be used to calculate the necessary correction that should be made in the main link's angle in order to bring the arm to the desired location. The same information is used to move the mating link in the opposite direction.

Results

To test this approach, a simple three degree of freedom manipulator was designed and constructed. The device consists of a set of main and mating links, stepper motors and stepper drivers, the error detection system, computer interface, and control software. The device differs in its operation from the above mentioned technique in that it has two independent drive actuators on each link, one for each mating link. This simplifies the design since no differential gearing is needed for the coupled motion, but is not as efficient as using a primary and secondary actuators. In this device, all actuators are located on the waist. The motion is transferred to each link and joint through chain drives and sprockets. Since the weight of the actuators are not carried by the links they can be designed very lightweight and relatively flexible. Thus, large deflections can be created under different loading conditions without having to use large weights.

The arms are made from hollow brass tubings. A photo diode and light source set is mounted at the tip of each main link while the mating links carry a small plate with a narrow slot. If the tips of the two links are exactly at the same location, the sensor is on, indicating that there is no error. If there is any deflection in the main link, then the sensor is off and the controller will try to correct the position until the error goes back to zero. Since the deflections are always downward the controller will always try to move the arm upward. The stepper motors with their reduction gears can provide a resolution of 0.02 inches on each 12 inch arm link. Stepper motors were chosen for their availability, simplicity, and high resolution.

The system has performed satisfactorily under different loads. When the arm is programmed to go to specific locations under different weights, the sensor will detect any deflection in the main arm as the arm is moving toward the desired position and will turn the main link until the error goes to zero.

Discussion

For this project an assumption is made that the deflections of the links will be relatively small so that a single rotation of the arm will be sufficient to correct the end position of the arm. If these deflections are large, the link will become a curved beam and the effective straight distance between the two ends of the link will be shorter than the original length. As a result, a simple rotation of the link will not suffice to bring the end point to the original location. In this case it will be necessary to rotate and extend the arm simultaneously to correct the error.

Although the main arms of the robot are made lighter and as a result more flexible, they are not to be so flexible as to cause large vibrations, overshoots, or other control problems. Flexible arm robots are a different class of manipulators and are not discussed here.

Other combinations and techniques can also be employed to achieve this active positional correction of the arm. For instance, the main arm can be forced to just follow the mating link while the mating link is controlled through the software. In other words, the controller will drive the mating link to the desired position while the tip of the main link is forced to follow the tip of the mating link in real time through the error detection system. Another possibility is to use a single drive motor which drives both the mating and the main links. The differential motion between the two links for error correction purposes is provided by a clutch and differential mounted between the two links.

The significance of the system is in its merit to reduce weight and inertia of the links without compromising the arm's accuracy, resulting in additional weight and power savings in the drive systems. It will separate the payload capacity of the arm and its positional accuracy such that the payload will be limited only by the physical capabilities of the arm. The additional cost of the correction system is minimal compared to the long term gains in the robot.

At the present time, there is no system that actively tries to close the kinematic loop of the arm. The customary technique is to have open or closed loop drive systems with a certain accuracy, and to rely on the strength of the links to minimize deflections. This scheme tries to effectively close the robot's kinematic chain.

Acknowledgements: The Author would like to thank the California State University for the State Faculty Support Grant program which provided funds for the project, and to thank Jon Conte and Cliff Dey for their contributions in making the arm and the controller.

Feedback Control of Robot End-Effector Probable Position Error

Y.C. PAO
  Engineering Mechanics
University of Nebraska
Lincoln, NE

L.C. CHANG
Systems Engineer
Electronic Data Systems Corporation
Troy, MI

Summary

In an earlier paper (Chang *et al.*, 1988), a **probabilistic model** has been proposed for estimating the effect of **process and measurement noises** on the robot end-effector's position error. In this paper, we show that a **Kalman filter** can be selected to estimate unmeasured states and the **separation theorem** (Brogan, 1985) can be applied for design of a **feedback controller** to stabilize the unstable perturbed state of the robot system. Error envelopes of two- and three-dimensional robots in the form of elliptical and ellipsoidal regions, respectively, have been derived for the cases where process and measurements noises at the robot joints are assumed to be **Gaussian-distributed, white noises** and not correlated. The developed methodology is for a general, n-link robot. As a numerical example, the probable error envelope is illustrated using a two-link robot that has two revolute joints and an end effector which undergoes planar motion. Calculations of the Kalman filter feedback, and of the shape and dimensions of the elliptical error envelope are explained in detail.

Method

Figure 1 shows a block diagram for the perturbed state of a robot, $\underline{\epsilon}_x$, subject to both the process noise $\underline{w}$ and measurement noise $\underline{v}$. The actually measured perturbed state is denoted as $\underline{z}$. The Kalman filter is the best linear estimator in the sense that it produces unbiased, minimum variance estimates (Kalman and Bucy, 1961; Brown, 1983). Let $\underline{e}_x(t)$ be the estimated perturbed state and $\delta\underline{e}_x(t)$ be the **residual** which is the difference between the true measured perturbed state, $\underline{z}(t)$, and the estimated perturbed state based on $\underline{e}_x(t)$, here denoted as $\underline{e}_z(t)$. It has already been shown (Lewis, 1986) that $\underline{e}_x$ satisfies a differential equation which can be schematically represented by the block diagram shown in Fig. 2 where K(t) is a Kalman filter **gain**. K(t) is to be calculated according to the equation

$$[K(t)] = [P(t)][C]^T[R]^{-1} \tag{1}$$

The measurement noise $\underline{v}$ is assumed as stationary, zero-mean, white noise of Gaussian distribution and assumed to have the auto-correlation

$$E\{\underline{v}(t)\underline{v}(t+\tau)\} = [R]\delta(\tau) \tag{2}$$

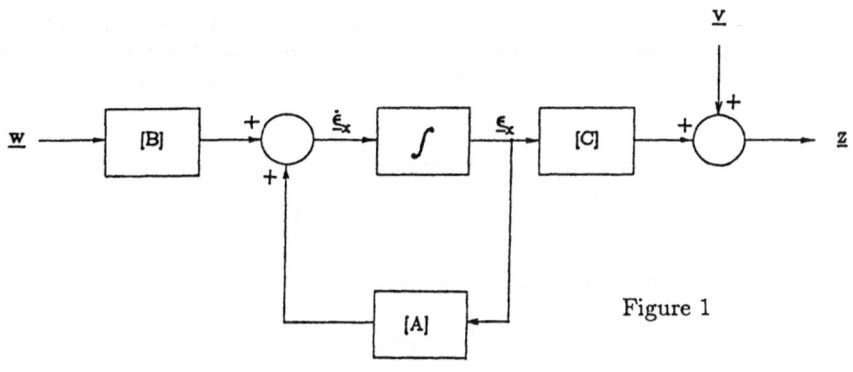

Figure 1

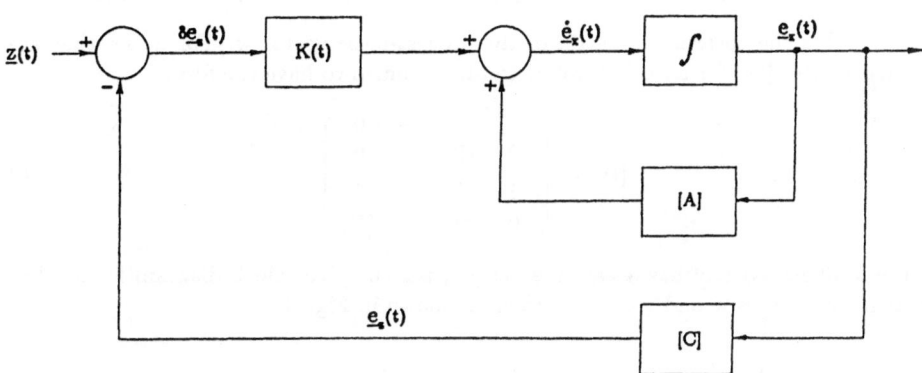

Figure 2

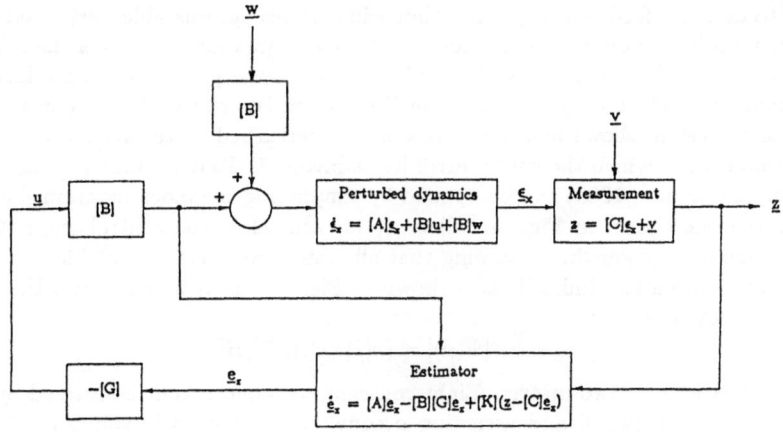

Figure 3

where [R] which appears in Eq. (1) is the covariance matrix, $\delta(\tau)$ is the unit impulse function, and τ is a time variable. If the variance of the ith joint of the robot system is denoted by R_i and if the noises are not correlated, the matrix [R] can be expressed as

$$[R] = \begin{bmatrix} R_1 & 0 & \cdots & 0 \\ 0 & R_2 & \cdots & 0 \\ \vdots & \vdots & & \vdots \\ 0 & 0 & \cdots & R_n \end{bmatrix} \tag{3}$$

In Eq. (1), the covariance matrix [P] is defined as

$$[P] = E\{(\underline{\epsilon}_x - \underline{e}_x)(\underline{\epsilon}_x - \underline{e}_x)^T\} \tag{4}$$

And, [P] satisfies the Riccati equation

$$[\dot{P}(t)] = [A][P(t)] + [P(t)][A]^T + [B][W][B]^T - [P(t)][C]^T[R]^{-1}[C][P(t)] \tag{5}$$

where [W] is the covariance matrix of the process-noise vector $\underline{w}$. For noises that are not correlated, [W] for an n-link robot can be assumed to have the form

$$[W] = \begin{bmatrix} W_1 & 0 & \cdots & 0 \\ 0 & W_2 & \cdots & 0 \\ \vdots & \vdots & \vdots & \vdots \\ 0 & 0 & \cdots & W_n \end{bmatrix} \tag{6}$$

If the feedback control has a gain matrix [G], the complete block diagram for a robot system with noises $\underline{w}$ and $\underline{v}$ can be drawn as shown in Fig. 3.

Design of Feedback Control

To design a feedback controller that will stabilize an unstable perturbed robot system, a feedback gain matrix should be added to insure that all poles of the robot system are in the left-half of the complex s-plane, s being the parameter of Laplace transformation. The block diagram shown in Fig. 3 consists of two major parts - the robot dynamic system shows in the top row and a perturbed state estimator shown in the bottom row, in which the gain matrix [G] is involved. To determine [G], the separation theorem (Brogan, 1985) can be applied by simplifying the block diagram shown in Fig. 3 to the one shown in Fig. 4. That is, we will select the control gains [G] using a pole-placement algorithm assuming that all states are directly available.

Let us use a two-link robot as shown in Fig. 5. First, we construct the **resolvent matrix** equation

$$[\Psi(p)] = (p[I] - [A])^{-1}[B] \tag{7}$$

$[\Psi(p)]$ is a four-by-two matrix. Eight constant vectors can thus be formed by using the two columns of $[\Psi(p)]$ and assigning p equal to the four pole values of the feedback system. Let these eight vectors be denoted by Ψ_{ij} which is identically equal to $\Psi_j(p_i)$ for i=1,2,3,4 indicating which pole value is used, and j=1,2 indicating which column of

$[\Psi(p)]$ has been selected to generate the vector. The gain $[G]$ is to be calculated from the equation

$$[G][\Psi] = -[I_{j1}\ I_{j2}\ I_{j3}\ I_{j4}] \tag{8}$$

where

$$[\Psi] = [\psi_{i1,j1}\ \psi_{i2,j2}\ \psi_{i3,j3}\ \psi_{i4,j4}] \tag{9}$$

Notice that $[G]$ is a two-by-four matrix , $[\Psi]$ is a square matrix of order four, and I_{jk}'s are column vectors of order two whose jkth element equal to one and the other element equal to zero. j1 through j4 can have value 1 or 2 whereas i1 through i4 can be either 1, or 2, or 3, or 4. Although any four of the eight vectors ψ_{ij} can potentially be selected but there is a restriction that $[\Psi]$ must be a nonsingular matrix.

After the addition of a feedback controller, the perturbed robot system is stabilized but should remain dominated by the dynamic characteristics of the uncompensated robot system rather than dominated by the filter effect. The filter pole locations are predetermined by the system properties and noise variances. The control poles can be arbitrarily positioned by choice of the gains $[G]$, and they should be selected somewhat to the right of the filter poles.

Numerical Example and Discussion

Let the lengths of the robot links shown in Fig. 5 be $\ell_1=75$ mm and $\ell_2=70$ mm and let both be of cylindrical shape of radii $r_1=10$ mm and $r_2=8$ mm and made of aluminum with a mass density equal to 2710 kg/m^3. If the orientations of the links are $\theta_1=30$ and $\theta_2=50$ degrees, it is easy to find that eigenvalues of $[A]$ are equal to ± 5.03254 and ± 3.18447 indicating that the perturbed state is unstable. Let us assume that the input torques have process noises of standard deviations of 1 N-m and .1 N-m, at the joints 1 and 2, respectively, and the measurement devices have errors of standard deviations .1 degree ($\pi/1800$ radian) for both θ_1 and θ_2. For $W_1=1$, $W_2=.01$, and $R_1=R_2=3.046$x10^{-6}, the poles of the Kalman filter are obtained to be $-10.66 \pm 9.700j$ and $-4.368 \pm 1.773j$ which indicate that the filter is a stable system. Knowing the poles of the filter, we may then select the poles for the feedback controlled robot perturbed system to be -2, -2.5, -3, and -3.5 (they are all on the right side of the filter poles.) Consequently, we can have $\psi_1(p_1) = [-.0190\ .0275\ .0380\ -.0550]^T$, $\psi_2(p_2) = [-.0160\ .0662\ .0402\ -.1655]^T$, $\psi_3(p_3) = [-.0938\ .1120\ .2815\ -.3362]^T$, $\psi_4(p_4) = [.0227\ -.0958\ -.0796\ .3353]^T$, and the first and second rows of $[G]$ are [212.229 51.334 65.711 15.827] and [73.904 44.899 17.492 8.985], respectively. Knowing $[G]$, we proceed to construct the error covariance matrix $[T_x]$ which then enables the geometrical shape and dimensions of the error envelope of the end effector to be determined (Chang et al., 1988).

Figure 6 shows the elliptical error envelope for the end effector of the two-link robot after being stabilized with a Kalman-filter feedback-controller. We have calculated that the probability of the end effector being inside an elliptical error envelope whose semi-major axis eqaul to 2.67 cm and semiminor axis equal to 1.86 cm is 39.35%, the probability of being inside an elliptical error envelope whose semi-major axis equal to 6.54 cm and semi-minor axis equal to 4.55 cm is 95.02%, and so on.

We thus have demonstrated the application of our proposed probabilistic error model for a two-dimensional robot. Even though the numerical example only covers a two-link analysis, the generalization of the methodology to n-link robots and for three-dimensional motions leading to ellipsoidal error envelopes is straight forward and is presented elsewhere (Chang, 1987). Our present research effort of extending this probabilistic model includes the incorporation of interactive block manipulation (Pao and Chang, 1985) for the feedback-control analysis on microcomputer, the software development of computer-aided prediction of error envelope also using microcomputers (Pao, 1984), and the study of the effect of elastic deformation of the robot links on the error envelope. In view of the complex geometrical shapes and material composition of the robot links, finite element techniques (Pao, 1986) need to be applied for determination of their elastic deformations. Knowing the effect of including the elastic deformation on the links' perturbed state is a prerequisite to the consideration of stabilizing the robot system affected by the process and measurement noises.

Acknowledgement

This work was supported, in part, by the General Motors Corporation through a CAD/CAM Fellowship Program and by the Engineering Research Center, University of Nebraska-Lincoln. The authors would like to thank their colleagues, Professors George R. Schade and William L. Brogan for enlightening discussions of probabilstic robotic analysis and feedback control.

References

1. Brogan, W. L. (1985) *Modern Control Theory*, Prentice-Hall, Englewood Cliffs, New Jersey.
2. Brown, R. G. (1983) *Introduction to Random Signal Analysis and Kalman Filter*, John Wiley, New York.
3. Chang, L. C. (1987) *Probabilistic Error Model of Robot End-Effector's Dynamic State*, Doctoral Dissertation, Department of Engineering Mechanics, University of Nebraska, Lincoln, Nebraska.
4. Chang, L. C., Schade, G. R., and Pao, Y. C. (1988) "Probabilistic Estimate of Robot End-Effector's Position Due to Process Noise," in *Robotics and Factories of the Future, '87*, edited by R. Radharamanan, Springer-Verlag, New York.
5. Kalman, R. E., and Bucy, R. S. (1961) "New Results in Linear Filtering and Prediction Theory," *Journal of Applied Mechanics*, V. 83, pp. 95-108.
6. Lewis, F. L. (1986) *Optimal Estimation*, John Wiley, New York.
7. Pao, Y. C. (1984) *Elements of Computer-Aided Design and Manufacturing, CAD/CAM*, John Wiley, New York.
8. Pao, Y. C., and Chang, L. C. (1985) "Manipulating Transfer Functions of Engineering System Using Interactive Graphics," *Computers in Mechanical Engineering*, V. 3, pp. 65-72.
9. Pao, Y. C. (1986) *A First Course in Finite Element Analysis*, Allyn and Bacon, Boston, Massachusetts.

10. Ross, S. R. (1985) *Introduction to Probability Models*, Academic Press, London, United Kingdom.

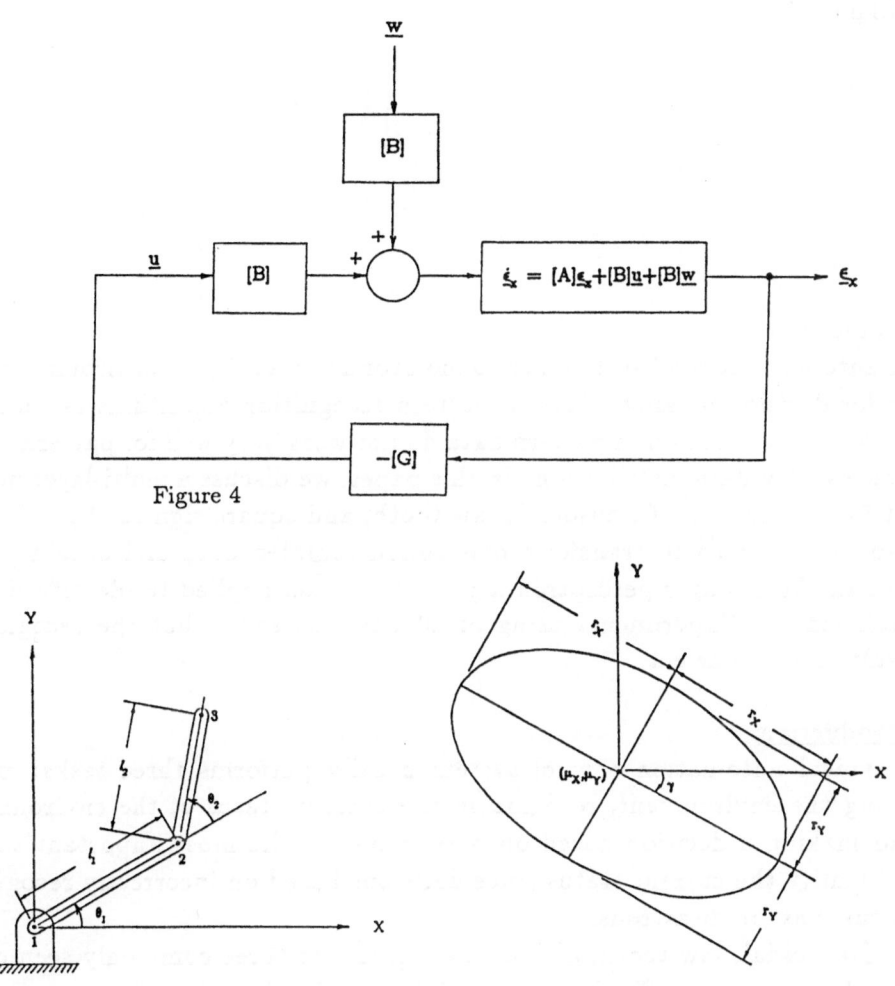

Figure 4

Figure 5

Figure 6

An Intelligent Signal Recognition System

T. HOU and L. LIN

Department of Industrial Engineering
State University of New York at Buffalo
Buffalo, NY

Summary

An automatic control system has to monitor and identify an environment status for decision making. Current pattern recognition algorithms assuming a Gaussian distribution in pattern data do not work very well for patterns with some overlay data distribution. In this paper, we discuss a multi-layer neural net for recognition of sinusoidal, saw tooth, and square signals. Fast Fourier transform is used to transform one period sampled data and obtain power spectra. Multi-layer perceptron algorithm is then applied to identify signals intelligently. Experiments using of 40 patterns show that the recognition results are all correct.

Introduction

A typical automation control system usually performs three tasks: monitoring the environment, recognizing the current status of the environment, and making a decision based on observations. The most important task is to identify the current status since decisions based on incorrectly recognized status may be disastrous.

Sinusoidal, saw tooth, and square signals are three commonly seen cyclic signals. It is not difficult to recognize them in time domain using Fourier series. The sinusoidal signal has only one dominating term in the series while the saw tooth and the square wave have many non-dominating terms with diminishing amplitude. However, it is time consuming to find all the coefficients from Fourier series representation.

The objective of the research is to develop an efficient and intelligent system for recognizing cyclic signals and to apply this system to these three signals.

Discrete Signal Processing

In an automatic control system, continuous signals are sampled and converted into discrete signals. A discrete periodic signal can be represented as a discrete Fourier series as follows:

$$x[n] = \sum_{k=<N>} a_k e^{jk(2\pi/N)n}, \quad k = 0, 1, 2, ..., N-1. \tag{1}$$

where a_k is the Fourier series coefficient and it can be determined from the following equation:

$$a_k = \frac{1}{N} \sum_n x[n] e^{-jk(2\pi/N)n} \tag{2}$$

Equations 1 and 2 are called discrete-time Fourier series pairs.

For a general aperiodic sequence $\hat{x}[n]$ with finite duration N, the Fourier transform of $\hat{x}[n]$ is defined as follows [3]:

$$X(k) = \frac{1}{N} \sum_{n=0}^{N-1} \hat{x}[n] e^{-jk(2\pi/N)n}, \quad k = 0, 1, 2, ... \tag{3}$$

It has been shown that $X(k) = a_k$ [3]. In other words, Fourier series coefficients of a periodic signal can be obtained from Fourier transform of one period of that sequence.

An Intelligent Signal Recognition System

The system shown in Figure 1 contains the following modules:

1. Signal collecting: signal generator is used to create sinusoidal, saw tooth, and square signals. Multiplexer and A/D converter are used to collect signal.

2. Signal processing: as can be seen from the previous section, only one period sampled data is needed for cyclic signal recognition. In this module, the sampled signals are processed to get one period data.

3. FFT processing: fast Fourier transform (FFT) is used to process the one period data and obtain the power spectrum.

4. Signal recognition: implementing a pattern recognition technique to identify signals. In this research, a multi-layer perceptron neural network based pattern recognition routine is used.

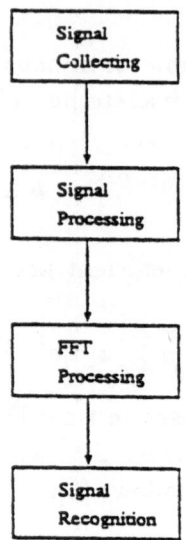

Figure 1: An intelligent signal recognition system

Topology of the multi-layer perceptron neural network contains an input layer, hidden layers, and an output layer [4]. The neural net uses a training data set as input,and a sigmoidal activation function and a learning algorithm to adjust the weights and the biases. After a few iterations the system will converge to a reasonable weight which is then used to identify a new incoming pattern. Generally there are two stages in the learning process: feed forward and back propagation [1].

In the feed forward stage the net input to each node is the sum of the weighted outputs of the nodes in the previous layer. The outputs of nodes in one layer are transmitted to nodes in another layers through links that amplify or attenuate or inhibit such outputs through weighting factors. Objectives of the stage are to obtain calculated outputs from the output layer and determine system errors between the calculated output and the desired output.

In the back propagation stage, it is to propagate the system errors back to the lower layers and to adjust the weights and biases in the output layer and hidden layer.

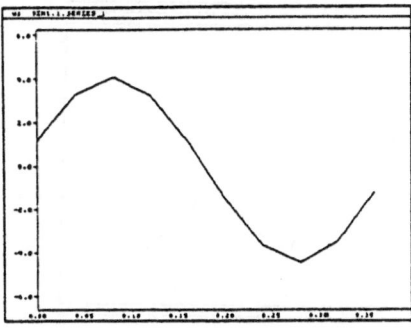 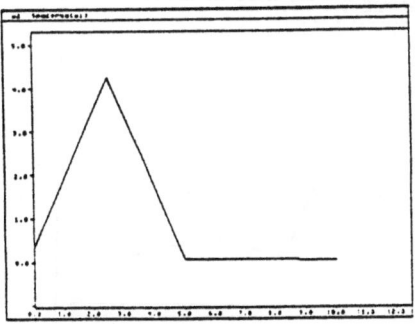

Figure 2: Time domain and power spectrum of a sinusoidal signal

Experiment and Results

In the signal collecting stage, a DYNASCAN 3010 function generator is used to generate sinusoidal, saw tooth, and square signals. The DATA TRANS-LATION PCLAB software [5] is used for multiplexing and A/D conversion. A total of 1024 samples are collected for each input signal. Signal frequency is set at 2.5 Hz. and sampling frequency is 25 Hz. It means that there are 10 sampled data in one signal period. These 10 sampled data has to be extracted from those 1024 samples. The system allows to start extracting at any sampled point.

FFT is then applied to the 10 sampled data to transform time domain data into frequency domain data and then obtain the power spectrum. Time domain data and power spectrum of one period data for a sinusoidal signal is shown in Figure 2. Figure 3 shows the power spectra of a saw tooth and a square signal.

After the FFT operation there are only 5 data in the power spectrum which are then used in the multi-layer perceptron neural net computation. A two layer neural net with 5 nodes in the input layer, 12 nodes in the hidden layer, and 3 nodes in the output layer is used.

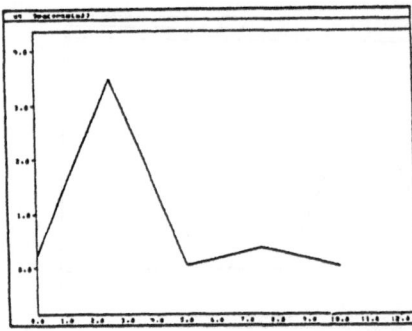 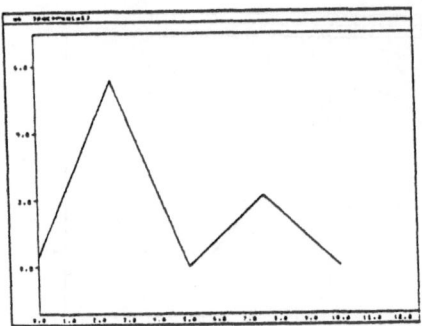

Figure 3: Power spectrum of a saw tooth and a square signals

In the training data set 9 patterns (three patterns in each signal) with learning rate 0.4, momentum 0.7 and scaling factor 1.0 are used. The multi-layer perceptron algorithm is programmed in FORTRAN. As the results show in Figure 4, the learning stage stopped after 451 iterations and the program started the pattern recognition stage with the weights obtained in the learning stage. A total of 40 patterns were tested using the recognition routine and the recognition results are all correct.

Conclusion

In this research the authors propose a system to identify cyclic signals and multi-layer perceptron neural net is applied to do the pattern recognition. Sinusoidal, saw tooth, and square signals are tested. Fourier transform of one period sampled data are used to represent a signal. This reduces much data processing time. After running the multi-layer perceptron program, it shows that the learning is very fast and the recognition result is excellent. It appears that the system is very useful for identifying cyclic signals. Potential applications of the system would be in cyclic signals recognition and machine faults diagnosis, such as faults in gears and rotors.

```
START THE LEARNING STAGE
        31    3.412826E-01      10.348600      1.661381
        61    1.194219E-01       6.064578      1.296881
        91    1.280643E-01       5.670607      3.147967
       121    2.833024E-01       6.163828      1.562475
       151    5.232073E-02       2.947738      1.408949
       181    2.616090E-01       6.257407      5.806849
       211    4.411582E-02       2.949939      9.762725E-01
       241    1.611302E-01       6.583477      2.722264
       271    7.564608E-02       3.743400      3.237298
       301    1.500261E-02       1.820072      2.943093E-01
       331    8.594129E-03       1.389130      1.803703E-01
       361    5.704645E-03       1.143592      1.290583E-01
       391    4.143907E-03       9.846560E-01   9.895636E-02
       421    3.197833E-03       8.731978E-01   7.944673E-02
       451    2.375489E-03       7.904000E-01   6.597544E-02
   LEARNING STAGE HAS FINISHED

START THE PATTERN RECOGNITION STAGE
Enter NEW pattern filename  -->     9.395765E-01    5.750675E-02     1.433609E-02
the pattern belongs to category  1
Do another pattern(Y/N)? Enter NEW pattern filename  -->     9.394671E-03     9.86
6891E-01     1.480577E-02
the pattern belongs to category  2
Do another pattern(Y/N)? Enter NEW pattern filename  -->     1.174292E-03     1.16
7005E-02     9.832875E-01
the pattern belongs to category  3
Do another pattern(Y/N)? Stop - Program terminated.
```

Figure 4: Running results of pattern recognition routine

References

1. Lippmann, R. P., "An Introduction to Computing with Neural Nets," IEEE ASAP Magazine, Vol. 4, No. 1, pp.4-22, April 1987.

2. Nilsson, N. J., Learning Machines: Foundations of Trainable Pattern Classifying Systems, McGraw-Hill, New York, 1965.

3. Oppenheim, A. V., Willsky, A. S. and Young, I. T., Signals and Systems, Prentice-Hall Inc., 1983.

4. Pao, Y. H., Adaptive Pattern Recognition and Neural Networks, Addison-Wesley Publishing Company, Inc. 1989.

5. PCLAB User Manual, Data Translation, Inc., 100 Locke Drive, Marlboro, MA 01752-1192.

State Estimation Under Unknown Noises - A Least-Squares Approach

CHUNG-WEN CHEN and JEN-KUANG HUANG

Mechanical Engineering and Mechanics
Old Dominion University
Norfolk, VA

Abstract

A simple filter for state estimation of linear systems under unknown noises is developed. Through state space model, the current state of the system is linked to previous measurement data and can be estimated by using least-squares techniques. Due to the unknown process noise, past data are regarded as of decaying importance in determining the current state and a forgetting factor is employed in the recursive least-squares to function fading memory. The relations between the fading memory filter and the Kalman filter are discussed. A numerical example is given to illustrate the feasibility of the approach.

Introduction

The Kalman filter is a widely used technique in state estimation [1]. However, a well-known limitation in applying Kalman filter is the requirement of a priori knowledge about the system space model and the noise statistics. This information, in practice, is either partially known or totally unknown. If the noise statistics are not known, experience-required guesses and/or time-consuming filter tuning process should be made before the filter can actually be used. Otherwise, more complicated adaptive filtering algorithm which can improve the initial guesses during the operating process should be adopted [2,3].

In this paper the estimation problem is solved from a point of view which is essentially different from the stochastic approach of the conventional Kalman filter. The time-variant state estimation problem is re-phrased into a time-invariant parameter estimation problem, and least-squares techniques, which is derived under deterministic framwork, are then used. The advantage of using least-squares approach is that it does not require a priori knowledge of the noise statistics, the initial values of the estimated state, and its corresponding error covariance. Close relations are found between the Kalman filter and the least-squares filter. Finally, a numerical example is provided to illustrate the feasibility of the proposed method.

Filters for Linear Systems

A discrete-time finite-dimensional linear system can be represented by the following state space model:

$$x_{k+1} = Ax_k + w_k, \qquad (1)$$

$$y_k = Cx_k + v_k, \tag{2}$$

where $A \in R^{n \times n}$ is the system matrix, $C \in R^{p \times n}$ the output matrix, $x \in R^{n \times 1}$ the state vector, $y \in R^{p \times 1}$ the measurement vector. The process noise sequence $\{w_k\}$ and the measurement noise sequence $\{v_k\}$ are assumed to be zero-meaned stationary Gaussian white noises with constant variance matrices Q and R, respectively. Based on this model the following eqution can be derived:

$$
\begin{bmatrix} v_k \\ v_{k-1} \\ \vdots \\ v_{k-m+2} \\ v_{k-m+1} \end{bmatrix}
=
\begin{bmatrix} C \\ CA^{-1} \\ \vdots \\ CA^{-m+2} \\ CA^{-m+1} \end{bmatrix} x_k
-
\begin{bmatrix} 0 & 0 & \cdots & 0 & 0 \\ 0 & CA^{-1} & \cdots & 0 & 0 \\ \vdots & \vdots & \ddots & \vdots & \vdots \\ 0 & CA^{-m+2} & \cdots & CA^{-1} & 0 \\ 0 & CA^{-m+1} & \cdots & CA^{-2} & CA^{-1} \end{bmatrix}
\begin{bmatrix} w_k \\ w_{k-1} \\ \vdots \\ w_{k-m+2} \\ w_{k-m+1} \end{bmatrix}
+
\begin{bmatrix} v_k \\ v_{k-1} \\ \vdots \\ v_{k-m+2} \\ v_{k-m+1} \end{bmatrix}
$$

or in short: $\tag{3}$

$$Y_m = U_m x_k + M W_m + V_m, \tag{4}$$

where m denotes the number of the successive data used in the equation relating $m - 1$ previous data to the current state. The second term in the right hand side arises because of the existence of process noise.

Filters for Systems of Negligible Process Noise

For an observable system with negligible process noise, if the number m is fixed and large enough to make the matrix U_m full-column-ranked, then x_k has a least-squares solution [4]:

$$\hat{x}_k = (U_m^T U_m)^{-1} U_m Y_m = U_m^\dagger Y_m, \tag{5}$$

where $\hat{x}_k$ is the estimate of the state vector x_k, and $U_m^\dagger$ the pseudoinverse of the observability-liked matrix U_m. The estimation is unbiased and, if all the measurement noises are uncorrelated and equally strong with variance σ^2, the error covariance of the estimation becomes $\Psi_k = \sigma^2 (U_m^T U_m)^{-1}$ [4]. Therefore, the quality of the estimate is directly proportioal to the variance of the measurement noise.

If the filter order m is not fixed but increases with time index k instead, a recursive least-squares filter can be derived from the batch type least-squares [4].

Fading Memory Least-Squares Filter (FMLS)

Now, consider the situation when the process noise is significant. Since the process noise term in Eq. (3) is correlated with the current state, which violates the "uncorrelated measurement noise" assumption of the least-squares method, the ordinary least-squares can not be applied in this case. However, the problem can be viewed in a different way.

Due to the effect of the process noise, the previous data provide less reliable information than the current data do for estimating the current state. According to this intuitive understanding, therefore, we can still use recursive least-squares technique but introducing a weighting factor to give less weight to the previous data in determing the current state. This method is called recursive weighted least-squares [5]. The weighting of the previous data will be exponentially reduced, therefore, the method has an ability to gradually "forget" the old data and emphasize on the newer ones. The recursive forgeting algorithm is derived as follows.

Suppose at time step k we have the estimation equation as:

$$\hat{x}_k = (U_k^T \bar{R}_k^{-1} U_k)^{-1} U_k \bar{R}_k^{-1} Y_k, \tag{6}$$

where $\bar{R}_k$ is a weighting matrix, then at time step k+1 by introducing the forgeting factor λ one can write:

$$
\begin{aligned}
\hat{x}_{k+1} &= (U_{k+1}^T R_{k+1}^{-1} U_{k+1})^{-1} U_{k+1} R_{k+1}^{-1} Y_{k+1} \\
&= \left([C^T \vdots A^{-T} U_k^T] \begin{bmatrix} I_p & 0 \\ 0 & \lambda R_k^{-1} \end{bmatrix} \begin{bmatrix} C \\ U_k A^{-1} \end{bmatrix} \right)^{-1} \times [C^T \vdots A^{-T} U_k^T] \begin{bmatrix} I_p & 0 \\ 0 & \lambda R_k^{-1} \end{bmatrix} \begin{bmatrix} v_{k+1} \\ Y_k \end{bmatrix} \\
&= [C^T C + \lambda A^{-T} U_k^T R_k^{-1} U_k A^{-1}]^{-1} [C^T \vdots A^{-T} U_k^T] \begin{bmatrix} I_p v_{k+1} \\ \lambda R_k^{-1} Y_k \end{bmatrix} \\
&= [\lambda^{-1} A (U_k^T R_k^{-1} U_k)^{-1} A^T - \lambda^{-1} A (U_k^T R_k^{-1} U_k)^{-1} A^T C^T (I_p + \lambda^{-1} C A (U_k^T R_k^{-1} U_k)^{-1} \\
&\quad \times A^T C^T)^{-1} \lambda^{-1} C A (U_k^T R_k^{-1} U_k)^{-1} A^T] [C^T v_{k+1} + \lambda A^{-T} U_k^T R_k^{-1} Y_k] \\
&= \cdots \\
&= \lambda^{-1} A (U_k^T R_k^{-1} U_k)^{-1} A^T C^T (I_p + \lambda^{-1} C A (U_k^T R_k^{-1} U_k)^{-1} A^T C^T)^{-1} v_{k+1} \\
&\quad + [A (U_k^T R_k^{-1} U_k)^{-1} U_k^T R_k^{-1} Y_k - \lambda^{-1} A (U_k^T R_k^{-1} U_k)^{-1} A^T C^T \\
&\quad \times (I_p + \lambda^{-1} C A (U_k^T R_k^{-1} U_k)^{-1} A^T C^T)^{-1} \times C A (U_k^T R_k^{-1} U_k)^{-1} U_k^T R_k^{-1} Y_k] \\
&= A \hat{x}_k + \Pi_{k+1} (v_{k+1} - C A \hat{x}_k) \tag{7}
\end{aligned}
$$

where the matrix inverse lemma [4,5] is used to expand the inverse term, and

$$
\begin{aligned}
\Pi_{k+1} &= \lambda^{-1} A (U_k^T R_k^{-1} U_k)^{-1} A^T C^T (I_p + \lambda^{-1} C A (U_k^T R_k^{-1} U_k)^{-1} A^T C^T)^{-1} \\
&= \Phi_{k+1} C^T (I_p + C \Phi_{k+1} C^T)^{-1} \tag{8}
\end{aligned}
$$

$$
\Phi_{k+1} = \lambda^{-1} A (U_k^T R_k^{-1} U_k)^{-1} A^T = \lambda^{-1} A \Psi_k A^T \tag{9}
$$

$$
\begin{aligned}
\Psi_{k+1} &= (U_{k+1}^T R_{k+1}^{-1} U_{k+1})^{-1} \\
&= \lambda^{-1} A (U_k^T R_k^{-1} U_k)^{-1} A^T - \lambda^{-1} A (U_k^T R_k^{-1} U_k)^{-1} A^T C^T \\
&\quad \times (I_p + \lambda^{-1} C A (U_k^T R_k^{-1} U_k)^{-1} A^T C^T)^{-1} \lambda^{-1} C A (U_k^T R_k^{-1} U_k)^{-1} A^T \\
&= \lambda^{-1} A \Psi_k A^T - \lambda^{-1} \Pi_{k+1} C A \Psi_k A^T \\
&= (I_n - \Pi_{k+1} C) \lambda^{-1} A \Psi_k A^T = (I_n - \Pi_{k+1} C) \Phi_{k+1} \tag{10}
\end{aligned}
$$

Eqs. (7)–(10) constitute the structures of the fading memory least-squares filter for state estimation, which are in the recursive form.

Parameter λ can be used to adjust the memory length. If $\lambda = 1$, which corresponds to infinite memory, the filter will weight all the data equally as in the negligible process noise case. Reduce λ will reduce the memory length and the filter will "forget" old data faster, which is suitable when the process noise is strong.

The question remained now is how to choose a proper value of λ. This problem is similar to how to determine the process noise covariance Q in the Kalman filter, except only one variable has to be determined in this case. The usual way is to "tune" the filter by observing its residual sequence, the difference between the real measurement and the estimated measurement by the filter. After all, residual sequences are the only information available for judging the performance of the filter. It is known that if the filter is optimal the residual sequences should be white and zero-meaned. In practice, therefore, we can adjust λ by monitoring the residual sequences. We may not be able to obtain white residuals by adjusting only one variable, however, if all the measurement residuals remain random enough we can expect satisfactory results.

Relations Between Kalman Filter and Fading Memory Least-Squares Filter

The optimal Kalman filter is derived under the optimality criterion of least-mean-square error of the state, while the fading memory least-squares filter (FMLS) is under least-squares error of the measurement. However, it is interesting to see that they turn out to have the same form of structures. Examing Eqs. (7)–(10), it is found that Φ_k and Ψ_k in FMLS are equivalent to the a priori state error covariance P_k^- and the posterior error covariance P_k, respectively, in the Kalman filter. The formula for P_{k+1}^- is

$$P_{k+1}^- = A P_k A^T + Q. \tag{11}$$

Comparing to Eq. (9) we note that instead of adding process noise covariance to the propagation of state error covariance matrix as in the Kalman filter, the FMLS simply multiplies a factor λ^{-1} which is larger than 1 to account for the effect of the process noise. Or by writing Eq. (9) as

$$\Phi_{k+1} = A \Psi_k A^T + (\lambda^{-1} - 1) A \Psi_k A^T, \tag{12}$$

it can be clearly seen that the FMLS method implicitly assigns the process noise covariance as $Q' = (\lambda^{-1} - 1) A \Psi_k A$. Therefore, the state with larger error covariance will be assigned larger process noise covariance, which makes sense intuitively. If the process noise is not significantly different from Q', the FMLS method can give reasonably good result.

Numerical Examples

A three-mode simulated lumped-mass system as shown in Fig. 1 is excited initially by a pulse input force applied on node 3 and then left to free decay. Two measurements are taken from nodes 1 and 2. The process noise is about 5% of the initial state and the measurement noise is also about 5% of the measuremen in variance ratio. system has three modal frequencies, 1.6369, 4.4719, and 6.1085 (rad/sec) with damping ratios, 0.63, 1.01, 1.30 (%), respectively.

For the optimal Kalman filter, the initial state is set to zero and the initial error covariance of the Kalman filter is set to $10 \times I_n$. The exact values of the noise covariances are assume known. The results of the estimation of the first state and the auto-correlation function of one of the residuals are shown in Fig. 2. Hereafter, the solid lines in the state estimation plots represent the true state histories and the dashed lines represent the estimated ones. For the optimal Kalman filter case these two are almost coincide. Theoretically the residual of the optimal Kalman filter is a while sequence and, therefore, its auto-correlation function should be zero everywhere except at $\tau = 0$. Note the auto-correlation figures are not normalized.

For the fading memory filter (FMLS), the results of three cases of different forgeting factor ($\lambda = 1.0$, 0.90, 0.50) are shown in Figs. 3–5. Comparing to the optimal Kalman filter we

610

can see that the estimates of the case $\lambda = 0.90$ are fairly well, which also can be expected by the comparative whiteness of the corresponding residual sequence. For the case of $\lambda = 1.0$, which corresponds to infinite memory, the estimates deteriorate gradually due to the effect of the process noise in the system and finally fail to track the states. For the case of $\lambda = 0.50$, the memory is short and the results, therefore, are sensitive to the current noise. The residual in this case is less white as can be seen in its auto-corrlation plot. Therefore, by monitoring the whiteness of the residual, a proper forgetting factor can be chosen.

Concluding Remarks

A simple filter for state estimation of systems under unknown noises is developed. Though the filter is suboptimal, the results are fairly accurate in most practical cases. With only one parameter needs to be adjusted, the fading memory filter is much simpler than the conventional Kalman filter in application.

References

[1] Kalman, R. E., "A new approach to linear filtering and prediction problems," Trans. ASME, J. Basic Eng., Vol 82, pp. 35-44, 1960.

[2] Mehra, R. K. "On the identification of variances and adaptive Kalman filtering," IEEE Transactions On Automatic Control, Apr. 1970, AC-15, pp. 175-184.

[3] Mehra, R. K. "Approach to adaptive filtering," IEEE Transactions On Automatic Control, Oct. 1972, AC-17, pp. 693-698.

[4] Hsia, T. C., "System Identification - Least-Squares Methods, " D. C. Heath and Company Lexington, Massachusetts Toronto, 1977.

[5] Haykin, S., "Adaptive Filter Theory," Prentice-Hall, Englewood Cliffs, New Jersey 07632, 1986.

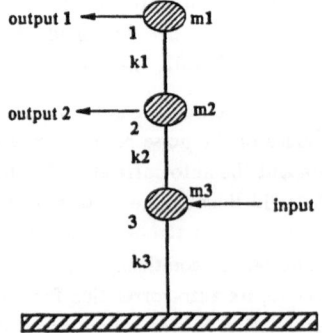

m1	0.5
m2	1.0
m3	1.0
k1	10.0
k2	10.0
k3	10.0

Fig. 1 Simulated Lumped-Mass System

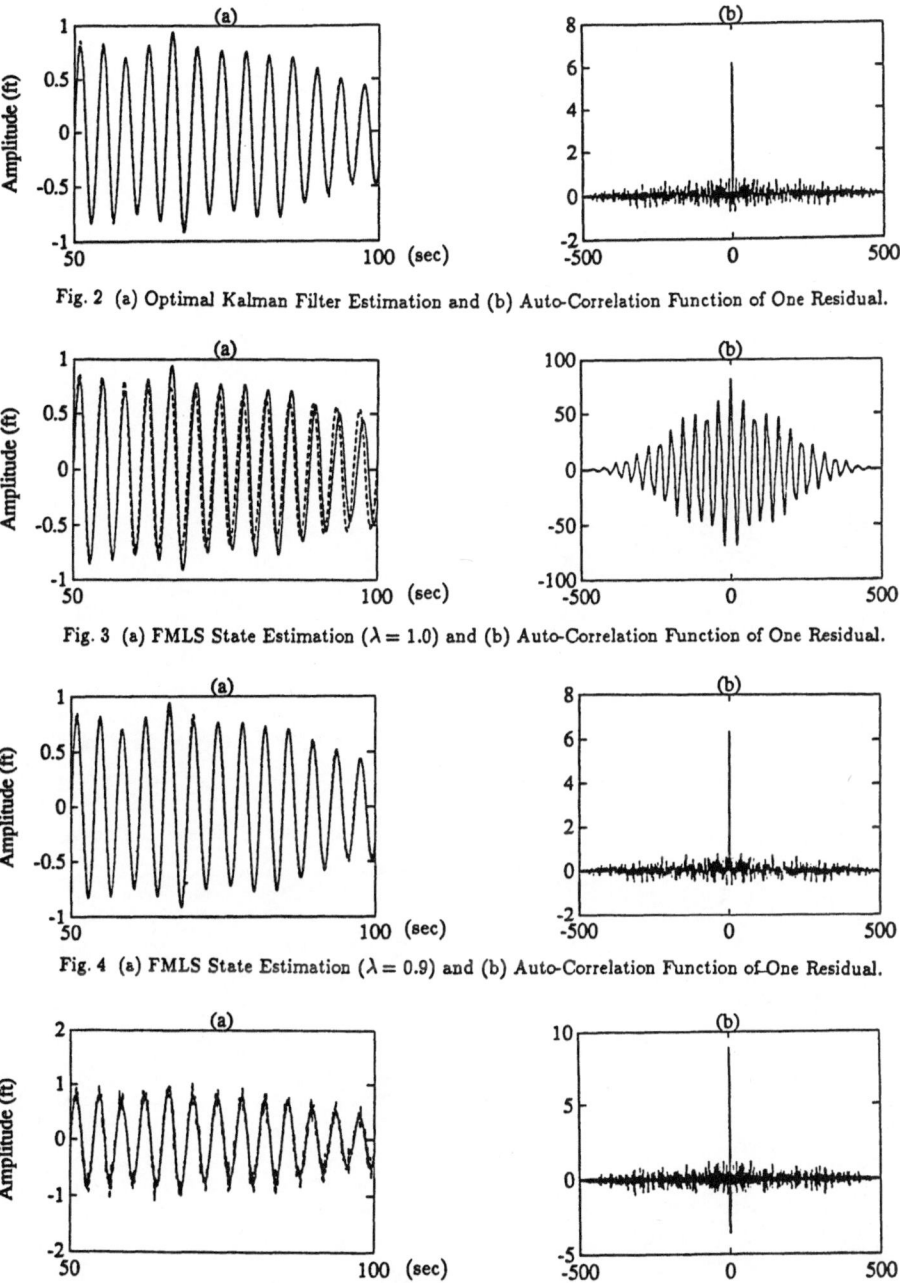

Fig. 2 (a) Optimal Kalman Filter Estimation and (b) Auto-Correlation Function of One Residual.

Fig. 3 (a) FMLS State Estimation ($\lambda = 1.0$) and (b) Auto-Correlation Function of One Residual.

Fig. 4 (a) FMLS State Estimation ($\lambda = 0.9$) and (b) Auto-Correlation Function of One Residual.

Fig. 5 (a) FMLS State Estimation ($\lambda = 0.5$) and (b) Auto-Correlation Function of One Residual.

Chapter X

Educational Endeavors

Introduction

Design and construction of an industrial robot and its control system served as an ideal project in the first paper for a combined group of electrical and mechanical engineering senior students. Students were forced to appreciate and respond to some of the larger issues of interdisciplinary project work. In the second paper, an automated manufacturing laboratory for teaching concepts and techniques of computer integrated manufacturing systems are described. The educational concepts and implementation approach for a graduate level manufacturing systems engineering program jointly developed by the University of Pittsburgh School of Engineering and several industrial partners over a six-year period are described in the third paper. An internship requires the student to address an applied research problem in an industrial setting, working with real-world processes and equipment. In the fourth paper, the extensive use of teach pendants to program robots in both technician training as well as workspace programming for extended periods of time is of human factors interest. Design improvements are offered based on survey results and human factor principles.

Combined EE-ME Senior Capstone Projects In Robotics at West Virginia University

NIGEL T. MIDDLETON
Department of Engineering
Colorado School of Mines
Golden, CO

LARRY BANTA
Department of Mechanical and Aerospace Engineering
West Virginaia University
Morgantown, WV

Summary

There are considerable pedagogical benefits in combining groups of electrical and mechanical engineering senior students in a capstone project. The total design and construction of an industrial robot and its control system serves as an ideal project for such a combined group of students. This paper describes two "Build a Robot" projects which have been carried out at West Virginia University.

Introduction

The Departments of Electrical and Computer Engineering (ECE) and Mechanical and Aerospace Engineering (MAE) at West Virginia University both have rigorous capstone senior design project courses which are conducted independently. The two departments have slightly different credit allocations (5 hrs. in ECE and 6 hrs. in MAE over 2 semesters), and slightly different perspectives on the organization of the courses. In ECE, the course is tightly structured in terms of curriculum and presentation requirements. Typically, groups of 3 to 5 ECE students work on a suitable electrical, electronic or computer related projects, under the supervision of 3 or 4 faculty supervisors. In MAE, the student groupings favor larger numbers directed at larger projects, like a Baja car, or a formula racer. The classes in MAE are therefore separated into different sections associated with each project, and are independently instructed by faculty members who lead activities in the different project areas.

Consistent with the national trend during the mid 1980's, a growing research and educational endeavor in robotics and automation has developed in WVU's College of Engineering. A consequence of this was the Fall 1987 offering by MAE of a "Build a Robot" senior capstone project. An implicit component of this project, however, was the need for an electronic control system. It was therefore proposed that this MAE project should be combined with one project in the ECE capstone course.

The 1987-88 "Build a Robot" project broke new ground in terms of inter-departmental cooperation at the undergraduate level. Because of the differences in the organization of the capstone courses in the two departments, the ECE and MAE students on the project were working according to different curricular and credit requirements, causing some administrative and motivational difficulties. Nevertheless, the 1987-88 project was considered to be a success, and so it was repeated in 1988-89 with much greater success.

The 1987-88 "Build a Robot" Project

The initial solicitation to interest students in this project was quite loosely phrased, with the intent that there was pedagogical advantage in having the project group develop their own objectives and design specifications. Some general guidance was offered by the faculty in selecting the intended application of the robot, and this was motivated by a possible need in the plant of a nearby glass manufacturer. After watching video recordings of a pick-and-place function in this glass plant, the group developed appropriate time budget, operating envelope, and degree-of-freedom requirements for a robot. Based on these, a more detailed design for what turned out to be an industrial 4 degree-of-freedom gantry robot was developed.

During the first semester of the course sequence, the ten ME students investigated all aspects of the kinematics, dynamics and material strength requirements of the design. Their deliverable at the end of the first semester was a technical description of the assembly and each mechanical part of the proposed robot, together with supporting design calculations and CAD developed shop drawings. At the same time, the five EE students submitted an engineering proposal for the design of the robot's control system. This proposal described the control system at a functional block- diagram level, and included all project planning and budget issues for the implementation phase.

During the second semester of the sequence, the robot parts were either fabricated in-house, or purchased, and the structure was assembled. The EE's developed their system-level design into circuit and software details, and then built and programmed the control system. They completely built a bus based multi-processor system, with one Intel 8085 microprocessor for digital PID control on each of the three cartesian axes, one 8085 to handle the gripper and sensor inputs, and one 8085 to coordinate communications on the bus and to a host personal computer. They also designed and built sensor interfaces and pulse-width-modulated power amplifiers for the dc motor actuators.

It turned out that the scope of work in the control system was too large for the EE students to complete fully during the second semester, although they were able to demonstrate a partially working robot. The mechanical structure appeared to be satisfactory, but, because the control system was not completed, it was never tested to its limits.

The 1988-89 "Build a Robot" Project

The scope and intent of the 1988-89 project was essentially the same as that set in 1987-88. The coordination between the MAE and ECE departments was not changed substantially, and the students were expected to deliver in much the same way as with the previous project. Once again, the goal was to build completely a pick-and-place industrial robot. Some aspects of the objectives were more ambitious than before. In particular, the system was to use vision equipment to enable it to identify and sort randomly placed parts on a moving conveyor. A complete set of MicroVax based vision processing hardware was loaned to the project by Lawrence Livermore Laboratory.

The EE group of seven correctly recognized that the downfall of the 1987-88 EE group had been the tremendous amount of hardware and Assembler software development needed. Therefore, they explored technological options very carefully, and were able to reduce the hardware complexity dramatically. Their hardware design amounted to a PC/AT bus compatible half-card, containing three LM629 programmable digital PID controllers (for each of the cartesian axes), and a Motorola 68HC11 microcontroller for the stepper motors in the twist and close functions of the gripper. The 68HC11 also handled some sensor inputs and coordinated control of the robot by a teach pendant. Programmable Logic Arrays were used for the bus interface logic. A separate 19" rack module was developed for power supplies and power amplifiers for the dc motor and stepper motor actuators. A system block diagram of the control system is shown in Figure 1.

An immediate consequence of this design was that, with the exception of the on-chip software for the 68HC11, the bulk of the programming of the system could be done in a high level language. 'C' source code was developed for all of the image processing routines running on the MicroVax, the MicroVax to PC/AT communications link, and the PC/AT based control system kernel. Compiled QuickBasic was used for the user interfaces and utilities, and the teach pendant operation. A large amount of code was therefore developed fairly quickly and with little error.

The mechanical side of the project proceeded with similar diligence. After a serious critique of the earlier project, a completely new gantry frame was designed and developed. The new design took advantage of plastics and was designed to be lighter yet be much more rigid than its predecessor, giving a mechanical repeatability in gripper position of about ±0.05".

The performance of the robot exceeded all expectations. It was able to successfully identify and sort known and unknown objects on a moving conveyor arriving at 2 second intervals. The students also set it up to do some effective demonstrations like trace the outline of a map (with a felt pen placed in the gripper), and play some simple tunes on a portable electronic keyboard. The project was demonstrated at an SME Student Robot

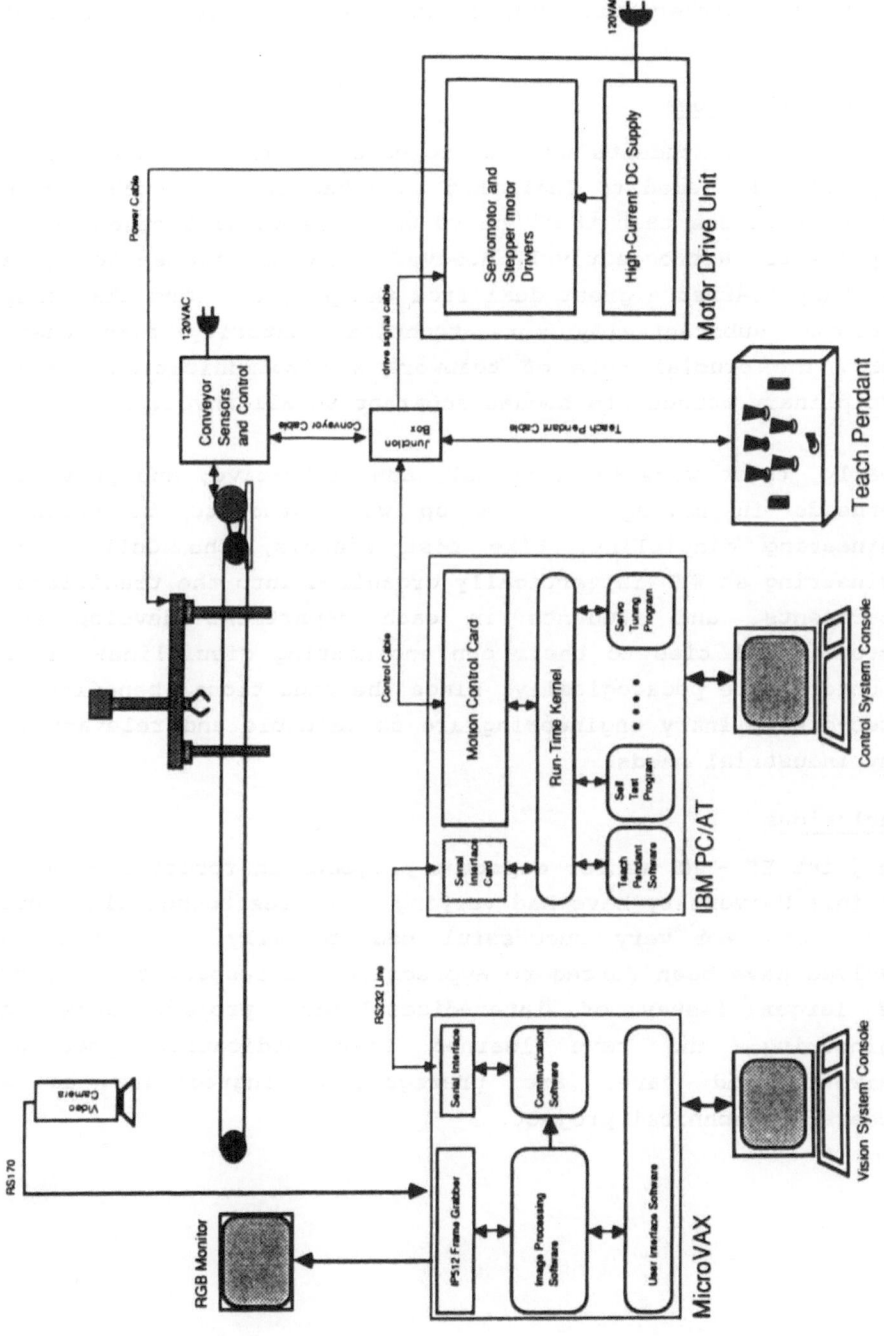

Figure 1. Control system block diagram for the 1988-89 robot.

Meeting in Gaithersburg, MD, in May, 1989, where it received much acclaim.

Pedagogical Issues

The EE and ME students who participated in the robot capstone projects all seemed to feel that they had to work much harder for their grades than their peers who were doing focused EE or ME projects. Retrospectively however, they all seemed to agree that they learned a great deal from the projects, and that they developed substantially more technical maturity than their peers. The crucial role of teamwork and communication across disciplinary boundaries became apparent to all involved.

Clearly, there were intellectual, administrative, and personal overheads in having to team up with students in another engineering discipline. Like most schools, the College of Engineering at WVU is vertically organized into the traditional departments, and students in each department develop and nurture loyal ties to their own engineering disciplines. This is unfortunate pedagogically, since the educational benefits of inter-disciplinary engineering are so valuable and relevant to many industrial needs.

Conclusions

Two joint EE - ME senior capstone projects in robotics at West Virginia University have had varying successes technically, but have both been very successful pedagogically. The students involved have been forced to appreciate and respond to some of the larger issues of inter-disciplinary project work in engineering, and have learned that dedication, people, teamwork, and care, are prerequisite ingredients to a successful technical project.

Automated Manufacturing at Western Kentucky University

R.I. EVERSOLL, H. T. LEEPER and L. T. ROSS

Department of Industrial Technology
Western Kentucky University
Bowling Green, KY

Summary

Faculty members in the Department of Industrial Technology
at Western Kentucky University have developed a model automated
manufacturing laboratory (AML) for the purpose of teaching
concepts and techniques of computer integrated manufacturing.
The facility is equipped with industrial grade machines
that offer the following capabilities: robotic materials
handling, automated inspection using a coordinate measuring
machine (CMM) and machine vision, computer numerical controlled
machining applications and materials handling using a programmable
logic controlled conveyor system (see Figure 1).

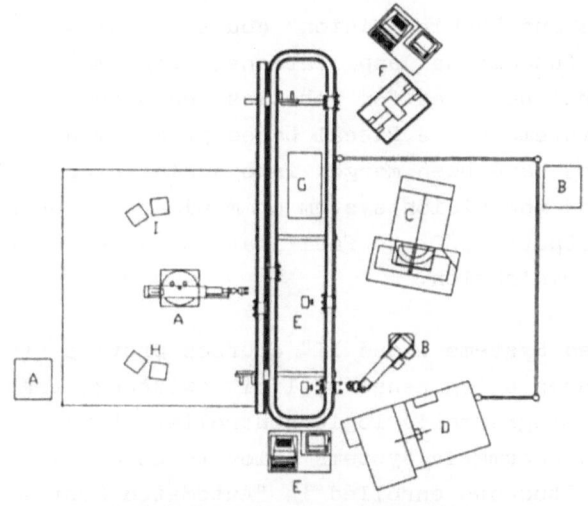

A - GMF M-100 ROBOT WITH A KAREL CONTROLLER
B - GMF S-10 ROBOT WITH A KAREL CONTROLLER
C - PRATT WHITNEY MACHINING CENTER WITH FANUC CONTROL
D - PRATT WHITNEY LATHE WITH FAPT AND FANUC CONTROL
E - ALLEN-BRADLEY EXPERT VISION SYSTEM WITH TWO CAMERAS
F - MITUTOYO COORDINATE MEASURING MACHINE
G - ALLEN-BRADLEY PLC 5/25
H - MILL AND LATHE LIFT STATIONS
I - PALLETIZED MILL AND LATHE STATIONS

Figure 1.   AML Layout and Equipment Descriptions

This paper describes the use of the facility for instructional purposes and the actual electrical/programming communications used to integrate the individual pieces of equipment into a total manufacturing system.

Instruction

The AML supports faculty instruction in seven different courses, and as a support facility, the equipment may be used on a stand-alone basis. The Mitutoyo Coordinate Measuring Machine of the AML is used by students in the "Metal Processes II" course for precision measurement experiments and to evaluate their final projects. The same machine is used in "Statistical Quality Control" for developing statistical process control (SPC) charts and for conducting process capability studies on the machine tools in the AML.

The "Robotics and Machine Vision" course enables students to learn the fundamental applications, programming, and lighting techniques used for robotics and machine vision. Since both systems use a Pascal based programming language, the two topics have been merged into a single course. The S-10 robot and the vision system communicate through digital inputs and outputs (I/O) to facilitate selected part assembly and part discrimination.

The "Automated Systems I and II" courses provide students with experiences using sensors, limit switches, traditional relay logic, programmable logic controller (PLC) programming, robotics, and pneumatic systems prior to conducting experiments in the AML. Students enrolled in "Automated Systems II" use the AML materials handling conveyor and the Allen-Bradley PLC 5/25 to write ladder logic programs to control the movement and transfer of part pallets and other peripherals of the system.

The "Computer Aided Manufacturing" (CAM) course requires students to use the support equipment in the AML for instruction in numerical control (NC) programming, graphically assisted programming using the Fanuc Automatic Programmed Tools (FAPT) language, and for making the CAD/CAM interface using SmartCam

software. NC programming is emphasized so that students
can correct and modify the "G, M, and T" coded information
for manual or software generated programs.

A capstone course, entitled "Senior Project in Manufacturing",
requires students to utilize the full capabilities of the
AML. Students organize themselves into manufacturing teams
with group leaders, and these teams then design lathe and
mill parts using concepts of group technology based upon
existing tooling and fixtures. The teams also program the
equipment in the AML and produce a number of parts to predetermined
specifications. The thrust of the course is to teach systems
integration under actual manufacturing conditions.

Communications

The diagram below shows the electrical and programming links
between the loading and unloading robot, the materials handling
conveyor and the machining cell. The communication is program
oriented and the physical connections between controllers
use digital I/O's via twisted pair cables (see Figure 2).

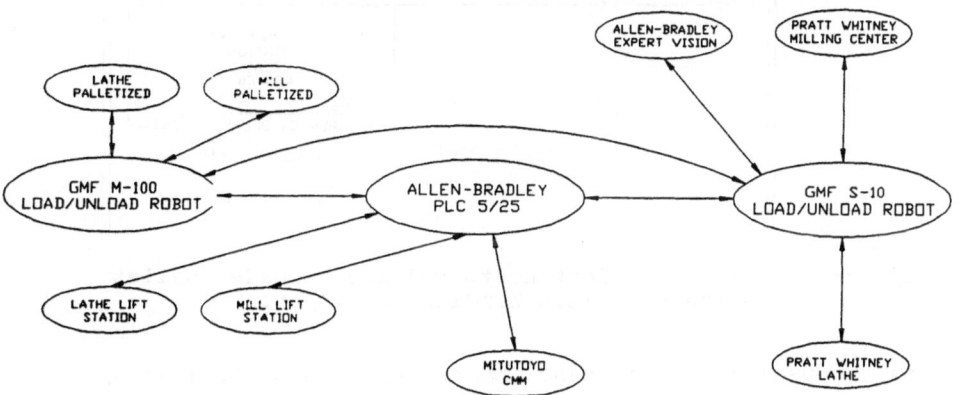

Figure 2.   Communication Linkages.

A Karel controller is responsible for controlling the General
Motors Fanuc (GMF) M-100 cylindrical coordinate pick and
place robot and the two palletizing stations for finished
parts. The controller also communicates with the materials

handling conveyor PLC and the machining cell. A program
has been written for the robot, identifying the conditions
by which it operates and reacts to the other controllers
or sensors of the system. The actual robot program is a
series of routines. Based upon inputs received, the robot
reacts in a predetermined manner. At the conclusion of
the robot movement, the controller sends an output signal
to one of the other system controllers to complete the cycle
of operation (see an example in Figure 3).

Robot Routine

```
Move to LCONV[5]            --Move to Load CONV pos 5
Close Hand(2)               --Close robot hand
Pulse DOUT[10] for 2000     --Pulse digital output 10
                              for 2 seconds to the PLC
                              5/25 to release the pallet
                              from station
Move to PERCH               --Move to PERCH position
```

PLC 5/25 Rung

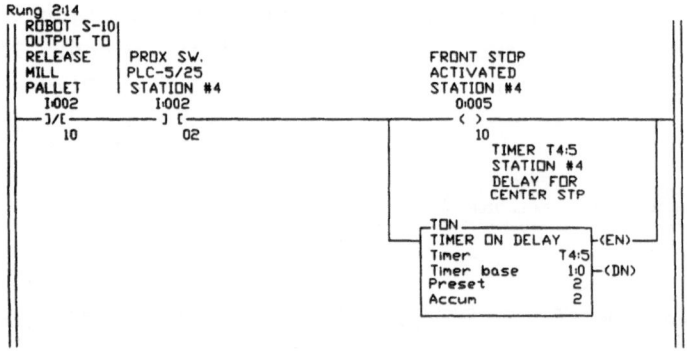

Figure 3.   Excerpt of Routine to release a loaded pallet
            on the materials handling conveyor.

The carousel materials handling conveyor is controlled by
an Allen-Bradley PLC 5/25. The conveyor is responsible
for: transporting unmachined and machined parts to the
loading and unloading stations, discrimination between lathe
and mill parts, controlling the inspection of parts by the
CMM and for controlling the function of the two raw material
lift stations. The PLC 5/25 communicates with the other
controllers of the AML to signal the arrival of parts at
the various queue stations (see Figure 4).

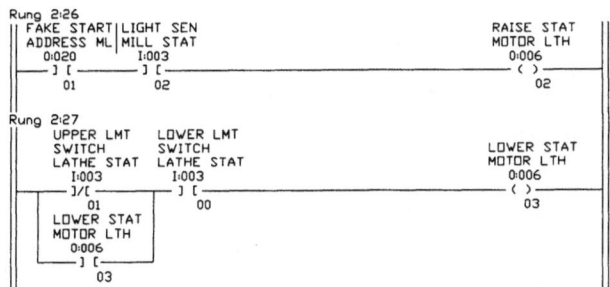

Figure 4. Excerpt of a ladder logic program for the lathe
lift station.

The machining cell consists of a GMF S-10 articulate coordinate, pick and place robot, two CNC machining centers and
an Allen-Bradley Expert Vision System. Using program routines,
the Karel controller for the S-10 robot controls: loading
and unloading of parts at the machining centers, external
functions and the cycle start of the machining centers,
and automated inspection of parts prior to and after machining.
The S-10 Karel controller also communicates with the materials
handling conveyor and the M-100 Karel controller. The direct
link with the M-100 controller enables the S-10 to issue
a service request to load a new unmachined part onto the
materials handling conveyor. The entire machining cell
operates on the "pull processing" philosophy. See Figure 5
below for an excerpt of a Karel controller routine.

```
Routine Mill _unload
  Begin
    Pulse DOUT[1] for 3000             --Blow coolant off part
    Pulse DOUT[2] for 4000             --Open mill doors
    Move to MILL[5];Move to MILL[7]    --Moving hand into mill
    Move to MVABV;Open hand(1)         --Open hand
    With $Speed=20 move to MILLDROP    --Move over mill part
    Close Hand(1):delay 3000           --Close hand pickup part
    Pulse DOUT[4] for 5000 NOWAIT      --Open vise for 5 sec
    Move to MVABV                      --Move above vise
    Move to MILL[7];Move to MILL[5]    --Move out of mill
    Pulse DOUT[3] for 3000             --Close mill doors
    Move to MCONV;Move to MCPICK       --Place part on pallet
    Open Hand(1):delay 3000            --Open hand release part
    Move to PERCH                      --Move to PERCH position
    Pulse DOUT[5] for 3000:NOWAIT      --Release pallet
  End Mill_unload
```

Figure 5. Routine to unload a finished mill part.

Manufacturing Systems Engineering Education at the University of Pittsburgh

JOHN H. MANLEY

Manufacturing Systems Engineering Program
School of Engineering
University of Pittsburgh
Pittsburgh, PA

Summary

An innovative manufacturing systems engineering education program to support factories of the future has been instituted at the University of Pittsburgh. The educational concepts and implementation approach were jointly developed by the School of Engineering and several industry partners over a six year period. The charter class was enrolled in August 1989, and the first Master of Science in Manufacturing Systems Engineering degrees are expected to be awarded in August 1990. This paper describes the program rationale, organization, and curriculum structure. Twelve newly-developed interdisciplinary courses, in conjunction with a thesis-level internship requirement, clearly distinguish this program from manufacturing engineering educational offerings at other leading research universities.

Background

> The mission of the Manufacturing Systems Engineering Program (MSEP) is two-fold. First, to become a nationally recognized leader in manufacturing systems engineering education. Second, to transfer the university's manufacturing systems research results into industrial practice through student internships.

To stand apart from other university programs, the new MSEP degree requires more than traditional classroom and laboratory work. In particular, a unique thesis-level internship fosters very close working relationships with industry partners, and permits the actual transfer of research results into real-world practice. The program also includes close collaboration with other University of Pittsburgh units, the Joseph M. Katz Graduate School of Business (KGSB), the University Center for International Studies (UCIS), the Center for Hazardous Materials Research (CHMR), and the National Center for Excellence in Metalworking Technology (NCEMT).

The first MSEP graduates are expected to receive their Master of Science in Manufacturing Systems Engineering degrees after completing their theses in

August 1990. It is anticipated that the current enrollment of 38 degree-seeking students will increase to over 50 in the 1990-91 academic year.

Approach to Manufacturing Engineering Education

The MSEP curriculum emphasizes a systems and integration approach to real world manufacturing engineering problem solving. The intention is to improve enterprise-wide manufacturing effectiveness by explaining how to:

- Link all phases of the manufacturing life cycle from product conceptualization through production to eventual product retirement
- Connect individual tools into integrated manufacturing systems
- Develop comprehensive information systems from factory floor data gathering through corporate strategic planning
- Recognize and factor in the human component of manufacturing from assembly-line worker through top management, especially with regard to quality-consciousness and team organization
- Apply the latest information related to engineering methods, tools, materials, and processes

Graduate Program Depth versus Breadth

When the interdisciplinary curriculum design issue of educational depth versus breadth was considered, it was decided that the MSEP graduate-level student must be provided with a balance of each. This is especially important if he or she is to be educated for the role of integrating a variety of manufacturing enterprise operations, such as linking engineering design to production and customer service. In this regard, traditional graduate engineering curricula are focused in a specific discipline and therefore must have more depth than the MSEP, i.e., total number of required courses in a defined subject area. On the other hand, a traditional curriculum is not able to offer as much breadth as an interdisciplinary program due to this same course offering restriction. If considerable breadth is desired in addition to an engineering-type discipline, then a predominantly business-oriented curriculum with a major or minor in manufacturing might be considered suitable. Since the MSEP curriculum requires some depth, but also some breadth, it has been positioned midway between the *traditional engineering depth* and the *business-oriented breadth* alternatives.

The MSEP educational philosophy is in full agreement with recommendations contained in an independent study conducted by A.T. Kearney, Inc. for the Society of Manufacturing Engineers [1]. One important part of the report

628

was a description of three distinct roles that manufacturing engineers will play in the next century, namely, *technical specialist*, *operations integrator*, and *manufacturing strategist*. A subsequent analysis of the Kearney report by Worthley recommended that manufacturing engineering education at the graduate level should focus on the operations integrator role [2] as depicted in Figure 1.

SOURCE [Worthley, 1990]

Fig. 1. Role of the MSEP-graduate manufacturing systems engineer

Engineering Degree Comparisons

Figure 2 illustrates additional important differences between the Master of Science in Manufacturing Systems Engineering degree and traditional graduate degrees offered to students within the School of Engineering.

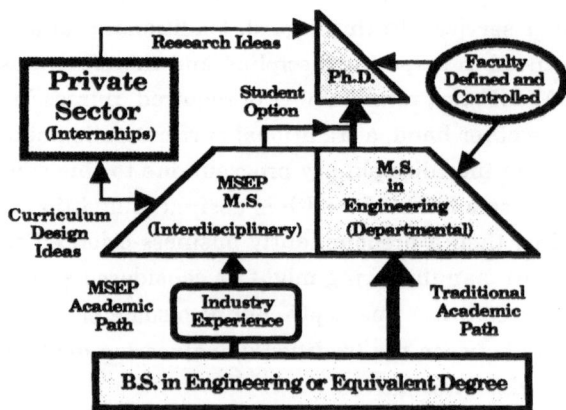

Fig. 2. MSEP versus traditional education paths

Students who wish to pursue graduate-level manufacturing engineering education can take manufacturing options within traditional departments, such as Industrial Engineering, or pursue the interdisciplinary degree within the MSEP program. The traditional education path leads toward the role of technical specialist at the master's level, or to a research career in education or industry at the doctoral level. The MSEP path provides education for the industry role as operations integrator, with a potential return to the traditional path and doctoral-level education in the student's original engineering discipline. These key differences are summarized below.

Interdisciplinary MSEP M.S. Degree	Traditional Engineering M.S. Degree
• Professional degree in manufacturing	• Option in manufacturing
• Industrial applied research	• Basic or applied engineering research
• Emphasis on operations integration	• Emphasis on traditional discipline
• Industry internship required	• No internship required
• Thesis required	• Thesis optional
• Student must have industry support	• Student support from industry optional
• Work experience strongly desired	• Work experience not required
• Interdisciplinary faculty	• Departmental faculty

MSEP Student Internship

Figure 3 illustrates that the required thesis-level industrial internship component of the MSEP is carried out in three major phases, normally during a single academic year beginning in the fall term.

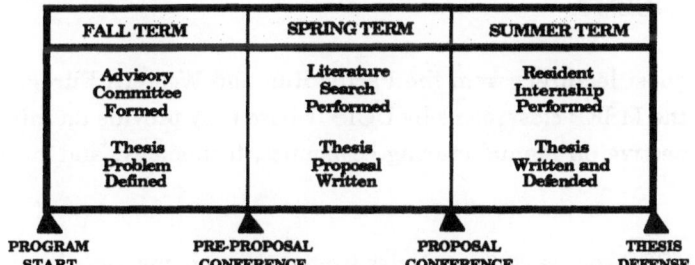

Fig. 3. MSEP internship life cycle

The student is first matched with a host company where he or she will later perform his or her resident internship. A thesis Advisory Committee is established that is composed of one or two faculty members, and one industry supervisor representing the host company. The student identifies an applied research problem of interest and presents it to the Advisory Committee for approval at a formal Pre-Proposal Conference. During the second phase of the MSEP internship process the student takes a research course which requires pre-internship research in the approved problem area. During this course, the student is advised on scientific method and guided in research methods by the faculty advisor. During the final phase, the thesis course requires an equivalent of four months of full-time work at the host company site on the approved problem. Finally, the student develops a written thesis report on his or her findings and defends it before the Advisory Committee during a formal Thesis Defense meeting.

Academic Program

The MSEP academic program consists of 36-credit hours combining class-room work with the industrial internship described above. The program is constructed to be completed in one calendar year of full-time study. The requirements for earning this degree are heavily engineering-oriented and are intended to be consistent in academic content and rigor with the School of Engineering's other masters degree programs. Multiple disciplines are integrated into the MSEP. These include not only the traditional engineering and materials science subjects, but also information systems engineering, safety engineering, human factors, engineering management, and other science and technology areas. Additional elective courses that emphasize international competitiveness have been developed for the MSEP by faculty in KGSB. Important specialized course modules have been developed jointly by School of Engineering faculty, industry experts, and researchers from the CHMR and NCEMT.

Renowned guest lecturers from the Pacific Rim and Western Europe are also brought to the MSEP classrooms by UCIS where they provide an international perspective on manufacturing economics, technology, and cultural differences.

Thus, the MSEP course content varies from a rigorous master's level in the form of both cross-listed and specialized MSEP elective courses, to high-level seminars that provide the student with a broad perspective and understanding of significant, current issues facing the manufacturing industry in a global, and highly competitive environment.

Current Results and Projections for the Future

The first class is highly qualified, averaging nine years of industrial experience, and possessing 14 different engineering and "hard science" degrees (mathematics, physics, chemistry, metallurgy, and computer science). All reports from faculty have been highly complimentary concerning student dedication and overall quality. Predicting the future of any new educational program is highly dependent upon expectations of student enrollments. The number of students enrolled in the MSEP was originally expected to increase from an initial class of five students to a "steady-state" complement of thirty over a five year development phase. However, our first-year expectations were greatly exceeded when five full-time and 28 part-time students enrolled in the first term. It seems that while the MSEP 12-month curriculum is oriented to the needs of full-time students from coast-to-coast, it has become especially attractive to part-time practicing engineers from western Pennsylvania and eastern Ohio. As a result, a new "steady state" five-year projection is to have a mix of 75 full- and part-time students enrolled, and graduate approximately 30 students per year.

Summary

An innovative, graduate-level, manufacturing systems engineering education program to support factories of the future has been instituted at the University of Pittsburgh. The educational concepts and implementation approach were jointly developed by the School of Engineering and several industry partners over a six year period. The MSEP curriculum design currently includes 12 newly-developed, rigorous, masters-level courses. The program requires eight months of classroom work, followed by a four-month industrial internship. The unique internship requires the student to address an applied research problem and write and defend a thesis to both an academic and an industry supervisor. The thesis work is carried out in an industrial setting, working with real-world processes and equipment. The charter class was enrolled in August 1989, and the first Master of Science in Manufacturing Systems Engineering degrees are expected to be awarded in August 1990.

References

1. "Countdown to the Future: The Manufacturing Engineer in the 21st Century," A.T. Kearney, Inc. report to the Society of Manufacturing Engineers, Fall 1988.

2. Worthley, Warren. W., "The Focus for Future Manufacturing Education," 1990 College Industry Education Conference, Session 333, American Society of Engineering Education, 1990.

Human Factor Considerations in the Design of a Teach Pendant

ALOK K. VERMA

Mechanical Engineering Technology
Old Dominion University
Norfolk, VA 23529

CHENG Y. LIN

Mechanical Engineering Technology
Old Dominion University
Norfolk, VA 23529

ABSTRACT Teach Pendants are extensively used for programming robots. This paper deals with the human factor considerations in the design of a teach pendant. Parameters affecting the human performance and ergonomic design principles are discussed. Some of these parameters include weight, size, location of control, color coding, symbolic coding and frequency of use.

INTRODUCTION Research effort in the Robotics area has largely tried to address and improve parameters like accuracy, response time of controllers ,multiple arms, flexible arms and weight of robots to reduce inertia force. It is not surprising that human factor aspects have largely been ignored during this developmental effort. A survey of existing teach pendants revealed several flaws violating ergonomic design principles.

1. Teach Pendant, too heavy for holding in one hand for extended duration of time.
2. The operation of Deadman's switch requires large amount of force.Since this switch must be depressed continuously while operating the robot, this often leads to sore thumb.
3. Physical dimensions, too large to be held comfortably for extended periods.
4. The digital displays used are of electronic register type and are only visible when looked head-on.
5. Numeral coding is used for the keys for each joint which means that the operator has to remember the joint number and the direction of positive and negative rotation.

ANALYSIS

a **Design Objectives** A good design must provide:

1. Ease of handling and gripping.
2. Ease of programming in terms of locating and identifying controls.
3. Ease of learning by incorporating symbolic and color coding of various controls.
4. Ease of usage for both right and left handed persons.
5. Light weight teach pendant.

b.   **Functional Analysis of Teach Pendants**

The teach pendant is generally designed to provide following functions :

1.   To manually move the arm in it's workspace.
2.   To record the location of the arm.
3.   To initialize the arm.
4.   To operate gripper or end-effector.
5.   To select coordinate system.
6.   To edit a program.
7.   To stop the arm in case of emergency.

Due to limitations on the number of keys that can be provided on the teach pendant and it's size, teach pendants do not have as much functional control on the robot as a keyboard. However they provide the programmer a means of operating and programming the robot from different locations around the work envelope as may be necessary. The controls on a teach pendant can be functionally classified into following groups.

1.   Controls to move different joints of arm.
2.   Controls to select coordinate system or mode of operation.
3.   Program run and record control.
4.   Emergency stop control.
5.   Control to edit programs.

c.   **Human Performance**

The size and weight of the teach pendant are important human factor considerations since they directly determine the length of time a person will be able to hold it comfortably.

1.   **Weight**

Programming   tasks may require varying amounts of time depending upon the complexity of task performed by the robot. Several studies have been done relating to the lifting tasks,method of lifting and energy consumption related to different methods of lifting (Davies [5] and NIOSH standards 1981). Studies have also been done for carrying tasks(Snook [6]) which explored effect of frequency and distance on the weight.None of the studies done in this area were done in a posture relevant to teach pendant usage. Figure 1 shows the most probable posture for the use of teach pendant with elbow angle in the range of 50-70 degrees.

2.   **Size**

The size of an object which can be comfortably held in the palm of person will depend upon

1.   Size of hand (Grip axis opening)

2.   Width of object
3.   Height of object

A study done by Greenberg and Chaffin [7] showed that the grip strength of the hand is related to the size of the object being gripped and that the maximum grip force is obtained when the grip axis opening is between 3 to 3.5 inch.. Although the amount of force is irrelevant to the present situation, it has indirect bearing on the size of the object being gripped. It is fair to assume that the grip axis corresponding to the maximum grip force will provide a comfortable grasp for holding the teach pendant for long durations. Based upon this assumption a 3 inch wide object will provide comfortable grip for most individuals including both males and females.

3.   Left or Right Hand Orientation

Majority of the existing designs are made for only right handed person and prove to be very awkward for a left handed person . With some modification and duplication of few controls these designs can be made easily operable by everyone. A right handed person would normally hold the teach pendant in left hand and operate it with right hand. Figure 2 shows the recommended shape. The symmetrical shape allows it to be operated by both left and right hand.

The deadman's switch with cover is on either side of the teach pendant. The speed control switch is on the side as shown.

DESIGN PRINCIPLES AND DATA

This section discusses the layout of the controls based on the ergonomic principles of design. The general arrangement of controls must be guided by considerations regarding their importance, frequency of usage, functional grouping and sequence of usage.

a.   **Importance Principle**   This principle deals with the degree to which the operation of specific control is vital to the whole system. Table 1. shows the controls listed in order of importance.

Controls In Descending Order of Importance

NO	CONTROLS
1.	Emergency Controls
2.	Modes of Operation Controls
3.	Program Run & Record Controls
4.	Controls to move joints
5.	Edit Controls

Table 1

b. **Frequency of Use Principle** This concept applies to the frequency of usage of the control and suggests that most frequently used controls should be placed in a more accessible location. Table 2 shows the list of controls used in descending order of usage frequency.

Usage Frequency of Controls

NO	CONTROLS
1.	Controls for joint movement
2.	Modes of operation controls
3.	Program run and record
4.	Edit controls
5.	Emergency controls

Table 2

c. **Functional Principle** This concept provides the grouping of components according to their function. Table 1,2 and 3 list the controls according to their functions.

d. **Sequence of Use Principle** In use of certain controls, sequence of patterns frequently occur and arranging the controls in that sequence may be helpful for the operator. For example the general sequence in which the controls on the majority of teach pendants are used is given in Table 3.

Sequence of Use

NO	CONTROLS
1.	Modes of Operation
2.	Controls to move joints
3.	Program run and record
4.	Edit controls
5.	Emergency controls

Table 3

e. **Discussion** A comparative study by Fowler and Young [8] showed that out of above four principles, the sequence of use principle is far superior to the other three principles. In their study, application of this principle lead to minimum response time.

Figure 2 shows the recommended location of all the controls. This layout conforms to the principle of sequence of operation, functional principle and importance principle.

f. **Types of Controls** Since most operations require high emergency action, high speed and high frequency of operation, push button type of controls are recommended. Among the pushbutton type, visco-elastic membrane switch will be preferred since it requires certain amount of force to activate it.(Inadvertent activation is minimized) For the speed selection switch which has three settings, a toggle switch will be preferable.

g. **Size and Displacement of Controls** The recommended size of pushbutton can be obtained from Handbooks on Ergonomics and is found to be .5 inch. Visco-elastic membrane switches will provide minimal displacement.

h. **Color Coding** It is recommended to color code the switches based upon their functional grouping

CONCLUSION

Figure 2 shows the final design which incorporates modifications based upon ergonomic considerations. This design satisfies the principles of arrangements of controls, is easy to operate and carry for extended periods. Two deadman's switches are installed with lower spring constant to provide ease of operation. Duplication of this switch and the symmetrical shape of the device makes it possible to operate by either hand.Lower weight makes it comfortable to carry for long periods.

REFERENCES

1. Bernotat and Gartner, "Display and Controls" Published by Swets and Zeitlinger, 1972.

2. Clark and Corlett, "The Ergonomics of Workspaces and Machines, A Design Manual" Published by Taylor and Francis,1984.

3. Johnson and Wilson, "Ergonomics matters in Advanced Manufacturing Technology" Published by Butterworths, 1987.

4. Lupton Edited by,"Proceeding of the 1st International Conference on Human Factors in Manufacturing" Published by IFS, 1984.

5. Greenberg,L., and Chaffin, D.(1977). Workers and their tools. Midland, MI: Pendell Publishing.

6. Davies, B.T.(1972). Moving loads manually. Applied Ergonomics, 3(4), 190-194.

7. Snook, S.H.(1978). The design of manual handling tasks. Ergonomics, 21(12), 963-985.

8. Fowler, R.L., Williams W.E., Fowler M.G.,and Young D.D.(1968, December). An investigation of the relationship between operator performance and operator panel layout for continuous tasks, Tech Rept. 68-170. U.S Air Force AMRL.

FIG. 1

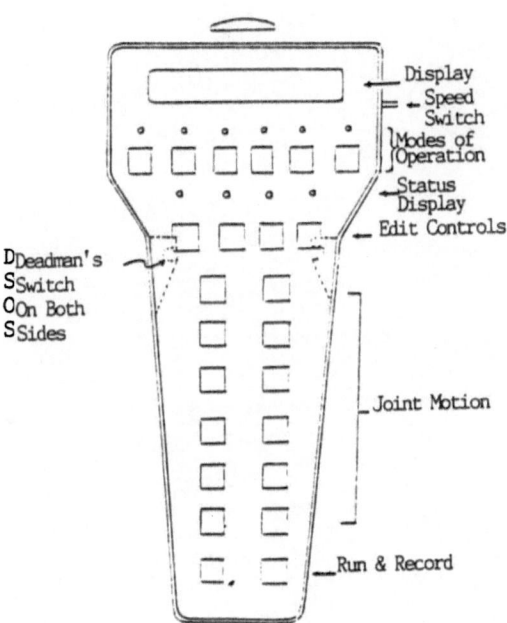

FIG. 2